# PROBABILITY AND PROBABILITY DISTRIBUTIONS

**Probability**

Bayes' theorem

$$P(B_i|A) = \frac{P(B_i) \cdot P(A|B_i)}{P(B_1) \cdot P(A|B_1) + P(B_2) \cdot P(A|B_2) + \cdots + P(B_k) \cdot P(A|B_k)}$$

Conditional probability

$$P(A|B) = \frac{P(A \cap B)}{P(B)}$$

General addition rule

$$P(A \cup B) = P(A) + P(B) - P(A \cap B)$$

General multiplication rule

$$P(A \cap B) = P(B) \cdot P(A|B) \quad \text{or} \quad P(A \cap B) = P(A) \cdot P(B|A)$$

Mathematical expectation

$$E = a_1 p_1 + a_2 p_2 + \cdots + a_k p_k$$

**Probability Distributions**

Binomial distribution

$$f(x) = \binom{n}{x} p^x (1 - p)^{n-x}$$

Mean of probability distribution

$$\mu = \sum x \cdot f(x)$$

Standard deviation of probability distribution

$$\sigma = \sqrt{\sum (x - \mu)^2 \cdot f(x)}$$

*Continued inside back cover*

# MODERN

# ELEMENTARY

# STATISTICS

# MODERN ELEMENTARY STATISTICS

## TWELFTH EDITION

**John E. Freund**
*Arizona State University*

**Benjamin M. Perles**
*Suffolk University*

PEARSON

Prentice
Hall

Upper Saddle River, New Jersey 07458

**Library of Congress Cataloging-in-Publication Data**

Freund, John E.
  Modern elementary statistics/John E. Freund, Benjamin M. Perles.—12 ed.
    p.cm.
  Includes bibliographical references and index.
  ISBN 0-13-187439-X
    1. Statistics. I. Perles, Benjamin M., II. Title.

QA276.12.F738 2007
519.5—dc22                                    2005054616

Executive Acquisitions Editor: *Petra Recter*
Editor in Chief: *Sally Yagan*
Project Manager: *Michael Bell*
Production Editors: *Raegan Keida and Barbara Mack*
Assistant Managing Editor: *Bayani Mendoza DeLeon*
Senior Managing Editor: *Linda Mihatov Behrens*
Executive Managing Editor: *Kathleen Schiaparelli*
Media Production Editor: *Jenelle Woodrup*
Managing Editor, Digital Supplements: *Nicole M. Jackson*
Manufacturing Buyer: *Lisa McDowell*
Director of Marketing: *Patrice Jones*
Marketing Assistant: *Jennifer de Leeuwerk*
Editorial Assistant/Print Supplements Editor: *Joanne Wendelken*
Art Director: *John Christiana*
Interior Designer: *511 Design*
Cover Designer: *Kiwi Design*
Art Editor: *Thomas Benfatti*
Director of Creative Services: *Paul Belfanti*
Permission Researcher: *Melinda Alexander*
Compositor & Art Studio: *Laserwords*

ISBN 0-13-187439-X

Pearson Education, Ltd., *London*
Pearson Education Australia PTY. Limited, *Sydney*
Pearson Education Singapore, Pte., Ltd
Pearson Education North Asia Ltd, *Hong Kong*
Pearson Education Canada, Ltd., *Toronto*
Pearson Education de Mexico, S.A. de C.V.
Pearson Education–Japan, *Tokyo*
Pearson Education Malaysia, Pte. Ltd

# CONTENTS†

■ **PREFACE**  xi

■ **CHAPTER 1**  **INTRODUCTION**  **1**

**1.1**  The Growth of Modern Statistics  2
**1.2**  Sources of Statistical Data  5
**1.3**  The Nature of Statistical Data  7
  Checklist of Key Terms  10
  References  10

■ **CHAPTER 2**  **SUMMARIZING DATA: LISTING AND GROUPING**  **12**

**2.1**  Listing Numerical Data  13
**2.2**  Stem-and-Leaf Displays  16
**2.3**  Frequency Distribution  21
**2.4**  Graphical Presentations  30
**2.5**  Summarizing Two-Variable Data  37
  Checklist of Key Terms  41
  References  41

■ **CHAPTER 3**  **SUMMARIZING DATA: MEASURES OF LOCATION**  **43**

**3.1**  Populations and Samples  44
**3.2**  The Mean  45
**3.3**  The Weighted Mean  49
**3.4**  The Median  54
**3.5**  Other Fractiles  57
**3.6**  The Mode  59
*3.7  The Description of Grouped Data  63
**3.8**  Technical Note (Summations)  70
  Checklist of Key Terms  73
  References  73

†All sections marked * are optional; they can be omitted without loss of continuity.

■ **CHAPTER 4   SUMMARIZING DATA: MEASURES OF VARIATION   74**

**4.1**   The Range   75
**4.2**   The Standard Deviation and the Variance   75
**4.3**   Applications of the Standard Deviation   79
*__4.4__   The Description of Grouped Data   86
**4.5**   Some Further Descriptions   88
           Checklist of Key Terms   93
           References   93

**Review Exercises for Chapters 1, 2, 3, & 4   94**

■ **CHAPTER 5   POSSIBILITIES AND PROBABILITIES   100**

**5.1**   Counting   101
**5.2**   Permutations   104
**5.3**   Combinations   107
**5.4**   Probability   114
           Checklist of Key Terms   122
           References   123

■ **CHAPTER 6   SOME RULES OF PROBABILITY   124**

**6.1**   Sample Spaces and Events   125
**6.2**   The Postulates of Probability   132
**6.3**   Probabilities and Odds   135
**6.4**   Addition Rules   139
**6.5**   Conditional Probability   145
**6.6**   Multiplication Rules   147
*__6.7__   Bayes' Theorem   152
           Checklist of Key Terms   157
           References   157

■ **CHAPTER 7   EXPECTATIONS AND DECISIONS   159**

**7.1**   Mathematical Expectation   160
*__7.2__   Decision Making   164
*__7.3__   Statistical Decision Problems   168
           Checklist of Key Terms   171
           References   171

**Review Exercises for Chapters 5, 6, & 7   172**

## ■ CHAPTER 8   PROBABILITY DISTRIBUTIONS   177

**8.1**   Random Variables   178
**8.2**   Probability Distributions   179
**8.3**   The Binomial Distribution   181
**8.4**   The Hypergeometric Distribution   189
**8.5**   The Poisson Distribution   193
*****8.6**   The Multinomial Distribution   197
**8.7**   The Mean of a Probability Distribution   198
**8.8**   The Standard Deviation of a Probability Distribution   201
         Checklist of Key Terms   205
         References   205

## ■ CHAPTER 9   THE NORMAL DISTRIBUTION   206

**9.1**   Continuous Distributions   207
**9.2**   The Normal Distribution   209
*****9.3**   A Check for Normality   218
**9.4**   Applications of the Normal Distribution   219
**9.5**   The Normal Approximation to the Binomial
         Distribution   222
         Checklist of Key Terms   228
         References   228

## ■ CHAPTER 10   SAMPLING AND SAMPLING DISTRIBUTIONS   229

**10.1**   Random Sampling   230
*****10.2**   Sample Designs   236
*****10.3**   Systematic Sampling   237
*****10.4**   Stratified Sampling   237
*****10.5**   Cluster Sampling   239
**10.6**   Sampling Distributions   242
**10.7**   The Standard Error of the Mean   245
**10.8**   The Central Limit Theorem   248
**10.9**   Some Further Considerations   249
*****10.10**  Technical Note (Simulation)   252
          Checklist of Key Terms   254
          References   255

### Review Exercises for Chapters 8, 9, & 10   256

## ■ CHAPTER 11   PROBLEMS OF ESTIMATION   262

**11.1**   The Estimation of Means   263
**11.2**   The Estimation of Means ($\sigma$ unknown)   268
**11.3**   The Estimation of Standard Deviations   274
**11.4**   The Estimation of Proportions   279
          Checklist of Key Terms   286
          References   286

■ **CHAPTER 12   TESTS OF HYPOTHESES: MEANS   287**

**12.1**   Tests of Hypotheses   287
**12.2**   Significance Tests   293
**12.3**   Tests Concerning Means   300
**12.4**   Tests Concerning Means ($\sigma$ unknown)   304
**12.5**   Differences Between Means   307
**12.6**   Differences Between Means ($\sigma$'s unknown)   311
**12.7**   Differences Between Means (Paired data)   313
          Checklist of Key Terms   315
          References   315

■ **CHAPTER 13   TESTS OF HYPOTHESES: STANDARD DEVIATIONS   317**

**13.1**   Tests Concerning Standard Deviations   317
**13.2**   Tests Concerning Two Standard Deviations   321
          Checklist of Key Terms   325
          References   325

■ **CHAPTER 14   TESTS OF HYPOTHESES BASED ON COUNT DATA   326**

**14.1**   Tests Concerning Proportions   327
**14.2**   Tests Concerning Proportions (Large Samples)   328
**14.3**   Differences between Proportions   329
**14.4**   The Analysis of an $r \times c$ Table   333
**14.5**   Goodness of Fit   345
          Checklist of Key Terms   350
          References   350

**Review Exercises for Chapters 11, 12, 13, & 14   351**

■ **CHAPTER 15   ANALYSIS OF VARIANCE   357**

**15.1**   Differences Among $k$ Means: An Illustration   358
**15.2**   The Design of Experiments: Randomization   362
**15.3**   One-Way Analysis of Variance   363
**15.4**   Multiple Comparisons   370
**15.5**   The Design of Experiments: Blocking   376
**15.6**   Two-Way Analysis of Variance   377
**15.7**   Two-Way Analysis of Variance without Interaction   378
**15.8**   The Design of Experiments: Replication   382
**15.9**   Two-Way Analysis of Variance with Interaction   382
**15.10** The Design of Experiments: Further Considerations   387
          Checklist of Key Terms   394
          References   395

■■ **CHAPTER 16    REGRESSION    396**

**16.1**    Curve Fitting    397
**16.2**    The Method of Least Squares    399
**16.3**    Regression Analysis    410
*   **16.4**    Multiple Regression    418
*   **16.5**    Nonlinear Regression    422
            Checklist of Key Terms    430
            References    430

■■ **CHAPTER 17    CORRELATION    431**

**17.1**    The Coefficient of Correlation    432
**17.2**    The Interpretation of $r$    438
**17.3**    Correlation Analysis    443
*   **17.4**    Multiple and Partial Correlation    447
            Checklist of Key Terms    450
            References    451

■■ **CHAPTER 18    NONPARAMETRIC TESTS    452**

**18.1**    The Sign Test    453
**18.2**    The Sign Test (Large Samples)    455
*   **18.3**    The Signed-Rank Test    458
*   **18.4**    The Signed-Rank Test (Large Samples)    462
**18.5**    The $U$ Test    465
**18.6**    The $U$ Test (Large Samples)    469
**18.7**    The $H$ Test    471
**18.8**    Tests of Randomness: Runs    475
**18.9**    Tests of Randomness: Runs (Large Samples)    476
**18.10**   Tests of Randomness: Runs Above and Below
            the Median    477
**18.11**   Rank Correlation    479
**18.12**   Some Further Considerations    483
**18.13**   Summary    483
            Checklist of Key Terms    484
            References    484

**Review Exercises for Chapters 15, 16, 17, & 18    486**

**STATISTICAL TABLES    493**

**ANSWERS TO ODD-NUMBERED EXERCISES    519**

**INDEX    557**

# PREFACE

*This edition of Modern Elementary Statistics, like all the preceding editions, is keynoted by the statement that "In general, statistical methods are nothing but a refinement of everyday thinking." These words paraphrase the statement made by the prominent scientist Albert Einstein that "the whole of science is nothing more than a refinement of everyday thinking."*

This book is intended primarily for use in a first course in statistics. There is a systematic academic approach in *Modern Elementary Statistics*, twelfth edition. Its emphasis is on instruction in meaningful, well-established statistical techniques. The future would-be medical doctor, business executive, scientist, teacher, or other professional specialist must comprehend and be skillful in the application of basic statistical tools and methodology. The student's knowledge of statistics is greatly enhanced by repeated exposure to statistical exercises.

Many changes to introductory statistics courses have been advocated in recent years, consisting of more emphasis on data analysis utilizing real data and decreasing the level of mathematical rigor, particularly in the area of probability. Our views have remained basically unchanged, and we present this new edition with the belief that this is an appropriate introduction to statistics that should be taught as a vital part of one's general education. It is difficult to deny the importance of emphasizing the analysis of real data, but it is more important to recommend that it be interesting real data. A classical example of the former, often used in the past, consists of the number of deaths resulting from the kick of a horse per year per army corps of the Prussian army! Such data fit the conditions for so-called Poisson data (see Section 8.5).

## EXERCISES

Many of the more than 1200 exercises are new or updated from previous editions. Among the new exercises in this edition, there are essentially two kinds. First, there is the conceptual kind that makes one think rather than devote one's time to procedures & calculations. The second kind serves to check whether a set of data satisfies the conditions required by a particular statistical procedure.

## EXAMPLES

Each chapter contains an abundant number of examples to help motivate the concepts and techniques that are presanted.

## OTHER FEATURES

The book also contains a large number of illustrations and examples, designed to provide guidance to readers. Important formulas, definitions, and rules are highlighted. Statistical tables are located in the back of the book, and frequently used formulas are repeated inside the front and back covers.

## COMPUTER PRINTOUTS

The purpose of the computer printouts and the graphing calculator reproductions is to make the reader aware of some of the most popular and some of the most up-to-date technologies that are available for work in statistics. The computer printouts from early editions have been replaced by new ones generated by MINITAB.

With one exception, neither computers nor graphing calculators are required for the use of our book. Indeed, the book can be used effectively by readers who do not possess or have easy access to computers and statistical software or to a graphing calculator. The one exception is Section 9.3, the check for normality.

Some of the exercises are labeled with the special icon 🖥 for the use of a computer and/or the icon ▦ for the use of a graphing calculator, but this is optional.

## SUPPLEMENTS FOR TEACHING AND LEARNING

### Instructor Supplements

Instructor's Solutions Manual **(ISBN 0-13-187443-8 for download only)**
Worked-out solutions to all of the exercises are provided in this manual. Careful attention has been paid to ensure that all methods of solution and notation are consistent with those used in the text.

Available for download from www.prenhall.com (then search for Freund/Perles, "Modern Elementary Statistics", and click on "Instructor Resources" in left column). This is a password protected site that requires instructors to set up an account.

Test Item File **(ISBN 0-13-187446-2 for download only)**
Hard copy of the computerized software testing materials provides a quick reference for the testing software.

TestGen withQuiz Master CD-ROM (Windows/Macintosh)
**(ISBN 0-13-187446-2)**
A computerized test bank which allows you to view and edit test bank questions, transfer them to exams, and print in a variety of formats. The program offers many options for organizing and displaying test banks and tests and a built-in random number and test generator allows instructors to create multiple versions of tests.

## Student Supplements

### Data CD
Available with the text, this CD includes the larger data sets used in the text problems and exercises, eliminating extensive data entry. The selected exercises included on the data CD are indicated by a computer icon.

### Student Solutions Manual (ISBN 0-13-187442-X)
The Student Solutions Manual contains worked-out solutions to all the text's odd-numbered exercises—a great resource for students as they study and work through the problem material. The Student Solutions Manual can be packaged with a new textbook. The package ISBN (Text & SSM) is 0-13-228287-9.

## For Instructors and Students

### Prentice Hall Companion Web site—www.prenhall.com/freund
Designed to complement and expand upon the text, the text Web site offers a variety of interactive teaching and learning tools, including links to related Web sites, quizzes, Syllabus Builder, and more! For more information, contact your local Prentice Hall representative.

For additional information on the supplements and special statistical software packages available from Prentice Hall, visit us online at http://www.prenhall.com.

## ACKNOWLEDGMENTS

Appreciation goes to my colleagues and students for their helpful suggestions and reviews of all of the editions of this text. I am grateful to:

Michael B. Dollinger, *Pacific Lutheran University*

Kathleen Ebert, *Alfred State College*

Lynn R. Eisenberg, *Durham Technical Community College*

J. S. Huang, *Columbia College*

Bob Jensen, *Pacific Lutharn University*

Lionel Mordecai, *Southwestern Community College*

I want to thank Douglas E. Freund for his help with the proofreading of prior editions.

The author is indebted to Prentice Hall for permission to reproduce the material in Table II; to the *Biometrika* trustees to reproduce the material in Tables III and IV; to the American Cyanamid Company to reproduce the material in Table VI; to the Addison-Wesley Publishing Company to reproduce the material in Table VII; to the editor of the *Annals of Mathematical Statistics* to reproduce the material in Table VIII; and to the Aerospace Research Laboratories, U.S. Air Force, to reproduce the material in Table IX.

Thanks also to Priscilla Gathoni for accuracy checking this book, and to the Prentice Hall team: Petra Recter, Joanne Wendelken, Patrice Jones, Linda Behrens, Tom Benfatti, John Christiana, Michael Bell, Bayani DeLeon, Barbara Mack, and Raegan Keida.

JOHN E. FREUND
BENJAMIN M. PERLES

# INTRODUCTION

**1.1** The Growth of Modern Statistics 2

**1.2** Sources of Statistical Data 5

**1.3** The Nature of Statistical Data 7

Checklist of Key Terms 10

References 10

Everything that deals even remotely with the collection, processing, interpretation, and presentation of data belongs to the domain of statistics, and so does the detailed planning that precedes all these activities. Indeed, statistics includes such diversified tasks as calculating the batting averages of baseball players; collecting and recording data on births, marriages, and deaths; evaluating the effectiveness of commercial products; and forecasting the weather. Even one of the most advanced branches of atomic physics goes by the name of quantum statistics.

The word "statistics" itself is used in various ways. It can be used, for example, to denote the mere tabulation of numerical data, as in reports of stock market transactions and in publications such as the *Statistical Abstract of the United States* or the *World Almanac*. It can also be used to denote the totality of methods that are employed in the collection, processing, and analysis of data, numerical and otherwise, and it is in this sense that "statistics" is used in the title of this book.

The word "statistician" is also used in several ways. It can be applied to those who simply collect information, as well as to those who prepare analyses or interpretations, and it is also applied to scholars who develop the mathematical theory on which the whole subject is based. Finally, the word "statistic" in the singular is used to denote a particular measure or formula, such as an average, a range of values, a growth rate such as an economic indicator, or a measure of the correlation (or relationship) between variables.

In Sections 1.1 and 1.2 we briefly discuss the history and recent growth of modern statistics with its ever-widening range. Some different types of data and their applications are explained. We also mention various sources of data.

1

In Section 1.3 we discuss the nature of statistical data and introduce additional terminology. The reader is warned against the indiscriminate application of some methods used in the analysis of data.

# 1.1 THE GROWTH OF MODERN STATISTICS

The origin of modern statistics can be traced to two areas that, on the surface, have very little in common: *government* (political science) and *games of chance.*

Governments have long used *censuses* to count persons and property. The ancient Romans used this technique to assist in the taxation of their subjects; indeed, the Bible tells how Mary and Joseph, subjects of Rome, went to Bethlehem to have their names listed in a census. Another famous census is reported in the *Domesday Book* of William of Normandy, completed in the year 1086. This census covered most of England, listing its economic resources, including property owners and the land which they owned. The United States Census of 1790 was the first "modern" census, but government agents merely counted the population. More recent censuses have become much more wide in scope, providing a wealth of information about the population and the economy. The United States Census is conducted every ten years (in the years that end in zero such as the years 2000, 2010, and 2020). The most recent one, the twenty-first decennial census, was conducted in the year 2000.

The problem of describing, summarizing, and analyzing census data led to the development of methods which, until recently, constituted almost all that there was to the subject of statistics. These methods, which originally consisted mainly of presenting the most important features of *data* by means of tables and charts, constitute what is now referred to as **descriptive statistics**. To be more specific, this term applies to anything done to data that does not infer anything which generalizes beyond itself.

If the government reports on the basis of census counts that the population of the United States was 248,709,873 in 1990 and 281,421,906 in 2000, this belongs to the field of descriptive statistics. This would also be the case if we calculated the corresponding percentage growth, which, as can easily be verified, was 13.2%, but not if we used these data to predict, say, the population of the United States in the year 2010 or 2020. Such a prediction goes beyond the available information.

The scope of statistics and the need to study statistics has grown enormously in the last few decades. One reason for this is that the amount of data that is collected, processed, and disseminated to the public for one purpose or another has increased almost beyond comprehension. To act as watchdogs, more and more persons with some knowledge of statistics are needed to take an active part in the collection of the data.

The second, and even more important, reason why the scope of statistics and the need to study statistics has grown so tremendously in recent years is the increasingly quantitative approach employed in business, economics, and industry as well as in the sciences and many other activities which directly affect our lives. Since most of the information required by this approach comes from *samples* (namely, from observations made on only part of a large set of items), its analysis requires **generalizations** which go beyond the data, and this is why there has been a pronounced shift in emphasis from **descriptive statistics** to **statistical inference**, or **inductive statistics**. In other words,

*Statistics has grown from the art of constructing charts and tables to the science of basing decisions on numerical data, or even more generally the science of decision making in the face of uncertainty.*

To mention a few examples, **generalizations** (that is, methods of statistical inference) are needed to estimate the number of long distance telephone calls which will be made in the United States ten years hence (on the basis of business trends and population projections); to determine the effect of newspaper advertising on the sales of supermarkets (on the basis of experiments conducted at a few markets); to evaluate conflicting and uncertain legal evidence; to predict the extent to which elevators will be used in a large office building which has not yet been built; to estimate the demand for a new computer which is being developed; or to rate the efficiency of salespersons based on fragmentary information about their performance. In each of these examples there are uncertainties, only partial or incomplete information, and it is here that we must use statistical methods which find their origin in games of chance.

Games of chance date back thousands of years, as is evidenced, for example, by the use of **astragali** (the forerunners of dice) in Egypt about 3500 B.C., but the mathematical study of such games began less than four centuries ago. The study of probability as a mathematical science began in the year 1654, when Blaise Pascal (a mathematician) wrote to Pierre de Fermat (another mathematician) with regard to a gambling problem. They solved the problem independently, using different mathematical methods. It may seem surprising that it took so long, but until then chance was looked upon as an expression of divine intent, and it would have been impious, or even sacrilegious, to analyze the "mechanics" of the supernatural through mathematics.

Although the mathematical study of games of chance, called **probability theory**, dates back to the seventeenth century, it was not until the early part of the nineteenth century that the theory developed for "heads or tails," for example, or "red or black" or "even or odd," was applied also to real-life nongambling situations where the outcomes were "boy or girl," "life or death," "pass or fail," and so forth. Thus probability theory was applied to many problems in the social as well as the physical sciences, and nowadays it provides an important tool for the analysis of any situation in business, in science, or in everyday life which in some way involves an element of uncertainty or risk. In particular, it provides the basis for methods which we use when we generalize from observed data, namely, when we use the methods of statistical inference to make predictions. This process is called **statistical inference**.

Many aspects of data collection are just common sense, as is illustrated by the following examples.

**EXAMPLE  1.1**   To determine public sentiment about the continuation of a government program, an interviewer asks "Do you feel that this wasteful program should be continued?" Explain why this question will probably not elicit a fair (objective) response.

**Solution**    The interviewer is *begging the question* by suggesting, in fact, that the program is wasteful.

**EXAMPLE  1.2**    To study consumer reaction to a new convenience food, a house-to-house survey is conducted during weekday mornings, with no provisions for return visits in case no one is home. Explain why this approach may well yield misleading information.

**Solution**    This survey will fail to reach those most likely to use the product—single persons and married couples with both spouses employed.

Although much of the aforementioned growth of statistics began prior to the "computer revolution," the widespread availability and use of computers has greatly accelerated the process. In particular, computers enable one to handle, analyze, and dissect large masses of data and to perform calculations that previously had been too cumbersome to contemplate. Let us point out, though, that access to a computer is not imperative for the study of statistics so long as one's main goal is to gain an understanding of the subject. *Some computer uses are illustrated in this textbook, but they are intended merely to make the reader aware of the technology that is available for work in statistics. Thus, computers are not required for the use of this textbook, but for those familiar with statistical software, there are some special exercises marked with an appropriate icon. Otherwise, none of the exercises require more than a simple hand-held calculator.*

The previous Examples 1.1 and 1.2 provide illustrations of **biased data**. The compilation of statistical data is biased when inappropriately (intentionally or unintentionally) certain persons or items are included or omitted from a sample. The results of very costly surveys can be completely useless if questions are ambiguous or asked in the wrong way, if they are asked of the wrong persons, in the wrong place, or at the wrong time. The following illustrates how questions may inadvertently be asked of the wrong persons: In a survey conducted by a manufacturer of an "instant" dessert, the interviewers, who worked from 9 A.M. to 5 P.M. on weekdays, got only the opinions of persons who happened to be at home during that time. Unfortunately, this did not include persons regularly employed and others who may have been especially interested in the manufacturer's product, and this made the whole survey worthless.

There are many subtle reasons for getting biased data. For one thing, many persons are reluctant to give honest answers to questions about their sanitary habits, say, how often they use a deodorant, bathe, or brush their teeth; they may be reluctant to return a mail questionnaire inquiring about their success in life unless they happen to be doing rather well; and they may talk about their "Xerox copies" even though the copies were produced on a copying machine made by another manufacturer. Another factor is the human element which may lead a poll taker to reduce the difficulty or expense of his assignment by interviewing the first persons who happen to come along, by interviewing persons who all live in the same apartment house or neighborhood, by falsely reporting interviews which were never held, or by spending an inordinate amount of time interviewing members of the opposite sex.

Then there are those annoying hidden biases which may be very difficult to foresee or detect. For instance, if the registrar of a large university wanted to select a "typical" group of students and happened to take a sample from the files of students whose last name begins with M, N, and O, he may well get a disproportionate number of students of Irish or Scottish descent: MacLeod, MacPherson, McDonald, O'Brien, O'Toole, and so forth. Similarly, if a quality control inspection procedure calls for a check of every tenth jar of jelly filled and sealed by a machine, the results would be disastrously misleading if it so happened that due to a defect in the machine, every fifth jar is improperly sealed. Needless to say, it can also happen that computers are programmed incorrectly or that information is transmitted or recorded incorrectly; *there is literally no end to what can conceivably cause trouble when it comes to the collecting, recording, and processing of data.* Of course, there are also such things as *intentional biases* and outright fraudulent misrepresentations (as in false advertising or when using "loaded" questions), but this is a matter of ethics rather than statistics.

# 1.2 SOURCES OF STATISTICAL DATA

Statistical data come from many sources, such as the day-to-day operation of businesses, hospitals, universities, government agencies, decennial censuses, experiments conducted by individuals or organizations, market analyses, and opinion polls.

Mainly in connection with business statistics, it has become the custom to make the following distinction: Data taken by a business firm or some organization from its *own* accounting records, payrolls, production records, inventories, sales vouchers, and the like, are called **internal data**, while data coming from outside the particular firm or organization, such as data from federal, state, or local government agencies, trade associations, private reporting organizations, or perhaps other firms, are called **external data**.

External data may be classified further as **primary data** or **secondary data**, depending on whether the data are collected and released (reported, or published) by the same organization. To be more specific, primary data are collected and released by the same organization, while secondary data are released by an organization other than the one by which they are collected. For instance, the Bureau of Labor Statistics collects the data required for its Consumer Price Index and publishes it in its *Monthly Labor Review*, which is thus the primary source for this important index. On the other hand, when the values of the Consumer Price Index are given in the *Federal Reserve Bulletin* or in the financial pages of newspapers or magazines, we refer to these publications as secondary sources of the index. Generally, primary sources are preferred to secondary sources because the possibility of errors of transcription is reduced, and also because primary sources are often accompanied by documentation and precise definitions.

The U.S. government is undoubtedly the largest publisher of both primary and secondary statistical data. For instance, the *Statistical Abstract of the United States*, published annually by the Bureau of the Census, contains a wide variety of information from various sources on the life of the nation; the *Survey of Current Business*, issued monthly by the Department of Commerce, contains a wealth of information about prices, production, inventories, income, sales, employment, wages, . . . , as well as important indicators of business conditions in general; and the *Monthly Labor Review*, published monthly by the Bureau of Labor Statistics, provides data concerning the labor force, employment, hours, wages, work stoppages, labor turnover, and other related activities. In view of the vast scope of statistics available, the *Statistical Masterfile*, published by the Congressional Information Service, is a valuable aid in locating certain types of statistical data. Their CD-ROM contains the American Statistics Index (ASI), and the Statistical Reference Index (SRI), and the Index to International Statistics (IS).

Some non–U.S. governmental sources include numerous other governments, a wealth of business and financial data, and an assortment of professional and scientific groups in the areas of education, medicine, psychology, chemistry, physics, sociology, engineering, and others. Experiments conducted by individuals or organizations, market analyses, and opinion polls may be published.

EXERCISES

**1.1**   Bad statistics may well result from asking questions in the wrong way, or of the wrong persons. Explain why the following may lead to useless data:

(a)  To study business executives' reaction to photocopying machines, the Xerox Corporation hires a research organization to ask business executives the question, "How do you like using Xerox copiers?"

(b)  To determine what the average person spends on a wristwatch, a researcher interviews only persons wearing a Rolex wristwatch.

(c)  To determine the average expenditures of a typical person on a vacation, a researcher interviews only passengers aboard a luxury ocean liner on a lengthy cruise.

**1.2**   Bad statistics may result from asking questions in the wrong place or at the wrong time. Explain why the following may lead to useless data:

(a)  To predict whether the House of Representatives of the U.S. Congress will vote in favor of a proposed law, a newspaper reporter for a Los Angeles paper interviews several Representatives from the State of California. *Note*: California has 52 Representatives (32 Democrats + 20 Republicans; the U.S. House of Representatives has 435 members (208 Democrats + 223 Republicans + 1 Independent + 3 vacancies) at the time of the interview.

(b)  To study the spending patterns of individuals, a survey is conducted during the first three weeks of December.

**1.3**   Explain why each of the following studies may fail to yield the desired information.

(a)  To ascertain facts about personal habits, a sample of adults are asked how often they take a bath or shower.

(b)  To determine the average annual income of its graduates 10 years after graduation, the alumni office of a university sent questionnaires in 2005 to all members of the class of 1995.

**1.4**   A person vacationing on the shores of a certain lake went fishing from his boat on four consecutive days and caught 3, 8, 10, and 12 fish. Which of the following conclusions may be obtained from these data by purely descriptive methods and which require generalizations? Explain your answers.

(a)  On only 3 days did the catch exceed 5 fish.

(b)  The vacationer's catch increased from each day to the next.

(c)  The vacationer learned more about the best fishing locations on the lake on each successive day.

(d)  On the fourth day the vacationer must have been lucky because he changed the type of bait used before going fishing.

**1.5**   Smith and Jones are medical doctors. On a recent day, Dr. Smith treated four male patients and two female patients, while Dr. Jones treated three male patients and three female patients. Which of the following conclusions can be obtained from these figures by purely descriptive methods and which require generalizations? Explain your answers.

(a)  On a given day, Doctors Smith and Jones treat equal numbers of patients.

(b)  Dr. Smith "always" treats more female patients than Dr. Jones.

(c)  Over a week, Dr. Smith averages three female patients per day.

(d)  The amount of time it takes Dr. Smith and Dr. Jones to treat a patient is about the same.

**1.6**   According to the Cellular Telecommunications and Internet Association (CTIA), Semi-Annual Wireless Survey, the number of U.S. cellular subscribers (in millions) for the five years 1997, 1998, 1999, 2000, and 2001 was 55.3, 69.2, 86.0, 109.5, and 128.4. Which of the following conclusions can be obtained from these data by purely descriptive methods, and which require generalizations?

(a)  The number of phones increases from one year to the next.

(b)  The number of telephone subscribers increased about 232%.

(c)  The number of telephone subscribers may increase to 150 or more in the year following the survey.

**1.7**   Driving the same model truck, 5 persons averaged 15.5, 14.7, 16.0, 15.5, and 14.8 miles per gallon of gasoline. Which of the following conclusions can be obtained from these data by purely descriptive methods and which require generalizations? Explain your answers.

(a)  The third driver must have driven mostly on rural roads.

(b)  The second driver must have driven faster than the other four.

(c)  More often than any other figure, the drivers averaged 15.5 miles per gallon.

(d)  None of the drivers averaged better than 16.0 miles per gallon.

**1.8**   An automobile dealer sells 3 automobiles on Monday, 4 automobiles on Tuesday, 5 automobiles on Wednesday, 6 automobiles on Thursday, and 7 on Friday. The average number of automobiles sold per day during this period was $\frac{3+4+5+6+7}{5} = 5$ automobiles per day. Was this a generalization or was this descriptive?

**1.9**   A statistically minded broker has her office on the third floor of a very tall office building, and whenever she leaves her office she records whether the first elevator that stops is going up or going down. Having done this for some time, she discovers that the vast majority of the time the first elevator that stops is going down. Comment on the following "conclusions."

(a)  Fewer elevators are going up than are going down.

(b)  The next time she leaves her office the first elevator that stops will be going down.

# 1.3 THE NATURE OF STATISTICAL DATA

Essentially, there are two kinds of statistical data: **numerical data** and **categorical data**. The former are obtained by measuring or counting, and they are also referred to as **quantitative data**. Such data may consist, for example, of the weights of the guinea pigs used in an experiment (obtained by measuring) or the daily absences from a class throughout a school year (obtained by counting). In contrast, categorical data result from descriptions, and they may consist, for example, of the blood types of hospital patients, their marital status, or their religious affiliation. Categorical data are also referred to as **qualitative data**. For ease in manipulating (recording or sorting) categorical data, they are often **coded** by assigning numbers to the different categories, thus converting the categorical data to numerical data in a trivial sense. For example, marital status might be coded by letting 1, 2, 3, and 4 denote a person's being single (never married), married, widowed, or divorced.

Numerical data are classified further as being **nominal data**, **ordinal data**, **interval data**, or **ratio data**. *Nominal data* are numerical in name only, as typified by the preceding example, where the numbers 1, 2, 3, and 4 were used to denote a person's being single (never married), married, widowed, or divorced. By saying "nominal data are numerical in name only" we mean that *they do not share any of the properties of the numbers we deal with in ordinary arithmetic.* With regard to the codes for marital status, we cannot write $3 > 1$ or $2 < 4$, and we cannot write $2 - 1 = 4 - 3, 1 + 3 = 4$, or $4 \div 2 = 2$. This illustrates how important it is always to check whether the mathematical treatment of statistical data is really legitimate.

Let us now consider some examples of ordinal data (data which can be rank-ordered) where data share some, but not necessarily all, of the properties of the numbers we deal with in ordinary arithmetic. For instance, in mineralogy the hardness of solids is sometimes determined by observing "what scratches what." If one mineral can scratch another it receives a higher hardness number, and on the Mohs scale the numbers from 1 to 10 are assigned, respectively, to talc, gypsum, calcite, fluorite, apatite, feldspar, quartz, topaz, sapphire, and diamond. With these numbers we can write $6 > 3$, for example, or $7 < 9$, since feldspar is harder than calcite and quartz is softer than sapphire. On the other hand, we cannot write $10 - 9 = 2 - 1$, for example, because the difference in hardness between diamond and sapphire is actually much greater than that between gypsum and talc. Also, it would be meaningless to say that topaz is twice as hard as fluorite simply because their respective hardness numbers on the Mohs scale are 8 and 4.

If we cannot do anything except set up inequalities, as was the case in the preceding example, we refer to the data as *ordinal* data. In connection with ordinal data, $>$ does not necessarily mean "greater than." It may be used to denote "happier than," "preferred to," "more difficult than," "tastier than," and so forth.

*If we can also form differences, but not multiply or divide,* we refer to the data as **interval data**. To give an example, suppose we are given the following temperature readings in degrees Fahrenheit: $63°$, $68°$, $91°$, $107°$, $126°$, and $131°$. Here we can write $107° > 68°$ or $91° < 131°$, which simply means that $107°$ is warmer than $68°$ and that $91°$ is colder than $131°$. Also, we can write $68° - 63° = 131° - 126°$, since equal temperature differences are equal in the sense that the same amount of heat is required to raise the temperature of an object from $63°$ to $68°$ as from $126°$ to $131°$. On the other hand, it would not mean much if we say that $126°$ is twice as hot as $63°$, even though $126 \div 63 = 2$. To show why, we have only to change to the Celsius scale, where the first temperature becomes $\frac{5}{9}(126 - 32) = 52.2°$, the second temperature becomes $\frac{5}{9}(63 - 32) = 17.2°$, and the first figure is now more than three times the second. This difficulty arises because the Fahrenheit and Celsius scales both have artificial origins (zeros). In other words, in neither scale is the number 0 indicative of the absence of the quantity (in this case temperature) which we are trying to measure.

*If we can also form quotients, we refer to the data as* **ratio data**, and such data are not difficult to find. They include all the usual measurements (or determinations) of length, height, money amounts, weight, volume, area, pressure, elapsed time (though not calendar time), sound intensity, density, brightness, velocity, and so on.

The distinction we have made here among nominal, ordinal, interval, and ratio data is important, for as we shall see, the nature of a set of data may suggest the use of particular statistical techniques. To emphasize the point that what we can and cannot do arithmetically with a given set of data depends on the nature of the data, consider the following scores that four students obtained in the three parts of a comprehensive history test:

| | American history | European history | Ancient history |
|---|---|---|---|
| Linda | 89 | 51 | 40 |
| Tom | 61 | 56 | 54 |
| Henry | 40 | 70 | 55 |
| Rose | 13 | 77 | 72 |

The totals for the four students are 180, 171, 165, and 162, so that Linda scored highest, followed by Tom, Henry, and Rose.

Suppose now that somebody proposes that instead of adding the scores obtained in the three parts of the test, we compare the overall performance of the four students by ranking their scores from a high score of one to a low score of four for each part of the test and then average their ranks (that is, add them and divide by 3). What we get is shown in the following table:

| | American history | European history | Ancient history | Average rank |
|---|---|---|---|---|
| Linda | 1 | 4 | 4 | 3 |
| Tom | 2 | 3 | 3 | $2\frac{2}{3}$ |
| Henry | 3 | 2 | 2 | $2\frac{1}{3}$ |
| Rose | 4 | 1 | 1 | 2 |

Now, if we look at the average ranks, we find that Rose came out best, followed by Henry, Tom, and Linda, so that the order has been reversed from what it was before. How can this be? Well, strange things can happen when we average ranks. For instance, when it comes to their ranks, Linda's outscoring Tom by 28 points in American history counts just as much as Tom's outscoring her by 5 points in European history, and Tom's outscoring Henry by 21 points in American history counts just as much as Henry's outscoring him by a single point in ancient history. We conclude that, perhaps, we should not have averaged the ranks, but it might also be pointed out that, perhaps, we should not even have totaled the original scores. The variation of the American history scores, which go from 13 to 89, is much greater than that of the other two kinds of scores, and this strongly affects the total scores and suggests a possible shortcoming of the procedure. These observations are thought provoking, but we shall not follow up on them as it has been our goal merely to warn the reader against the indiscriminate use of statistical techniques; that is, to show how the choice of a statistical technique may be dictated by the nature of the data.

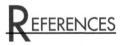

**1.10** Do we get ordinal or nominal data if students were asked whether a certain final examination which they had recently completed was easy, difficult, or very difficult, and these alternatives are coded 1, 2, and 3?

**1.11** What kind of data do we get if the bottles on a chemistry laboratory shelf are numbered 1, 2, 3, 4, and 5 representing sulfuric acid, hydrochloric acid, nitric acid, sodium hydroxide, and potassium hydroxide?

**1.12** Are the following data nominal, ordinal, interval, or ratio data?

(a) The number of White Siamese cats that are in a cage containing cats in a veterinarian's office

(b) The numbers on the jerseys of a team of football players

(c) Numbered musical exercises in a book of musical exercises

(d) The length and width of an official United States flag

**1.13** Are the following data nominal, ordinal, interval or ratio data?

(a) Presidential election years in the United States

(b) Checking account numbers

(c) Comparing the weight of a 2-pound pumpkin with that of a 5-pound pumpkin

**1.14** In two major golf tournaments one professional golfer finished second and ninth, while another finished sixth and fifth. Comment on the argument that since $2 + 9 = 6 + 5$, the overall performance of the two golfers in these two tournaments was equally good.

## CHECKLIST OF KEY TERMS (with page references to their definitions)

Astragali, 3
Biased data, 4
Categorical data, 7
Coded, 8
Descriptive statistics, 2
External data, 5
Generalizations, 2
Inductive statistics, 3
Internal data, 5
Nominal data, 8

Numerical data, 7
Ordinal data, 8
Primary data, 5
Probability theory, 3
Qualitative data, 8
Quantitative data, 7
Ratio data, 8
Scaling, 11
Secondary data, 5
Statistical inference, 3

## REFERENCES

*Brief and informal discussions of what statistics is and what statisticians do may be found in pamphlets titled* Careers in Statistics *and* Statistics as a Career: Women at Work, *which*

*are published by the American Statistical Association. They may be obtained by writing to this organization at 1429 Duke Street, Alexandria, Virginia 22314-3402.*
*Among the few books on the history of statistics, on the elementary level, is*

WALKER, H. M., *Studies in the History of Statistical Method.* Baltimore: The Williams & Wilkins Company, 1929.

*and on the more advanced level*

KENDALL, M. G., and PLACKETT, R. L., eds., *Studies in the History of Statistics and Probability,* Vol. II. New York: Macmillan Publishing Co., Inc., 1977.

PEARSON, E. S., and KENDALL, M. G., eds., *Studies in the History of Statistics and Probability.* New York: Hafner Press, 1970.

STIGLER, S. M., *The History of Statistics.* Cambridge, Mass.: Harvard University Press, 1986.

*A more detailed discussion of the nature of statistical data and the general problem of **scaling** (namely, the problem of constructing scales of measurement or assigning scale scores) may be found in*

HILDEBRAND, D. K., LAING, J. D., and ROSENTHAL, H., *Analysis of Ordinal Data.* Beverly Hills, Calif.: Sage Publications, Inc., 1977.

REYNOLDS, H. T., *Analysis of Nominal Data.* Beverly Hills, Calif.: Sage Publications, Inc., 1977.

SIEGEL, S., *Nonparametric Statistics for the Behavioral Sciences.* New York: McGraw-Hill Book Company, 1956.

*The following are some titles from the ever-growing list of books on statistics that are written for the layperson:*

BROOK, R. J., ARNOLD, G. C., HASSARD, T. H., and PRINGLE, R. M., eds., *The Fascination of Statistics.* New York: Marcel Dekker, Inc., 1986.

CAMPBELL, S. K., *Flaws and Fallacies in Statistical Thinking.* Englewood Cliffs, N.J.: Prentice Hall, 1974.

FEDERER, W. T., *Statistics and Society: Data Collection and Interpretation.* New York: Marcel Dekker, Inc., 1991.

GONICK, L., and SMITH, W., *The Cartoon Guide to Statistics.* New York: HarperCollins, 1993.

HOLLANDER, M., and PROSCHAN, F., *The Statistical Exorcist: Dispelling Statistics Anxiety.* New York: Marcel Dekker, Inc., 1984.

HOOKE, R., *How to Tell the Liars from the Statisticians.* New York: Marcel Dekker, Inc., 1983.

HUFF, D., *How to Lie with Statistics.* New York: W. W. Norton & Company, Inc., 1993.

JAFFE, A., and SPIRER, H. F., *Misused Statistics: Straight Talk for Twisted Numbers,* New York: Marcel Dekker, Inc., 1987.

MOSTELLER, F., PIETERS, R. S., KRUSKAL, W. H., RISING, G. R., LINK, R. F., CARLSON, R. and ZELINKA, M., *Statistics by Example.* Reading, Mass.: Addison-Wesley Publishing, Inc., 1973.

PAULOS, J. A., *Innumeracy: Mathematical Illiteracy and Its Consequences.* New York: Hill and Wang, 2001.

WANG, C., *Sense and Nonsense of Statistical Inference.* New York: Marcel Dekker, Inc., 1993.

WEAVER, W., *Lady Luck: The Theory of Probability.* New York: Dover Publications Inc., 1982.

# SUMMARIZING DATA: LISTING AND GROUPING†

**2.1** Listing Numerical Data 13

**2.2** Stem-and-Leaf Displays 16

**2.3** Frequency Distribution 21

**2.4** Graphical Presentations 30

**2.5** Summarizing Two-Variable Data 37

Checklist of Key Terms 41

References 41

S tatistical data that directly affect our lives are often disseminated in summarized form. The whole process of putting large masses of data into a usable form has increased greatly in the last few decades. This has been due to the development of computer capabilities and partly due to the deluge of data generated by the increasingly quantitative approach of the behavioral and social sciences, in which many aspects of life are measured in some way.

In Sections 2.1 and 2.2 we shall present ways of listing data so that they present a good overall picture and, hence, are easy to use. By **listing** we are referring to any kind of treatment that preserves the identity of each value (or item). In other words, we rearrange but do not change. A speed of 63 mph remains a speed of 63 mph, a salary of $75,000 remains a salary of $75,000, and when sampling public opinion, a Republican remains a Republican and a Democrat remains a Democrat. In Sections 2.3 and 2.4 we shall discuss ways of **grouping** data into a number of classes, intervals, or categories and presenting the result in the form of a table or a chart. This will leave us with data in a relatively compact and easy-to-use form, but it does entail a substantial loss of information. Instead of a person's weight, we may know only that he or she weighs anywhere from 160 to 169 pounds, and instead of an actual pollen count we may know only that it is "medium" (11–25 parts per cubic meter).

† Since computer printouts and a reproduction from the display screen of a graphing calculator appear first in this chapter, let us repeat from the Preface that the purpose of the printouts and the graphing calculator reproductions is to make the reader aware of the existence
*(cont. on next page)*

# 2.1 LISTING NUMERICAL DATA

Listing and, thus, organizing the data is usually the first task in any kind of statistical analysis. As a typical situation, consider the following data, representing the lengths (in centimeters) of 60 sea trout caught by a commercial trawler in Delaware Bay:

| | | | | | | | | | |
|---|---|---|---|---|---|---|---|---|---|
| 19.2 | 19.6 | 17.3 | 19.3 | 19.5 | 20.4 | 23.5 | 19.0 | 19.4 | 18.4 |
| 19.4 | 21.8 | 20.4 | 21.0 | 21.4 | 19.8 | 19.6 | 21.5 | 20.2 | 20.1 |
| 20.3 | 19.7 | 19.5 | 22.9 | 20.7 | 20.3 | 20.8 | 19.8 | 19.4 | 19.3 |
| 19.5 | 19.8 | 18.9 | 20.4 | 20.2 | 21.5 | 19.9 | 21.7 | 19.5 | 20.9 |
| 18.1 | 20.5 | 18.3 | 19.5 | 18.3 | 19.0 | 18.2 | 21.9 | 17.0 | 19.7 |
| 20.7 | 21.1 | 20.6 | 16.6 | 19.4 | 18.6 | 22.7 | 18.5 | 20.1 | 18.6 |

The mere gathering of this information is no small task, but it should be clear that more must be done to make the numbers comprehensible.

What can be done to make this mass of information more usable? Some persons may find it interesting to locate the extreme values, which are 16.6 and 23.5 for this list. Occasionally, it is useful to sort the data in an ascending or descending order. The following list gives the lengths of the trout arranged in an ascending order:

| | | | | | | | | | |
|---|---|---|---|---|---|---|---|---|---|
| 16.6 | 17.0 | 17.3 | 18.1 | 18.2 | 18.3 | 18.3 | 18.4 | 18.5 | 18.6 |
| 18.6 | 18.9 | 19.0 | 19.0 | 19.2 | 19.3 | 19.3 | 19.4 | 19.4 | 19.4 |
| 19.4 | 19.5 | 19.5 | 19.5 | 19.5 | 19.5 | 19.6 | 19.6 | 19.7 | 19.7 |
| 19.8 | 19.8 | 19.8 | 19.9 | 20.1 | 20.1 | 20.2 | 20.2 | 20.3 | 20.3 |
| 20.4 | 20.4 | 20.4 | 20.5 | 20.6 | 20.7 | 20.7 | 20.8 | 20.9 | 21.0 |
| 21.1 | 21.4 | 21.5 | 21.5 | 21.7 | 21.8 | 21.9 | 22.7 | 22.9 | 23.5 |

Sorting a large set of numbers manually in an ascending or descending order can be a surprisingly difficult task. It is simple, though, if we can use a computer or a graphing calculator. In that case, entering the data is the most tedious part. With a graphing calculator we then press **STAT** and **2**, fill in the list where we put the data, press **ENTER**, and the display screen spells out **DONE**.

If a set of data consists of relatively few values, many of which are repeated, we simply count how many times each value occurs and then present the result in the form of a table or a **dot diagram**. In such a diagram we indicate by means of dots how many times each value occurs.

EXAMPLE **2.1**   **Dot diagrams** are used to summarize visually data which have been tallied, thus indicating by means of dots how many times each value (or item) has occurred.

of these technologies for work in statistics. Let us make it clear, however, that neither computers nor graphing calculators are required for the use of our text. Indeed, this book can be used effectively by readers who do not possess or have easy access to computers and statistical software or to graphing calculators. Some of the exercises are labeled with special icons for the use of a computer or a graphing calculator. Everything marked with an asterisk, text material and corresponding exercises, is optional.

Suppose, for instance, that the following data pertain to the number of times per week that the departure of an airline's 48 daily flights are delayed:

$$
\begin{array}{cccccccccccc}
2 & 1 & 5 & 0 & 1 & 3 & 2 & 0 & 7 & 1 & 3 & 4 \\
2 & 4 & 1 & 2 & 2 & 5 & 1 & 3 & 4 & 3 & 1 & 1 \\
3 & 2 & 6 & 4 & 1 & 0 & 2 & 2 & 3 & 5 & 2 & 3 \\
0 & 2 & 4 & 1 & 1 & 3 & 2 & 3 & 5 & 2 & 4 & 4
\end{array}
$$

Counting how many times each value occurs, we get the dot diagram of Figure 2.1. The figure gives a clearer picture of the situation than the original numbers shown in four rows of 12 numbers each.

## Solution

**Figure 2.1**
Dot diagram of weekly departure delays.

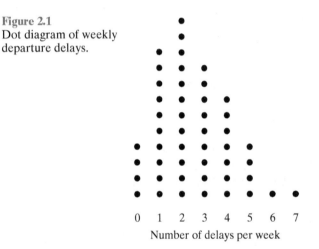

0   1   2   3   4   5   6   7

Number of delays per week

EXAMPLE **2.2** There are various ways in which dot diagrams can be modified. For instance, instead of the dots we can use other symbols, such as x's or ⋆'s. Also, we could align the dots horizontally rather than vertically. Suppose, for example, that thirty persons were asked to name their favorite color, and that their responses were

| | | | | | |
|---|---|---|---|---|---|
| blue | red | green | blue | red | blue |
| brown | blue | red | red | red | yellow |
| white | red | blue | green | blue | blue |
| orange | green | blue | blue | blue | red |
| green | blue | red | blue | brown | green |

Counting the number of times each color occurs, we get the dot diagram of Figure 2.2. In this chart, the colors have been ordered according to their frequencies.

## Solution

**Figure 2.2**
Dot diagram of color preferences.

```
Blue   │ *  *  *  *  *  *  *  *  *  *  *  *
Red    │ *  *  *  *  *  *  *  *
Green  │ *  *  *  *  *
Brown  │ *  *
White  │ *
Orange │ *
Yellow │ *
```

---

**EXAMPLE 2.3**  Another way of modifying dot diagrams is to represent the frequencies of the various numbers (or items) by means of rectangles whose lengths are proportional to the respective frequencies. Such diagrams are referred to as **bar charts** or, more specifically, as **vertical bar charts** or **horizontal bar charts**. The rectangles are usually supplemented with the corresponding frequencies, as shown in Figure 2.3. This figure displays the vertical and horizontal bar charts of the color preferences on which we based Figure 2.3. The bars here are plotted in descending order of frequency. This emphasizes the fact that twice as many of the respondents prefer the colors blue and red to all of the other colors combined. Diagrams like those shown in Figures 2.2 and 2.3 are also referred to as **Pareto diagrams**, named after Vilfredo Pareto, an Italian economist who used them to illustrate that roughly 80% of the wealth of a country is owned by about 20% of the people.

## Solution

**Figure 2.3**
Bar charts of color preferences: (a) vertical bar chart, (b) horizontal bar chart.

(a)

Figure 2.3
(*continued*).

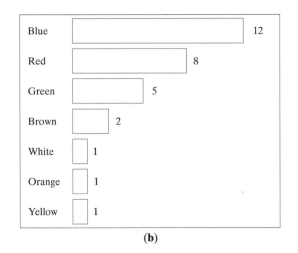

**(b)**

## 2.2 STEM-AND-LEAF DISPLAYS

Dot diagrams are impractical and ineffective when a set of data contains many different values or categories, or when some of the values or categories require too many dots to yield a coherent picture. To give an example, consider the first-round scores in a PGA tournament, where the lowest score was a 62, the highest score was an 88, and 27 of the 126 golfers shot a par 72. This illustrates both of the reasons cited previously for not using dot diagrams. There are too many different values from 62 to 88, and at least one of them, 72, requires too many dots.

In recent years, an alternative method of listing data has been proposed for the exploration of relatively small sets of numerical data. It is called a **stem-and-leaf display** and it also yields a good overall picture of the data without any appreciable loss of information. Again, each value retains its identity, and the only information we lose is the order in which the data were obtained.

To illustrate this technique, consider the following data on the number of rooms occupied each day in a resort hotel during a recent month of June:

$$\begin{array}{cccccccccc}
55 & 49 & 37 & 57 & 46 & 40 & 64 & 35 & 73 & 62 \\
61 & 43 & 72 & 48 & 54 & 69 & 45 & 78 & 46 & 59 \\
40 & 58 & 56 & 52 & 49 & 42 & 62 & 53 & 46 & 81
\end{array}$$

Let us combine all the values beginning with a 3, all those beginning with a 4, all those beginning with a 5, and so on. This would yield

$$\begin{array}{ccccccccccc}
37 & 35 \\
49 & 46 & 40 & 43 & 48 & 45 & 46 & 40 & 49 & 42 & 46 \\
55 & 57 & 54 & 59 & 58 & 56 & 52 & 53 \\
64 & 62 & 61 & 69 & 62 \\
73 & 72 & 78 \\
81
\end{array}$$

This arrangement is quite informative, but it is not the kind of diagram we use in actual practice. To simplify it further, we show the first digit only once for each row, on the left and separated from the other digits by means of a vertical line. This leaves us with

```
3 | 7 5
4 | 9 6 0 3 8 5 6 0 9 2 6
5 | 5 7 4 9 8 6 2 3
6 | 4 2 1 9 2
7 | 3 2 8
8 | 1
```

and this is what we refer to as a **stem-and-leaf display**. In this arrangement, each row is called a **stem**, each number on a stem to the left of the vertical line is called a **stem label**, and each number on a stem to the right of the vertical line is called a **leaf**. As we shall see later, there is a certain advantage to arranging the leaves on each stem according to size, and for our data this would yield

```
3 | 5 7
4 | 0 0 2 3 5 6 6 6 8 9 9
5 | 2 3 4 5 6 7 8 9
6 | 1 2 2 4 9
7 | 2 3 8
8 | 1
```

A stem-and-leaf display is actually a hybrid kind of arrangement, obtained in part by grouping and in part by listing. The values are grouped into the six stems, and yet each value retains its identity. Thus, from the preceding stem-and-leaf display, we can reconstruct the original data as 35, 37, 40, 40, 42, 43, 45, 46, 46, 46, 48, 49, 49, 52, 53, . . . , and 81, though not in their original order.

There are various ways in which stem-and-leaf displays can be modified. For instance, the stem labels or the leaves could be two-digit numbers, so that

$$24 \mid 0 \quad 2 \quad 5 \quad 8 \quad 9$$

would represent the numbers 240, 242, 245, 248, and 249; and

$$2 \mid 31 \quad 45 \quad 70 \quad 88$$

would represent the numbers 231, 245, 270, and 288.

Now suppose that in the room occupancy data; we had wanted to use more than six stems. Using each stem label twice, if necessary, once to hold the leaves from 0 to 4 and once to hold the leaves from 5 to 9, we would get

```
3 | 5 7
4 | 0 0 2 3
4 | 5 6 6 6 8 9 9
5 | 2 3 4
5 | 5 6 7 8 9
6 | 1 2 2 4
6 | 9
7 | 2 3
7 | 8
8 | 1
```

and this is called a **double-stem display**. Other modifications of stem-and-leaf displays also exist.

Had we wanted to use a computer in the preceding example, MINITAB would have yielded the standard and double-stem displays shown in Figure 2.4.

In case the reader is curious about the figures in the columns on the left, they are simply the accumulated numbers of items (leaves) counted from either end. The 8 and the 3 in parentheses tell us that the middle of the data falls on the respective stems and that these stems have, respectively, 8 and 3 leaves. Finding the middle position of a set of data will be illustrated in Sections 3.4 and 3.5.

We shall not discuss stem-and-leaf displays in any great detail, as it has been our objective mainly to present one of the relatively new techniques, which come under the general heading of **exploratory data analysis**. These techniques are really quite simple and straightforward, and as we have seen, the work can be simplified even further by using a computer and appropriate software.

**Figure 2.4**
Computer printouts of a standard stem-and-leaf display and a double-stem display of the room occupancy data.

**Stem-and-Leaf Display: Room Occupancy**

```
Stem-and-leaf of Room occ
N = 30
Leaf Unit = 1.0

    2     3 57
   13     4 00235666899
   (8)    5 23456789
    9     6 12249
    4     7 238
    1     8 1
```

**Stem-and-Leaf Display: Room Occupancy**

```
Stem-and-leaf of Room occ
N = 30
Leaf Unit = 1.0

    2     3  57
    6     4  0023
   13     4  5666899
   (3)    5  234
   14     5  56789
    9     6  1224
    5     6  9
    4     7  23
    2     7  8
    1     8  1
```

**EXERCISES**

**2.1**  According to the U.S. National Oceanic and Atmospheric Administration, 12, 11, 6, 7, 12, 11, 14, 8, 7, 8, and 7 North Atlantic tropical storms and hurricanes reached the U.S. coast in 11 consecutive years. Construct a dot diagram.

**2.2**  At a busy intersection with photo radar in a town in Arizona, 0, 1, 0, 0, 0, 1, 0, 0, 2, 1, 0, 0, 1, 0, 0, 0, 0, 0, 1, 0, 0, 2, 2, 0, 0, 0, 1, 0, 1, 0, 3, 0, 0, 0, 0, and 0 cars entered the intersection after the light had turned red on 36 randomly chosen occasions. Construct a dot diagram.

**2.3**  On 40 business days, a pharmacy filled 7, 4, 6, 9, 5, 8, 8, 7, 6, 10, 7, 7, 6, 9, 6, 8, 4, 9, 8, 7, 5, 8, 7, 5, 8, 10, 6, 9, 7, 7, 8, 10, 6, 6, 7, 8, 7, 9, 7, and 8 prescriptions for AMBIEN® sleeping pills.

    (a)  Construct a table showing on how many days the pharmacy filled 4, 5, 6, 7, 8, 9, and 10 prescriptions for this sleeping pill.

    (b)  Construct a dot diagram for these data, using asterisks instead of dots.

**2.4**  In a special sale, a used car dealer advertised the following secondhand Buicks, identified by years and models: '02 Le Sabre, '05 Regal, '04 LeSabre, '04 Park Avenue, 01 Regal, '02 Skylark, '04 Le Sabre, '01 Skylark, '01 Century, '05 Skylark, '04 Skylark, '05 LeSabre, '03 Regal, '04 Skylark, '03 Century, '05 Regal, '04 LeSabre, '03 Le Sabre, '05 Park Ave, '04 Skylark, '04 Riviera, '01 Le Sabre, '03 Regal, '03 Century, and '05 Le Sabre.

    (a)  Construct a dot diagram showing how these cars are distributed according to model year.

    (b)  Construct a dot diagram showing how these cars are distributed according to model name.

**2.5**  At a dog show, an interviewer asked 30 persons to name their favorite breed of dog in the Hound Group. Their replies were dachshund, greyhound, basset, beagle, afghan, afghan, beagle, dachshund, beagle, afghan, dachshund, greyhound, beagle, greyhound, dachshund, dachshund, afghan, dachshund, greyhound, beagle, afghan, greyhound, beagle, dachshund, dachshund, beagle, bloodhound, greyhound, basset, and beagle. Construct a dot diagram for these categorical data.

**2.6**  Draw a bar chart with horizontal bars for the categorical data of Exercise 2.5.

**2.7**  On Wednesdays, mutual funds are denoted by the letter A in the financial pages of the Arizona Republic if they are in the top 20% among funds with the same investment objectives, by the letter B if they are in the next 20%, ..., and by the letter E if they are in the bottom 20%. On the third Wednesday in July, 2001, the 20 Vanguard Index Funds listed were denoted by

A  C  A  NA  B  B  E  C  A  A  B  B  D  A  C  A  A  B  D  C

where NA means "not available." Construct a dot diagram of the 19 rated funds.

**2.8**  If the categories in a dot diagram are arranged in descending order according to their frequencies (numbers of dots), such a dot diagram is also referred to as a Pareto diagram. Present the data of Exercise 2.5 in the form of a Pareto diagram.

**2.9**  Pareto diagrams are often used in industrial quality control to illustrate the relative importance of different kinds of defects. Denoting broken parts, paint defects, missing parts, faulty connections, and all other defects by the codes 0, 1, 2, 3, and 4, respectively, a quality control inspector observed the following kinds of defects in a large production run of cellular phones:

    3  3  2  3  2  2  0  3  3  4  1  3  2  0  2  0  3  3
    2  0  1  2  3  4  3  3  0  2  3  3  1  3  2  3  3

Present these defects in the form of a Pareto diagram.

**2.10**   List the data that correspond to the following stems of stem-and-leaf displays:

(a)  1 | 4 7 0 1 5

(b)  4 | 2 0 3 9 8

(c)  7 | 3 5 1 1 6

**2.11**   List the data that correspond to the following stems of stem-and-leaf displays:

(a)  3 | 6 1 7 5 2

(b)  4 | 15 38 50 77

(c)  25 | 4 4 0 3 9

**2.12**   List the data that correspond to the following stems of double-stem displays:

(a)  5 | 3  0  4  4  1  2

     5 | 9  9  7  5  8  6

(b)  6 | 7  8  5  9  6

     7 | 1  1  0  4  3

     7 | 5  5  8  9  6

**2.13**   Following are the lengths of young-of-the-year freshwater drum (in millimeters), caught near Rattlesnake Island in Lake Erie: 79, 77, 65, 78, 71, 66, 95, 86, 84, 83, 88, 72, 81, 64, 71, 58, 60, 81, 73, 67, 85, 89, 75, 80, and 56. Construct a stem-and-leaf display with the stem labels 5, 6, 7, 8, and 9.

**2.14**   Convert the stem-and-leaf display obtained in Exercise 2.13 into a double-stem display.

**2.15**   On page 13 we gave the following lengths (in centimeters) of 60 sea trout caught by a commercial trawler in Delaware Bay:

| | | | | | | | | | |
|---|---|---|---|---|---|---|---|---|---|
| 19.2 | 19.6 | 17.3 | 19.3 | 19.5 | 20.4 | 23.5 | 19.0 | 19.4 | 18.4 |
| 19.4 | 21.8 | 20.4 | 21.0 | 21.4 | 19.8 | 19.6 | 21.5 | 20.2 | 20.1 |
| 20.3 | 19.7 | 19.5 | 22.9 | 20.7 | 20.3 | 20.8 | 19.8 | 19.4 | 19.3 |
| 19.5 | 19.8 | 18.9 | 20.4 | 20.2 | 21.5 | 19.9 | 21.7 | 19.5 | 20.9 |
| 18.1 | 20.5 | 18.3 | 19.5 | 18.3 | 19.0 | 18.2 | 21.9 | 17.0 | 19.7 |
| 20.7 | 21.1 | 20.6 | 16.6 | 19.4 | 18.6 | 22.7 | 18.5 | 20.1 | 18.6 |

Construct a stem-and-leaf display with the stem labels 16., 17., ... and 23., and the leaves 0, 1, ..., and 9.

**2.16**   A city engineer counted the number of trucks that passed a certain point on a highway in 24 consecutive business days. The numbers were

| | | | | | | | |
|---|---|---|---|---|---|---|---|
| 168 | 195 | 227 | 193 | 207 | 189 | 176 | 216 |
| 164 | 199 | 198 | 203 | 214 | 191 | 171 | 200 |
| 197 | 195 | 184 | 202 | 188 | 173 | 197 | 181 |

Construct a stem-and-leaf display with the stem labels 16, 17, 18, 19, 20, 21, and 22.

**2.17**   Following are the lifetimes of 25 electronic components sampled from a production lot: 834, 919, 784, 865, 839, 912, 888, 783, 655, 831, 886, 842, 760, 854, 939, 961, 826, 954, 866, 675, 760, 865, 901, 632, and 718. Construct a stem-and-leaf display with one-digit stem labels and two-digit leaves. (Data are in hours of continuous use.)

**2.18**   Following are the low temperatures recorded at the Phoenix Sky Harbor Airport during a recent month of February: 46, 43, 54, 53, 43, 42, 47, 46, 46, 45, 43, 39, 52,

51, 48, 42, 43, 47, 49, 54, 53, 45, 50, 52, 53, 49, 35, and 34. Construct a double-stem display. Arrange the leaves according to size.

**2.19**   Following are measurements (to the nearest hundredth of a second) of the time required for sound to travel between two points: 1.53, 1.66, 1.42, 1.54, 1.37, 1.44, 1.60, 1.68, 1.72, 1.59, 1.54, 1.63, 1.58, 1.46, 1.52, 1.58, 1.53, 1.50, 1.49, and 1.62. Construct a stem-and-leaf display with the stem labels 1.3, 1.4, 1.5, 1.6, and 1.7, and one-digit leaves.

**2.20**   Convert the stem-and-leaf display obtained in Exercise 2.19 into a double-stem display.

**2.21**   Following are the IQs of 24 persons empaneled for jury duty by a municipal court: 108, 97, 103, 122, 84, 105, 101, 113, 127, 103, 124, 97, 88, 109, 103, 115, 96, 110, 104, 92, 105, 106, 93, and 99. Construct a stem-and-leaf display with the stem labels 8, 9, 10, 11, and 12, and one-digit leaves.

**2.22**   Following are the numbers of outpatients seeking treatment at a hospital during the 28 days of February:

| 78 | 66 | 54 | 62 | 67 | 68 | 62 |
|----|----|----|----|----|----|----|
| 60 | 71 | 67 | 80 | 60 | 56 | 61 |
| 63 | 65 | 52 | 69 | 59 | 65 | 76 |
| 68 | 64 | 60 | 71 | 57 | 56 | 76 |

Construct a double-stem display for these values.

# 2.3 FREQUENCY DISTRIBUTION

When we deal with large sets of data, and sometimes even when we deal with not so large sets of data, it can be quite a problem to get a clear picture of the information that they convey. As we saw in Sections 2.1 and 2.2, this usually requires that we rearrange and/or display the **raw** (untreated) **data** in some special form. Traditionally, this involves a **frequency distribution** or one of its **graphical presentations**, where we group or classify the data into a number of categories or classes.

Following are two examples. A recent study of their total billings (rounded to the nearest dollar) yielded data for a sample of 4,757 law firms. Rather than providing printouts of the 4,757 values, the information is disseminated by means of the following table:

| Total billings | Number of law firms |
|----------------|---------------------|
| Less than $300,000 | 2,405 |
| $300,000 to $499,999 | 1,088 |
| $500,000 to $749,999 | 271 |
| $750,000 to $999,999 | 315 |
| $1,000,000 or more | 678 |
| Total | 4,757 |

This does not show much detail, but it may well be adequate for some special purposes. This should also be the case with the following table, provided by the Office of Consumer Affairs of the U.S. Department of Transportation, which summarizes consumer complaints against U.S. airlines in a recent year.

| Nature of complaint | Number of complaints |
|---|---|
| Cancellations and delays | 1,586 |
| Customer service | 805 |
| Baggage handling | 761 |
| Ticketing and boarding | 598 |
| Refunds | 393 |
| Bumping | 301 |
| Information about fares | 267 |
| Other | 468 |
| Total | 5,179 |

When data are grouped according to numerical size, as in the first example, the resulting table is called a **numerical** or **quantitative distribution**. When they are grouped into nonnumerical categories, as in the second example, the resulting table is called a **categorical** or **qualitative distribution**. In either case we refer to them as **frequency distributions**.

Frequency distributions present data in a relatively compact form, give a good overall picture, and contain information that is adequate for many purposes, but, as we said previously, there is some loss of information. Some things that can be determined from the original data cannot be determined from a distribution. For instance, in the first example the distribution does not tell us the exact size of the lowest and the highest billings, nor does it provide the total of the billings of the 4,757 law firms. Similarly, in the second example we cannot tell how many of the complaints about baggage handling pertain to physical damage and how many of the complaints pertain to its loss or its delay in transit.

Nevertheless, frequency distributions present information in a more handy form, and the price we pay for this—the loss of certain information—is usually a fair exchange.

The construction of a frequency distribution consists essentially of three steps:

1. Choosing the **classes** (intervals or categories)
2. Sorting or tallying the data into these classes
3. Counting the number of items in each class

Since the second and third steps are purely mechanical, we concentrate here on the first, namely, that of choosing a suitable classification.

For numerical distributions, this consists of deciding how many classes we are going to use and from where to where each class should go. Both of these choices are essentially arbitrary, but the following rules are usually observed:

> **We seldom use fewer than 5 or more than 15 classes; the exact number we use in a given situation depends largely on how many measurements or observations there are.**

Clearly, we would lose more than we gain if we group five observations into 12 classes with most of them empty, and we would probably discard too much information if we group a thousand measurements into three classes.

---

**We always make sure that each item (measurement or observation) goes into one and only one class.**

---

To this end we must make sure that the smallest and largest values fall within the classification, that none of the values can fall into a gap between successive classes, and that the classes do not overlap, namely, that successive classes have no values in common.

---

**Whenever possible, we make the classes cover equal ranges of values.**

---

Also, if we can, we make these ranges multiples of numbers that are easy to work with, such as 5, 10, or 100, since this will tend to facilitate the construction and the use of a distribution.

If we assume that the law firm billings were all rounded to the nearest dollar, only the third of these rules was violated in the construction of the distribution on page 21. However, had the billings been given to the nearest cent, then a billing of, say, $499,999.54 would have fallen between the second class and the third class, and we would also have violated the second rule. The third rule was violated because the classes do not all cover equal ranges of values; in fact, the first class and the last class have, respectively, no specified lower and upper limits.

Classes of the "less than," "or less," "more than," or "or more" variety are referred to as **open classes**, and they are used to reduce the number of classes that are needed when some of the values are much smaller than or much greater than the rest. Generally, open classes should be avoided, however, because they make it impossible to calculate certain values of interest, such as averages or totals.

Insofar as the second rule is concerned, we have to watch whether the data are given to the nearest dollar or to the nearest cent, whether they are given to the nearest inch or to the nearest tenth of an inch, whether they are given to the nearest ounce or to the nearest hundredth of an ounce, and so on. For instance, if we want to group the weights of certain animals, we might use the first of the following classifications when the weights are given to the nearest kilogram, the second when the weights are given to the nearest tenth of a kilogram, and the third when the weights are given to the nearest hundredth of a kilogram:

| Weight (kilograms) | Weight (kilograms) | Weight (kilograms) |
|---|---|---|
| 10–14 | 10.0–14.9 | 10.00–14.99 |
| 15–19 | 15.0–19.9 | 15.00–19.99 |
| 20–24 | 20.0–24.9 | 20.00–24.99 |
| 25–29 | 25.0–29.9 | 25.00–29.99 |
| 30–34 | 30.0–34.9 | 30.00–34.99 |
| etc. | etc. | etc. |

To illustrate what we have been discussing in this section, let us now go through the actual steps of grouping a set of data into a frequency distribution.

EXAMPLE **2.4**    Based on information supplied by the Chief Park Naturalist of Yellowstone National Park and updated in accordance with information about the trend of such data provided on the Internet, the following are 110 simulated "waiting times" (in minutes) between eruptions of the Old Faithful geyser:

| 81 | 83 | 94 | 73 | 78 | 94 | 73 | 89 | 112 | 80 |
|----|----|----|----|----|----|----|----|-----|----|
| 94 | 89 | 35 | 80 | 74 | 91 | 89 | 83 | 80 | 82 |
| 91 | 80 | 83 | 91 | 89 | 82 | 118 | 105 | 64 | 56 |
| 76 | 69 | 78 | 42 | 76 | 82 | 82 | 60 | 73 | 69 |
| 91 | 83 | 67 | 85 | 60 | 65 | 69 | 85 | 65 | 82 |
| 53 | 83 | 62 | 107 | 60 | 85 | 69 | 92 | 40 | 71 |
| 82 | 89 | 76 | 55 | 98 | 74 | 89 | 98 | 69 | 87 |
| 74 | 98 | 94 | 82 | 82 | 80 | 71 | 73 | 74 | 80 |
| 60 | 69 | 78 | 74 | 64 | 80 | 83 | 82 | 65 | 67 |
| 94 | 73 | 33 | 87 | 73 | 85 | 78 | 73 | 74 | 83 |
| 83 | 51 | 67 | 73 | 87 | 85 | 98 | 91 | 73 | 108 |

As can be verified easily, the smallest value is 33 and the largest value is 118, so that a convenient choice for grouping these data would be the nine classes 30–39, 40–49, 50–59, 60–69, 70–79, 80–89, 90–99, 100–109, and 110–119. These classes will accommodate all of the data, they do not overlap, and they are all of the same size. There are other possibilities (for instance, 25–34, 35–44, 45–54, 55–64, 65–74, 75–84, 85–94, 95–104, 105–114, and 115–124), but it should be apparent that our first choice will facilitate the tally. Use the original classification to construct a frequency distribution of the given "waiting time" data.

Solution    Tallying the 110 values, we get the results shown in the following table:

| Waiting times between eruptions (minutes) | Tally | Frequency |
|------------------------------------------|-------|-----------|
| 30–39 | \|\| | 2 |
| 40–49 | \|\| | 2 |
| 50–59 | \|\|\|\| | 4 |
| 60–69 | ЖҤ ЖҤ ЖҤ \|\|\|\| | 19 |
| 70–79 | ЖҤ ЖҤ ЖҤ ЖҤ \|\|\|\| | 24 |
| 80–89 | ЖҤ ЖҤ ЖҤ ЖҤ ЖҤ ЖҤ ЖҤ \|\|\|\| | 39 |
| 90–99 | ЖҤ ЖҤ ЖҤ | 15 |
| 100–109 | \|\|\| | 3 |
| 110–119 | \|\| | 2 |
| | Total | 110 |

The numbers given in the right-hand column of this table, which show how many values fall into each class, are called the **class frequencies**. The smallest and largest values that can go into any given class are called its **class limits**, and for the distribution of the waiting times between eruptions they are 30 and 39, 40 and 49, 50 and 59, . . . , and 110 and 119. More specifically, 30, 40, 50, . . . , and 110 are called the **lower class limits**, and 39, 49, 59, . . . , and 119 are called the **upper class limits**.

The amounts of time that we grouped in our example were all given to the nearest minute, so that 30 actually includes everything from 29.5 to 30.5, 39 includes everything from 38.5 to 39.5, and the class 30–39 includes everything from 29.5 to 39.5. Similarly, the second class includes everything from 39.5 to 49.5, . . . , and the class at the bottom of the distribution includes everything from 109.5 to 119.5. It is customary to refer to 29.5, 39.5, 49.5, . . . , and 119.5 as the **class boundaries** or the **real class limits** of the distribution. Although 39.5 is the **upper boundary** of the first class and also the **lower boundary** of the second class, 49.5 is the upper boundary of the second class and also the lower boundary of the third class, and so forth, there is no cause for alarm. The class boundaries are by choice *impossible values* that cannot occur among the data being grouped. If we assume again that the law firm billings grouped in the distribution on page 21 were all rounded to the nearest dollar, the class boundaries $299,999.50, $499,999.50, $749,999.50, and $999,999.50 are also impossible values.

We emphasize this point because, to avoid gaps in the continuous number scale, some statistics texts, some widely used computer programs, and some graphing calculators (MINITAB, for example, and the TI-83) include in each class its lower boundary, and the highest class also includes its upper boundary. They would include 29.5 but not 39.5 in the first class of the preceding distribution of waiting times between eruptions of Old Faithful. Similarly, they would include 39.5 but not 49.5 in the second class, . . . , but 109.5 as well as 119.5 in the highest class of the distribution. All this is immaterial, of course, so long as the class boundaries are impossible values that cannot occur among the data being grouped.

Numerical distributions also have what we call **class marks** and **class intervals**. Class marks are simply the midpoints of the classes, and they are found by adding the lower and upper limits of a class (or its lower and upper boundaries) and dividing by 2. A class interval is merely the length of a class, or the range of values it can contain, and it is given by the difference between its boundaries. If the classes of a distribution are all equal in length, their common class interval, which we call the **class interval of the distribution**, is also given by the difference between any two successive class marks. Thus, the class marks of the waiting-time distribution are 34.5, 44.5, 54.5, . . . , and 114.5, and the class intervals and the class interval of the distribution are all equal to 10.

There are essentially two ways in which frequency distributions can be modified to suit particular needs. One way is to convert a distribution into a **percentage distribution** by dividing each class frequency by the total number of items grouped, and then multiplying by 100.

**EXAMPLE 2.5**   Convert the waiting-time distribution obtained in Example 2.4 into a percentage distribution.

**Solution**  The first class contains $\frac{2}{110} \cdot 100 = 1.82\%$ of the data (rounded to two decimals), and so does the second class. The third class contains $\frac{4}{110} \cdot 100 = 3.64\%$ of the data, the fourth class contains $\frac{19}{110} \cdot 100 = 17.27\%$ of the data, ..., and the bottom class again contains 1.82% of the data. These results are shown in the following table:

| Waiting times between eruptions (minutes) | Percentage |
|---|---|
| 30–39 | 1.82 |
| 40–49 | 1.82 |
| 50–59 | 3.64 |
| 60–69 | 17.27 |
| 70–79 | 21.82 |
| 80–89 | 35.45 |
| 90–99 | 13.64 |
| 100–109 | 2.73 |
| 110–119 | 1.82 |

The percentages total 100.01, with the difference, of course, due to rounding.

The other way of modifying a frequency distribution is to convert it into a "less than," "or less," "more than," or "or more" **cumulative distribution**. To construct a cumulative distribution, we simply add the class frequencies, starting either at the top or at the bottom of the distribution.

**EXAMPLE 2.6**  Convert the waiting-time distribution obtained in Example 2.4 into a cumulative "less than" distribution.

**Solution**  Since none of the values is less than 30, two $(0+2)$ of the values are less than 40, four $(0+2+2)$ of the values are less than 50, eight $(0+2+2+4)$ of the values are less than 60, ..., and all 110 of the values are less than 120, we get

| Waiting times between eruptions (minutes) | Cumulative frequency |
|---|---|
| Less than 30 | 0 |
| Less than 40 | 2 |
| Less than 50 | 4 |
| Less than 60 | 8 |
| Less than 70 | 27 |
| Less than 80 | 51 |
| Less than 90 | 90 |
| Less than 100 | 105 |
| Less than 110 | 108 |
| Less than 120 | 110 |

Note that instead of "less than 30" we could have written "29 or less," instead of "less than 40" we could have written "39 or less," instead of "less than 50" we could have written "49 or less," and so forth. Of course, we would then have referred to the distribution as a cumulative "or less" distribution.

In the same way we can also convert a percentage distribution into a **cumulative percentage distribution**. We simply add the percentages instead of the frequencies, starting either at the top or at the bottom of the distribution.

So far we have discussed only the construction of numerical distributions, but the general problem of constructing categorical (or qualitative) distributions is about the same. Here again we must decide how many categories (classes) to use and what kind of items each category is to contain, making sure that all the items are accommodated and that there are no ambiguities. Since the categories must often be chosen before any data are actually collected, it is usually prudent to include a category labeled "others" or "miscellaneous."

For categorical distributions, we do not have to worry about such mathematical details as class limits, class boundaries, and class marks. On the other hand, there is often a serious problem with ambiguities and we must be very careful and explicit in defining what each category is to contain. For instance, if we had to classify items sold at a supermarket into "meats," "frozen foods," "baked goods," and so forth, it would be difficult to decide, for example, where to put frozen beef pies. Similarly, if we had to classify occupations, it would be difficult to decide where to put a farm manager, if our table contained (without qualification) the two categories "farmers" and "managers." For this reason, it is advisable, where possible, to use standard categories developed by the Bureau of the Census and other government agencies.

EXERCISES

**2.23**  The weights of 125 rats used in medical research vary from 231 grams to 365 grams. Show the class limits of a table with eight classes into which these weights (rounded to the nearest gram) could conveniently be grouped.

**2.24**  The burning times of certain solid-fuel rockets vary from 3.2 to 5.9 seconds. Show the class limits of a table with six classes into which these burning times could be grouped.

**2.25**  The average monthly electric bills of the residents of a retirement community vary from $37.65 to $184.66, with the large variation being due to the high cost of air conditioning during the summer months. Show the class limits of a table into which these figures could be grouped, if it is to contain
(a)  only four classes;
(b)  six classes;
(c)  eight classes.

**2.26**  Decide for each of the following whether it can be determined from the distribution of law firm billings on page 21; if possible, give a numerical answer:
(a)  The number of law firms with billings exceeding $300,000.
(b)  The number of law firms with billings exceeding $749,999.
(c)  The number of law firms with billings less than $250,000.
(d)  The number of law firms with billings less than $500,000.

**2.27**  Following is the distribution of the weights of 133 mineral specimens collected on a field trip:

| Weight (grams) | Number of specimens |
|---|---|
| 5.0–19.9 | 8 |
| 20.0–34.9 | 27 |
| 35.0–49.9 | 42 |
| 50.0–64.9 | 31 |
| 65.0–79.9 | 17 |
| 80.0–94.9 | 8 |

Find
(a)  the lower class limits;
(b)  the upper class limits;
(c)  the class boundaries;
(d)  the class intervals.

**2.28**   To group data on the number of rainy days reported by a newspaper for the month of May, we plan to use the classes 1–9, 10–19, 20–25, and 25–30. Explain where difficulties might arise.

**2.29**   To group sales invoices ranging from $12.00 to $79.00, a department store's accountant uses the following classes: 10.00–29.99, 30.00–49.99, 60.00–79.99, and 70.00–99.99. Explain where difficulties might arise.

**2.30**   Difficulties can also arise when we choose inappropriate classes for grouping categorical data. If men's shirts are classified according to the fibers of which they are made, explain where difficulties might arise if we include only the three categories: wool, silk, and synthetic fibers.

**2.31**   Explain where difficulties might arise if, in a study of the nutritional value of desserts, we use the categories pie, cake, fruit, pudding, and ice cream.

**2.32**   Temperature readings, rounded to the nearest degree Fahrenheit, are grouped into a distribution with the classes 55–60, 61–66, 67–72, 73–78, and 79–84. Find
(a)  the class boundaries of this distribution;
(b)  the class marks.

**2.33**   Measurements given to the nearest centimeter are grouped into a table having the class boundaries 19.5, 24.5, 29.5, 34.5, 39.5, and 44.5. Find
(a)  the class limits of these five classes;
(b)  their class marks;
(c)  their class intervals.

**2.34**   The class marks of a distribution of retail food prices (in cents) are 27, 42, 57, 72, 87, and 102. Find the corresponding
(a)  class boundaries;
(b)  class limits.

**2.35**   The wingspans of certain birds are grouped into a distribution with the class boundaries 59.95, 74.95, 89.95, 104.95, 119.95, and 134.95 centimeters. Find the corresponding
(a)  class limits;
(b)  class marks.

**2.36**   Following are the percent shrinkages on drying of 40 plastic clay specimens:

| | | | | | | | |
|---|---|---|---|---|---|---|---|
| 20.3 | 16.8 | 21.7 | 19.4 | 15.9 | 18.3 | 22.3 | 17.1 |
| 19.6 | 21.5 | 21.5 | 17.9 | 19.5 | 19.7 | 13.3 | 20.5 |
| 24.4 | 19.8 | 19.3 | 17.9 | 18.9 | 18.1 | 23.5 | 19.8 |
| 20.4 | 18.4 | 19.5 | 18.5 | 17.4 | 18.7 | 18.3 | 20.4 |
| 20.1 | 18.5 | 17.8 | 17.3 | 20.0 | 19.2 | 18.4 | 19.0 |

Group these percentages into a frequency distribution with the classes 13.0–14.9, 15.0–16.9, 17.0–18.9, 19.0–20.9, 21.0–22.9, and 23.0–24.9.

**2.37**  Convert the distribution obtained in Exercise 2.36 into a percentage distribution.

**2.38**  Convert the distribution obtained in Exercise 2.36 into a cumulative "less than" distribution.

**2.39**  Following are 60 measurements (in 0.00001 inch) of the thickness of an aluminum alloy plating obtained in the analysis of an anodizing process:

$$
\begin{array}{cccccccccccc}
24 & 24 & 41 & 36 & 32 & 33 & 22 & 34 & 39 & 25 & 21 & 32 \\
36 & 26 & 43 & 28 & 30 & 27 & 38 & 25 & 33 & 42 & 30 & 32 \\
31 & 34 & 21 & 27 & 35 & 48 & 35 & 26 & 21 & 30 & 37 & 39 \\
25 & 33 & 36 & 27 & 29 & 28 & 26 & 22 & 23 & 30 & 43 & 20 \\
31 & 22 & 37 & 23 & 30 & 29 & 31 & 28 & 36 & 38 & 20 & 24 \\
\end{array}
$$

Group these thicknesses into a distribution with the classes 20–24, 25–29, 30–34, 35–39, 40–44, and 45–49.

**2.40**  Convert the distribution obtained in Exercise 2.39 into a percentage distribution.

**2.41**  Convert the distribution obtained in Exercise 2.40 into a cumulative "or less" percentage distribution.

**2.42**  Following are the lengths of root penetrations (in feet) of 120 crested wheatgrass seedlings one month after planting:

$$
\begin{array}{cccccccccc}
0.95 & 0.88 & 0.90 & 1.23 & 0.83 & 0.67 & 1.41 & 1.04 & 1.01 & 0.81 \\
0.78 & 1.21 & 0.80 & 1.43 & 1.27 & 1.16 & 1.06 & 0.86 & 0.70 & 0.80 \\
0.71 & 0.93 & 1.00 & 0.62 & 0.80 & 0.81 & 0.75 & 1.25 & 0.86 & 1.15 \\
0.91 & 0.62 & 0.84 & 1.08 & 0.99 & 1.38 & 0.98 & 0.93 & 0.80 & 1.25 \\
0.82 & 0.97 & 0.85 & 0.79 & 0.90 & 0.84 & 0.53 & 0.83 & 0.83 & 0.60 \\
0.95 & 0.68 & 1.27 & 0.97 & 0.80 & 1.13 & 0.89 & 0.83 & 1.47 & 0.96 \\
1.34 & 0.87 & 0.75 & 0.95 & 1.13 & 0.95 & 0.85 & 1.00 & 0.73 & 1.36 \\
0.94 & 0.80 & 1.33 & 0.91 & 1.03 & 0.93 & 1.34 & 0.82 & 0.82 & 0.95 \\
1.11 & 1.02 & 1.21 & 0.90 & 0.80 & 0.92 & 1.06 & 1.17 & 0.85 & 1.00 \\
0.88 & 0.86 & 0.64 & 0.96 & 0.88 & 0.95 & 0.74 & 0.57 & 0.96 & 0.78 \\
0.89 & 0.81 & 0.89 & 0.88 & 0.73 & 1.08 & 0.87 & 0.83 & 1.19 & 0.84 \\
0.94 & 0.70 & 0.76 & 0.85 & 0.97 & 0.86 & 0.94 & 1.06 & 1.27 & 1.09 \\
\end{array}
$$

Group these lengths into a table having the classes 0.50–0.59, 0.60–0.69, 0.70–0.79, ..., and 1.40–1.49.

**2.43**  Convert the distribution obtained in Exercise 2.42 into a cumulative "more than" distribution.

**2.44**  The alumni association of a university sponsors monthly outings for its single members. Its records show that during the most recent four years these outings have been attended by

$$
\begin{array}{cccccccccccc}
28 & 51 & 31 & 38 & 27 & 35 & 33 & 40 & 37 & 28 & 33 & 27 \\
33 & 31 & 41 & 46 & 40 & 36 & 53 & 23 & 33 & 27 & 40 & 30 \\
33 & 22 & 37 & 38 & 36 & 48 & 22 & 36 & 45 & 34 & 26 & 28 \\
40 & 42 & 43 & 41 & 35 & 50 & 31 & 48 & 38 & 33 & 39 & 35 \\
\end{array}
$$

single members of the alumni association of the university. Group these figures into a frequency distribution with the classes 20–24, 25–29, 30–34, 35–39, 40–44, 45–49, and 50–54.

**2.45**   Convert the distribution obtained in Exercise 2.44 into an "or more" percentage distribution.

**2.46**   During the Major League baseball season, use the sports pages of a Monday newspaper to list the number of runs scored by the winning teams in all Major League games played on the preceding Sunday. Provided that not too many games rained out, construct a distribution of these winning numbers of runs with four or five classes.

**2.47**   With the permission of the manager, station yourself near one of the cash registers and record the total amounts spent by 50 customers. Also, group these amounts into a distribution with six or seven classes.

**2.48**   Locate a site near you where free blood pressure tests are offered. With their permission, record the ages of forty persons having their blood pressure checked and construct a frequency distribution.

# 2.4 GRAPHICAL PRESENTATIONS

When frequency distributions are constructed mainly to condense large sets of data and present them in an "easy to digest" form, it is usually most effective to display them graphically. As the saying goes, a picture speaks louder than a thousand words, and this was true even before the current proliferation of computer graphics. Nowadays, each statistical software package strives to outdo its competitors by means of more and more elaborate pictorial presentations of statistical data.

For frequency distributions, the most common form of graphical presentation is the **histogram**, like the one shown in Figure 2.5. Histograms are constructed by representing the measurements or observations that are grouped (in Figure 2.5 the waiting times in minutes between eruptions of Old Faithful) on a horizontal scale, and drawing rectangles whose bases equal the class intervals and whose heights are the corresponding class frequencies. The markings on the horizontal scale of a histogram can be the class marks as in Figure 2.5, class limits, the class boundaries, or arbitrary key values. For practical reasons, it is generally preferable to show the class limits, even though the rectangles actually go from one class boundary to the next. After all, the class limits tell us *what values go into each class*. Note that histograms cannot be drawn for distributions with open classes and that they require special care when the class intervals are not all equal (see Exercise 2.57 on page 35).

The histogram shown in Figure 2.5 was obtained with the use of a computer, even though the data had already been grouped on page 24, and it would have been very easy to draw it manually. In actual practice, just entering the data in a computer can be more work than tallying the data and then drawing the rectangles.

Our definition of "histogram" may be referred to as traditional, for nowadays the term is applied much more loosely to all sorts of graphical presentations of frequency distributions, where the class frequencies are not necessarily represented by rectangles. For instance, Figure 2.6 shows an older MINITAB-generated

**Figure 2.5**
Computer printout of a histogram for the waiting times between eruptions of Old Faithful geyser.

**Figure 2.6**
Computer printout of histogram obtained with MINITAB 10Xtra.

```
MTB > GStd.
MTB > Histogram cl;
SUBC>    Start 34.5 114.5;
SUBC>    Increment 10.



Histogram of Cl      N = 110

Midpoint     Count
    34.5         2 **
    44.5         2 **
    54.5         4 ****
    64.5        19 *******************
    74.5        24 ************************
    84.5        39 ***************************************
    94.5        15 ***************
   104.5         3 ***
   114.5         2 **
```

histogram, which really looks more like a dot diagram (see Section 2.1), except that the dots aligned at the class marks represent the various values in the corresponding classes rather than repeated identical values.

Also referred to at times as histograms are bar charts (see Section 2.1), such as the one shown in Figure 2.7. The heights of the rectangles, or bars, again represent the class frequencies, but there is no pretense of having a continuous horizontal scale.

Another less widely used form of graphical presentation of a frequency distribution is the **frequency polygon**, as illustrated by Figure 2.8. Here, the class frequencies are plotted at the class marks and the successive points are connected by straight lines. Note that we added classes with zero frequencies at both ends of the distribution to "tie down" the graph to the horizontal scale.

**Figure 2.7**
Bar chart of the distribution of waiting times between eruptions of Old Faithful geyser.

**Figure 2.8**
Frequency polygon of the distribution of waiting times between eruptions of Old Faithful geyser.

If we apply a similar technique to a cumulative distribution, usually a "less than" distribution, we obtain what is called an **ogive** (which rhymes with five or jive). However, in an ogive the cumulative frequencies are plotted at the class boundaries instead of the class marks—it stands to reason that the cumulative frequency corresponding to, say, "less than 60" should be plotted at the class boundary 59.5, since "less than 60" actually includes everything up to 59.5. Figure 2.9 shows an ogive of the "less than" distribution of the waiting times obtained on page 26.

Although the visual appeal of histograms, bar charts, frequency polygons, and ogives is a marked improvement over that of mere tables, there are various ways in which distributions can be presented even more dramatically and often more effectively. An example of such a pictorial presentation (often seen in newspapers, magazines, and reports of various sorts) is the **pictogram** shown in Figure 2.10.

Categorical distributions are often presented graphically as **pie charts**, like the one shown in Figure 2.11, where a circle is divided into sectors—pie-shaped pieces—that are proportional in size to the corresponding frequencies or percentages. To construct a pie chart, we first convert the distribution into a percentage distribution. Then, since a complete circle corresponds to 360 degrees, we obtain the central angles of the various sectors by multiplying the percentages by 3.6.

**Figure 2.9**
Ogive of the distribution of waiting times between eruptions of Old Faithful geyser.

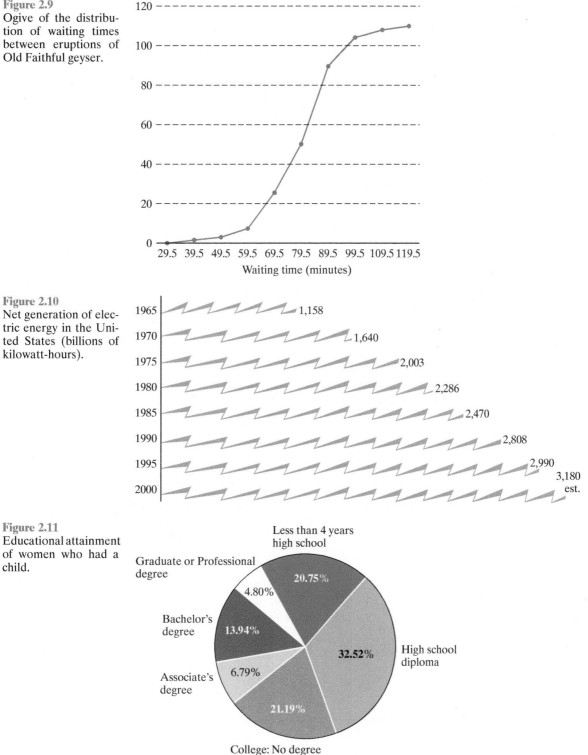

Waiting time (minutes)

**Figure 2.10**
Net generation of electric energy in the United States (billions of kilowatt-hours).

1965    1,158
1970    1,640
1975    2,003
1980    2,286
1985    2,470
1990    2,808
1995    2,990
2000    3,180 est.

**Figure 2.11**
Educational attainment of women who had a child.

Less than 4 years high school

Graduate or Professional degree

4.80%    20.75%

Bachelor's degree    13.94%

32.52%    High school diploma

Associate's degree    6.79%

21.19%

College: No degree

EXAMPLE **2.7**   The following table, showing the educational attainment of women who had a child in a certain year is based on *The Statistical Abstract of the United States*. The data are given in thousands.

| | |
|---|---|
| Less than 4 years of high school | 12,159 |
| High school diploma | 19,063 |
| College: No degree | 12,422 |
| Associate's degree | 3,982 |
| Bachelor's degree | 8,173 |
| Graduate or Professional degree | 2,812 |
| | 58,611 |

Construct a pie chart.

**Solution**   The percentages corresponding to the six categories are $\frac{12,159}{58,611} \cdot 100 = 20.75\%$, $\frac{19,063}{58,611} \cdot 100 = 32.52\%$, $\frac{12,422}{58,611} \cdot 100 = 21.19\%$, $\frac{3,982}{58,611} \cdot 100 = 6.79\%$, $\frac{8,173}{58,611} \cdot 100 = 13.94\%$, and $\frac{2,812}{58,611} \cdot 100 = 4.80\%$ rounded to two decimals. Multiplying these percentages by 3.6, we find that the central angles of the six sectors are 74.7, 117.1, 76.3, 24.4, 50.2, and 17.3 degrees. Rounding the angles to the nearest degree and using a protractor, we get the pie chart shown in Figure 2.11.   ▪◼

Many computers are programmed so that, once the data have been entered, a simple command will produce a pie chart, or a variation thereof. Some computer-generated pie charts use color, some are three dimensional, some cut out sectors (like pieces of pie) for emphasis, and some shade or tint the various sectors.

EXERCISES

**2.49**   Draw a histogram of the following distribution of the frequencies with which rifle shots hit the respective distances from the center of a target. Indicate the class limits that correspond to each rectangle of the histogram.

| Distance (centimeters) | Frequency |
|---|---|
| 0.0–1.9 | 23 |
| 2.0–3.9 | 18 |
| 4.0–5.9 | 12 |
| 6.0–7.9 | 9 |
| 8.0–9.9 | 5 |
| 10.0–11.9 | 2 |
| 12.0–13.9 | 1 |

**2.50**   Construct a histogram of the distribution of the weights of mineral specimens given in Exercise 2.27.

**2.51**    Construct a histogram of whichever data you grouped among those of Exercise 2.36 or Exercise 2.39.

**2.52**    Construct a frequency polygon of whichever data you grouped among those of Exercise 2.36 or Exercise 2.39.

**2.53**    Following is the distribution of the number of fish tacos served for lunch by a Mexican restaurant on 60 weekdays:

| Number of fish tacos | Number of weekdays |
|:---:|:---:|
| 30–39 | 4 |
| 40–49 | 23 |
| 50–59 | 28 |
| 60–69 | 5 |

Construct a

(a) histogram;       (c) frequency polygon;

(b) bar chart;       (d) ogive.

**2.54**    Following are 80 measurements of the iron-solution index of tin-plate specimens, designed to measure the corrosion resistance of tin-plated steel:

| | | | | | | | | | |
|---|---|---|---|---|---|---|---|---|---|
| 0.78 | 0.65 | 0.48 | 0.83 | 1.43 | 0.92 | 0.92 | 0.72 | 0.48 | 0.96 |
| 0.72 | 0.48 | 0.83 | 0.49 | 0.78 | 0.96 | 1.06 | 0.83 | 0.78 | 0.82 |
| 1.12 | 0.78 | 1.03 | 0.88 | 1.23 | 0.28 | 0.95 | 1.16 | 0.47 | 0.55 |
| 0.97 | 1.20 | 0.77 | 0.72 | 0.45 | 1.36 | 0.65 | 0.73 | 0.39 | 0.94 |
| 0.79 | 1.26 | 1.06 | 0.90 | 0.77 | 0.45 | 0.78 | 0.77 | 1.09 | 0.73 |
| 0.64 | 0.91 | 0.95 | 0.71 | 1.20 | 0.88 | 0.83 | 0.78 | 1.04 | 1.33 |
| 0.52 | 0.32 | 0.54 | 0.63 | 0.44 | 0.92 | 1.00 | 0.79 | 0.63 | 1.23 |
| 0.65 | 0.64 | 0.48 | 0.79 | 0.99 | 0.57 | 0.91 | 1.12 | 0.70 | 1.05 |

Group these measurements into a table with the class interval 0.20 and draw its histogram. The smallest class is 0.20–0.39.

**2.55**    Convert the distribution obtained in Exercise 2.54 into a cumulative "less than" distribution and draw its ogive.

**2.56**    Draw a frequency polygon of the distribution obtained in Exercise 2.54.

**2.57**    Figure 2.12 shows the distribution of the scores of 80 incoming college freshmen on a French language placement examination. Explain why it might give a misleading impression and indicate how it might be improved.

**Figure 2.12**
Distribution of scores on a French language placement examination.

Scores on French language placement examination

**2.58**   Combine the second and third classes of the distribution of Exercise 2.49 and draw a histogram in which the areas of the rectangles are proportional to the class frequencies.

**2.59**   Construct a pie chart of the distribution of consumer complaints against U.S. airlines given on page 22.

**2.60**   Observe 80 automobiles in the lane nearest to where you are located along a fairly busy street and check off each one as being a sedan, a coupe, a convertible, a minivan, an SUV, a pickup, or some other commercial vehicle. Construct a pie chart of these categorical data.

**2.61**   Observe what dessert, if any, was chosen by 60 customers in a cafeteria and check off each one as none, a piece of pie, a piece of cake, a pudding, some ice cream or a sherbert, a French pastry, or a fruit cup. Construct a pie chart of these categorical data.

**2.62**   Use a computer and appropriate software to produce a pie chart for the categorical data obtained in Exercise 2.61. (Note that the TI-83 is not programmed for pie charts. An earlier model, the TI-73, was programmed for this purpose, but since its pie charts were not particularly attractive, they were not included in the TI-83.)

**2.63**   Asked to rate the maneuverability of a new model car as excellent, very good, good, fair, or poor, 50 drivers responded as follows: very good, good, good, fair, excellent, good, good, good, very good, poor, good, good, good, good, very good, good, fair, good, good, poor, very good, fair, good, good, excellent, very good, good, good, good, fair, fair, very good, good, very good, excellent, very good, fair, good, very good, good, fair, good, good, excellent, very good, fair, fair, good, very good, and good. Construct a pie chart showing the percentages corresponding to these ratings.

**2.64**   The pictogram of Figure 2.13 is intended to illustrate that per capita personal income in the United States has doubled from $10,000 in 1980 to $20,000 in 1992. Explain why this pictogram conveys a misleading impression and indicate how it might be modified.

**Figure 2.13**
Per capita personal income.

**2.65**    Following are the scores which 150 applicants for secretarial positions in a government agency obtained in an achievement test:

| | | | | | | | | | | | | | | |
|---|---|---|---|---|---|---|---|---|---|---|---|---|---|---|
| 62 | 37 | 49 | 56 | 89 | 52 | 41 | 70 | 80 | 28 | 54 | 45 | 95 | 52 | 66 |
| 43 | 59 | 56 | 70 | 64 | 55 | 62 | 79 | 48 | 26 | 61 | 56 | 62 | 49 | 71 |
| 58 | 77 | 74 | 63 | 37 | 68 | 41 | 52 | 60 | 69 | 58 | 73 | 14 | 60 | 84 |
| 55 | 44 | 63 | 47 | 28 | 83 | 46 | 55 | 53 | 72 | 54 | 83 | 70 | 61 | 36 |
| 46 | 50 | 35 | 56 | 43 | 61 | 76 | 63 | 66 | 42 | 50 | 65 | 41 | 62 | 74 |
| 45 | 60 | 47 | 72 | 87 | 54 | 67 | 45 | 76 | 52 | 57 | 32 | 55 | 70 | 44 |
| 81 | 72 | 54 | 57 | 92 | 61 | 42 | 30 | 57 | 58 | 62 | 86 | 45 | 63 | 28 |
| 57 | 40 | 44 | 55 | 36 | 55 | 44 | 40 | 57 | 28 | 63 | 45 | 86 | 61 | 51 |
| 68 | 56 | 47 | 86 | 52 | 70 | 59 | 40 | 71 | 56 | 34 | 62 | 81 | 58 | 43 |
| 46 | 60 | 45 | 69 | 74 | 42 | 55 | 46 | 50 | 53 | 77 | 70 | 49 | 58 | 63 |

Construct a histogram of the distribution of these scores with the classes 10–19, 20–29, 30–39, . . . , and 90–99, using

(a) a computer with appropriate software;

(b) a graphing calculator.

# 2.5 SUMMARIZING TWO-VARIABLE DATA

So far we have dealt only with situations involving one variable—the room occupancies in Section 2.2, the waiting times between eruptions of Old Faithful in Example 2.4, the plating thicknesses of Exercise 2.39, the root penetrations of Exercise 2.42, and so on. In actual practice, many statistical methods apply to situations involving two variables, and some of them apply even when the number of variables cannot be counted on one's fingers and toes. Not quite so extreme would be a problem in which we want to study the values of one-family homes, taking into consideration their age, their location, the number of bedrooms, the number of baths, the size of the garage, the type of roof, the number of fireplaces, the lot size, the value of nearby properties, and the accessibility of schools.

Leaving some of this work to later chapters and, in fact, most of it to advanced courses in statistics, we shall treat here only the display, listing, and grouping of data involving two variables; that is, problems dealing with the display of paired data. In most of these problems the main objective is to see whether there is a relationship, and if so what kind of relationship, so that we can predict one variable, denoted by the letter $y$, in terms of the other variable, denoted by the letter $x$. For instance, the $x$'s might be family incomes and the $y$'s might be family expenditures on medical care, they might be annealing temperatures and the hardness of steel, or they might be the time that has elapsed since the chemical treatment of a swimming pool and the remaining concentration of chlorine.

Pairs of values of $x$ and $y$ are usually referred to as **data points**, denoted by $(x, y)$, in the same way in which we denote points in the plane, with $x$ and $y$ being their $x$- and $y$-coordinates. When we actually plot the points corresponding to paired values of $x$ and $y$, we refer to the resulting graph as a **scatter diagram**, a **scatter plot**, or a **scattergram**. As their name implies, such graphs are useful tools in the analysis of whatever relationship there may exist between the $x$'s and the $y$'s, namely, in judging whether there are any discernible patterns.

**EXAMPLE 2.8** Raw materials used in the production of synthetic fiber are stored in a place that has no humidity control. Following are measurements of the relative humidity in the storage place, $x$, and the moisture content of a sample of the raw material, $y$, on 15 days:

| $x$ (Percent) | $y$ (Percent) |
|:---:|:---:|
| 36 | 12 |
| 27 | 11 |
| 24 | 10 |
| 50 | 17 |
| 31 | 10 |
| 23 | 12 |
| 45 | 18 |
| 44 | 16 |
| 43 | 14 |
| 32 | 13 |
| 19 | 11 |
| 34 | 12 |
| 38 | 17 |
| 21 | 8 |
| 16 | 7 |

Construct a scattergram.

**Solution** Scattergrams are easy enough to draw, yet the work can be simplified by using appropriate computer software or a graphing calculator. The one shown in Figure 2.14 was reproduced from the display screen of a TI-83 graphing calculator.

As can be seen from the diagram, the points are fairly widely scattered, yet there is evidence of an upward trend; that is, increases in the water content of the raw material seem to go with increases in humidity. In Figure 2.14 the dots are squares with their centers removed, but they can also be circles, ×'s, dots, or other kinds of symbols. (The units are not marked on either scale, but on the horizontal axis the tick marks are at 10, 20, 30, 40, and 50, and on the vertical axis they are at 5, 10, 15, and 20.) ■

Some difficulties arise when two or more of the data points are identical. In that case, the TI-83 graphing calculator shows only one point and the printouts

**Figure 2.14**
Scattergram of the humidity and water content data.

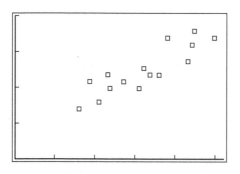

obtained with some statistical software. However, MINITAB has a special scattergram to take care of situations like this. Its so-called **character plot** prints the number 2 instead of the symbol $\times$ or * to indicate that there are two identical data points, and it would print a 3 if there were three. This is illustrated by the following example.

EXAMPLE **2.9**    The following data were obtained in a study of the relationship between the resistance (in ohms) and the failure time (in minutes) of certain overloaded resistors.

| Resistance $x$ | Failure time $y$ |
|---|---|
| 33 | 39 |
| 36 | 36 |
| 30 | 34 |
| 44 | 51 |
| 34 | 36 |
| 25 | 21 |
| 40 | 45 |
| 28 | 25 |
| 40 | 45 |
| 46 | 36 |
| 42 | 39 |
| 48 | 41 |
| 47 | 45 |
| 25 | 21 |

Construct a scattergram.

**Solution**    As can be seen, there are two duplicates among the data points: (40, 45) appears twice, and so does (25, 21). A scattergram that shows the number 2 instead of the symbol * at these two points is given in the computer printout of Figure 2.15.

**2.66**    The following data pertain to the chlorine residue in a swimming pool at various times after it has been treated with chemicals:

| Number of hours | Chlorine residual (parts per million) |
|---|---|
| 2 | 1.8 |
| 4 | 1.5 |
| 6 | 1.4 |
| 8 | 1.1 |
| 10 | 1.1 |
| 12 | 0.9 |

Draw a scatter diagram and describe what relationship, if any, it seems to indicate.

EXERCISES

**Figure 2.15**
Scattergram of the resistance and failure time data.

**2.67**   Following are the high school averages, $x$, and the first-year college grade-point indexes, $y$, of ten students:

| $x$ | $y$ |
|-----|-----|
| 3.0 | 2.6 |
| 2.7 | 2.4 |
| 3.8 | 3.9 |
| 2.6 | 2.1 |
| 3.2 | 2.6 |
| 3.4 | 3.3 |
| 2.8 | 2.2 |
| 3.1 | 3.2 |
| 3.5 | 2.8 |
| 3.3 | 2.5 |

Draw a scattergram and describe what kind of relationship, if any, it seems to display.

**2.68**   Following are the drying times of a certain varnish and the amount of a certain chemical that has been added:

| Amount of additive (grams) $x$ | Drying time (hours) $y$ |
|--------------------------------|-------------------------|
| 1 | 7.2 |
| 2 | 6.7 |
| 3 | 4.7 |
| 4 | 3.7 |
| 5 | 4.7 |
| 6 | 4.2 |
| 7 | 5.2 |
| 8 | 5.7 |

Draw a scatter diagram and describe what kind of relationship, if any, it seems to display.

# CHECKLIST OF KEY TERMS (with page references to their definitions)

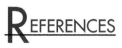

Bar chart, 15
   Horizontal, 15
   Vertical, 15
Categorical distribution, 22
Character plot, 39
Class, 22
   boundary, 25
   frequency, 24
   interval, 25
   limit, 25
   mark, 25
Cumulative distribution, 26
Cumulative percentage distribution, 27
Data point, 37
Dot diagram, 13
Double-stem display, 18
Exploratory data analysis, 18
Frequency distribution, 21
Frequency polygon, 31
Graphical presentation, 21
Grouping, 12
Histogram, 30
Interval of distribution, 25

Leaf, 17
Listing, 12
Lower boundary, 25
Lower class limit, 25
Numerical distribution, 22
Ogive, 32
Open class, 23
Pareto diagram, 15
Percentage distribution, 25
Pictogram, 32
Pie chart, 32
Qualitative distribution, 22
Quantitative distribution, 22
Raw data, 21
Real class limits, 25
Scattergram, 37
Scatter diagram, 37
Scatter plot, 37
Stem, 17
Stem-and-leaf display, 16
Stem label, 17
Upper boundary, 25
Upper class limit, 25

# REFERENCES

*Detailed information about statistical charts may be found in*

CLEVELAND, W. S., *The Elements of Graphing Data.* Monterey, Calif: Wadsworth Advanced Books and Software, 1985.

SCHMID, C. F., *Statistical Graphics: Design Principles and Practices.* New York: John Wiley & Sons, Inc., 1983.

TUFTE, E. R., *The Visual Display of Quantitative Information.* Cheshire, Conn.: Graphics Press, 1985.

*and some interesting information about the history of the graphical presentation of statistical data is given in an article by E. Royston in*

PEARSON, E. S., and KENDALL, M. G., eds., *Studies in the History of Statistics and Probability.* New York: Hafner Press, 1970.

*Discussions of what not to do in the presentation of statistical data may be found in*

CAMPBELL, S. K., *Flaws and Fallacies in Statistical Thinking.* Upper Saddle River, N.J.: Prentice Hall, Inc., 1974.

HUFF, D., *How to Lie with Statistics.* New York: W.W. Norton & Company, Inc., 1954.

REICHMAN, W. J., *Use and Abuse of Statistics.* New York: Penguin Books, 1971.

SPIRER, H. E., SPIRER, L., and JAFFE, A. J., *Misused Statistics*, 2nd ed. New York: Marcel Dekker, Inc., 1998.

*Useful references to lists of standard categories are given in*

HAUSER, P. M., and LEONARD, W. R., *Government Statistics for Business Use*, 2nd ed. New York: John Wiley & Sons, Inc., 1956.

*For further information about exploratory data analysis and stem-and-leaf displays in particular, see*

HARTWIG, F., and DEARING, B. E., *Exploratory Data Analysis.* Beverly Hills, Calif.: Sage Publications, Inc., 1979.

HOAGLIN, D. C., MOSTELLER, F., and TUKEY, J. W., *Understanding Robust and Exploratory Data Analysis.* New York: John Wiley & Sons, Inc., 1983.

KOOPMANS, L. H., *An Introduction to Contemporary Statistics.* North Scituate, Mass.: Duxbury Press, 1981.

TUKEY, J. W., *Exploratory Data Analysis.* Reading, Mass.: Addison-Wesley Publishing Company, Inc., 1977.

VELLEMAN, P. F., and HOAGLIN, D. C., *Applications, Basics, and Computing for Exploratory Data Analysis.* North Scituate, Mass.: Duxbury Press, 1980.

# SUMMARIZING DATA: MEASURES OF LOCATION

**3.1**  Populations and Samples 44

**3.2**  The Mean 45

**3.3**  The Weighted Mean 49

**3.4**  The Median 54

**3.5**  Other Fractiles 57

**3.6**  The Mode 59

**\*3.7**  The Description of Grouped Data 63

**3.8**  Technical Note (Summations) 70

Checklist of Key Terms 73

References 73

When we are about to describe a set of data, we are always well advised to say neither too little nor too much. Depending on the nature of the data and the purpose for which we want them described, statistical descriptions can be very brief or very elaborate. Sometimes we present data just as they are and let them speak for themselves. Sometimes we just group them and present their distribution in tabular or graphical form. Most of the time, however, we describe data in various other ways. Indeed, we usually describe data by means of a few well-chosen numbers that, in their way, summarize the entire set. Exactly what sort of numbers we choose depends on the particular aspects of the data that we want to describe. In one study we may be interested in a value that describes the middle or the most typical of a set of data; in another we may be interested in the value that is exceeded by 25% of the data; and in still another we may be interested in the interval between the smallest and the largest values among the data. The statistical measures asked for in the first two situations come under the heading of **measures of location** and the one asked for in the third situation comes under the heading of a **measure of variation**.

In this chapter we shall concentrate on measures of location, and in particular on **measures of central location**, which in some way describe the center or the middle of a set of data. Measures of variation and some other kinds of statistical descriptions will be discussed in Chapter 4.

# 3.1 POPULATIONS AND SAMPLES

When we said that the choice of a statistical description may depend on the nature of the data, we were referring among other things to the following distinction: If a set of data consists of all conceivably possible (or hypothetically possible) observations of a given phenomenon, we call it a **population**; if a set of data consists of only a part of a population, we call it a **sample**.

Here we added the phrase "hypothetically possible" to take care of such clearly hypothetical situations as where we look at the outcomes (heads or tails) of 12 flips of a coin as a sample from the potentially unlimited number of flips of the coin, where we look at the weights of ten 30-day-old lambs as a sample of the weights of all (past, present, and future) 30-day-old lambs raised at a certain farm, or where we look at four determinations of the uranium content of an ore as a sample of the many determinations that could conceivably be made. In fact, we often look at the results of an experiment as a sample of what we might get if the experiment were repeated over and over again.

Originally, statistics dealt with the description of human populations, census counts, and the like; but as it grew in scope, the term "population" took on the much wider connotation given to it in the preceding distinction between populations and samples. Whether or not it sounds strange to refer to the heights of all the trees in a forest or the speeds of all the cars passing a checkpoint as populations is beside the point. In statistics, "population" is a technical term with a meaning of its own.

Although we are free to call any group of items a population, what we do in practice depends on the context in which the items are to be viewed. Suppose, for instance, that we are offered a lot of 400 ceramic tiles, which we may or may not buy depending on their strength. If we measure the breaking strength of 20 of these tiles in order to estimate the average breaking strength of all the tiles, these 20 measurements are a sample from the population that consists of the breaking strengths of the 400 tiles. In another context, however, if we consider entering into a long-term contract calling for the delivery of tens of thousands of such tiles, we would look upon the breaking strengths of the original 400 tiles only as a sample. Similarly, the complete figures for a recent year, giving the elapsed times between the filing and disposition of divorce suits in San Diego County, can be looked upon as either a population or a sample. If we are interested only in San Diego County and that particular year, we would look upon the data as a population; on the other hand, if we want to generalize about the time that is required for the disposition of divorce suits in the entire United States, in some other county, or in some other year, we would look upon the data as a sample.

As we have used it here, the word "sample" has very much the same meaning as it has in everyday language. A newspaper considers the attitudes of 150 readers toward a proposed school bond to be a sample of the attitudes of all its readers toward the bond; and a consumer considers a box of Mrs. See's candy a sample of the firm's product. Later, we shall use the word "sample" only when referring to data that can reasonably serve as the basis for valid generalizations about the populations from which they came. In this more technical sense, many sets of data that are popularly called samples are not samples at all.

In this chapter and in Chapter 4 we shall describe things statistically without making any generalizations. For future reference, though, it is important to distinguish even here between populations and samples. Thus, we shall use different symbols depending on whether we are describing populations or samples.

# 3.2 THE MEAN

The most popular measure of central location is what the layperson calls an "average" and what the statistician calls an **arithmetic mean**, or simply a **mean**.[†] It is defined as follows:

> **The mean of $n$ numbers is their sum divided by $n$.**

It is all right to use the word "average," and on occasion we shall use it ourselves, but there are other kinds of averages in statistics and we cannot afford to speak loosely when there is any risk of ambiguity.

**EXAMPLE 3.1**    The mean of a set of $n$ numbers is very simply their sum divided by $n$. For instance, if a service station repaired 18, 15, 12, 20, 19, 11, 14, 38, 18, and 17 flat tires on ten consecutive days, the average number of flat tires repaired, namely, the mean, is given by

$$\frac{18 + 15 + 12 + 20 + 19 + 11 + 14 + 38 + 18 + 17}{10} = 18.2$$

**EXAMPLE 3.2**    Following are the number of packages delivered by a truck during a ten-day period. The numbers are

110, 112, 120, 128, 115, 150, 151, 91, 88, and 162

Find the mean of the daily number of packages delivered during this ten-day period.

**Solution**    Since    $110 + 112 + 120 + 128 + 115 + 150 + 151 + 91 + 88 + 162 = 1,227$,    the mean daily number of packages is

$$\frac{1,227}{10} = 122.7$$

or 123, rounded to the nearest package.

To give a formula for the mean, which is applicable to any kind of data, it will be necessary to represent them by means of symbols such as $x$, $y$, or $z$. For instance, in the tire-repairs example we might have used the letter $x$ and denoted

---

[†]The term "arithmetic mean" is used mainly to distinguish the mean from the **geometric mean** and the **harmonic mean**, two other kinds of averages used only in very special situations (see Exercises 3.15 and 3.16).

the ten values by $x_1$ (*x sub-one*), $x_2$ (*x sub-two*) ..., and $x_{10}$ (*x sub-ten*). More generally, if we have $n$ values, which we denote by $x_1, x_2, \ldots,$ and $x_n$, their mean is

$$\frac{x_1 + x_2 + \cdots + x_n}{n}$$

If these $x$'s constitute a sample, we denote their mean by $\bar{x}$ (*x bar*). Of course, had we referred to the data as $y$'s or $z$'s, we would have denoted their mean by $\bar{y}$ or $\bar{z}$. Furthermore, let us introduce the symbol $\Sigma$ (capital **sigma**, the Greek letter for $S$), which is simply a mathematical shorthand notation indicating the process of summation, or addition. When we write $\Sigma x$, this means literally "the sum of the $x$'s," and we can thus write the formula for a sample mean as

**SAMPLE MEAN**

$$\bar{x} = \frac{\sum x}{n}$$

Following the practice of using Greek letters to denote descriptions of populations, we denote the mean of a finite population of size $N$ by $\mu$ (the Greek letter for $m$). Thus, when the $x$'s constitute a population, we write

**POPULATION MEAN**

$$\mu = \frac{\sum x}{N}$$

with the reminder that $\sum x$ is now the sum of all $N$ values of $x$ that constitute the population.[†]

In general, to distinguish between descriptions of samples and descriptions of populations, we not only use different symbols such as $\bar{x}$ and $\mu$, but we refer to descriptions of samples as **statistics** and descriptions of populations as **parameters**. Parameters are usually denoted by Greek letters.

To illustrate the terminology and the notation just introduced, suppose that we are interested in the mean lifetime of a production lot of $N = 40{,}000$ light bulbs. Obviously, we cannot test all of the light bulbs for there would be none left to use or sell, so we take a sample, calculate $\bar{x}$, and use this quantity as an estimate of $\mu$.

**EXAMPLE 3.3**  If the light bulbs in the sample last 967, 949, 952, 940, and 922 hours of continuous use, what can we conclude about the mean lifetime of the 40,000 light bulbs in the production lot?

**Solution**  Since $n = 5$, the mean of this sample is

$$\bar{x} = \frac{967 + 949 + 952 + 940 + 922}{5} = 946 \text{ hours}$$

[†] When the **population size** is unlimited, population means cannot be defined in this way. Definitions of the mean of infinite populations can be found in most textbooks on mathematical statistics.

and, assuming that the data constitute a sample in the technical sense (namely, a set of data from which valid generalizations can be made), we estimate the mean of all the 40,000 light bulbs in the production lot as $\mu = 946$ hours.    ■

For nonnegative data, the mean not only describes their middle, but it also puts some limitation on their size. If we multiply by $n$ on both sides of the equation $\bar{x} = \frac{\sum x}{n}$, we find that $\sum x = n \cdot \bar{x}$ and, hence, that no part, or subset of the data can exceed $n \cdot \bar{x}$.

**EXAMPLE 3.4**    If the average (mean) salary paid to three NBA players for the season was $3,650,000, can

(a) any one of them have received $6,000,000?
(b) any two of them have each received $6,000,000?

**Solution**    The combined salaries of the three players added up to $3(3,650,000) = \$10,950,000$.

(a) If one of them had received a salary of $6,000,000, this would have left $10,950,000 - 6,000,000 = 4,950,000$ for the other two players. So this would have been possible.
(b) If two of them had each received a salary of $6,000,000, this would have required $2(6,000,000) = \$12,000,000$. Since this would have required more than the total paid to the three players, it would not have been possible.    ■

**EXAMPLE 3.5**    If six high school juniors averaged 57 on the verbal part of a standardized test, at most how many of them could have scored 72 or better on the test?

**Solution**    Since $n = 6$ and $\bar{x} = 57$, it follows that their combined scores total $6(57) = 342$. Since $342 = 4 \cdot 72 + 54$, we find that at most four of the six students could have scored 72 or more.    ■

The popularity of the mean as a measure of the "center" or "middle" of a set of data is not just accidental. Any time we use a single number to describe a whole set of data, there are certain requirements (namely, certain desirable properties) we must keep in mind. Thus, some of the noteworthy properties of the mean are as follows:

---

**It is familiar to most persons, although they may not call it by that name.**

**It always exists, that is, it can be calculated for any kind of data.**

**It is always unique, or in other words, a set of data has one and only one mean.**

**It takes into account each individual item.**

**It is relatively reliable in the sense that for sample data it is generally not as strongly affected by chance as some of the other measures of location.**

---

This question of reliability is of fundamental importance when it comes to problems of estimation, hypothesis testing, and making predictions.

Whether the fourth property which we have listed is actually desirable is open to some doubt; a single extreme (very small or very large) value can affect the mean to such an extent that it is debatable whether it is really "representative" or "typical" of the data it is supposed to describe. To give an example, suppose that in copying the five mean lifetimes of the light bulbs in Example 3.3 on page 46 we actually made a mistake—the second value should of been 499 instead of 949. This means that the mean would have been

$$\frac{967 + 499 + 952 + 940 + 922}{5} = 856 \text{ hours}$$

instead of 18.2, and this illustrates how one careless mistake can have a pronounced effect on the mean.

**EXAMPLE 3.6**    The editor of a book on nutritional values needs a figure for the calorie count of a slice of a 12-inch pepperoni pizza. Letting a laboratory with a calorimeter do the job, she gets the following figures for a slice of pizza from six different fast-food chains: 265, 332, 340, 225, 238, and 346.

(a) Calculate the mean, which the editor will report in her book.
(b) Suppose that when calculating the mean, the editor makes the mistake of entering 832 instead of 238 in her calculator. How much of an error would this make in the figure that she reports in her book?

**Solution**    (a) The correct mean is

$$\bar{x} = \frac{265 + 332 + 340 + 225 + 238 + 346}{6}$$

$$= 291$$

(b) The incorrect mean is

$$\bar{x} = \frac{265 + 332 + 340 + 225 + 832 + 346}{6}$$

$$= 390$$

and the error would be a disastrous $390 - 291 = 99$.

**EXAMPLE 3.7**    The ages of nine students from a high school who went on a field trip to a local zoo were 18, 16, 16, 17, 18, 15, 17, 17, and 17, and the age of the Biology teacher who accompanied them was 49. What was the mean age of the ten persons on their field trip?

**Solution**    Substituting the data into the formula for $\bar{x}$, we get

$$\bar{x} = \frac{18 + 16 + \cdots + 17 + 49}{10}$$

$$= 20$$

Note, however, that any statement to the effect that the average (mean) age of those on the trip was 20 can easily be misinterpreted. The students were teenagers and the teacher was 49 years old.

To avoid the possibility of being misled by an extreme (very small or very large) value, it may be advisable to omit such an **outlier** (the 49 in our example) or to use a statistical measure other than the mean. Perhaps, the **median** (see Section 3.4), which, as we shall see, is not as sensitive to an outlier as is the mean.

# 3.3 THE WEIGHTED MEAN

When we calculate a mean and the quantities we are averaging are not all of equal importance or of equal significance, we may not be getting a statistical measure that tells us what we had hoped to describe. In other words, we may get a result that is totally useless. Consider the following example:

**EXAMPLE  3.8**    Each Wednesday, a market advertises its specials for the week, and this week the specials include chuck steaks for $3.99 per pound, T-bone steaks for $7.99 per pound, and filet mignons for 11.85 per pound. Calculate the mean price per pound for these kinds of steaks.

**Solution**    As we said on page 47, the mean can be calculated for any kind of numerical data, and here we get

$$\bar{x} = \frac{3.99 + 7.99 + 11.85}{3}$$

$$= 7.9433$$

rounded to 4 decimals or $7.94 rounded to the nearest cent.

What we have done here is precisely what we were asked to do in Example 3.8, but what, if anything does it mean to a customer of the market, or to its management? Only if the customer had wanted to buy exactly one of the three kinds of steak (or exactly equal numbers of pounds of the three kinds of steak) would the result of Example 3.8 have been of any value. However, since butchers really have no choice but to cut the steaks first and then weigh and price them, it would have been just about impossible to get this done. The management would be interested mainly in the total receipts for the three kinds of steaks, and they could get this figure from the tape of the cash register. The $7.94 average obtained in Example 3.8 would be of no value, but the total receipts could be calculated if one knew how many pounds were sold of each of the three kinds of steaks. Given that the market had sold 83.52 pounds of the chuck steaks, 140.72 pounds of the T-bone steaks, and 35.60 pounds of the filet mignons, it would have followed that the total receipts had been $(83.52)(3.99) + (140.72)(7.99) + (35.60)(11.85) = 1,879.46$ rounded to the nearest cent. Thus, the $83.52 + 140.72 + 35.60$ pounds of steaks had sold at an average price of $\frac{1,879.46}{259.84} = \$7.23$ per pound, rounded to the nearest cent.

To obtain this meaningful average, it was necessary to give each of the prices a **relative importance weight** and then calculate a **weighted mean**. In general, the weighted mean $\bar{x}_w$ of a set of numbers $x_1, x_2, x_3, \ldots$, and $x_n$ whose relative importance is expressed numerically by a corresponding set of numbers $w_1, w_2, w_3, \ldots$, and $w_n$ is given by

# WEIGHTED MEAN

$$\bar{x}_w = \frac{w_1 x_1 + w_2 x_2 + \cdots + w_n x_n}{w_1 + w_2 + \cdots + w_n} = \frac{\sum w \cdot x}{\sum w}$$

Here $\sum w \cdot x$ is the sum of the products obtained by multiplying each $x$ by the corresponding weight, and $\sum w$ is simply the sum of the weights. Note that when the weights are all equal, the formula for the weighted mean reduces to that for the ordinary (arithmetic) mean.

**EXAMPLE 3.9**    To consider another example, let us determine the combined batting averages of the five leading batters of the Boston Red Sox baseball team. Data are from the American League Final Standings, 2004 (published in the *World Almanac*, 2005).

| Batters | Batting Averages | Times at Bat |
|---------|------------------|--------------|
| N. Garciaparra | .321 | 156 |
| M. Ramirez | .308 | 568 |
| J. Damon | .304 | 621 |
| D. Ortiz | .301 | 582 |
| K. Millar | .297 | 508 |

**Solution**    Using the Times at Bat as weights, we get

$$\bar{x}_w = \frac{(156)(.321) + (568)(.308) + (621)(.304) + (582)(.301) + (508)(.297)}{156 + 568 + 621 + 582 + 508}$$

$$= \frac{739.862}{2,435} = 0.304$$

The choice of the weights did not pose any particular problems in Examples 3.8 and 3.9, but there are situations in which their selection is not quite so obvious. For instance, if we wanted to construct a cost-of-living index, we would have to worry about the relative importance of such things as food, rent, entertainment, medical care, and so on, in the average person's budget. To give one general rule, the weights that are commonly used to average prices are, as in Example 3.8, the corresponding quantities sold, produced or consumed.

A special application of the formula for the weighted mean arises when we must find the overall mean, or **grand mean**, of $k$ sets of data having the means $\bar{x}_1, \bar{x}_2, \bar{x}_3, \ldots,$ and $\bar{x}_k$, and consisting of $n_1, n_2, n_3, \ldots,$ and $n_k$ measurements or observations. The result denoted by $\bar{\bar{x}}$ is given by

# GRAND MEAN OF COMBINED DATA

$$\bar{\bar{x}} = \frac{n_1 \bar{x}_1 + n_2 \bar{x}_2 + \cdots + n_k \bar{x}_k}{n_1 + n_2 + \cdots + n_k} = \frac{\sum n \cdot \bar{x}}{\sum n}$$

where the weights are the sizes of the samples, the numerator is the total of all the measurements or observations, and the denominator is the number of items in the combined samples.

EXAMPLE **3.10**    In a psychology class there are 14 freshmen, 25 sophomores, and 16 juniors. Given that the freshmen averaged 76 in the final examination, the sophomores averaged 83, and the juniors averaged 89, what is the mean grade for the entire class?

**Solution**    Substituting $n_1 = 14$, $n_2 = 25$, $n_3 = 16$, $\bar{x}_1 = 76$, $\bar{x}_2 = 83$, and $\bar{x}_3 = 89$ into the formula for the grand mean of combined data, we get

$$\bar{\bar{x}} = \frac{14 \cdot 76 + 25 \cdot 83 + 16 \cdot 89}{14 + 25 + 16}$$

$$= 82.96$$

rounded to two decimal places, or approximately 83.

EXERCISES

**3.1**    Suppose that a cardiologist takes the blood pressure of all patients in a cardiac ward. Give one illustration each of a situation where these data would be looked upon as
(a)  a population;
(b)  a sample.

**3.2**    Suppose that we are given the daily amount of rain at SEATAC (The Seattle–Tacoma airport) during April of the year 2005. Give one illustration each of a situation where these data would be looked on as
(a)  a population;
(b)  a sample.

**3.3**    Suppose that we are told, yes or no, whether there was at least one rain delay on each day of the tennis tournament at Wimbledon. Describe one situation where this information would be looked on as a sample, and one situation where it would be looked on as a population.

**3.4**    A bookstore keeps track of the daily sales of three new hard-covers by very popular mystery writers. Do these data constitute a population or a sample if the bookstore wants to
(a)  use the data for tax purposes;
(b)  use this information in advertising to whet potential customers' appetite;
(c)  determine the size of subsequent printings of the mysteries as paperbacks;
(d)  decide which of the three mysteries should be awarded a place of honor as best mystery of the month.

**3.5**    Following are the numbers of seconds that 12 insects survived after having been sprayed with a certain insecticide: 112, 83, 102, 84, 105, 121, 76, 110, 98, 91, 103, and 85. Find the mean of these survival times.

**3.6**    Following are the numbers of twists that were required to break ten forged alloy bars: 23, 14, 37, 25, 29, 45, 19, 30, 36, and 42. Calculate the mean of this set of data.

**3.7**    A 10-ml pipette was calibrated and found to deliver the following volumes: 9.96, 9.98, 9.92, 9.98, and 9.96. Find the mean of these data and use it to estimate by how much the calibration is off.

**3.8**    Following are the speeds (miles per hour) at which 20 cars were timed on a highway in late-evening traffic: 77, 69, 82, 76, 69, 71, 80, 66, 70, 77, 72, 73, 80, 86, 74, 77, 69, 89, 74, and 75.
(a)  Find the mean of these speeds.

(b) Find the mean of these speeds after subtracting 75 from each value and then adding 75 to the result. What general simplification does this suggest for the calculation of means?

**3.9** An elevator in a department store is designed to carry a maximum load of 3,200 pounds. If it is loaded with 18 persons having a mean weight of 166 pounds, is there any danger of it being overloaded?

**3.10** The cargo bay of an airplane is designed to carry a maximum load of 15,000 pounds. If it is loaded with 214 crates weighing on the average 65 pounds, is there any real danger of the cargo bay being overloaded?

**3.11** The hours of sleep that a student had during each of the eight nights prior to a final examination were 7, 6, 7, 0, 7, 9, 6, and 0 hours. Calculate the mean and discuss whether this relatively low figure might account for the student's relatively poor performance in this final examination.

**3.12** A carload of stone taken from a quarry by the plug-and-feather method (where stone is split into definite shapes using wedges and other tools) contains pieces of stone have a mean weight of 1,500 pounds and a total weight of 36,000 pounds. Another carload of stone produced by the explosive method (where stone is blasted into random shapes by explosives) contains pieces of stone that have a mean weight of 600 pounds and a total weight of 30,000 pounds. What is the mean weight of all the pieces of stone combined?

**3.13** A classroom of 32 students of statistics received grades averaging 78 points on a standardized test, and another section of 48 students of statistics received grades averaging 84 points on the same test. What is the overall average of these grades rounded to the nearest point?

**3.14** A pharmaceutical manufacturer utilized 867, 849, 840, 852, and 822 mice in five trials of the effectiveness of a new drug.

(a) Determine the mean number of mice used in the five trials.

(b) Recalculate the mean if the second trial had been incorrectly recorded as 489 instead of 849.

**3.15** Generalizing the argument of Examples 3.4 and 3.5, it can be shown that for any set of nonnegative data with the mean $\bar{x}$, the fraction of the data that are greater than or equal to any positive constant $k$ cannot exceed $\bar{x}/k$. Use this result, called **Markov's theorem**, in the following problems:

(a) If the mean breaking strength of certain linen threads is 33.5 ounces, at most what fraction of the threads can have a breaking strength of 50.0 ounces or more?

(b) If the diameters of the orange trees in an orchard have a mean of 17.2 cm, at most what fraction of the trees can have a diameter of 20.0 cm or more?

**3.16** Records show that in Phoenix, Arizona, the normal daily maximum temperature for each month is 65, 69, 74, 84, 93, 102, 105, 102, 98, 88, 74, and 66 degrees Fahrenheit. Verify that the mean of these figures is 85 and comment on the claim that, in Phoenix, the average daily maximum temperature is a very comfortable 85 degrees.

**3.17** The **geometric mean** of $n$ positive numbers is the $n$th root of their product. For example, the geometric mean of 3 and 12 is $\sqrt{3 \cdot 12} = \sqrt{36} = 6$ and the geometric mean of 1, 3, and 243 is $\sqrt[3]{1 \cdot 3 \cdot 243} = \sqrt[3]{729} = 9$.

(a) Find the geometric mean of 9 and 36.

(b) Find the geometric mean of 1, 2, 8, and 81.

(c) During a flu epidemic, 12 cases were reported on the first day, 18 on the second day, and 48 on the third day. Thus, from the first day to the second day the number of cases reported was multiplied by $\frac{18}{12} = \frac{3}{2}$, and from the second day to the third day the number of cases was multiplied by $\frac{48}{18} = \frac{8}{3}$. Find the geometric mean of these two growth rates and, assuming that the growth pattern continues, predict the number of cases that will be reported on the fourth and fifth days.

**3.18**  The **harmonic mean** of $n$ positive numbers $x_1, x_2, x_3, \ldots$, and $x_n$ is defined as the reciprocal of the mean of their reciprocals. Its usefulness is limited, but it is appropriate in some special situations. For instance, if someone drives 10 miles on a highway at 60 mph and on the way back at 30 mph, he will not have averaged $\frac{60+30}{2} = 45$ mph. He will have driven 20 miles in 30 minutes, which makes his average speed 40 mph.

(a) Verify that the harmonic mean of 60 and 30 is 40, so that it gives the appropriate average for the preceding example.

(b) If an investor buys $18,000 worth of a company's stock at $45 a share and another $18,000 at $36 a share, she is buying

$$\frac{18,000}{45} + \frac{18,000}{36} = 900$$

shares for $36,000 and she is paying $\frac{36,000}{900} = \$40$ per share. Verify that 40 is, in fact, the harmonic mean of 45 and 36.

**3.19**  Having received a bonus of $20,000 for accepting early retirement a company's sales representative invested $6,000 in a municipal bond paying 3.75%, $10,000 in a mutual fund paying 3.96%, and $4,000 in a certificate of deposit paying 3.25%. Using the respective amounts invested as weights, find the weighted mean of the three percentages. Does this figure equal the actual total return on the three investments?

**3.20**  An instructor counts the final examination in a course four times as much as each of three one-hour examinations. Which of two students has a higher weighted average score, the one who received scores of 72, 80, and 65 in the one-hour examinations and an 82 in the final examination, or the one who received scores of 81, 87, and 75 in the one-hour examinations and a 78 in the final examination?

**3.21**  Among the students receiving bachelor's degrees from a certain university in 2005, 382 majoring in the humanities had salary offers averaging $33,373, 450 majoring in the social sciences had salary offers averaging $31,684, and 113 majoring in computer science had salary offers averaging $40,329. What was the average salary offered to these 945 graduates?

**3.22**  A home appliance center advertised the following refrigerators, of which it had, respectively, 18, 12, 9, 14, and 25 in stock.

| Brand | Size | Price |
|-------|------|-------|
| Company A | 15 cu. ft. | $416 |
| Company B | 21 cu. ft. | $549 |
| Company C | 19 cu. ft. | $649 |
| Company D | 21 cu. ft. | $716 |
| Company E | 24 cu. ft. | $799 |

(a) What is the average size of these refrigerators?

(b) What is the average price of these refrigerators?

 **3.23** Use a computer, a graphing calculator, or, for that matter, any calculator to find the mean of the 110 waiting times between eruptions of Old Faithful geyser given in Example 2.4 on page 24.

 **3.24** Use a computer, a graphing calculator, or, for that matter, any calculator to find the mean of the 150 test scores of Exercise 2.65 on page 36.

# 3.4 THE MEDIAN

To avoid the possibility of being misled by one or a few very small or very large values, we sometimes describe the "middle" or "center" of a set of data with statistical measures other than the mean. One of these, the **median** of $n$ values, requires that we first arrange the data according to size. Then it is defined as follows:

> **The median is the value of the middle item when $n$ is odd, and the mean of the two middle items when $n$ is even.**

In either case, when no two values are alike, the median is exceeded by as many values as it exceeds. When some of the values are alike, this may not be the case.

**EXAMPLE  3.11**   In five recent weeks, a town in England reported 14, 17, 20, 22, and 17 burglaries. Find the median number of burglaries in that town for these weeks.

**Solution**   The median is not 20, the third (middle) value, since the data are not arranged according to size. Doing so, we get

$$14 \quad 17 \quad 17 \quad 20 \quad 22$$

and it can be seen that the median is 17.

Note that there are two 17s among the data and that we did not refer to either of them as *the median*—the median is a number and not necessarily any particular measurement or observation.

**EXAMPLE  3.12**   In some cities, persons cited for minor traffic violations can attend a class in defensive driving in lieu of paying a fine. Given that in one week in a certain city 12 such classes were attended by 37, 32, 28, 40, 35, 38, 40, 24, 30, 37, 32, and 40 persons, find the median of these data.

**Solution**   Ranking these attendance figures according to size, from low to high, we get

$$24 \quad 28 \quad 30 \quad 32 \quad 32 \quad 35 \quad 37 \quad 37 \quad 38 \quad 40 \quad 40 \quad 40$$

and we find that the median is the mean of the two values nearest the middle, namely, $\frac{35+37}{2} = 36$.

Some of the values were alike in this example, but this did not affect the median, which exceeds six of the values and is exceeded by equally many. The situation is quite different, however, in the example that follows.

**EXAMPLE  3.13**    On the seventh hole of a golf course in Palm Springs, California, nine golfers scored *par*, *birdie* (one below par), *par*, *par*, *bogey* (one above par), *eagle* (two below par), *par*, *birdie*, and *birdie*. Find the median.

**Solution**    Ranking these scores from low to high, we get *eagle*, *birdie*, *birdie*, *birdie*, *par*, *par*, *par*, *par*, and *bogey*, and it can be seen that the fifth score, the median, is *par*. ∎

Note that in this example the median, *par*, exceeds four of the scores but is exceeded by only one. This may seem misleading, but by definition the median is *par*.

The symbol that we use for the median of *n* sample values $x_1, x_2, x_3, \ldots$, and $x_n$ is $\tilde{x}$ (and, hence, $\tilde{y}$ or $\tilde{z}$ if we refer to the values of *y*'s or *z*'s). If a set of data constitutes a population, we denote its median by $\tilde{\mu}$.

Thus, we have a symbol for the median, but no formula; there is only a formula for the **median position**. Referring again to data arranged according to size, usually ranked from low to high, we can write

**MEDIAN POSITION**    *The median is the value of the $\frac{n+1}{2}$th item.*

**EXAMPLE  3.14**    Find the median position for

**(a)** $n = 17$;

**(b)** $n = 41$.

**Solution**    With the data arranged according to size (and counting from either end)

**(a)** $\frac{n+1}{2} = \frac{17+1}{2} = 9$ and the median is the value of the 9th item;

**(b)** $\frac{n+1}{2} = \frac{41+1}{2} = 21$ and the median is the value of the 21st item. ∎

**EXAMPLE  3.15**    Find the median position for

**(a)** $n = 16$;

**(b)** $n = 50$.

**Solution**    With the data arranged according to size (and counting from either end)

**(a)** $\frac{n+1}{2} = \frac{16+1}{2} = 8.5$ and the median is the mean of the values of the 8th and 9th items;

**(b)** $\frac{n+1}{2} = \frac{50+1}{2} = 25.5$ and the median is the mean of the values of the 25th and 26th items. ∎

It is important to remember that $\frac{n+1}{2}$ is the formula for the median position and not a formula for the median, itself. It is also worth mentioning that determining the median can usually be simplified, especially for large sets of data, by first presenting the data in the form of a stem-and-leaf display.

**EXAMPLE  3.16**    In the printout of Figure 2.2 on page 15, we gave the following stem-and-leaf display of the room occupancy data of a resort during the month of June:

|      |   |         |
|------|---|---------|
| 2    | 3 | 57      |
| 6    | 4 | 0023    |
| 13   | 4 | 5666899 |
| (3)  | 5 | 234     |
| 14   | 5 | 56789   |
| 9    | 6 | 1224    |
| 5    | 6 | 9       |
| 4    | 7 | 23      |
| 2    | 7 | 8       |
| 1    | 8 | 1       |

Use this double-stem display to find the median of these data.

**Solution**   The original display in Figure 2.2 was actually a computer printout, and we explained that the figures in the column on the left were the accumulated numbers of leaves counted from either end, and that the parentheses around the three were meant to indicate that the middle of the data was on that stem. Having defined the median, we can now substitute "median" for "middle," and use these features in determining the median of the data.

Since $n = 30$ for the given table, the median position is $\frac{30+1}{2} = 15.5$, so that the median is the mean of the fifteenth and sixteenth largest values among the data. Since $2 + 4 + 7 = 13$ of the values are represented by leaves on the first three stems, the median is the mean of the values represented by the second and third leaves on the fourth stem. These are 53 and 54, and hence the median of the room-occupancy data is $\frac{53+54}{2} = 53.5$. Note that this illustrates why it is generally advisable to rank the leaves on each stem from low to high.

As a matter of interest, let us also point out that the mean of the room-occupancy data is 54.4 and that this figure differs from 53.5, the value we obtained for the corresponding median. It really should not come as a surprise that the median we obtained here does not equal the mean of the same data—the median and the mean define the middle of a set of data in different ways. The median is average in the sense that it divides the data into two parts so that, unless there are duplicates, there are equally many values above and below the median. The mean, in the other hand, is average in the sense that if each value is replaced by some constant k while the total remains unchanged, this number k will have to be the mean. (This follows directly from the relationship $n \cdot \bar{x} = \sum x$.) In this sense, the mean has also been likened to a center of gravity.

The median shares some, but not all, of the properties of the mean listed on page 47. Like the mean, the median always exists and it is unique for any set of data. Also like the mean, the median is simple enough to find once the data have been arranged according to size, but as we indicated earlier, sorting a set of data manually can be a surprisingly difficult task.

Unlike the mean, the medians of several sets of data cannot generally be combined into an overall median of all the data, and in problems of statistical inference the median is usually less reliable than the mean. This is meant to say that the medians of repeated samples from the same population will usually vary more widely than the corresponding means (see Exercises 3.34 and 4.20). On the other hand, sometimes the median may be preferable to the mean because it is not so easily, or not at all, affected by extreme (very small or very large) values. For instance, in Example 3.6 we showed that incorrectly entering 832 instead of 238 into a calculator caused an error of 99 in the mean. As the reader will be asked to verify in Exercise 3.31, the corresponding error in the median would have been only 37.5.

Finally, also unlike the mean, the median can be used to define the middle of a number of objects, properties, or qualities that can be ranked, namely, when we deal with ordinal data. For instance, we might rank a number of tasks according to their difficulty and then describe the middle (or median) one as being of "average difficulty." Also, we might rank samples of chocolate fudge according to their consistency and then describe the middle (or median) one as having "average consistency."

Besides the median and the mean there are several other measures of central location; for example, the **midrange** described in Exercise 3.37 and the **midquartile** defined on page 58. Each describes the "middle" or "center" of a set of data in its own way, and it should not come as a surprise that their values may well all be different. Then there is also the **mode** described in Section 3.6.

# 3.5 OTHER FRACTILES

The median is but one of many **fractiles** that divide data into two or more parts, as nearly equal as they can be made. Among them we also find **quartiles**, **deciles**, and **percentiles**, which are intended to divide data into four, ten, and a hundred parts. Until recently, fractiles were determined mainly for distributions of large sets of data, and in this connection we shall study them in Section 3.7.

In this section we shall concern ourselves mainly with a problem that has arisen in **exploratory data analysis**—in the preliminary analysis of relatively small sets of data. It is the problem of dividing such data into four nearly equal parts, where we say "nearly equal" because there is no way in which we can divide a set of data into four equal parts for, say, $n = 27$ or $n = 33$. Statistical measures designed for this purpose have traditionally been referred to as the three quartiles, $Q_1$, $Q_2$, and $Q_3$, and there is no argument about $Q_2$, which is simply the median. On the other hand, there is some disagreement about the definition of $Q_1$ and $Q_3$.

As we shall define them, the quartiles divide a set of data into four parts such that there are as many values less than $Q_1$ as there are between $Q_1$ and $Q_2$, between $Q_2$ and $Q_3$, and greater than $Q_3$. Assuming that no two values are alike, this is accomplished by letting

> $Q_1$ be the median of all the values less than the median of the whole set of data,

and

> $Q_3$ be the median of all the values greater than the median of the whole set of data.

It remains to be shown that with this definition there are, indeed, as many values less than $Q_1$ as there are between $Q_1$ and the median, between the median and $Q_3$, and greater than $Q_3$. We shall demonstrate this here for the four cases where $n = 4k$, $n = 4k + 1$, $n = 4k + 2$, and $n = 4k + 3$, with $k = 3$; namely, for $n = 12$, $n = 13$, $n = 14$, and $n = 15$.

**EXAMPLE 3.17**   Verify that there are three values less than $Q_1$, between $Q_1$ and the median, between the median and $Q_3$, and greater than $Q_3$ for

**(a)** $n = 12$;       **(d)** $n = 14$;

**(b)** $n = 13$;       **(e)** $n = 15$.

**Solution**   **(a)** For $n = 12$ the median position is $\frac{12+1}{2} = 6.5$ and for the six values below the median the $Q_1$ position is $\frac{6+1}{2} = 3.5$, and for the six values above the median the $Q_3$ position is 3.5 from the other end; namely, 9.5. As can be seen from Figure 3.1, there are three values less than $Q_1$, between $Q_1$ and the median, between the median and $Q_3$, and greater than $Q_3$,

**(b)**, **(c)**, and **(d)**, In each case the median and quartile positions are obtained in the same way, and as can be seen from Figure 3.1, in each case there are three values less than $Q_1$, between $Q_1$ and the median, between the median and $Q_3$, and greater than $Q_3$.   ■

If some of the values are alike, we modify the definitions of $Q_1$ and $Q_3$ by replacing "less than the median" by "to the left of the median position" and "greater than the median" by "to the right of the median position.

Quartiles are not meant to be descriptive of the "middle" or "center" of a set of data, and we have given them here mainly because, like the median, they are fractiles and they are determined in more or less the same way. The **midquartile** $\frac{Q_1+Q_3}{2}$, has been used on occasion as another measure of central location.

The information provided by the median, the quartiles $Q_1$ and $Q_3$, and the smallest and largest values is sometimes presented in the form of a **boxplot**.

**Figure 3.1**
The $Q_1$, median, and $Q_3$ positions for $n = 12$, $n = 13$, $n = 14$, and $n = 15$.

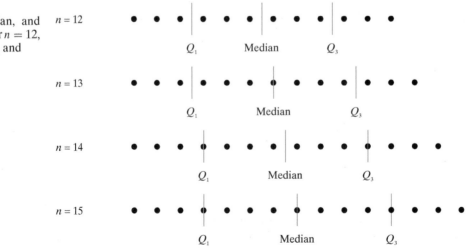

Originally referred to somewhat whimsically as a **box-and-whisker plot**, such a display consists of a rectangle that extends from $Q_1$ to $Q_3$, lines drawn from the smallest value to $Q_1$ and from $Q_3$ to the largest value, and a line at the median that divides the rectangle into two parts. In practice, boxplots are sometimes embellished with other features, but the simple form shown here is adequate for most purposes.

**EXAMPLE 3.18**    In Example 3.16 we used the following double-stem display to show that the median of the room-occupancy data, originally given on page 18, is 53.5. For convenience, the double-stem display of Example 3.16 is repeated below.

| | | |
|---|---|---|
| 2 | 3 | 57 |
| 6 | 4 | 0023 |
| 13 | 4 | 5666899 |
| (3) | 5 | 234 |
| 14 | 5 | 56789 |
| 9 | 6 | 1224 |
| 5 | 6 | 9 |
| 4 | 7 | 23 |
| 2 | 7 | 8 |
| 1 | 8 | 1 |

(a) Find the smallest and largest values.
(b) Find $Q_1$ and $Q_3$.
(c) Draw a boxplot.

**Solution**    (a) As can be seen by inspection, the smallest value is 35 and the largest value is 81.
(b) For $n = 30$ the median *position is* $\frac{30+1}{2} = 15.5$ and, hence, for the 15 *values* below 53.5 the median *position* is $\frac{15+1}{2} = 8$. It follows that $Q_1$, the eighth *value*, is 46. Similarly, $Q_3$, the eighth *value* from the other end, is 62.
(c) Combining all this information, we obtain the boxplot shown in Figure 3.2.

**Figure 3.2**
Boxplot of the room occupancy data (manually prepared).

Boxplots can also be produced with appropriate computer software or a graphing calculator. Using the data from Example 3.18, we obtained the computer printout shown in Figure 3.3.

# 3.6 THE MODE

Another statistical measure that is sometimes used to describe the middle or center of a set of data is the **mode**, which is defined simply as the value or

**Figure 3.3**
Boxplot of the room
occupancy data (com-
puter printout).

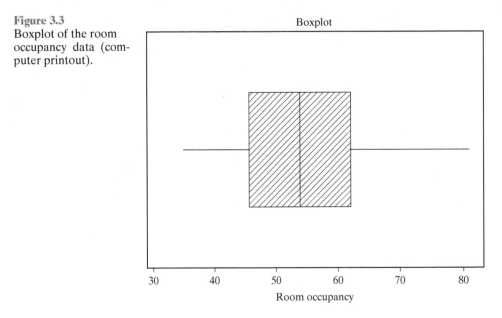

Boxplot

Room occupancy

category that occurs with the highest frequency and more than once. The two
main advantages of the mode are that it requires no calculations, only counting,
and that it can be determined for numerical as well as categorical data.

**EXAMPLE 3.19**   The 21 meetings of a square dance club were attended by 22, 24, 23, 24, 27, 25,
24, 20, 24, 26, 28, 26, 23, 21, 24, 25, 23, 28, 24, 26, and 25 of its members. Find the
mode.

**Solution**   Among these numbers, 20, 21, 22, and 27 each occurs once, 28 occurs twice, 23,
25, and 26 each occurs three times, and 24 occurs six times. Thus, the modal
attendance at the club's dances is 24.   ◼

**EXAMPLE 3.20**   In Example 3.13 we gave the scores of nine golfers on a par-four hole as eagle,
birdie, birdie, birdie, par, par, par, par, and bogey. Find the mode.

**Solution**   Since these data have already been arranged according to size, it can easily be
seen that *par*, which occurs four times, is the modal score.   ◼

As we have seen in this chapter, there are various measures of central location
that describe the middle of a set of data. What particular "average" should be used
in any given situation can depend on many different things (see, for example, the
optional Section 7.3) and the choice may be difficult to make. Since the selection
of statistical descriptions often contains an element of arbitrariness, some persons
believe that the magic of statistics can be used to prove nearly anything. Indeed,
a famous nineteenth-century British statesman is often quoted as having said that
there are three kinds of lies: *lies, damned lies, and statistics.* Exercises 3.36 and
3.37 describe a situation where this kind of criticism is well justified.

EXERCISES

**3.25** Find the median position for
(a) $n = 55$;
(b) $n = 34$.

**3.26** Find the median position for
(a) $n = 33$;
(b) $n = 45$.

**3.27** Find the median of the following 12 figures on the average monthly percent of sunshine in Pittsburgh, Pennsylvania, as reported by the U.S. Weather Bureau: 38, 40, 53, 53, 57, 65, 66, 63, 68, 59, 50, and 40.

**3.28** On 15 days, a restaurant served breakfast to 38, 50, 53, 36, 38, 56, 46, 54, 54, 58, 35, 61, 44, 48, and 59 persons. Find the median.

**3.29** Thirty-two NBA games lasted 138, 142, 113, 164, 159, 157, 135, 122, 126, 139, 140, 142, 157, 121, 143, 140, 169, 130, 142, 146, 155, 117, 158, 148, 145, 151, 137, 128, 133, 150, 134, and 147 minutes. Determine the median length of these games.

**3.30** In a study of the stopping ability of standard passenger cars on dry, clean, level pavement, 21 drivers going 30 mph were able to stop in 78, 69, 79, 91, 66, 72, 74, 85, 84, 66, 76, 67, 79, 83, 70, 77, 67, 79, 79, 77, and 67 feet. Find the median of these stopping distances.

**3.31** With reference to Example 3.6, suppose that the editor of the book on nutrition had used the median instead of the mean to average the respective calorie counts. Show that with the median the error of using 832 instead of 238 would have been only 37.5.

**3.32** Following are the miles per gallon obtained with 30 tankfuls of gas:

| | | | | | | | | | |
|---|---|---|---|---|---|---|---|---|---|
| 24.1 | 24.9 | 25.2 | 23.8 | 24.7 | 22.9 | 25.0 | 24.1 | 23.6 | 24.5 |
| 23.7 | 24.4 | 24.7 | 23.9 | 25.1 | 24.6 | 23.3 | 24.3 | 24.8 | 22.8 |
| 23.9 | 24.2 | 24.7 | 24.9 | 25.0 | 24.8 | 24.5 | 23.4 | 24.6 | 25.3 |

Construct a double-stem display and use it to determine the median of these data.

**3.33** Following are the body weights of 60 small lizards used in a study of vitamin deficiencies (in grams):

| | | | | | | | | | |
|---|---|---|---|---|---|---|---|---|---|
| 125 | 128 | 106 | 111 | 116 | 123 | 119 | 114 | 117 | 143 |
| 136 | 92 | 115 | 118 | 121 | 137 | 132 | 120 | 104 | 125 |
| 119 | 115 | 101 | 129 | 87 | 108 | 110 | 133 | 135 | 126 |
| 127 | 103 | 110 | 126 | 118 | 82 | 104 | 137 | 120 | 95 |
| 146 | 126 | 119 | 113 | 105 | 132 | 126 | 118 | 100 | 113 |
| 106 | 125 | 117 | 102 | 146 | 129 | 124 | 113 | 95 | 148 |

Construct a stem-and-leaf display with the stem labels 8, 9, 10, 11, 12, 13, and 14, and use it to determine the median of these data.

**3.34** To verify the claim that the mean is generally more reliable than the median (namely, that the mean is subject to smaller chance fluctuations), a student conducts an experiment consisting of 12 tosses of three dice. The following are his results:
2, 4, and 6,  5, 3, and 5,  4, 5, and 3,  5, 2, and 3,  6, 1, and 5,  3, 2, and 1
3, 1, and 4,  5, 5, and 2,  3, 3, and 4,  1, 6, and 2,  3, 3, and 3,  4, 5, and 3
(a) Calculate the 12 medians and the 12 means.
(b) Group the medians and the means obtained in part (a) into separate distributions having the classes

$$1.5-2.5, 2.5-3.5, 3.5-4.5, 4.5-5.5$$

(Note that there will be no ambiguities since the medians of three whole numbers and the means of three whole numbers cannot equal 2.5, 3.5, or 4.5.)

(c) Draw histograms of the two distributions obtained in part (b) and explain how they illustrate the claim that the mean is generally more reliable than the median.

**3.35** Repeat Exercise 3.34 with your own data by repeatedly rolling three dice (or one die three times) and constructing distributions of the medians and the means.

**3.36** A consumer testing service obtains the following mileages per gallon during five test runs performed with each of three compact cars:

| | | | | | |
|---|---|---|---|---|---|
| Car A: | 27.9 | 30.4 | 30.6 | 31.4 | 31.7 |
| Car B: | 31.2 | 28.7 | 31.3 | 28.7 | 31.3 |
| Car C: | 28.6 | 29.1 | 28.5 | 32.1 | 29.7 |

(a) If the manufacturers of car A want to advertise that their car performed best in this test, which of the "averages" discussed so far in this chapter, could they use to substantiate this claim?

(b) If the manufacturers of car B want to advertise that their car performed best in this test, which of the "averages" discussed so far in this chapter could they use to substantiate this claim?

**3.37** With reference to Exercise 3.36, suppose that the manufacturers of car C hire an unscrupulous statistician and instruct him or her to find some kind of "average" that will substantiate that their car performed best in the test. Show that the **midrange** (the mean of the smallest and largest values) will serve this purpose.

**3.38** Find the positions of the median, $Q_1$, and $Q_3$ for
(a) $n = 32$;
(b) $n = 35$.

**3.39** Find the positions of the median, $Q_1$, and $Q_3$ for
(a) $n = 41$;
(b) $n = 50$.

**3.40** Find the positions of the median, $Q_1$, and $Q_3$ for $n = 21$ and verify that when no two values are alike, there are as many values below $Q_1$ as there are between $Q_1$ and the median, between the median and $Q_3$, and above $Q_3$.

**3.41** Find the positions of the median, $Q_1$, and $Q_3$ for $n = 34$ and verify that even though some of the values may be alike, there are as many values to the left of the $Q_1$ position as there are to the right of the $Q_1$ position and to the left of the median position, to the right of the median position and to the left of the $Q_3$ position, and to the right of the $Q_3$ position.

**3.42** Following are the thicknesses of grease coatings (in microns) produced on twenty steel rods: 41, 51, 63, 57, 57, 66, 63, 60, 44, 41, 46, 43, 53, 55, 48, 49, 65, 61, 58, and 66. Find the median, $Q_1$, and $Q_3$. Also verify that there are as many values below $Q_1$ as there are between $Q_1$ and the median, between the median and $Q_3$, and above $Q_3$.

**3.43** With reference to Exercise 3.42, determine the smallest and the largest values among the data, and use the information obtained here and in Exercise 3.42 to draw a boxplot.

**3.44** Following are fourteen temperature readings taken at different locations in a large kiln: 409, 412, 439, 411, 432, 432, 405, 411, 422, 417, 440, 427, 411, and 417. Find the median, $Q_1$, and $Q_3$. Also verify that there are as many values to the left of the

$Q_1$ position as there to the right of the $Q_1$ position and to the left of the median position, to the right of the median position and to the left of the $Q_3$ position, and to the right of the $Q_3$ position.

**3.45** With reference to Exercise 3.44, determine the smallest and the largest values among the data, and use the information obtained here and in Exercise 3.44 to draw a boxplot.

**3.46** With reference to Example 2.4, use a computer package or a graphing calculator to sort the $n = 110$ waiting times between eruptions of Old Faithful in ascending order, and use this arrangement to determine the median, $Q_1$, and $Q_3$.

**3.47** Use the arrangement obtained in Exercise 3.46 to read off the smallest and the largest values, and use the information obtained here and in Exercise 3.46 to draw a boxplot.

**3.48** With reference to Exercise 3.33, use a computer package or a graphing calculator to sort the data, ranked from low to high, and use this arrangement to determine the median, $Q_1$, and $Q_3$ for the body weights of the 60 lizards.

**3.49** Use the arrangement obtained in Exercise 3.48 to read off the smallest and the largest of the body weights, and use the information obtained here and in Exercise 3.48 to draw a boxplot.

**3.50** Find the mode (if it exists) for each of the following sets of data:
(a) 6, 8, 6, 5, 5, 7, 7, 9, 7, 6, 8, 4, and 7;
(b) 57, 39, 54, 30, 46, 22, 48, 35, 27, 31, and 23;
(c) 11, 15, 13, 14, 13, 12, 10, 11, 12, 13, 11, and 13.

**3.51** Following are the numbers of chicken dinners served by a restaurant on 40 Sundays: 41, 52, 46, 42, 46, 36, 46, 61, 58, 44, 49, 48, 48, 52, 50, 45, 68, 45, 48, 47, 49, 57, 44, 48, 49, 45, 47, 48, 43, 45, 45, 56, 48, 54, 51, 47, 42, 53, 48, and 41. Find the mode.

**3.52** Following are the numbers of blossoms on 50 cacti in a desert botanical garden: 1, 0, 3, 0, 4, 1, 0, 1, 0, 0, 1, 6, 1, 0, 0, 0, 3, 3, 0, 1, 1, 5, 0, 2, 0, 3, 1, 1, 0, 4, 0, 0, 1, 2, 1, 1, 2, 0, 1, 0, 3, 0, 0, 1, 5, 3, 0, 0, 1, and 0. Find the mode.

**3.53** On a seven-day cruise, twenty passengers complained of seasickness on 0, 4, 5, 1, 0, 0, 5, 4, 5, 5, 0, 2, 0, 0, 6, 5, 4, 1, 3, and 2 days. Find the mode and explain why it may very well give a misleading picture of the actual situation.

**3.54** When there is more than one mode, this is often construed as an indication that the data actually consist of a combination of several distinct sets of data. Reanalyze the data of Exercise 3.53 after changing the fourth value from 1 to 5.

**3.55** Asked whether they ever go to the opera, 40 persons in the age group from 20 to 29 replied as follows: rarely, occasionally, never, occasionally, occasionally, occasionally, rarely, rarely, never, occasionally, never, rarely, occasionally, frequently, occasionally, rarely, never, occasionally, occasionally, rarely, rarely, never, occasionally, occasionally, rarely, frequently, rarely, occasionally, occasionally, never, rarely, frequently, never, rarely, occasionally, occasionally, rarely, rarely, occasionally and never. What is their modal reply?

**3.56** With reference to Exercise 2.5, what is the modal reply of the 30 persons interviewed at the dog show?

# *3.7 THE DESCRIPTION OF GROUPED DATA

In the past, considerable attention was paid to the description of grouped data, because it usually simplified matters to group large sets of data before calculating various statistical measures. This is no longer the case, since the necessary calculations can now be made in a matter of seconds with the use of computers or even hand-held calculators. Nevertheless, we shall devote this section and Section 4.4 to the description of grouped data, since many kinds of data (for example, those reported in government publications) are available only in the form of frequency distributions.

As we have already seen in Chapter 2, the grouping of data entails some loss of information. Each item loses its identity, so to speak; we know only how many values there are in each class or in each category. This means that we shall have to be satisfied with approximations. *Sometimes we treat our data as if all the values falling into a class were equal to the corresponding class mark, and we shall do so to define the mean of a frequency distribution. Sometimes we treat our data as if all the values falling into a class are spread evenly throughout the corresponding class interval, and we shall do so to define the median of a frequency distribution.* In either case, we get good approximations since the resulting errors will tend to average out.

To give a general formula for the mean of a distribution with $k$ classes, let us denote the successive class marks by $x_1, x_2, \ldots,$ and $x_k$, and the corresponding class frequencies by $f_1, f_2, \ldots,$ and $f_k$. Then, the sum of all the measurements is approximated by

$$x_1 \cdot f_1 + x_2 \cdot f_2 + \cdots + x_k \cdot f_k = \sum x \cdot f$$

and the mean of the distribution is given by

**M**EAN OF
GROUPED DATA

$$\overline{x} = \frac{\sum x \cdot f}{n}$$

Here $f_1 + f_2 + f_3 + \cdots + f_k = n$, the size of the sample. To write a corresponding formula for the mean of a population, we substitute $\mu$ for $\overline{x}$ and $N$ for $n$, obtaining

$$\mu = \frac{\sum x \cdot f}{N}$$

**EXAMPLE 3.21**  Find the mean of the distribution of waiting times between eruptions of Old Faithful geyser that was obtained in Example 2.4 on page 24.

**Solution**  To get $\sum x \cdot f$, we perform the calculations shown in the following table, where the first column contains minutes, the second column consists of the class marks, the third column contains frequencies, and the fourth column contains the products $x \cdot f$.

| Minutes | Class Mark $x$ | Frequency $f$ | $x \cdot f$ |
|---------|-----------|-----------|------|
| 30–39   | 34.5      | 2         | 69.0 |
| 40–49   | 44.5      | 2         | 89.0 |
| 50–59   | 54.5      | 4         | 218.0 |
| 60–69   | 64.5      | 19        | 1,225.5 |
| 70–79   | 74.5      | 24        | 1,788.0 |
| 80–89   | 84.5      | 39        | 3,295.5 |
| 90–99   | 94.5      | 15        | 1,417.5 |
| 100–109 | 104.5     | 3         | 313.5 |
| 110–119 | 114.5     | 2         | 229.0 |
|         |           | 110       | 8,645.0 |

Then, substitution into the formula yields $\bar{x} = \dfrac{8,645.0}{110} = 78.5909 = 78.59$ rounded to two decimals.   ■

To check on the **grouping error**, namely, the error introduced by replacing each value within a class by the corresponding class mark, we can calculate $\bar{x}$ for the original data given on page 24, or use the same computer software that led to Figure 2.4. Having already entered the data, we simply change the command to MEAN C1 and we get 78.273, or 78.27 rounded to two decimals. Thus, the grouping error is only $78.59 - 78.27 = 0.32$, which is fairly small.

When dealing with grouped data, we can determine most other statistical measures besides the mean, but we may have to make different assumptions and/or modify the definitions. For instance, for the median of a distribution we use the second of the assumptions mentioned on page 64 (namely, the assumption that the values within a class are spread evenly throughout the corresponding class interval). Thus, with reference to a histogram

> **The median of a distribution is such that the total area of the rectangles to its left equals the total area of the rectangles to its right.**

The method for finding the median location of grouped data is slightly different from that of ungrouped data. It is simply $\dfrac{n}{2}$. For example, if there were 100 frequencies in a distribution, the 50th value would be the median location. If in cumulating the frequencies, 46 were cumulated up to the median class, only 4 more would be needed in the median class to reach the median value.

To find the dividing line between the two halves of a histogram (each of which represents $\frac{n}{2}$ of the items grouped), we must count $\frac{n}{2}$ of the items starting at either end of the distribution. How this is done is illustrated by the following example and Figure 3.4:

**EXAMPLE 3.22**   Find the median of the distribution of the waiting time between eruptions of Old Faithful geyser.

**Figure 3.4**
The median of the distribution of the waiting times between eruptions of Old Faithful.

**Solution**  Since $\frac{n}{2} = \frac{110}{2} = 55$, we must count 55 of the items starting at either end. Starting at the bottom of the distribution (that is, beginning with the smallest values), we find that $2 + 2 + 4 + 19 + 24 = 51$ of the values fall into the first five classes. Therefore, we must count $55 - 51 = 4$ more values from among the values in the sixth class. Based on the assumption that the 39 values in the sixth class are spread evenly throughout that class, we accomplish this by adding $\frac{4}{39}$ of the class interval of 10 to 79.5, which is its lower class boundary. This yields

$$\tilde{x} = 79.5 + \frac{4}{39} \cdot 10 = 80.53$$

rounded to two decimals.                                                              ■

In general, if $L$ is the lower boundary of the class into which the median must fall, $f$ is its frequency, $c$ is its class interval, and $j$ is the number of items we still lack when we reach $L$, then the median of the distribution is given by

Median of
grouped data

$$\tilde{x} = L + \frac{j}{f} \cdot c$$

If we prefer, we can find the median of a distribution by starting to count at the other end (beginning with the largest values) and subtracting an appropriate fraction of the class interval from the upper boundary of the class into which the median must fall.

**EXAMPLE 3.23**  Use this formula to find the median of the waiting times between eruptions of Old Faithful geyser.

**Solution** Since $2 + 3 + 15 = 20$ of the values fall above 89.5, we need $55 - 20 = 35$ of the 39 values in the next class to reach the median. Thus, we write

$$\tilde{x} = 89.5 - \frac{35}{39} \cdot 10 = 80.53$$

and the result is, of course, the same.

Note that the median of a distribution can be found regardless of whether the class intervals are all equal. In fact, it can be found even when either or both classes at the top and at the bottom of a distribution are open, so long as the median does not fall into either class (see Exercise 3.57.)

The method by which we found the median of a distribution can also be used to determine other fractiles. For instance, $Q_1$ and $Q_3$ are defined for grouped data so that 25% of the total area of the rectangles of the histogram lies to the left of $Q_1$ and 25% lies to the right of $Q_3$. Similarly, the nine deciles (which are intended to divide a set of data into ten equal parts) are defined for grouped data so that 10% of the total area of the rectangles of the histogram lies to the left of $D_1$, 10% lies between $D_1$ and $D_2, \ldots$, and 10% lies to the right of $D_9$. And finally, the ninety-nine percentiles (which are intended to divide a set of data into a hundred equal parts) are defined for grouped data so that 1% of the total area of the rectangles of the histogram lies to the left of $P_1$, 1% lies between $P_1$ and $P_2, \ldots$, and 1% lies to the right of $P_{99}$. Note that $D_5$ and $P_{50}$ are equal to the median and that $P_{25}$ equals $Q_1$ and $P_{75}$ equals $Q_3$.

**EXAMPLE 3.24** Find the quartiles $Q_1$ and $Q_3$ for the distribution of the waiting times between eruptions of Old Faithful geyser.

**Solution** To find $Q_1$ we must count $\frac{110}{4} = 27.5$ of the items starting at the bottom (with the smallest values) of the distribution. Since there are $2 + 2 + 4 + 19 = 27$ values in the first four classes, we must count $27.5 - 27 = 0.5$ of the 24 values in the fifth class to reach $Q_1$. This yields

$$Q_1 = 69.5 + \frac{0.5}{24} \cdot 10 \approx 69.71$$

Since $2 + 3 + 15 = 20$ of the values fall into the last three classes, we must count $27.5 - 20 = 7.5$ of the 39 values in the next class to reach $Q_3$. Thus, we write

$$Q_3 = 89.5 - \frac{7.5}{39} \cdot 10 \approx 87.58$$

**EXAMPLE 3.25** Find the decile $D_2$ and the percentile $P_5$ for the distribution of the waiting times between eruptions of Old Faithful geyser.

**Solution** To find $D_2$ we must count $110 \cdot \frac{2}{10} = 22$ of the items starting at the bottom of the distribution. Since there are $2 + 2 + 4 = 8$ values in the first three classes, we must count $22 - 8 = 14$ of the 19 values in the fourth class to reach $D_2$. This yields

$$D_2 = 59.5 + \frac{14}{19} \cdot 10 \approx 66.87$$

Since $2 + 3 + 15 = 20$ of the values fall into the last three classes, we must count $22 - 20 = 2$ of the 39 values in the next class to reach $P_8$. Thus, we write

$$P_8 = 89.5 - \frac{2}{39} \cdot 10 \approx 88.99$$

Note that when we determine a fractile of a distribution, the number of items we have to count and the quantity $j$ in the formula on page 66 need not be a whole number.

The calculation of the mean of a distribution can usually be simplified by replacing the class marks with consecutive integers. This process is referred to as **coding**; when the class intervals are all equal, and only then, we assign the value 0 to a class mark near the middle of the distribution and code the class marks $\ldots, -3, -2, -1, 0, 1, 2, 3, \ldots$. Denoting the coded class marks by the letter $u$, we then use the formula

MEAN OF
GROUPED DATA
(WITH CODING)

$$\bar{x} = x_0 + \frac{\sum u \cdot f}{n} \cdot c$$

where $x_0$ is the class mark in the original scale to which we assign 0 in the new scale, $c$ is the class interval, $n$ is the total number of items grouped, and $\sum u \cdot f$ is the sum of the products obtained by multiplying each of the new class marks by the corresponding class frequency.

**(a)** the mean asked for in Example 3.21;

Using this method of coding, we calculate the mean asked for in Example 3.21, obtaining the same answer as before.

| Class Mark | Frequency | Coded Class Mark | $u \cdot f$ |
|---|---|---|---|
| 34.5 | 2 | −4 | −8 |
| 44.5 | 2 | −3 | −6 |
| 54.5 | 4 | −2 | −8 |
| 64.5 | 19 | −1 | −19 |
| 74.5 | 24 | 0 | 0 |
| 84.5 | 39 | 1 | 39 |
| 94.5 | 15 | 2 | 30 |
| 104.5 | 3 | 3 | 9 |
| 114.5 | 2 | 4 | 8 |
| | $n = 110$ | | $\sum u \cdot f = 45$ |

$$\bar{x} = x_0 + \frac{\sum u \cdot f}{n} \cdot c = 74.5 + \frac{45}{110} \cdot 10 = 74.5 + 4.0909 = 78.59$$

rounded to two decimals

EXERCISES

**3.57** For each of the following distributions, determine whether it is possible to find the mean and/or the median.

(a)

| Grade | Frequency |
|-------|-----------|
| 40–49 | 5 |
| 50–59 | 18 |
| 60–69 | 27 |
| 70–79 | 15 |
| 80–89 | 6 |

(b)

| IQ | Frequency |
|----|-----------|
| Less than 90 | 3 |
| 90–99 | 14 |
| 100–109 | 22 |
| 110–119 | 19 |
| More than 119 | 7 |

(c)

| Weight | Frequency |
|--------|-----------|
| 100 or less | 41 |
| 101–110 | 13 |
| 111–120 | 8 |
| 121–130 | 3 |
| 131–140 | 1 |

**3.58** Following is the distribution of the percentages of the students in 50 elementary schools who are bilingual:

| Percentage | Number of schools |
|------------|-------------------|
| 0–4 | 18 |
| 5–9 | 15 |
| 10–14 | 9 |
| 15–19 | 7 |
| 20–24 | 1 |

Find the mean and the median.

**3.59** Following is a distribution of the compressive strength (in 1,000 psi) of 120 samples of concrete:

| Compressive strength | Frequency |
|----------------------|-----------|
| 4.20–4.39 | 6 |
| 4.40–4.59 | 12 |
| 4.60–4.79 | 23 |
| 4.80–4.99 | 40 |
| 5.00–5.19 | 24 |
| 5.20–5.39 | 11 |
| 5.40–5.59 | 4 |
| | 120 |

Find the mean and the median.

**3.60** With reference to the distribution of Exercise 3.59, find

(a) $Q_1$ and $Q_3$;

(b)  $D_1$ and $D_9$;

(c)  $P_{15}$ and $P_{85}$.

**3.61**  In Exercise 2.27 we gave the following distribution of the weights of 133 mineral specimens collected on a field trip:

| Weight (grams) | Number of specimens |
|---|---|
| 5.0–19.9 | 8 |
| 20.0–34.9 | 27 |
| 35.0–49.9 | 42 |
| 50.0–64.9 | 31 |
| 65.0–79.9 | 17 |
| 80.0–94.9 | 8 |

Find the mean and the median.

**3.62**  In Exercise 2.53 we gave the following distribution of the number of fish tacos served for lunch by a Mexican restaurant on 60 weekdays:

| Number of fish tacos | Number of weekdays |
|---|---|
| 30–39 | 4 |
| 40–49 | 23 |
| 50–59 | 28 |
| 60–69 | 5 |

Find

(a)  the mean and the median;

(b)  $Q_1$ and $Q_3$.

**3.63**  With reference to the preceding exercise, could we have found the percentile $P_{95}$ if the fourth class had been "60 or more?"

**3.64**  Use the distribution obtained in Exercise 2.36 to determine the mean, the median, $Q_1$, and $Q_3$ for the percent shrinkages of the plastic clay specimens.

**\*3.65**  Use the distribution obtained in Exercise 2.42 to determine the mean, the median, $Q_1$, and $Q_3$ for the root penetrations of the 120 crested wheatgrass seedlings one month after planting.

# 3.8 TECHNICAL NOTE (SUMMATIONS)

In the notation introduced on page 46, $\sum x$ does not tell us which, or how many, values of $x$ we must add. This is taken care of by the more explicit notation

$$\sum_{i=1}^{n} x_i = x_1 + x_2 + \cdots + x_n$$

where it is made clear that we are adding the $x$'s whose subscripts $i$ are $1, 2, \ldots$, and $n$. We are not using the more explicit notation in this text to simplify the overall appearance of the formulas, assuming that it is clear in each case what $x$'s we are referring to and how many there are.

Using the $\sum$ notation, we shall also have occasion to write such expressions as $\sum x^2, \sum xy, \sum x^2 f, \ldots$, which (more explicitly) represent the sums

$$\sum_{i=1}^{n} x_i^2 = x_1^2 + x_2^2 + x_3^2 + \cdots + x_n^2$$

$$\sum_{j=1}^{m} x_j y_j = x_1 y_1 + x_2 y_2 + \cdots + x_m y_m$$

$$\sum_{i=1}^{n} x_i^2 f_i = x_1^2 f_1 + x_2^2 f_2 + \cdots + x_n^2 f_n$$

Working with two subscripts, we shall also have the occasion to evaluate **double summations** such as

$$\sum_{j=1}^{3} \sum_{i=1}^{4} x_{ij} = \sum_{j=1}^{3} (x_{1j} + x_{2j} + x_{3j} + x_{4j})$$

$$= x_{11} + x_{21} + x_{31} + x_{41} + x_{12} + x_{22} + x_{32} + x_{42}$$

$$+ x_{13} + x_{23} + x_{33} + x_{43}$$

To verify some of the formulas involving summations that are stated but not proved in the text, the reader will need the following rules:

**R**ULES FOR SUMMATIONS

$$Rule\ A: \sum_{i=1}^{n} (x_i \pm y_i) = \sum_{i=1}^{n} x_i \pm \sum_{i=1}^{n} y_i$$

$$Rule\ B: \sum_{i=1}^{n} k \cdot x_i = k \cdot \sum_{i=1}^{n} x_i$$

$$Rule\ C: \sum_{i=1}^{n} k = k \cdot n$$

The first of these rules states that the summation of the sum (or difference) of two terms equals the sum (or difference) of the individual summations, and it can be extended to the sum or difference of more than two terms. The second rule states that we can, so to speak, factor a constant out of a summation, and the third rule states that the summation of a constant is simply $n$ times that constant. All of these rules can be proved by actually writing out in full what each of the summation represents.

**3.66** Write each of the following in full; that is, without summation signs:

(a) $\displaystyle\sum_{i=1}^{6} x_i$;

(d) $\displaystyle\sum_{j=1}^{8} x_j f_j$;

(b) $\displaystyle\sum_{i=1}^{5} y_i$;

(e) $\displaystyle\sum_{i=3}^{7} x_i^2$;

(c) $\displaystyle\sum_{i=1}^{3} x_i y_i$;

(f) $\displaystyle\sum_{j=1}^{4} (x_j + y_j)$.

**3.67** Write each of the following as a summation; that is, in the $\sum$ notation:

(a) $z_1 + z_2 + z_3 + z_4 + z_5$;

(b) $x_5 + x_6 + x_7 + x_8 + x_9 + x_{10} + x_{11} + x_{12}$;

(c) $x_1 f_1 + x_2 f_2 + x_3 f_3 + x_4 f_4 + x_5 f_5 + x_6 f_6$;

(d) $y_1^2 + y_2^2 + y_3^2$;

(e) $2x_1 + 2x_2 + 2x_3 + 2x_4 + 2x_5 + 2x_6 + 2x_7$;

(f) $(x_2 - y_2) + (x_3 - y_3) + (x_4 - y_4)$;

(g) $(z_2 + 3) + (z_3 + 3) + (z_4 + 3) + (z_5 + 3)$;

(h) $x_1 y_1 f_1 + x_2 y_2 f_2 + x_3 y_3 f_3 + x_4 y_4 f_4$.

**3.68** Given $x_1 = 3, x_2 = 2, x_3 = -2, x_4 = 5, x_5 = -1, x_6 = 3, x_7 = 2,$ and $x_8 = 4,$ find

(a) $\displaystyle\sum_{i=1}^{8} x_i$;

(b) $\displaystyle\sum_{i=1}^{8} x_i^2$.

**3.69** Given $x_1 = 2, \ x_2 = 3, \ x_3 = 4, \ x_4 = 5, \ x_5 = 6, \ f_1 = 2, \ f_2 = 8, \ f_3 = 9, \ f_4 = 3,$ and $f_5 = 2,$ find

(a) $\displaystyle\sum_{i=1}^{5} x_i$;

(c) $\displaystyle\sum_{i=1}^{5} x_i f_i$;

(b) $\displaystyle\sum_{i=1}^{5} f_i$;

(d) $\displaystyle\sum_{i=1}^{5} x_i^2 f_i$.

**3.70** Given $x_1 = 4, x_2 = -2, x_3 = 3, x_4 = -1, y_1 = 5, y_2 = -2, y_3 = 4,$ and $y_4 = -1,$ find

(a) $\displaystyle\sum_{i=1}^{4} x_i$;

(d) $\displaystyle\sum_{i=1}^{4} y_i^2$;

(b) $\displaystyle\sum_{i=1}^{4} y_i$;

(e) $\displaystyle\sum_{i=1}^{4} x_i y_i$.

(c) $\displaystyle\sum_{i=1}^{4} x_i^2$;

**3.71** Given $x_{11} = 4, x_{12} = 2, x_{13} = -1, x_{14} = 3, x_{21} = 2, x_{22} = 5, x_{23} = -1, x_{24} = 6, x_{31} = 4, x_{32} = -1, x_{33} = 3,$ and $x_{34} = 4,$ find

(a) $\displaystyle\sum_{i=1}^{3} x_{ij}$ separately for $j = 1, 2, 3,$ and $4$;

(b) $\displaystyle\sum_{j=1}^{4} x_{ij}$ separately for $i = 1, 2,$ and $3$.

**3.72** With reference to Exercise 3.71, evaluate the double summation $\sum_{i=1}^{3} \sum_{j=1}^{4} x_{ij}$ using

    (a) the results of part (a) of that exercise;

    (b) the results of part (b) of that exercise.

**3.73** Is it true in general that $\left(\sum_{i=1}^{n} x_i\right)^2 = \sum_{i=1}^{n} x_i^2$? (*Hint*: Check whether the equation holds for $n = 2$.)

# CHECKLIST OF KEY TERMS (with page references to their definitions)

Arithmetic mean, 45
Boxplot, 58
Box-and-whisker plot, 59
*Coding, 68
Decile, 57
Double summation, 71
Exploratory data analysis, 57
Fractile, 57
Geometric mean, 45, 52
Grand mean, 50
*Grouping error, 65
Harmonic mean, 45, 53
Markov's theorem, 52
Mean, 45
Measures of central location, 43
Measures of location, 43
Measures of variation, 43

Median, 49, 54
Median position, 55
Midquartile, 57, 58
Midrange, 57, 62
Mode, 57, 59
Outlier, 49
Parameter, 46
Percentile, 57
Population, 44
Population size, 46
Quartile, 57
Relative importance weight, 49
Sample, 44
Statistic, 46
Sigma notation, 46
Weighted mean, 49

# REFERENCES

*Informal discussions of the ethics involved in choosing among averages and other questions of ethics in statistics in general are given in*

    HOOKE, R., *How to Tell the Liars from the Statisticians*. New York: Marcel Dekker, Inc., 1983.

    HUFF, D., *How to Lie with Statistics*. New York: W. W. Norton & Company, Inc., 1954.

*For further information about the use and interpretation of hinges, see the books on exploratory data analysis referred to on page 42.*

# SUMMARIZING DATA: MEASURES OF VARIATION

**4.1**  The Range 75

**4.2**  The Standard Deviation and the Variance 75

**4.3**  Applications of the Standard Deviation 79

*__4.4__  The Description of Grouped Data 86

**4.5**  Some Further Descriptions 88

Checklist of Key Terms 93

References 93

One aspect of most sets of data is that the values are not all alike; indeed, the extent to which they are unalike, or vary among themselves, is of basic importance in statistics. Consider the following examples:

In a hospital where each patient's pulse rate is taken three times a day, that of patient *A* is 72, 76, and 74, while that of patient *B* is 72, 91, and 59. The mean pulse rate of the two patients is the same, 74, but observe the difference in variability. Whereas patient *A*'s pulse rate is stable, that of patient *B* fluctuates widely.

A supermarket stocks certain 1-pound bags of mixed nuts, which on the average contain 12 almonds per bag. If all the bags contain anywhere from 10 to 14 almonds, the product is consistent and satisfactory, but the situation is quite different if some of the bags have no almonds while others have 20 or more.

Measuring variability is of special importance in statistical inference. Suppose, for instance, that we have a coin that is slightly bent and we wonder whether there is still a fifty–fifty chance for heads. What if we toss the coin 100 times and get 28 heads and 72 tails? Does the shortage of heads—only 28 where we might have expected 50—imply that the count is not "fair"? To answer such questions we must have some idea about the magnitude of the fluctuations, or variations, that are brought about by chance when coins are tossed 100 times.

We have given these three examples to show the need for measuring the extent to which data are dispersed, or spread out; the corresponding measures that provide this information are called **measures of variation**. In Sections 4.1 through 4.3 we present the most widely used measures of variation and some of their special applications. Some statistical descriptions other than measures of location and measures of variation are discussed in Section 4.5.

# 4.1 THE RANGE

To introduce a simple way of measuring variability, let us refer to the first of the three examples cited previously, where the pulse rate of patient $A$ varied from 72 to 76 while that of patient $B$ varied from 59 to 91. These extreme (smallest and largest) values are indicative of the variability of the two sets of data, and just about the same information is conveyed if we take the differences between the respective extremes. So, let us make the following definition:

> **The range of a set of data is the difference between the largest value and the smallest.**

For patient $A$ the pulse rates had a range of $76 - 72 = 4$ and for patient $B$ they had a range of $91 - 59 = 32$. Also, for the lengths of the trout on page 13, the range was $23.5 - 16.6 = 6.9$ centimeters; and for the waiting times between eruptions of Old Faithful in Example 2.4, the range was $118 - 33 = 85$ minutes.

Conceptually, the range is easy to understand, its calculation is very easy, and there is a natural curiosity about the smallest and largest values. Nevertheless, it is not a very useful measure of variation—its main shortcoming being that it does not tell us anything about the dispersion of the values that fall between the two extremes. For example, each of the following three sets of data

$$\begin{array}{lllllllllll} \text{Set A:} & 5 & 18 & 18 & 18 & 18 & 18 & 18 & 18 & 18 & 18 \\ \text{Set B:} & 5 & 5 & 5 & 5 & 5 & 18 & 18 & 18 & 18 & 18 \\ \text{Set C:} & 5 & 6 & 8 & 9 & 10 & 12 & 14 & 15 & 17 & 18 \end{array}$$

has a range of $18 - 5 = 13$, but their dispersions between the first and last values are totally different.

In actual practice, the range is used mainly as a "quick and easy" measure of variability; for instance, in industrial quality control it is used to keep a close check on raw materials and products on the basis of small samples taken at regular intervals of time.

Whereas the range covers all the values in a sample, a similar measure of variation covers (more or less) the middle 50%. It is the **interquartile range** $Q_3 - Q_1$, where $Q_1$ and $Q_3$ may be defined as in Section 3.5 or in Section 3.7. Some statisticians use the **semi-interquartile range** $\frac{1}{2}(Q_3 - Q_1)$, which is also referred to as the **quartile deviation**.

# 4.2 THE STANDARD DEVIATION AND THE VARIANCE

To define the **standard deviation**, by far the most generally useful measure of variation, let us observe that the dispersion of a set of data is small if the values are closely bunched about their mean, and that it is large if the values are scattered widely about their mean. It would seem reasonable, therefore, to measure the variation of a set of data in terms of the amounts by which the values deviate from their mean. If a set of numbers

$$x_1, x_2, x_3, \ldots, \quad \text{and} \quad x_n$$

constitutes a sample with the mean $\bar{x}$, then the differences

$$x_1 - \bar{x}, x_2 - \bar{x}, x_3 - \bar{x}, \ldots, \quad \text{and} \quad x_n - \bar{x}$$

are called the **deviations from the mean**, and we might use their average (that is, their mean) as a measure of the variability of the sample. Unfortunately, this will not do. Unless the $x$'s are all equal, some of the deviations from the mean will be positive, some will be negative. The sum of the deviations from the mean, $\sum(x - \bar{x})$, is always equal to zero.

Since we are really interested in the magnitude of the deviations, and not in whether they are positive or negative, we might simply ignore the signs and define a measure of variation in terms of the absolute values of the deviations from the mean. Indeed, if we add the deviations from the mean as if they were all positive or zero and divide by $n$, we obtain the statistical measure that is called the **mean deviation**. This measure has intuitive appeal, but because of the absolute values it leads to serious theoretical difficulties in problems of inference, and it is rarely used.

An alternative approach is to work with the squares of the deviations from the mean, as this will also eliminate the effect of the signs. Squares of real numbers cannot be negative; in fact, squares of the deviations from a mean are all positive (unless a value happens to coincide with the mean). Then, if we average the squared deviations from the mean and take the square root of the result (to compensate for the fact that the deviations were squared), we get

$$\sqrt{\frac{\sum(x - \bar{x})^2}{n}}$$

and this is how, traditionally, the standard deviation used to be defined. Expressing literally what we have done here mathematically, it is also called the **root-mean-square deviation**.

Nowadays, it is customary to modify this formula by dividing the sum of the squared deviations from the mean by $n - 1$ instead of $n$. Following this practice, which will be explained later, let us define the **sample standard deviation**, denoted by $s$, as

**SAMPLE STANDARD DEVIATION**

$$s = \sqrt{\frac{\sum(x - \bar{x})^2}{n - 1}}$$

The square of the sample standard deviation is called the **sample variance**, and it is appropriately denoted by $s^2$. There is no need to list a separate formula for the sample variance, which is simply like that for the sample standard deviation with the square root sign deleted. Correspondingly, $\sigma^2$ is the **population variance**.

These formulas for the standard deviation and the variance apply to samples, but if we substitute $\mu$ for $\bar{x}$ and $N$ for $n$, we obtain analogous formulas for the standard deviation and the variance of a population. It is customary to denote the **population standard deviation** by $\sigma$ (*sigma*, the Greek letter for lowercase $s$) when we divide by $N$, and by $S$ when we divide by $N - 1$.

**POPULATION STANDARD DEVIATION**

$$\sigma = \sqrt{\frac{\sum(x - \mu)^2}{N}}$$

Ordinarily, the purpose of calculating a sample statistic (such as the mean, the standard deviation, or the variance) is to estimate the corresponding population parameter. If we actually took many samples from a population that has the mean $\mu$, calculated the sample means $\bar{x}$, and then averaged all these estimates of $\mu$, we should find that their average is very close to $\mu$. However, if we calculated the variance of each sample by means of the formula $\dfrac{\sum(x - \bar{x})^2}{n}$ and then averaged all these supposed estimates of $\sigma^2$, we would probably find that their average is less than $\sigma^2$. Theoretically, it can be shown that we can compensate for this by dividing by $n - 1$ instead of $n$ in the formula for $s^2$. Estimators having the desirable property that their values will on the average equal the quantity they are supposed to estimate are said to be **unbiased**; otherwise, they are said to be **biased**. So, we say that $\bar{x}$ is an unbiased estimator of the population mean $\mu$, and that $s^2$ is an unbiased estimator of the population variance $\sigma^2$. It does not follow from this that $s$ is also an unbiased estimator of $\sigma$, but when $n$ is large the bias is small and can usually be ignored.

In calculating the sample standard deviation using the formula by which it is defined, we must (1) find $\bar{x}$, (2) determine the $n$ deviations from the mean $x - \bar{x}$, (3) square these deviations, (4) add all the squared deviations, (5) divide by $n - 1$, and (6) take the square root of the result arrived at in step 5. In actual practice, this method is rarely used—there are various shortcuts—but we shall illustrate it here to emphasize what is really measured by a standard deviation.

**EXAMPLE 4.1** A pathology lab found 8, 11, 7, 13, 10, 11, 7, and 9 bacteria of a certain kind in cultures from eight otherwise healthy persons. Calculate $s$.

**Solution** First calculating the mean, we get

$$\bar{x} = \frac{8 + 11 + 7 + 13 + 10 + 11 + 7 + 9}{8} = 9.5$$

and then the work required to find $\sum(x - \bar{x})^2$ may be arranged as in the following table:

| $x$ | $x - \bar{x}$ | $(x - \bar{x})^2$ |
|---|---|---|
| 8 | $-1.5$ | 2.25 |
| 11 | 1.5 | 2.25 |
| 7 | $-2.5$ | 6.25 |
| 13 | 3.5 | 12.25 |
| 10 | 0.5 | 0.25 |
| 11 | 1.5 | 2.25 |
| 7 | $-2.5$ | 6.25 |
| 9 | $-0.5$ | 0.25 |
| 76 | 0.0 | 32.00 |

Finally, dividing 32.00 by $8 - 1 = 7$ and taking the square root (using a simple handheld calculator), we get

$$s = \sqrt{\frac{32.00}{7}} = \sqrt{4.57} = 2.14$$

rounded to two decimals.   ◼

Note in the preceding table that the total of the middle column is zero; since this must always be the case, it provides a convenient check on the calculations.

It was easy to calculate $s$ in this example because the data were whole numbers and the mean was exact to one decimal. Otherwise, the calculations required by the formula defining $s$ can be quite tedious, and, unless we can get $s$ directly with a statistical calculator or a computer, it helps to use the formula

COMPUTING
FORMULA FOR THE
SAMPLE STANDARD
DEVIATION

$$s = \sqrt{\frac{S_{xx}}{n - 1}} \quad \text{where} \quad S_{xx} = \sum x^2 - \frac{(\sum x)^2}{n}$$

EXAMPLE 4.2   Use this computing formula to rework Example 4.1.

Solution   First we calculate $\sum x$ and $\sum x^2$, getting

$$\sum x = 8 + 11 + 7 + \cdots + 7 + 9$$
$$= 76$$

and

$$\sum x^2 = 64 + 121 + 49 + 169 + 100 + 121 + 49 + 81$$
$$= 754$$

Then, substituting these totals and $n = 8$ into the formula for $S_{xx}$, and $n - 1 = 7$ and the value obtained for $S_{xx}$ into the formula for $s$, we get

$$S_{xx} = 754 - \frac{(76)^2}{8} = 32$$

and, hence, $s = \sqrt{\frac{32}{7}} = 2.14$ rounded to two decimals. This agrees, as it should, with the result obtained before.   ∎

As should have been apparent from these two examples, the advantage of the computing formula is that we got the result without having to determine $\bar{x}$ and work with the deviations from the mean. Incidentally, the computing formula can also be used to find $\sigma$, with the $n$ in the formula for $S_{xx}$ and the $n - 1$ in the formula for $s$ replaced by $N$.

# 4.3 APPLICATIONS OF THE STANDARD DEVIATION

In subsequent chapters, sample standard deviations will be used primarily to estimate population standard deviations in problems of inference. Meanwhile, to provide the reader with more of a feeling of what a standard deviation really measures, we shall devote this section to some applications.

In the argument that led to the definition of the standard deviation, we observed that the dispersion of a set of data is small if the values are bunched closely about their mean, and that it is large if the values are scattered widely about their mean. Correspondingly, we can now say that if the standard deviation of a set of data is small, the values are concentrated near the mean, and if the standard deviation is large, the values are scattered widely about the mean. This idea is expressed more formally by the following theorem, called **Chebyshev's theorem** after the Russian mathematician P. L. Chebyshev (1821–1894):

**CHEBYSHEV'S THEOREM**

*For any set of data (population or sample) and any constant $k$ greater than 1, the proportion of the data that must lie within $k$ standard deviations on either side of the mean is at least*

$$1 - \frac{1}{k^2}$$

It may be surprising that we can make such definite statements, but it is a certainty that at least $1 - \frac{1}{2^2} = \frac{3}{4}$, or 75%, of the values in *any* set of data must lie within two standard deviations on either side of the mean, at least $1 - \frac{1}{5^2} = \frac{24}{25}$, or 96%, must lie within five standard deviations on either side of the mean, and at least $1 - \frac{1}{10^2} = \frac{99}{100}$, or 99%, must lie within ten standard deviations on either side of the mean. Here we arbitrarily let $k = 2, 5$, and 10.

**EXAMPLE 4.3** A study of the nutritional value of a certain kind of reduced-fat cheese showed that on the average a one-ounce slice contains 3.50 grams of fat with a standard deviation of 0.04 gram of fat.

(a) According to Chebyshev's theorem, at least what percent of the one-ounce slices of this kind of cheese must have a fat content between 3.38 and 3.62 grams of fat?

(b) According to Chebyshev's theorem, between what values must be the fat content of at least 93.75% of the one-ounce slices of this kind of cheese?

**Solution**

(a) Since $3.62 - 3.50 = 3.50 - 3.38 = 0.12$, we find that $k(0.04) = 0.12$ and, hence, $k = \frac{0.12}{0.04} = 3$. It follows that at least $1 - \frac{1}{3^2} = \frac{8}{9}$, or approximately 88.9%, of the one-ounce slices of the cheese have a fat content between 3.38 and 3.62 grams of fat.

(b) Since $1 - \frac{1}{k^2} = 0.9375$, we find that $\frac{1}{k^2} = 1 - 0.9375 = 0.0625$, $k^2 = \frac{1}{0.0625} = 16$, and $k = 4$. It follows that 93.75% of the one-ounce slices of this cheese contain between $3.50 - 4(0.04) = 3.34$ and $3.50 + 4(0.04) = 3.66$ grams of fat.

■

Chebyshev's theorem applies to any kind of data, but it has its shortcomings. Since it tells us merely "at least what proportion" of a set of data must lie between certain limits (that is, it provides only a lower limit to the actual proportion), it has few practical applications. Indeed, we have presented it here only to provide the reader with some idea of how to relate the standard deviation to the spread, or dispersion, of a set of data, and vice versa.

For distributions having the general shape of the cross section of a bell (see Figure 4.1), we can make the following much stronger statements:

---

**About 68% of the values will lie within one standard deviation of the mean, that is, between $\bar{x} - s$ and $\bar{x} + s$.**

**About 95% of the values will lie within two standard deviations of the mean, that is, between $\bar{x} - 2s$ and $\bar{x} + 2s$.**

**About 99.7% of the values will lie within three standard deviations of the mean, that is, between $\bar{x} - 3s$ and $\bar{x} + 3s$.**

---

This result is sometimes referred to as the **empirical rule**, presumably because such percentages are observed in practice. Actually, it is a theoretical result based

**Figure 4.1**
Bell-shaped distribution.

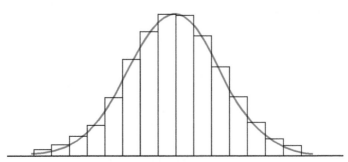

on the normal distribution, which we shall study in Chapter 9 (in particular, see Exercise 9.10).

**EXAMPLE 4.4**  In Example 3.21 we calculated the mean of the grouped waiting times between eruptions of Old Faithful, getting 78.59, and in Example 4.7 we will show that the corresponding standard deviation is 14.35. Use these figures to determine from the original data given in Example 2.4 what percentage of the values falls within three standard deviations of the mean.

**Solution**  Since $\bar{x} = 78.59$ and $s = 14.35$, we shall have to determine what percentage of the values falls between $78.59 - 3(14.35) = 35.54$ and $78.59 + 3(14.35) = 121.64$. Counting two of the values, 33 and 35, below 35.54 and none above 121.64, we find that $110 - 2 = 108$ of the values and, hence

$$\frac{108}{110} \cdot 100 = 98.2\%$$

of the original waiting times fall within three standard deviations of the mean. This is fairly close to the expected 99.7%, but then the distribution of the waiting times is not really perfectly bell shaped.

In the introduction to this chapter we gave three examples in which knowledge about the variability of the data was of special importance. This is also the case when we want to compare numbers belonging to different sets of data. To illustrate, suppose that the final examination in a French course consists of two parts, vocabulary and grammar, and that a certain student scored 66 points in the vocabulary part and 80 points in the grammar part. At first glance it would seem that the student did much better in grammar than in vocabulary, but suppose that all the students in the class averaged 51 points in the vocabulary part with a standard deviation of 12, and 72 points in the grammar part with a standard deviation of 16. Thus, we can argue that the student's score in the vocabulary part is $\frac{66-51}{12} = 1.25$ standard deviations above the average for the class, while the score in the grammar part is only $\frac{80-72}{16} = 0.50$ standard deviation above the average for the class. Whereas the original scores cannot be meaningfully compared, these new scores, expressed in terms of standard deviations, can. Clearly, the given student rates much higher on her command of French vocabulary than on the student's knowledge of French grammar, compared to the rest of the class.

What we have done here consists of converting the grades into **standard units** or **z-scores**. In general, if $x$ is a measurement belonging to a set of data having the mean $\bar{x}$ (or $\mu$) and the standard deviation $s$ (or $\sigma$), then its value in standard units, denoted by $z$, is

**FORMULA FOR CONVERTING TO STANDARD UNITS**

$$z = \frac{x - \bar{x}}{s} \quad \text{or} \quad z = \frac{x - \mu}{\sigma}$$

depending on whether the data constitute a sample or a population. In these units, $z$ tells us how many standard deviations a value lies above or below the mean of the set of data to which it belongs. Standard units will be used frequently in later chapters.

**EXAMPLE 4.5**   Mrs. Clark belongs to an age group for which the mean weight is 112 pounds with a standard deviation of 11 pounds, and Mr. Clark, her husband, belongs to an age group for which the mean weight is 163 pounds with a standard deviation of 18 pounds. If Mrs. Clark weighs 132 pounds and Mr. Clark weighs 193 pounds, which of the two is relatively more overweight compared to his/her age group?

**Solution**   Mr. Clark's weight is $193 - 163 = 30$ pounds above average while Mrs. Clark's weight is "only" $132 - 112 = 20$ pounds above average, yet in standard units we get $\frac{193-163}{18} \approx 1.67$ for Mr. Clark and $\frac{132-112}{11} \approx 1.82$ for Mrs. Clark. Thus, relative to their age groups Mrs. Clark is somewhat more overweight than Mr. Clark. ■

A serious disadvantage of the standard deviation as a measure of variation is that it depends on the units of measurement. For instance, the weights of certain objects may have a standard deviation of 0.10 ounce, but this really does not tell us whether it reflects a great deal of variation or very little variation. If we are weighing the eggs of quails, a standard deviation of 0.10 ounce would reflect a considerable amount of variation, but this would not be the case if we are weighing, say, 100-pound bags of potatoes. What we need in a situation like this is a **measure of relative variation** such as the **coefficient of variation**, defined by the following formula:

**C**OEFFICIENT OF VARIATION

$$V = \frac{s}{\bar{x}} \cdot 100 \quad \text{or} \quad V = \frac{\sigma}{\mu} \cdot 100$$

The coefficient of variation expresses the standard deviation as a percentage of what is being measured, at least on the average.

**EXAMPLE 4.6**   Several measurements of the diameter of a ball bearing made with one microm-eter had a mean of 2.49 mm and a standard deviation of 0.012 mm, and several measurements of the unstretched length of a spring made with another microme-ter had a mean of 0.75 in. with a standard deviation of 0.002 in. Which of the two micrometers is relatively more precise?

**Solution**   Calculating the two coefficients of variation, we get

$$\frac{0.012}{2.49} \cdot 100 \approx 0.48\% \quad \text{and} \quad \frac{0.002}{0.75} \cdot 100 \approx 0.27\%$$

Thus, the measurements of the length of the spring are relatively less variable, which means that the second micrometer is more precise. ■

**4.1**  Following are four determinations of the specific gravity of aluminum: 2.64, 2.70, 2.67, and 2.63. Find

(a) the range;

(b) the standard deviation (using the formula that defines $s$).

**4.2**  It has been claimed that for samples of size $n = 4$, the range should be roughly twice as large as the standard deviation. Use the results of Exercise 4.1 to check on this claim.

**4.3**  The ten employees of a nursing home, having been given a course in cardiopulmonary resuscitation (CPR), scored 17, 20, 12, 14, 18, 23, 17, 19, 18, and 15 on a test administered after the completion of the course. Find

(a) the range;

(b) the standard deviation (using the computing formula for $s$).

**4.4**  It has been claimed that for samples of size $n = 10$, the range should be roughly three times as large as the standard deviation. Use the results of Exercise 4.3 (a) to check this claim.

**4.5**  With reference to Exercise 4.3, use the result of part (a) of that exercise and compare it with twice the interquartile range. Should it come as a surprise that the range is the bigger of the two?

**4.6**  The response times for a sample of six switches designed to activate an alarm system upon receiving a certain stimulus are 9, 8, 5, 11, 7, and 5 milliseconds. Use the computing formula for $s$ to calculate the standard deviation.

**4.7**  Use a computer or a graphing calculator to determine the sample standard deviation for the six switches of Exercise 4.6.

**4.8**  In Exercise 3.5 we gave the numbers of seconds that 12 insects survived after having been sprayed with a certain insecticide as 112, 83, 102, 84, 105, 121, 76, 110, 98, 91, 103, and 85. Find the sample standard deviation using

(a) the formula that defines $s$;

(b) the computing formula for $s$.

**4.9**  In a recent year there were 7, 8, 4, 11, 13, 15, 6, and 4 Sunday newspapers in the eight mountain states. Since these data are a population, calculate $\sigma$, the population standard deviation.

**4.10**  A chemist used a finely graduated burette, in an experiment, to deliver very small amounts of acid to neutralize a chemical solution. The burette delivered 8.96, 8.92, 8.98, 8.96, and 8.93 ml of acid to the solution.

(a) Calculate $s$ for these five amounts;

(b) subtract 8.90 from each of these five amounts and then recalculate $s$, the sample standard deviation, for the resulting values.

**4.11**  Calculate the ranges for the preceding Exercises 4.10(a) and 4.10(b). In this exercise the standard deviation of part (a) was unaffected by the subtraction of the constant 8.90. Were the ranges affected by the subtraction of this constant value?

**4.12**  On a rainy day during the monsoon season, 0.13, 0.05, 0.26, 0.41, 0.57, 0.02, 0.25, 0.10, 0.60, and 0.18 inches of rain were recorded in ten cities in Arizona.

(a) Calculate $s$ for these data.

(b) Multiply each of these figures by 100, calculate $s$ for the resulting data, and compare the result with that obtained in part (a).

What possible simplification does this suggest for the calculation of a standard deviation?

**4.13**    Following are the numbers of fire insurance claims submitted to a casualty insurance company on six consecutive business days: 6, 13, 7, 4, 14, and 10.

(a)  Find the range.

(b)  Find $s$ using the formula for the sample standard deviation.

(c)  Find $s$ using the computing formula for the sample standard deviation.

**4.14**    If each item in a set of data has the same constant value added to it, the mean of this new set equals the mean of the original set plus the constant $a$, but the range and standard deviation remain unchanged. In Exercise 4.13 the data are 6, 13, 7, 4, 14, and 10.

(a)  Add five fire insurance claims to each of the six daily values. Verify that the range remains unchanged, as indicated above.

(b)  Add five fire insurance claims to each of the six daily values. Verify that the new mean equals the original mean, plus the constant value of five.

(c)  Verify that the sample standard deviation, $s$, remains unchanged.

**4.15**    Use a computer package or a graphing calculator to verify that the standard deviation of the waiting times between eruptions of Old Faithful given in Example 2.4 is $s = 14.666$ rounded to three decimals.

**4.16**    Use a computer package or a graphing calculator to determine the standard deviation of the lengths of the 60 sea trout given in the beginning of Section 2.1.

**4.17**    With reference to Exercise 2.44, use a computer package or a graphing calculator to determine the standard deviation of the attendance figures.

**4.18**    According to Chebyshev's theorem, what can we assert about the proportion of any set of data that must lie within $k$ standard deviations on either side of the mean when

(a)  $k = 6$;

(b)  $k = 12$;

(c)  $k = 21$?

**4.19**    According to Chebyshev's theorem, what can we assert about the proportion of any set of data that must lie within $k$ standard deviations on either side of the mean when

(a)  $k = 2.5$;

(b)  $k = 16$?

**4.20**    Hospital records show that a certain surgical procedure takes on the average 111.6 minutes with a standard deviation of 2.8 minutes. At least what percentage of these surgical procedures take anywhere between

(a)  106.0 and 117.2 minutes;

(b)  97.6 and 125.6 minutes?

**4.21**    With reference to Exercise 4.20, between how many minutes must be the lengths of

(a)  at least 35/36 of these surgical procedures;

(b)  at least 99% of these surgical procedures?

**4.22**    Having kept records for several months, Ms. Lewis knows that it takes her on the average 47.7 minutes with a standard deviation of 2.46 minutes to drive to work from her suburban home. If she always starts out exactly one hour before she has to arrive at work, at most what percentage of the time will she arrive late?

**4.23**   Following are the amounts of sulfur oxides (in tons) emitted by an industrial plant on 80 days:

| | | | | | | | | | |
|---|---|---|---|---|---|---|---|---|---|
| 15.8 | 26.4 | 17.3 | 11.2 | 23.9 | 24.8 | 18.7 | 13.9 | 9.0 | 13.2 |
| 22.7 | 9.8 | 6.2 | 14.7 | 17.5 | 26.1 | 12.8 | 28.6 | 17.6 | 23.7 |
| 26.8 | 22.7 | 18.0 | 20.5 | 11.0 | 20.9 | 15.5 | 19.4 | 16.7 | 10.7 |
| 19.1 | 15.2 | 22.9 | 26.6 | 20.4 | 21.4 | 19.2 | 21.6 | 16.9 | 19.0 |
| 18.5 | 23.0 | 24.6 | 20.1 | 16.2 | 18.0 | 7.7 | 13.5 | 23.5 | 14.5 |
| 14.4 | 29.6 | 19.4 | 17.0 | 20.8 | 24.3 | 22.5 | 24.6 | 18.4 | 18.1 |
| 8.3 | 21.9 | 12.3 | 22.3 | 13.3 | 11.8 | 19.3 | 20.0 | 25.7 | 31.8 |
| 25.9 | 10.5 | 15.9 | 27.5 | 18.1 | 17.9 | 9.4 | 24.1 | 20.1 | 28.5 |

(a)  Group these data into the classes 5.0–8.9, 9.0–12.9, 13.0–16.9, 17.0–20.9, 21.0–24.9, 25.0–28.9, and 29.0–32.9. Also, draw a histogram of this distribution and judge whether it may well be described as bell shaped.

(b)  Use a computer package or a graphing calculator to determine the values of $\bar{x}$ and $s$ for the ungrouped data.

(c)  Use the results of part (b) to determine the values of $\bar{x} \pm s$, $\bar{x} \pm 2s$, and $\bar{x} \pm 3s$.

(d)  Use the results of part (c) to determine what percent of the original data falls within one standard deviation on either side of the mean, what percent falls within two standard deviations on either side of the mean, and what percent falls within three standard deviations on either side of the mean.

(e)  Compare the percentages obtained in part (d) with the 68, 95, and 99.7% claimed by the empirical rule.

**4.24**   The applicants to one branch of a state university have a mean ACT English score of 19.4 with a standard deviation of 3.1, while the applicants to another branch of the state university have a mean ACT English score of 20.1 with a standard deviation of 2.8. If an applicant applied to both branches, with respect to which of the two branches is he or she in a relatively better position with

(a)  an ACT English score of 24;

(b)  an ACT English score of 29?

**4.25**   An investment service reports for each stock on its list the price at which it is currently selling, its average price over the last six months, and a measure of its variability. Stock A, it reports, is currently selling at $76.75 and averaged $58.25 over the last six months with a standard deviation of $11.00. Stock B is currently selling at $49.50 and averaged $37.50 over the last six months with a standard deviation of $4.00. Leaving all other considerations aside, which of the two stocks is relatively more overpriced at this time?

**4.26**   In 10 rounds of golf, one golfer averaged 76.2 with a standard deviation of 2.4, while another golfer averaged 84.9 with a standard deviation of 3.5. Which of the two golfers is relatively more consistent?

**4.27**   If five specimens of hard yellow brass had shearing strengths of 49, 52, 51, 53, and 55 thousand psi, and on four Sundays the rainfall at a marina amounted to 0.22, 0.18, 0.16, and 0.24 inches, which of these two sets of data is relatively more variable?

**4.28**   According to their medical records, one person's blood glucose level, measured before breakfast over several months, averaged 118.2 with a standard deviation of 4.8, while that of another person, also measured before breakfast over several months, averaged 109.7 with a standard deviation of 4.7. Which of the two persons' blood glucose level was relatively more variable?

**4.29**   An alternative measure of relative variation is the **coefficient of quartile variation**, which is defined as the ratio of the semi-interquartile range to the midquartile multiplied by 100, namely, as

$$\frac{Q_3 - Q_1}{Q_3 + Q_1} \cdot 100$$

Find the coefficient of quartile variation for the numbers of seconds that the insects of Exercise 4.8 survived after being sprayed with the insecticide.

**4.30**   Find the coefficient of quartile variation for the 14 temperature readings of Exercise 3.44.

**4.31**   Which of the data, those of Exercise 4.29 or those of Exercise 4.30, are

(a)  more variable?

(b)  relatively more variable?

# *4.4  THE DESCRIPTION OF GROUPED DATA

As we saw in Chapter 2 and then again in Section 3.7, the grouping of data entails some loss of information. Each item has lost its identity, and we know only how many values there are in each class or in each category. To define the standard deviation of a distribution, we shall have to be satisfied with an approximation and, as we did in connection with the mean, we shall treat our data as if all the values falling into a class were equal to the corresponding class mark. Thus, letting $x_1, x_2, \ldots,$ and $x_k$ denote the class marks, and $f_1, f_2, \ldots,$ and $f_k$ the corresponding class frequencies, we approximate the actual sum of all the measurements or observations with

$$\sum x \cdot f = x_1 f_1 + x_2 f_2 + \cdots + x_k f_k$$

and the sum of their squares with

$$\sum x^2 \cdot f = x_1^2 f_1 + x_2^2 f_2 + \cdots + x_k^2 f_k$$

Then we write the computing formula for the standard deviation of grouped sample data as

$$s = \sqrt{\frac{S_{xx}}{n-1}} \quad \text{where} \quad S_{xx} = \sum x^2 \cdot f - \frac{\left(\sum x \cdot f\right)^2}{n}$$

which is very similar to the corresponding computing formula for $s$ for ungrouped data. To obtain a corresponding computing formula for $\sigma$, we replace $n$ by $N$ in the formula for $S_{xx}$ and $n - 1$ by $N$ in the formula for $s$.

When the class marks are large numbers or given to several decimals, we can simplify things further by using the coding suggested on page 68. When the class intervals are all equal, and only then, we replace the class marks with consecutive integers, preferably with 0 at or near the middle of the distribution. Denoting the coded class marks by the letter $u$, we then calculate $S_{uu}$ and substitute into the formula

$$s_u = \sqrt{\frac{S_{uu}}{n-1}}$$

This kind of coding is illustrated by Figure 4.2, where we find that if $u$ varies (is increased or decreased) by 1, the corresponding value of $x$ varies (is increased or decreased) by the class interval $c$. Thus, to change $s_u$ from the $u$-scale to the original scale of measurement, the $x$-scale, we multiply it by $c$.

**Figure 4.2**
Coding the class marks
of a distribution.

| $x - 2c$ | $x - c$ | $x$ | $x + c$ | $x + 2c$ | $x$-scale |
|---|---|---|---|---|---|
| $-2$ | $-1$ | $0$ | $1$ | $2$ | $u$-scale |

**EXAMPLE 4.7** With reference to the distribution of the waiting times between eruptions of Old Faithful geyser shown in Example 2.4 and also in Example 3.21, calculate the standard deviation

(a) without coding;
(b) with coding.

**Solution** (a)

| $x$ | $f$ | $x \cdot f$ | $x^2 \cdot f$ |
|---|---|---|---|
| 34.5 | 2 | 69.0 | 2,380.50 |
| 44.5 | 2 | 89.0 | 3,960.50 |
| 54.5 | 4 | 218.0 | 11,881.00 |
| 64.5 | 19 | 1,225.5 | 79,044.75 |
| 74.5 | 24 | 1,788.0 | 133,206.00 |
| 84.5 | 39 | 3,295.5 | 278,469.75 |
| 94.5 | 15 | 1,417.5 | 133,953.75 |
| 104.5 | 3 | 313.5 | 32,760.75 |
| 114.5 | 2 | 229.0 | 26,220.50 |
| | 110 | 8,645.0 | 701,877.50 |

so that

$$S_{xx} = 701,877.5 - \frac{(8,645)^2}{110} \approx 22,459.1$$

and

$$s = \sqrt{\frac{22,459.1}{109}} \approx 14.35$$

(b)

| $u$ | $f$ | $u \cdot f$ | $u^2 \cdot f$ |
|---|---|---|---|
| $-4$ | 2 | $-8$ | 32 |
| $-3$ | 2 | $-6$ | 18 |
| $-2$ | 4 | $-8$ | 16 |
| $-1$ | 19 | $-19$ | 19 |
| 0 | 24 | 0 | 0 |
| 1 | 39 | 39 | 39 |
| 2 | 15 | 30 | 60 |
| 3 | 3 | 9 | 27 |
| 4 | 2 | 8 | 32 |
| | 110 | 45 | 243 |

so that

$$S_{uu} = 243 - \frac{(45)^2}{110} \approx 224.59$$

and

$$s_u = \sqrt{\frac{224.59}{109}} \approx 1.435$$

Finally, $s = 10(1.435) = 14.35$, which agrees, as it should, with the result obtained in part (a). This clearly demonstrates how the coding simplified the calculations.

## 4.5 SOME FURTHER DESCRIPTIONS

So far we have discussed only statistical descriptions that come under the general heading of measures of location or measures of variation. Actually, there is no limit to the number of ways in which statistical data can be described, and statisticians continually develop new methods of describing characteristics of numerical data that are of interest in particular problems. In this section we shall consider briefly the problem of describing the overall shape of a distribution.

Although frequency distributions can take on almost any shape or form, most of the distributions we meet in practice can be described fairly well by one or another of a few standard types. Among these, foremost in importance is the aptly described symmetrical **bell-shaped distribution** shown in Figure 4.1. The two distributions shown in Figure 4.3 can, by a stretch of the imagination, be described as bell shaped, but they are not symmetrical. Distributions like these, having a "tail" on one side or the other, are said to be **skewed**; if the tail is on the left we say that they are **negatively skewed** and if the tail is on the right we say that they are **positively skewed**. Distributions of incomes or wages are often positively skewed because of the presence of some relatively high values that are not offset by correspondingly low values.

**Figure 4.3**
Skewed distributions.

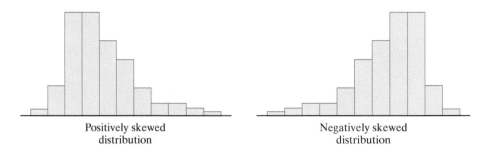

Positively skewed
distribution

Negatively skewed
distribution

The concepts of symmetry and skewness apply to any kind of data, not only distributions. Of course, for a large set of data we may just group the data and draw and study a histogram, but if that is not enough, we can use any one of several statistical **measures of skewness**. A relatively easy one is based on the fact that when there is perfect symmetry, as in the distribution shown in Figure 4.1,

**Figure 4.4**
Mean and median of positively skewed distribution.

Median Mean

the mean and the median will coincide. When there is positive skewness and some of the high values are not offset by correspondingly low values, as in Figure 4.4, the mean will be greater than the median; when there is a negative skewness and some of the low values are not offset by correspondingly high values, the mean will be smaller than the median.

This relationship between the median and the mean can be used to define a relatively simple measure of skewness, called the **Pearsonian coefficient of skewness**. It is given by

**PEARSONIAN COEFFICIENT OF SKEWNESS**

$$SK = \frac{3(mean - median)}{standard\ deviation}$$

For a perfectly symmetrical distribution, such as the one pictured in Figure 4.1, the mean and the median coincide and $SK = 0$. In general, values of the Pearsonian coefficient of skewness must fall between $-3$ and 3, and it should be noted that division by the standard deviation makes $SK$ independent of the scale of measurement.

**EXAMPLE 4.8**    Calculate $SK$ for the distribution of the waiting times between eruptions of Old Faithful geyser, using the results of Examples 3.21, 3.22, and 4.7, where we showed that $\bar{x} = 78.59$, $\tilde{x} = 80.53$, and $s = 14.35$.

**Solution**    Substituting these values into the formula for $SK$, we get

$$SK = \frac{3(78.59 - 80.53)}{14.35} \approx -0.41$$

which shows that there is a definite, though modest, negative skewness. This is also apparent from the histogram of the distribution, shown originally in Figure 2.3 and here again in Figure 4.5, reproduced from the display screen of a TI-83 graphing calculator. ■

When a set of data is so small that we cannot meaningfully construct a histogram, a good deal about its shape can be learned from a boxplot (defined originally on page 58). Whereas the Pearsonian coefficient is based on the difference between the mean and the median, with a boxplot we judge the

**Figure 4.5**
Histogram of the dis-
tribution of the waiting
times between erup-
tions of Old Faithful.

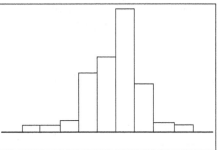

symmetry or skewness of a set of data on the basis of the position of the median relative to the two quartiles, $Q_1$ and $Q_3$. In particular, if the line at the median is at or near the center of the box, this is an indication of the symmetry of the data; if it is appreciably to the left of center, this is an indication that the data are positively skewed; and if it is appreciably to the right of center, this is an indication that the data are negatively skewed. The relative length of the two "whiskers," extending from the smallest value to $Q_1$ and from $Q_3$ to the largest value, can also be used as an indication of symmetry or skewness.

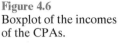

**EXAMPLE 4.9**   Following are the annual incomes of 15 certified public accountants (CPAs) in thousands of dollars: 88, 77, 70, 80, 74, 82, 85, 96, 76, 67, 80, 75, 73, 93, and 72. Draw a boxplot and use it to judge the symmetry or the lack of it for the data.

**Solution**   Arranging the data according to size, we get 67, 70, 72, 73, 74, 75, 76, 77, 80, 80, 82, 85, 88, 93, and 96, and it can be seen that the smallest value is 67; the largest value is 96; the median is the eighth value from either side that is 77; $Q_1$ is the fourth value from the left that is 73; and $Q_3$ is the fourth value from the right that is 85. All this information is summarized by the MINITAB printout of the boxplot shown in Figure 4.6. As can be seen, there is a strong indication that the

**Figure 4.6**
Boxplot of the incomes
of the CPAs.

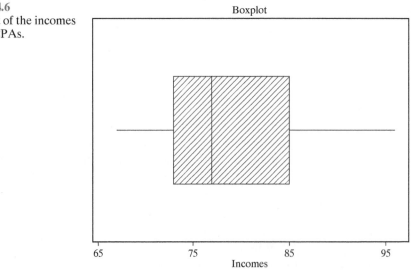

data are positively skewed. The line at the median is well to the left of the center of the box, and the "whisker" on the right is quite a bit longer than the one on the left. ▪

**Figure 4.7**
Reverse J-shaped and U-shaped distributions.

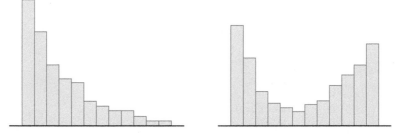

Besides the distributions we have discussed in this section, two others sometimes met in practice are the **reverse J-shaped** and **U-shaped** distributions shown in Figure 4.7. As can be seen from this figure, the names of these distributions literally describe their shape. Examples of such distributions may be found in Exercises 4.45 and 4.47.

## EXERCISES

**\*4.32** In a factory or office, the time during working hours in which a machine is not operating as a result of breakage or failure is called a *downtime*. The following distribution shows a sample of the length of the downtimes of a certain machine (rounded to the nearest minute).

| Downtime (minutes) | Frequency |
|---|---|
| 0–9 | 2 |
| 10–19 | 15 |
| 20–29 | 17 |
| 30–39 | 13 |
| 40–49 | 3 |

Find the standard deviation of this distribution.

**\*4.33** With reference to Exercise 3.58, find the standard deviation of the distribution of the students who are bilingual in the 50 elementary schools.

**\*4.34** Use the results of Exercises 3.58 and 4.33 to calculate the Pearsonian coefficient of skewness for the distribution of the percentages of bilingual students. Discuss the symmetry or skewness of this distribution.

**\*4.35** With reference to Exercise 3.59, find the standard deviation of the distribution of the compressive strengths of the 120 concrete samples.

**\*4.36** Use the results of Exercises 3.59 and 4.35 to calculate the Pearsonian coefficient of skewness for the distribution of the compressive strengths.

**\*4.37** Following is the distribution of the grades that 500 students received in a geography test:

| Grade | Number of Students |
|-------|--------------------|
| 10–24 | 44 |
| 25–39 | 70 |
| 40–54 | 92 |
| 55–69 | 147 |
| 70–84 | 115 |
| 85–99 | 32 |

Calculate

(a)  the mean and the median;

(b)  the standard deviation.

*4.38   Use the results of Exercise 4.37 to calculate the Pearsonian coefficient of skewness for the given data, and discuss the symmetry (or the lack of it) of their distribution.

*4.39   With reference to the distribution of Exercise 3.62, find the Pearsonian coefficient of skewness for the numbers of fish tacos served by the Mexican restaurant.

*4.40   Following are the numbers of accidents that occurred in July of 1999 in a certain town at 20 intersections without left-turn arrows:

$$25 \quad 30 \quad 32 \quad 22 \quad 26 \quad 10 \quad 2 \quad 32 \quad 6 \quad 13$$
$$27 \quad 22 \quad 18 \quad 12 \quad 28 \quad 35 \quad 8 \quad 29 \quad 31 \quad 8$$

Find the median, $Q_1$, and $Q_3$.

*4.41   Use the results of Exercise 4.40 to construct a boxplot, and use it to discuss the symmetry or skewness of these accident data.

*4.42   Following are the response times of 30 integrated circuits (in picoseconds):

$$3.7 \quad 4.1 \quad 4.5 \quad 4.6 \quad 4.4 \quad 4.8 \quad 4.3 \quad 4.4 \quad 5.1 \quad 3.9$$
$$3.3 \quad 3.4 \quad 3.7 \quad 4.1 \quad 4.7 \quad 4.6 \quad 4.2 \quad 3.7 \quad 4.6 \quad 3.4$$
$$4.6 \quad 3.7 \quad 4.1 \quad 4.5 \quad 6.0 \quad 4.0 \quad 4.1 \quad 5.6 \quad 6.0 \quad 3.4$$

Construct a stem-and-leaf display using the units digits as the stem labels and the tenths digits as the leaves. Use this stem-and-leaf display to judge the symmetry, or the lack of it, of these data.

*4.43   With reference to Exercise 4.42, find all the information that is required to draw a boxplot. Then use it to draw a boxplot and judge the symmetry or skewness of the response times of these integrated circuits.

*4.44   With reference to Exercise 3.29, find all the information that is necessary to construct a boxplot, draw the boxplot, and use it to judge the symmetry or the lack of it of the lengths of the NBA games.

*4.45   Following are the numbers of 3s obtained in fifty rolls of four dice: 0, 0, 1, 0, 0, 0, 2, 0, 0, 1, 0, 0, 0, 0, 1, 1, 0, 1, 2, 0, 0, 1, 0, 0, 0, 1, 1, 0, 1, 0, 0, 1, 2, 1, 0, 0, 3, 1, 1, 0, 4, 0, 0, 1, 2, 1, 0, 0, 1, and 1. Construct a frequency distribution and use it to judge the overall shape of the data.

*4.46   With reference to Exercise 4.45, find all the information that is needed to draw a boxplot. Then draw the boxplot and use it to describe the symmetry, or the lack of it, of the data. Also discuss the overall shape of the data.

*4.47   If a coin is flipped five times in a row, the result can be represented by means of a sequence of H's and T's (for example, HHTTH), where H stands for heads and T for tails. Having obtained such a sequence of H's and T's, we can then check after each successive flip whether the number of heads exceeds the number of tails. For instance, for HHTTH heads is ahead after the first flip, after the second flip, after the third flip, not after the fourth flip, but again after the fifth flip. Altogether, heads is ahead four times. Actually we repeated this "experiment" sixty times and found that heads was ahead

```
1  1  5  0  0  5  0  1  2  0  1  0  5  1  0
0  5  0  0  0  0  1  0  0  5  0  2  0  1  0
5  5  0  5  4  3  5  0  5  0  1  5  0  1  5
3  1  5  5  2  1  2  4  2  3  0  5  5  0  0
```

times. Construct a frequency distribution and discuss the overall shape of the data.

**\*4.48**  With reference to Exercise 4.47, find all the information that is needed to construct a boxplot. What features of the boxplot suggest that the data have a very unusual shape?

# CHECKLIST OF KEY TERMS (with page references to their definitions)

Bell-shaped distribution, 88
Biased estimator, 77
Chebyshev's theorem, 79
Coefficient of quartile variation, 86
Coefficient of variation, 82
Deviation from mean, 76
Empirical rule, 80
Interquartile range, 75
Mean deviation, 76
Measure of relative variation, 82
Measures of skewness, 88
Measures of variation, 75
Negatively skewed distribution, 88
Pearsonian coefficient of skewness, 89
Population standard deviation, 77
Population variance, 77

Positively skewed distribution, 88
Quartile deviation, 75
Range, 75
Reverse J-shaped distribution, 91
Root-mean-square deviation, 76
Sample standard deviation, 76
Sample variance, 77
Semi-interquartile range, 75
Skewed distribution, 88
Standard deviation, 75
Standard units, 81
Unbiased estimator, 77
U-shaped distribution, 91
Variance, 77
$z$-scores, 81

# REFERENCES

*A proof that division by $n - 1$ makes the sample variance an unbiased estimator of the population variance may be found in most textbooks on mathematical statistics, for instance, in*

MILLER, I., and MILLER, M., *John E. Freund's Mathematical Statistics*, 6th ed. Upper Saddle River, N.J.: Prentice Hall, 1998.

*Some information about the effect of grouping on the calculation of various statistical descriptions may be found in some of the older textbooks on statistics, for instance, in*

MILLS, F. C., *Introduction to Statistics*. New York: Holt, Rinehart and Winston, 1956.

**R.1** The numbers of artifacts uncovered each day at an archaeological dig are to be grouped into a table with the classes 0–4, 5–14, 15–24, 23–35, and 40 or more. Explain where difficulties might arise.

**R.2** An intercity bus was traveling down a highway when its driver was questioned, on a cell phone, by the starter at the next terminal. Which of the driver's answers would result in numerical data, and which would result in categorical data?

(a) How many passengers are aboard the bus?

(b) How many passengers are seated, and how many passengers are standing?

(c) How many miles is the bus from the terminal?

(d) What is the length of the bus?

(e) From what city did the trip originate?

(f) How many gallons of fuel do you think that the bus will need when it gets to the terminal?

**R.3** List the measurements that are grouped in the following stem-and-leaf display with unit leaves.

```
12 | 3  5
13 | 0  4  7  8
14 | 1  3  4  6  6  9
15 | 0  2  2  5  8
16 | 1  7
```

**R.4** Twenty pilots were tested in a flight simulator. Following are the times (in seconds) it took them to take corrective action to an emergency situation: 4.9, 10.1, 6.3, 8.5, 7.7, 6.3, 3.9, 6.5, 6.8, 9.0, 11.3, 7.5, 5.8, 10.4, 8.2, 7.4, 4.6, 5.3, 9.7, and 7.3. Find

(a) the median;

(b) $Q_1$ and $Q_3$.

**R.5** Use the results obtained in Exercise R.4 to construct a boxplot for the given data.

**R.6** Following is the distribution of a sample of the number of voter contacts made by volunteers assisting a candidate for election to a political office.

| Voter contacts | Volunteer campaign workers |
|---|---|
| 5–9 | 8 |
| 10–14 | 11 |
| 15–19 | 19 |
| 20–24 | 25 |
| 25–29 | 18 |
| 30–34 | 14 |
| 35–39 | 10 |

Calculate

(a) the mean;

(b) the median;

(c) quartiles $Q_1$ and $Q_3$;

(d) the standard deviation.

**R.7**   Following is a distribution of the numbers of mistakes that 80 graduate students made in translating a passage from French to English as part of the language requirement for an advanced degree:

| Number of mistakes | Number of students |
|---|---|
| 0–4 | 34 |
| 5–9 | 20 |
| 10–14 | 15 |
| 15–19 | 9 |
| 20–24 | 2 |

Calculate

(a) the mean;

(b) the median;

(c) the standard deviation;

(d) the Pearsonian coefficient of skewness.

**R.8**   Convert the distribution of Exercise R.7 into a cumulative "or less" distribution.

**R.9**   A fishery expert found the following concentrations of mercury, in parts per million, in 32 fish caught in a certain stream:

```
0.045   0.063   0.049   0.062   0.065   0.054   0.050   0.048
0.072   0.060   0.062   0.054   0.049   0.055   0.058   0.067
0.055   0.058   0.061   0.047   0.063   0.068   0.056   0.057
0.072   0.052   0.058   0.046   0.052   0.057   0.066   0.054
```

(a)  Construct a stem-and-leaf display with the stem labels 0.04, 0.05, 0.06, and 0.07.

(b)  Use the stem-and-leaf display obtained in part (a) to determine the median, $Q_1$, and $Q_3$.

(c)  Draw a boxplot and use it to describe the overall shape of the given data.

**R.10**   According to Chebyshev's theorem, what can we assert about the percentage of any set of data that must lie within $k$ standard deviations of the mean for (a) $k = 3.5$; (b) $k = 4.5$?

**R.11**   A meteorologist has complete data for the last 10 years on how many days in June the maximum temperature in Palm Springs, California, exceeded 110 degrees. Give one example each of a situation in which the meteorologist would look on these data as

(a)  a population;

(b)  a sample.

**R.12**   A 12-meeting sample of a lengthy series of seminars on musicology were attended by 22, 16, 20, 20, 15, 16, 12, 14, 16, 14, 11, and 16 musicians. Find

(a)  the mean number of musicians;

(b)  the median number of musicians;

(c)  the modal number of musicians;

(d)  the standard deviation of the 12-meeting sample;

(e)  the Pearsonian coefficient of skewness.

**R.13**   During the past few months, one computer repair service averaged 10 service calls per week with a standard deviation of 2 service calls. Another computer repair service averaged 25 service calls a week (during the same period) with a standard deviation of 3 service calls. Which of the two companies is relatively more constant in the number of service calls made weekly? Solve by comparing their coefficients of variation.

**R.14**  Following are the numbers of articles that 40 college professors have published in professional journals: 12, 8, 22, 45, 3, 27, 18, 12, 6, 32, 15, 17, 4, 19, 10, 2, 9, 16, 21, 17, 18, 11, 15, 2, 13, 15, 27, 16, 1, 5, 6, 15, 11, 32, 16, 10, 18, 4, 18, and 19. Determine

(a)  the mean;

(b)  the median.

**R.15**  Certain mass-produced metal shafts have a mean diameter of 24.00 mm with a standard deviation of 0.03 mm. At least what percentage of the shafts have diameters between 23.91 and 24.09 mm?

**R.16**  Among the students graduating from a university, 45 majoring in computer science had job offers averaging $31,100 (rounded to the nearest $100), 63 majoring in mathematics had job offers averaging $30,700, 112 majoring in engineering had job offers averaging $35,000, and 35 majoring in chemistry had job offers averaging $30,400. Find the mean job offer received by these 255 students.

**R.17**  From the distribution of Exercise R.7, can we determine how many of the 80 graduate students made

(a)  more than 14 mistakes;

(b)  anywhere from 5 to 19 mistakes;

(c)  exactly 17 mistakes;

(d)  anywhere from 10 to 20 mistakes?

If possible, give numerical answers.

**R.18**  Following are the numbers of whales seen breaching on 60 whale-watching trips off the coast of Baja California:

| 10 | 18 | 14 | 9 | 7 | 3 | 14 | 16 | 15 | 8 | 12 | 18 |
|----|----|----|----|----|----|----|----|----|----|----|----|
| 13 | 6 | 11 | 22 | 18 | 8 | 22 | 13 | 10 | 14 | 8 | 5 |
| 8 | 12 | 16 | 21 | 13 | 10 | 7 | 3 | 15 | 24 | 16 | 18 |
| 12 | 18 | 10 | 8 | 6 | 13 | 12 | 9 | 18 | 23 | 15 | 11 |
| 19 | 10 | 11 | 15 | 12 | 6 | 4 | 10 | 13 | 27 | 14 | 6 |

Group the data into a distribution with the classes 0–4, 5–9, 10–14, 15–19, 20–24, and 25–29.

**R.19**  Determine the mean, the median, and the standard deviation of the data of Exercise R.18.

**R.20**  Use the results of Exercise R.19 to calculate the coefficient of variation.

**R.21**  Following is the distribution of the total finance charges that 200 customers paid on their budget accounts at a department store.

| Amount (dollars) | Frequency |
|------------------|-----------|
| 1–20 | 18 |
| 21–40 | 62 |
| 41–60 | 63 |
| 61–80 | 43 |
| 81–100 | 14 |
| Total | 200 |

(a)  Draw a histogram of this distribution.

(b)  Draw a bar chart of this distribution.

**R.22**  The daily number of orders shipped by a mail-order company are grouped into a table having the classes 0–49, 50–99, 100–149, and 150–199. Find

(a)  the class boundaries;

(b)  the class marks;

(c)  the class interval.

**R.23**  The class limits of a distribution of weights (in ounces) are 10–29, 30–49, 50–69, 70–89, and 90–109. Find

(a)  the class boundaries;

(b)  the class marks;

(c)  the class interval of the distribution.

**R.24**  For a given set of data, the smallest value is 5.0, the largest value is 65.0, the median is 15.0, the first quartile is 11.5, and the third quartile is 43.5. Draw a boxplot and discuss the symmetry or skewness of the set of data.

**R.25**  Given $x_1 = 3.5$, $x_2 = 7.2$, $x_3 = 4.4$, and $x_4 = 2.0$, find

(a)  $\sum x$;

(b)  $\sum x^2$;

(c)  $\left( \sum x \right)^2$.

**R.26**  If a population consists of the integers 1, 2, 3, ..., and $k$, its variance is $\sigma^2 = \frac{k^2-1}{12}$. Verify this formula for

(a)  $k = 3$;

(b)  $k = 5$.

**R.27**  For a given set of data, the mean is 19.5 and the coefficient of variation is 32%. Find the standard deviation.

**R.28**  The manager of a bakery wishes to advertise a price based on the time that a job takes for icing an 8-inch cake and providing custom lettering and decorations. He knows from past experience that the simplest job of this sort takes at least 5 minutes and the mean time to complete the job is 15 minutes with a standard deviation of 1 minute. He plans to add an extra charge for jobs estimated to take more than 25 minutes. What is the smallest percentage of the jobs that can be performed at the advertised price?

**R.29**  In the construction of a categorical distribution, men's shirts are classified according to whether they are made of wool, silk, linen, or synthetic fibers. Explain where difficulties might arise.

**R.30**  On 30 days, the numbers of registered nurses present at a nursing home were 2, 3, 1, 1, 3, 0, 0, 2, 1, 2, 2, 3, 0, 1, 2, 3, 2, 2, 2, 1, 1, 0, 2, 3, 2, 2, 2, 1, 0, and 2. Construct a dot diagram.

**R.31**  The daily numbers of persons attending an art museum are grouped into a distribution with the classes 0–29, 30–59, 60–89, and 90 or more. Can the resulting distribution be used to determine on how many days

(a)  at least 89 persons attended the museum;

(b)  more than 89 persons attended the museum;

(c)  anywhere from 30 to 89 persons attended the museum;

(d)  more than 100 persons attended the museum?

**R.32**  If a set of measurements has the mean $\bar{x} = 45$ and the standard deviation $s = 8$, convert each of the following values of $x$ into standard units:

(a)  $x = 65$;

(b)  $x = 39$;

(c)  $x = 55$.

**R.33**  Explain why each of the following data may well fail to yield the desired information.

(a)  To determine public sentiment about certain import restrictions, an interviewer asks voters, "Do you feel that this unfair practice should be stopped?"

(b)  To predict an election for the governor of a state, a public opinion poll interviews persons selected haphazardly from a city's telephone directory.

**R.34**  Following are the numbers of false alarms that a security service received on twenty evenings: 9, 8, 4, 12, 15, 5, 5, 9, 3, 2, 6, 12, 5, 17, 6, 3, 7, 10, 8, and 4. Construct a boxplot and discuss the symmetry or skewness of the data.

**R.35**  The following distribution was obtained in a two-week study of the productivity of 100 workers:

| Number of acceptable pieces produced | Number of workers |
|---|---|
| 15–29 | 3 |
| 30–44 | 14 |
| 45–59 | 18 |
| 60–74 | 26 |
| 75–89 | 20 |
| 90–104 | 12 |
| 105–119 | 7 |

Find

(a)  the class boundaries;

(b)  the class marks;

(c)  the class interval.

**R.36**  Draw a histogram of the distribution of Exercise R.35.

**R.37**  Convert the distribution of Exercise R.35 into a cumulative "less than" distribution and draw an ogive.

**R.38**  Calculate the mean, the median, and the standard deviation of the distribution of Exercise R.35. Also, determine the Pearsonian coefficient of skewness.

**R.39**  Following are the systolic blood pressures of 22 hospital patients:

151   173   142   154   165   124   153   155   146   172   162
182   162   135   159   204   130   162   156   158   149   130

Construct a stem-and-leaf display with the stem labels 12, 13, 14, . . ., and 20.

**R.40**  Use the stem-and-leaf display obtained in Exercise R.39 to get the information needed for the construction of a boxplot. Draw a boxplot and discuss the symmetry or skewness of the data.

**R.41**  Based on past experience, it is known that the bus that leaves downtown Phoenix at 8:05 A.M. takes on the average 42 minutes with a standard deviation of 2.5 minutes to reach the Arizona State University campus. At least what percentage of the time will the bus reach the A.S.U. campus between 8:37 A.M. and 8:57 A.M.?

**R.42** An official of a symphony orchestra reported that its five concerts were attended by 462, 480, 1,455, 417, and 432 patrons.

(a) Calculate the mean and the median of these attendance figures.

(b) Discovering that the third value was printed incorrectly and should have been just 455, recalculate the mean and the median of the corrected attendance figures.

(c) Compare the effect of this printing error on the mean and on the median.

**R.43** The 30 pages of a preliminary printout of a manuscript were proofread for typographical errors, yielding the following numbers of mistakes:

$$2 \ 0 \ 3 \ 1 \ 0 \ 0 \ 0 \ 5 \ 0 \ 1 \ 2 \ 1 \ 4 \ 0 \ 1$$
$$0 \ 1 \ 3 \ 1 \ 2 \ 0 \ 1 \ 0 \ 3 \ 1 \ 2 \ 0 \ 1 \ 0 \ 2$$

Construct a dot diagram.

**R.44** An experiment was performed by scientists to estimate the average (mean) increase in the pulse rate of astronauts performing a certain task in outer space. Simulating weightlessness, they obtained the following data (increase in pulse rate in beats per minute) for 33 persons who performed the given task:

$$34 \ \ 26 \ \ 22 \ \ 24 \ \ 23 \ \ 18 \ \ 21 \ \ 27 \ \ 33 \ \ 26 \ \ 31$$
$$28 \ \ 29 \ \ 25 \ \ 13 \ \ 22 \ \ 21 \ \ 15 \ \ 30 \ \ 24 \ \ 23 \ \ 37$$
$$26 \ \ 22 \ \ 27 \ \ 31 \ \ 25 \ \ 28 \ \ 20 \ \ 25 \ \ 27 \ \ 24 \ \ 18$$

Calculate

(a) the mean and the median;

(b) the standard deviation;

(c) the Pearsonian coefficient of skewness.

**R.45** In an air pollution study, eight different samples of air yielded 2.2, 1.8, 3.1, 2.0, 2.4, 2.0, 2.1, and 1.2 micrograms of suspended benzene-soluble organic matter per cubic meter. Calculate the coefficient of variation for these data.

*****R.46** Following are the scores obtained by 44 cadets firing at a target from a kneeling position, $x$, and from a standing position, $y$:

| x | y | x | y | x | y | x | y |
|---|---|---|---|---|---|---|---|
| 81 | 83 | 81 | 76 | 94 | 86 | 77 | 83 |
| 93 | 88 | 96 | 81 | 86 | 76 | 97 | 86 |
| 76 | 78 | 86 | 91 | 91 | 90 | 83 | 78 |
| 86 | 83 | 91 | 76 | 85 | 87 | 86 | 89 |
| 99 | 94 | 90 | 81 | 93 | 84 | 98 | 91 |
| 98 | 87 | 87 | 85 | 83 | 87 | 93 | 82 |
| 82 | 77 | 90 | 89 | 83 | 81 | 88 | 78 |
| 92 | 94 | 98 | 91 | 99 | 97 | 90 | 93 |
| 95 | 94 | 94 | 94 | 90 | 96 | 97 | 92 |
| 98 | 84 | 75 | 76 | 96 | 86 | 89 | 87 |
| 91 | 83 | 88 | 88 | 85 | 84 | 88 | 92 |

Use a computer or a graphing calculator to produce a scattergram and describe the relationship, if any, between the cadets' scores in the two positions.

**R.47** If students calculate their grade-point indexes (that is, average their grades) by counting A, B, C, D, and F as 1, 2, 3, 4, and 5, what does this assume about the nature of the grades?

# POSSIBILITIES AND PROBABILITIES

**5.1** Counting 101

**5.2** Permutations 104

**5.3** Combinations 107

**5.4** Probability 114

Checklist of Key Terms 122

References 123

t was not until the advent of scientific thought, with its emphasis on observation and experimentation, that the study of uncertainty or chance developed as the **theory of probability** or (as it is sometime called) the **mathematics of chance**. We divide the subject matter of probability into three chapters. An informal introduction is presented in Chapter 5. Chapter 6 continues the subject in a somewhat more rigorous fashion. Chapter 7 discusses problems of decision making where there are uncertainties about the outcomes.

In Chapter 5 we shall see how uncertainties can actually be measured, how they can be assigned numbers, and how these numbers are to be interpreted. In subsequent chapters we shall see how these numbers, called **probabilities**, can be used to live with uncertainties—how they can be used to make choices or decisions that promise to be most profitable, or otherwise most desirable.

In Sections 5.1 through 5.3 we present mathematical preliminaries dealing with the question of "what is possible" in given situations. After all, we can hardly predict the outcome of a football game unless we know which teams are playing, and we cannot very well predict what will happen in an election unless we know which candidates are running for office. Then, in Section 5.4 we shall learn how to judge also "what is probable"; that is, we shall learn about several ways in which probabilities are defined, or interpreted, and their values are determined.

# 5.1 COUNTING

In contrast to the high-powered methods used nowadays in science, in business, and even in everyday life, the simple process of counting still plays an important role. We still have to count $1, 2, 3, 4, 5, \ldots$, for example, to determine how many persons take part in a demonstration, the size of the response to a questionnaire, the number of damaged cases in a shipment of wines from Portugal, or how many times the temperature in Phoenix, Arizona, went over 100 degrees in a given month. Sometimes, the process of counting can be simplified by using mechanical devices (for instance, when counting spectators passing through turnstiles), or by performing counts indirectly (for instance, by subtracting the serial numbers of invoices to determine the total number of sales). At other times, the process of counting can be simplified greatly by means of special mathematical techniques, such as the ones given in this section.

In the study of "what is possible," there are essentially two kinds of problems. There is the problem of listing everything that can happen in a given situation, and then there is the problem of determining how many different things can happen (without actually constructing a complete list). The second kind of problem is especially important, because there are many situations in which we do not need a complete list, and hence can save ourselves a great deal of work. Although the first kind of problem may seem straightforward and easy, the following example illustrates that this is not always the case.

**EXAMPLE 5.1** A restaurant offers three kinds of house wine by the glass—a red wine, a white wine, and a rosé. List the number of ways in which three dinner guests can order three glasses of wine, without taking into account who gets which wine.

**Solution** Evidently, there are many different possibilities. Guests might order three glasses of red wine; they might order two glasses of red wine and one glass of rosé; they might order one glass of white wine and two glasses of rosé; they might order one glass of each kind; and so forth. Continuing this way carefully, we may be able to list all ten possibilities, but there is a good chance that we will miss one or two. ◼

Problems like this can be handled systematically by drawing a **tree diagram** such as the one pictured in Figure 5.1. This diagram shows that there are four possibilities (four branches) corresponding to 0, 1, 2, or 3 glasses of red wine. Then, for white wine there are four branches coming from the top branch (0 glasses of red wine), three branches coming from the next branch (1 glass of red wine), two branches coming from the following branch (2 glasses of red wine), and only one branch coming from the bottom branch (3 glasses of red wine). After that, there is only one possibility for the number of glasses of rosé, since the numbers of glasses must always add up to three. Thus, we find that altogether there are 10 possibilities.

**Figure 5.1**
Tree diagram for
Example 5.1.

| Red | White | Rosé |
|-----|-------|------|

---

<table>
<tr><td></td><td></td><td>3</td></tr>
<tr><td></td><td>0</td><td>2</td></tr>
<tr><td></td><td>1</td><td>1</td></tr>
<tr><td></td><td>2</td><td></td></tr>
<tr><td>0</td><td>3</td><td>0</td></tr>
</table>

**EXAMPLE 5.2**  In a medical study, patients are classified according to whether they have blood type A, B, AB, or O, and also according to whether their blood pressure is low, normal, or high. In how many different ways can a patient thus be classified?

**Solution**  As is apparent from the tree diagram of Figure 5.2, the answer is 12. Starting at the top, the first path along the "branches" corresponds to a patient having blood type A and low blood pressure, the second path corresponds to a patient having blood type A and normal blood pressure, ..., and the twelfth path corresponds to a patient having blood type O and high blood pressure.  ▪

The answer we got in Example 5.2 is $4 \cdot 3 = 12$, namely, the product of the number of blood types and the number of blood pressure levels. Generalizing from this example, let us state the following rule:

**MULTIPLICATION OF CHOICES**

> *If a choice consists of two steps, of which the first can be made in m ways and for each of these the second can be made in n ways, then the whole choice can be made in m · n ways.*

We refer to this rule as the **multiplication of choices**, as indicated in the margin. To prove this rule, we need only draw a tree diagram like that of Figure 5.2. First there are *m* branches corresponding to the *m* possibilities in the first step, and then there are *n* branches emanating from each of these branches corresponding to the *n* possibilities in the second step. This leads to $m \cdot n$ paths along the branches of the tree diagram, and hence to $m \cdot n$ possibilities.

**Figure 5.2**
Tree diagram for
Example 5.2.

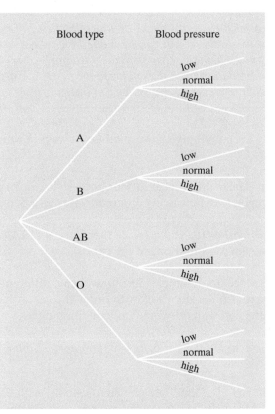

---

EXAMPLE **5.3**  If a research worker wants to experiment with one of 12 new medications for sinusitis, trying it on mice, guinea pigs, or rats, in how many different ways can she choose one of the medications and one of the three kinds of laboratory animals?

**Solution**  Since $m = 12$ and $n = 3$, there are $12 \cdot 3 = 36$ different ways in which the experiment can be arranged.    ■

EXAMPLE **5.4**  If a physics department schedules four lecture sections and 15 laboratory sections for its freshman course, in how many different ways can a student choose one of each? Also, how many choices are left, if two of the lecture sections and four of the laboratory sections are full by the time the student gets to enroll?

**Solution**  Since $m = 4$ and $n = 15$, there are $4 \cdot 15 = 60$ different ways in which a student can choose one of each. If two of the lecture sections and four of the laboratory sections are already full, $m = 4 - 2 = 2$ and $n = 15 - 4 = 11$ and the number of choices is reduced to $2 \cdot 11 = 22$.    ■

By using appropriate tree diagrams, we can easily generalize the rule for the multiplication of choices so that it will apply to choices involving more than two steps. For $k$ steps, where $k$ is a positive integer, we get the following rule:

MULTIPLICATION OF CHOICES (GENERALIZED)

> *If a choice consists of k steps, of which the first can be made in $n_1$ ways, for each of these the second can be made in $n_2$ ways, for each combination of choices made in the first two steps the third can be made in $n_3$ ways, ..., and for each combination of choices made in the first $k-1$ steps the kth can be made in $n_k$ ways, then the whole choice can be made in $n_1 \cdot n_2 \cdot n_3 \cdots \cdots n_k$ ways.*

We simply keep multiplying the numbers of ways in which the different steps can be made.

**EXAMPLE 5.5**   A new-car dealer offers a car in four body styles, in 10 colors, and with a choice of three engines. In how many different ways can a person order one of the cars?

**Solution**   Since $n_1 = 4, n_2 = 10$, and $n_3 = 3$, there are $4 \cdot 10 \cdot 3 = 120$ different ways in which a person can order one of the cars.   ■

**EXAMPLE 5.6**   With reference to Example 5.5, how many different choices are there if a person must also decide whether to order the car with or without an automatic transmission and with or without air conditioning?

**Solution**   Since $n_1 = 4, n_2 = 10, n_3 = 3, n_4 = 2$, and $n_5 = 2$, there are $4 \cdot 10 \cdot 3 \cdot 2 \cdot 2 = 480$ different choices.   ■

**EXAMPLE 5.7**   A test consists of 15 multiple choice questions, with each question having four possible answers. In how many different ways can a student check off one answer to each question?

**Solution**   Since $n_1 = n_2 = n_3 = \ldots = n_{15} = 4$, there are altogether $4 \cdot 4 \cdot 4 \cdot 4 \cdot 4 \cdot 4 \cdot 4 \cdot 4 \cdot 4 \cdot 4 \cdot 4 \cdot 4 \cdot 4 \cdot 4 \cdot 4 = 1{,}073{,}741{,}824$ different ways in which a student can check off one answer to each question. (Note that in only one of these possibilities are all the answers correct.)   ■

# 5.2 PERMUTATIONS

The rule for the multiplication of choices and its generalization is often used when several choices are made from one and the same set and we are concerned with the order in which the choices are made.

**EXAMPLE 5.8**   If 20 women entered the Miss Oregon contest, in how many different ways can the judges choose the winner and the first runner-up?

**Solution**   Since the winner can be chosen in $m = 20$ ways and the first runner-up must be one of the other $n = 19$, there are altogether $20 \cdot 19 = 380$ ways in which the judges can make their selection.   ■

**EXAMPLE 5.9**   In how many different ways can the 48 members of a labor union choose a president, a vice-president, a secretary, and a treasurer?

**Solution**   Since $n_1 = 48$, $n_2 = 47$, $n_3 = 46$, and $n_4 = 45$ (regardless of which officer is chosen first, second, third, and fourth), there are altogether $48 \cdot 47 \cdot 46 \cdot 45 = 4,669,920$ different possibilities. ◼

In general, if $r$ objects are selected from a set of $n$ distinct objects, any particular arrangement (order) of these objects is called a **permutation**. For instance, 4 1 2 3 is a permutation of the first four positive integers; Maine, Vermont, and Connecticut is a permutation (a particular ordered arrangement) of three of the six New England states; and

*UCLA, Stanford, USC, and Washington*

*ASU, Oregon, WSU, and California*

are two different permutations (ordered arrangements) of four of the ten universities with football teams in the PAC-10 conference.

**EXAMPLE 5.10**   Determine the number of different permutations of two of the five vowels $a, e, i, o, u$, and list them all.

**Solution**   Since $m = 5$ and $n = 4$, there are $5 \cdot 4 = 20$ different permutations, and they are *ae, ai, ao, au, ei, eo, eu, io, iu, ou, ea, ia, oa, ua, ie, oe, ue, oi, ui,* and *uo*. ◼

In general, it would be desirable to have a formula for the total number of permutations of $r$ objects selected from a set of $n$ distinct objects, such as the four universities chosen from among the 10 universities with football teams in the PAC-10. To this end, observe that the first selection is made from the whole set of $n$ objects, the second selection is made from the $n - 1$ objects that remain after the first selection has been made, the third selection is made from the $n - 2$ objects that remain after the first two selections have been made, ..., and the $r$th and final selection is made from the $n - (r - 1) = n - r + 1$ objects that remain after the first $r - 1$ selections have been made. Therefore, direct application of the generalized rule for the multiplication of choices yields the result that the total number of permutations of $r$ objects selected from a set of $n$ distinct objects, which we shall denote by $_nP_r$, is

$$n(n - 1)(n - 2) \cdot \cdots \cdot (n - r + 1)$$

Since products of consecutive integers arise in many problems relating to permutations and other kinds of special arrangements or selections, it is convenient to introduce here the **factorial notation**. In this notation, the product of all positive integers less than or equal to the positive integer $n$ is called "$n$ factorial" and denoted by n!. Thus,

$$1! = 1$$
$$2! = 2 \cdot 1 = 2$$
$$3! = 3 \cdot 2 \cdot 1 = 6$$
$$4! = 4 \cdot 3 \cdot 2 \cdot 1 = 24$$
$$5! = 5 \cdot 4 \cdot 3 \cdot 2 \cdot 1 = 120$$
$$6! = 6 \cdot 5 \cdot 4 \cdot 3 \cdot 2 \cdot 1 = 720$$

$$\cdots \cdots$$

and in general

$$n! = n(n-1)(n-2) \cdots \cdots 3 \cdot 2 \cdot 1$$

Also, to make various formulas more generally applicable, we let $0! = 1$ by definition.

Since factorials grow very rapidly, it has been claimed that the exclamation mark reflects one's surprise. Indeed, the value of 10! exceeds three million and 70! exceeds the memory limit of hand-held calculators.

To express the formula for $_nP_r$ more compactly in terms of factorials, we note, for example, that $12 \cdot 11 \cdot 10! = 12!$, that $9 \cdot 8 \cdot 7 \cdot 6! = 9!$, and that $37 \cdot 36 \cdot 35 \cdot 34 \cdot 33! = 37!$. Similarly,

$$_nP_r \cdot (n-r)! = n(n-1)(n-2) \cdots \cdots (n-r+1) \cdot (n-r)!$$

$$= n!$$

so that $_nP_r = \frac{n!}{(n-r)!}$. To summarize,

**NUMBER OF PERMUTATIONS OF $n$ OBJECTS TAKEN $r$ AT A TIME**

> *The number of permutations of $r$ objects selected from a set of $n$ distinct objects is*
>
> $$_nP_r = n(n-1)(n-2) \cdots \cdots (n-r+1)$$
>
> $$_nP_r = \frac{n!}{(n-r)!}$$

where either formula can be used for $r = 1, 2, \ldots$, or $n$. (The second formula, but not the first, could be used also for $r = 0$, for which we would get the trivial result that there is

$$_nP_0 = \frac{n!}{(n-0)!} = 1$$

way of selecting none of $n$ objects.) The first formula is generally easier to use because it requires fewer steps, but many students find the one in factorial notation easier to remember.

**EXAMPLE 5.11**   Find the number of permutations of $r = 4$ objects selected from a set of $n = 12$ distinct objects (say, the number of ways in which four of twelve new movies can be ranked first, second, third, and fourth by a panel of critics).

**Solution**   For $n = 12$ and $r = 4$, the first formula yields

$$_{12}P_4 = 12 \cdot 11 \cdot 10 \cdot 9 = 11{,}880$$

and the second formula yields

$$_{12}P_4 = \frac{12!}{(12-4)!} = \frac{12!}{8!} = \frac{12 \cdot 11 \cdot 10 \cdot 9 \cdot 8!}{8!} = 11,880$$

Essentially, the work is the same, but the second formula requires a few extra steps. ■

To find the formula for the number of permutations of $n$ distinct objects taken all together, we substitute $r = n$ into either formula for $_nP_r$ and get

**NUMBER OF PERMUTATIONS OF n OBJECTS TAKEN ALL TOGETHER**

$$_nP_n = n!$$

**EXAMPLE 5.12**  In how many different ways can eight teaching assistants be assigned to eight sections of a course in economics?

**Solution**  Substituting $n = 8$ into the formula for $_nP_n$, we get $_8P_8 = 8! = 40,320$. ■

Throughout this discussion it has been assumed that the $n$ objects are all distinct. When this is not the case, the formula for $_nP_n$ must be modified, and we shall illustrate how this is done for some special cases in Exercises 5.36 and 5.37.

# 5.3 COMBINATIONS

There are many problems in which we want to know the number of ways in which $r$ objects can be selected from a set of $n$ objects, but we do not care about the order in which the selection is made. For instance, we may want to know in how many ways a committee of four can be selected from among the 45 members of a college fraternity, or the number of ways in which the IRS can choose five of 36 tax returns for a special audit. To derive a formula that applies to problems like these, let us first examine the following 24 permutations of three of the first four letters of the alphabet:

| | | | | | |
|---|---|---|---|---|---|
| abc | acb | bac | bca | cab | cba |
| abd | adb | bad | bda | dab | dba |
| acd | adc | cad | cda | dac | dca |
| bcd | bdc | cbd | cdb | dbc | dcb |

If we do not care about the order in which the three letters are chosen from among the four letters $a$, $b$, $c$, and $d$, there are only four ways in which the selection can be made: $abc$, $abd$, $acd$, and $bcd$. Note that these are the groups of letters shown in the first column of the table, and that each row contains the $_3P_3 = 3! = 6$ permutations of the three letters in the first column.

In general, there are $_rP_r = r!$ permutations of $r$ distinct objects, so that the $_nP_r$ permutations of $r$ objects selected from among $n$ distinct objects contain each group of $r$ objects $r!$ times. (In our example, the $_4P_3 = 4 \cdot 3 \cdot 2 = 24$ permutations of three letters selected from among the first four letters of the alphabet contain each group of three letters $_3P_3 = 3! = 6$ times.) Therefore, to get a formula for the number of ways in which $r$ objects can be selected from a set of $n$ distinct objects *without regard to their order*, we divide $_nP_r$ by $r!$. Referring to such a selection as a **combination** of $n$ objects taken $r$ at a time, we denote the number of combinations of $n$ objects taken $r$ at a time by $_nC_r$ or $\binom{n}{r}$, and write

---

**NUMBER OF COMBINATIONS OF $n$ OBJECTS TAKEN $r$ AT A TIME**

*The number of ways in which $r$ objects can be selected from a set of $n$ distinct object is*

$$\binom{n}{r} = \frac{n(n-1)(n-2) \cdot \cdots \cdot (n-r+1)}{r!}$$

*or, in factorial notation,*

$$\binom{n}{r} = \frac{n!}{r!(n-r)!}$$

---

Like the two formulas for $_nP_r$, either formula can be used for $r = 1, 2, \ldots,$ or $n$, but the second one only for $r = 0$. Again, the first formula is generally easier to use because it requires fewer steps, but many students find the one in factorial notation easier to remember.

For $n = 0$ to $n = 20$, the values of $\binom{n}{r}$ may be read from Table XI on page 516, where these quantities are referred to as **binomial coefficients**. The reason for this is explained in Exercise 5.38.

**EXAMPLE 5.13**    On page 107 we asked for the number of ways in which a committee of four can be selected from among the 45 members of a college fraternity. Since the order in which the committee members are chosen does not matter, we can now ask for the value of $_nC_r$ or for the binomial coefficient $\binom{n}{r}$.

**Solution**    Substituting $n = 45$ and $r = 4$ into the first of the two formulas for $_nC_r$, we get
$$_{45}C_4 = \frac{45 \cdot 44 \cdot 43 \cdot 42}{4!} = 148,995.$$

**EXAMPLE 5.14**    On page 107 we also asked for the number of ways in which an IRS auditor can choose 5 of 36 tax returns for a special audit. Again, the order in which the choices are made does not matter, so we can now ask for the value of $_nC_r$ or for the binomial coefficient $\binom{n}{r}$.

**Solution**   Substituting $n = 36$ and $r = 5$ into the first of the two formulas for $\binom{n}{r}$, we get

$$\binom{36}{5} = \frac{36 \cdot 35 \cdot 34 \cdot 33 \cdot 32}{5!} = 376{,}992.$$

**EXAMPLE 5.15**   In how many different ways can a person choose three books from a list of 10 best-sellers, assuming that the order in which the books are chosen is of no consequence?

**Solution**   Substituting $n = 10$ and $r = 3$ into the first of the two formulas for $_nC_r$, we get

$$\binom{10}{3} = \frac{10 \cdot 9 \cdot 8}{3!} = 120$$

Similarly, substitution into the second formula yields

$$\binom{10}{3} = \frac{10!}{3!7!} = \frac{10 \cdot 9 \cdot 8 \cdot 7!}{3!7!} = \frac{10 \cdot 9 \cdot 8}{3!} = 120$$

Essentially, the work is the same, but the first formula required fewer steps.

**EXAMPLE 5.16**   In how many different ways can the director of a research laboratory choose two chemists from among seven applicants and three physicists from among nine applicants?

**Solution**   The two chemists can be selected in $\binom{7}{2}$ ways, the three physicists can be selected in $\binom{9}{3}$ ways, so that by the multiplication of choices, all five of them can be selected in

$$\binom{7}{2} \cdot \binom{9}{3} = 21 \cdot 84 = 1{,}764$$

ways. The values of the two binomial coefficients were obtained from Table XI.

In Section 5.2 we gave the special formula $_nP_n = n!$, but there is no need to do so here; substitution of $r = n$ into either formula for $\binom{n}{r}$ yields $\binom{n}{n} = 1$. In other words, there is one and only one way in which we can select all $n$ of the elements that constitute a set.

When $r$ objects are selected from a set of $n$ distinct objects, $n - r$ of the objects are left, and consequently, there are as many ways of leaving (or selecting) $n - r$ objects from a set of $n$ distinct objects as there are ways of selecting $r$ objects. Symbolically, we write

$$\binom{n}{r} = \binom{n}{n-r} \quad \text{for } r = 0, 1, 2, \ldots, n$$

Sometimes this rule serves to simplify calculations and sometimes it is needed in connection with Table XI.

**EXAMPLE  5.17**  Determine the value of $\binom{75}{72}$.

**Solution**  Instead of having to write down the product $75 \cdot 74 \cdot 73 \cdot \cdots \cdot 4$ and then to cancel $72 \cdot 71 \cdot 70 \cdot \cdots \cdot 4$, we write directly

$$\binom{75}{72} = \binom{75}{3} = \frac{75 \cdot 74 \cdot 73}{3!} = 67,525 \qquad \blacksquare$$

**EXAMPLE  5.18**  Find the value of $\binom{19}{13}$.

**Solution**  $\binom{19}{13}$ cannot be looked up directly in Table XI, but by making use of the fact that

$\binom{19}{13} = \binom{19}{19-13} = \binom{19}{6}$, we look up $\binom{19}{6}$ and obtain 27,132. For $r = 0$ we cannot substitute into the first of the two formulas for $\binom{n}{r}$, but if we substitute into the second formula, or if we write

$$\binom{n}{0} = \binom{n}{n-0} = \binom{n}{n}$$

we get $\binom{n}{0} = 1$. Evidently, there are as many ways of selecting none of the elements in a set as there are ways of choosing the $n$ elements that are left.  $\blacksquare$

**5.1**  A seafood restaurant stocks two live lobsters, reordering two more at the end of each day (for delivery early the next morning) if and only if both have been served. Draw a tree diagram to show that if it gets two live lobsters on a Monday morning, there are altogether eight different ways in which this seafood restaurant can serve such lobsters on Monday and Tuesday.

**5.2**  With reference to Exercise 5.1, in how many different ways can the seafood restaurant serve two or three such lobsters on these two days?

**5.3**  In the World Series of baseball, the winner is the first team to win four games. Suppose that the American League champions lead the National League champions by three games to one. Construct a tree diagram showing the number of ways that these teams can continue to the completion of the series.

**5.4**  In how many different ways can the judges choose a winner and first runner-up from the 10 finalists in a student essay contest?

**5.5**    In how many ways can a student organization with 50 members choose a president, a vice president, a secretary, and a treasurer?

**5.6**    Determine the number of possible permutations of two of the five vowels: *a, e, i, o,* and *u*. List them all.

**5.7**    In factorial notation the product of all positive integers less than or equal to the positive integer *n* is called "*n* factorial" and is denoted by *n*!. List and determine the values of 6!, 5!, 4!, and 3!.

**5.8**    The number of permutations of *r* objects taken from a set of *n* distinct objects is $_nP_r = \dfrac{n!}{(n-r)!}$. Find the number of ways that three Olympic skaters can be ranked first (gold medal), second (silver medal), and third (bronze medal) in a speed skating contest with 10 contestants.

**5.9**    Suppose that a merchant has 12 identical electric light bulbs for sale, from which a customer may select her choice. In how many ways can the customer choose 4 light bulbs from the 12?

**5.10**    An insurance company has a branch in each of the six New England states (Massachusetts, Rhode Island, Connecticut, Maine, New Hampshire, and Vermont). Find the number of ways in which six branch managers can be assigned to manage an office in each of the six states.

**5.11**    A person with $2 in his pocket bets $1, even money, on the flip of a coin, and he continues to bet $1 so long as he has any money left. Draw a tree diagram to show the various things that can happen in the first three flips of the coin (provided, of course, that there will be a third flip). In how many of the cases will he be

(a)  exactly $1 ahead;

(b)  exactly $1 behind?

**5.12**    On the faculty of a university there are three professors named Smith: Adam Smith, Brett Smith, and Craig Smith. Draw a tree diagram to show the various ways in which the payroll department can distribute their paychecks so that each of them receives a check made out to himself or to one of the other two Smiths. In how many of these possibilities will

(a)  only one of them get his own check;

(b)  at least one of them get his own check?

**5.13**    An artist has two paintings in an art exhibit that lasts two days. Draw tree diagrams to show in how many different ways he can make sales on the two days if

(a)  we are interested only in how many of his paintings are sold on each day;

(b)  we do care which painting is sold on what day.

**5.14**    In an election at a bank, Mr. Jones, Ms. Black, and Ms. Humphry are running for chairman of the board, and Mr. Arnold, Ms. Roberts, and Mr. Smith are running for general manager. Construct a tree diagram showing the nine possible outcomes, and use it to determine the number of ways in which the elected officials will not be of the same sex.

**5.15**    In a political science survey voters are classified into six categories according to income and into four categories according to education. In how many different ways can a voter thus be classified?

**5.16**    In a traffic court, violators are classified according to whether they are properly licensed, whether their violations are major or minor, and whether or not they have committed any other violations during the preceding 12 months.

(a) Construct a tree diagram showing the various ways in which a violator can be classified by this court.

(b) If there are 20 violators in each of the eight categories obtained on part (a) and the judge gives each violator who is not properly licensed a stern lecture, how many of the violators will receive a stern lecture?

(c) If the judge gives an $80 fine to everyone who has committed a major violation and/or another violation in the preceding 12 months, how many of the violators will receive an $80 fine?

(d) How many of the violators will receive a stern lecture as well as an $80 fine?

**5.17**   A chain of convenience stores has four warehouses and 32 retail outlets. In how many different ways can it ship a carton of maple syrup from one of the warehouses to one of the retail outlets?

**5.18**   A travel agent places his orders to cruise lines by telephone, by fax, by e-mail, by priority mail, or by an express carrier, and he requests that his orders be confirmed by telephone, by fax, or by priority mail. In how many different ways can one of his orders to a cruise line be placed and confirmed?

**5.19**   There are four different trails to the top of a mountain. In how many different ways can a person hike up and down the mountain if

(a) he must take the same trail both ways;

(b) he can, but need not, take the same trail both ways;

(c) he does not want to take the same trail both ways?

**5.20**   A student can study 1 or 2 hours for an astronomy test on any given night. Draw a tree diagram to find the number of ways in which the student can study altogether

(a) five hours on three consecutive nights;

(b) at least five hours on three consecutive nights.

**5.21**   In an optics kit there are five concave lenses, five convex lenses, two prisms, and three mirrors. In how many different ways can a person choose one of each kind?

**5.22**   A cafeteria offers 10 different soups or salads, eight entrees, and six desserts. In how many different ways can a customer choose a soup or salad, an entree, and a dessert?

**5.23**   A multiple choice test consists of 10 questions, with each question permitting a choice of three answers.

(a) In how many different ways can one choose an answer to each question?

(b) In how many different ways can one choose an answer to each question and get them all wrong? (It is assumed here that each question has only one correct answer.)

**5. 24**   A true-false test consists of 15 questions. In how many different ways can a person mark each question "true" or "false?"

**5.25**   Determine for each of the following whether it is true or false:

(a) $19! = 19 \cdot 18 \cdot 17 \cdot 16!$;

(b) $\dfrac{12!}{3!} = 4!$;

(c) $3! + 0! = 7$;

(d) $6! + 3! = 9!$;

(e) $\dfrac{9!}{7!2!} = 36$;

(f) $15! \cdot 2! = 17!$.

**5.26**   In how many different ways can a laboratory technician choose four of sixteen white mice? (No specific order is needed.) Use Table XI to verify your result.

**5.27**   In how many different ways can a television director schedule a sponsor's six different commercials in time spots allocated for them during the first half of a college football game?

**5.28**  Determine for each of the following whether it is true or false:

(a)  $\dfrac{1}{3!} + \dfrac{1}{4!} = \dfrac{5}{24}$;

(b)  $0! \cdot 8! = 0$;

(c)  $5 \cdot 4! = 5!$;

(d)  $\dfrac{16!}{12!} = 16 \cdot 15 \cdot 14$.

**5.29**  A motel chain wants to inspect 5 of its 32 franchised operations. If the order of the inspections does not matter, in how many different ways can it plan this series of inspections?

**5.30**  If the drama club of a college wants to present four of ten half-hour skits on one evening between 8 and 10 P.M., in how many different ways can it arrange its schedule?

**5.31**  In how many different ways can the curator of a museum arrange five of eight paintings horizontally on a wall?

**5.32**  In how many different ways can five graduate students choose one of ten research projects, if no two of them can choose the same project?

**5.33**  In how many different ways can the manager of a baseball team arrange the batting order of the nine players in the starting lineup?

**5.34**  Four married couples have bought eight seats in a row for a football game. In how many different ways can they be seated if

(a)  each husband is to sit to the left of his wife;

(b)  all the men are to sit together and all the women are to sit together?

**5.35**  A psychologist preparing four-letter nonsense words for use in a memory test chooses the first letter from among the consonants $q$, $w$, $x$, and $z$; the second letter from among the vowels $a$, $i$, and $u$; the third letter from among the consonants $c$, $f$, and $p$; and the fourth letter from among the vowels $e$ and $o$.

(a)  How many different four-letter nonsense words can this psychiatrist construct?

(b)  How many of these four-letter nonsense words will begin with the letter $q$?

(c)  How many of these four-letter nonsense words will begin with the letter $z$ and end with the letter $o$?

**5.36**  If among $n$ objects $r$ are alike, and the others are all distinct, the number of permutations of these $n$ objects taken all together is $\frac{n!}{r!}$.

(a)  How many permutations are there of the letters in the word "class"?

(b)  In how many ways (according to manufacturer only) can five cars place in a stock-car race if three of the cars are Fords, one is a Chevrolet, and one is a Dodge?

(c)  In how many ways can the television director of Exercise 5.27 fill the six time slots allocated to commercials if she has four different commercials, of which a given one is to be shown three times, while each of the others is to be shown once?

(d)  Present an argument to justify the formula given in this exercise.

**5.37**  If among $n$ objects $r_1$ are identical, another $r_2$ are identical, and the rest (if any) are all distinct, the number of permutations of these $n$ objects taken all together is $\frac{n!}{r_1! \cdot r_2!}$.

(a)  How many permutations are there of the letters in the word "greater"?

(b)  In how many ways can the television director of Exercise 5.27 fill the six time slots allocated to commercials if she has only two different commercials, each of which is to be shown three times?

(c)  Generalize the formula so that it applies if among $n$ objects $r_1$ are identical, another $r_2$ are identical, another $r_3$ are identical, and the rest (if any) are all distinct. In how many ways can the television director of Exercise 5.27 fill the six time slots allocated to commercials if she has three different commercials, each of which is to be shown twice?

**5.38**   Calculate the number of ways in which a chain of computer stores can choose three of 15 locations for new franchises.

**5.39**   An office supplies store carries 15 kinds of ball-point pens.

(a)   Calculate the number of ways in which an office manager can order a dozen each of three different kinds.

(b)   Use Table XI to verify the result obtained in part (a).

**5.40**   To develop the mathematics department of a new branch of a state university, the newly elected head of the department must choose two full professors from among six applicants, two associate professors from among ten applicants, and six assistant professors from among sixteen applicants. In how many different ways can she make this choice?

**5.41**   A student is required to report on three of eighteen books on a reading list. Calculate the number of ways in which a student can choose the three books, and verify the answer in Table XI.

**5.42**   A carton of 12 transistor batteries includes one that is defective. In how many different ways can an inspector choose three of the batteries and

(a)   get the one that is defective;

(b)   not get the one that is defective?

# 5.4 PROBABILITY

So far in this chapter we have studied only what is possible in a given situation. In some instances we listed all the possibilities and in others we merely determined how many different possibilities there are. Now we shall go one step further and judge also what is probable and what is improbable.

The most common way of measuring the uncertainties connected with events (say, the outcome of a presidential election, the side effects of a new medication, the durability of an exterior paint, or the total number of points we may roll with a pair of dice) is to assign them **probabilities** or to specify the **odds** at which it would be fair to bet that the events will occur. In this section we shall learn how probabilities are interpreted and how their numerical values are determined; odds will be discussed in Section 6.3.

Historically, the oldest way of measuring uncertainties is the **classical probability concept**. It was developed originally in connection with games of chance, and it lends itself most readily to bridging the gap between possibilities and probabilities. The classical probability concept applies only when all possible outcomes are equally likely, in which case we say that

**T**HE CLASSICAL PROBABILITY CONCEPT[†]

> *If there are n equally likely possibilities, of which one must occur and s are regarded as favorable, or as a "success," then the probability of a "success" is $\frac{s}{n}$.*

[†] In this discussion of probability, *s* denotes "success." In Chapter 4, *s* is the symbol for standard deviation of a sample.

In the application of this rule, the terms "favorable" and "success" are used rather loosely—what is favorable to one player is unfavorable to his opponent, and what is a success from one point of view is a failure from another. Thus, the terms "favorable" and "success" can be applied to any particular kind of outcome, even if "favorable" means that a television set does not work, or "success" means that someone catches the flu. This usage dates back to the days when probabilities were quoted only in connection with games of chance.

**EXAMPLE 5.19**    What is the probability of drawing an ace from a well-shuffled deck of 52 playing cards?

**Solution**    By "well-shuffled" we mean that each card has the same chance of being drawn, so that the classical probability concept can be applied. Since there are $s = 4$ aces among the $n = 52$ cards, we find that the probability of drawing an ace is

$$\frac{s}{n} = \frac{4}{52} = \frac{1}{13}$$

**EXAMPLE 5.20**    What is the probability of rolling a 3, 4, 5, or 6 with a balanced die?

**Solution**    By "balanced" we mean that each face of the die has the same chance, so that the classical probability concept applies. Since $s = 4$ and $n = 6$, we find that the probability of rolling a 3, 4, 5, or 6 is

$$\frac{s}{n} = \frac{4}{6} = \frac{2}{3}$$

**EXAMPLE 5.21**    If H stands for heads and T for tails, the eight possible outcomes for three flips of a balanced coin are HHH, HHT, HTH, THH, HTT, THT, TTH, and TTT. What are the probabilities of getting two heads or three heads?

**Solution**    Again, by "balanced" we mean that all possible outcomes have the same chance, so that the classical probability concept can be applied. Counting the possibilities, we find that for two heads $s = 3$ and $n = 8$, and that for three heads $s = 1$ and $n = 8$. Thus, for two heads the probability is $\frac{s}{n} = \frac{3}{8}$ and for three heads it is $\frac{s}{n} = \frac{1}{8}$.

Although equally likely possibilities are found mostly in games of chance, the classical probability concept applies also in a great variety of situations where gambling devices are used to make **random selections**—say, when offices are assigned to research assistants by lot, when laboratory animals are chosen for an experiment so that each one has the same chance of being selected (perhaps, by the method that is described in Section 10.1), when each family in a township has the same chance of being included in a survey, or when machine parts are chosen for inspection so that each part produced has the same chance of being selected.

**EXAMPLE 5.22**    If 3 of 20 weight lifters have used steroids and four of them are randomly tested for the use of steroids, what is the probability that one of the three weight lifters who have used steroids will be included in the tests?

**Solution** There are $n = \binom{20}{4} = \dfrac{20 \cdot 19 \cdot 18 \cdot 17}{4!} = 4{,}845$ ways of choosing the four weight lifters to be tested, and these possibilities may be regarded as equally likely by virtue of the random selection. The number of "favorable" outcomes is the number of ways in which one of the three weight lifters who have used steroids and three of the 17 weight lifters who have not used steroids will be selected, namely, $s = \binom{3}{1}\binom{17}{3} = 3 \cdot 680 = 2{,}040$, where the values of the binomial coefficients were obtained from Table XI. It follows that the probability that one and only one of the weight lifters who have used steroids will be caught is

$$\frac{s}{n} = \frac{2{,}040}{4{,}845} = \frac{8}{19}$$

or approximately 0.42. ∎

A major shortcoming of the classical probability concept is its limited applicability, because there are many situations in which the various possibilities cannot all be regarded as equally likely. This would be the case, for example, if we are concerned whether an experiment will support or refute a new theory; whether an expedition will be able to locate a shipwreck; whether a person's performance will justify a raise; or whether the Dow Jones Index will go up or down.

Among the various probability concepts, most widely held is the **frequency interpretation**, according to which probabilities are interpreted as follows:

**T**HE FREQUENCY INTERPRETATION OF PROBABILITY

> *The probability of an event (happening or outcome) is the proportion of the time that events of the same kind will occur in the long run.*

If we say that the probability is 0.78 that a jet from San Francisco to Phoenix will arrive on time, we mean that such flights arrive on time 78% of the time. Also, if the Weather Service predicts that there is a 40% chance for rain (that the probability is 0.40 that it will rain), they mean that under the same weather conditions it will rain 40% of the time. More generally, we say that an event has a probability of, say, 0.90, in the same sense in which we might say that our car will start in cold weather 90% of the time. We cannot guarantee what will happen on any particular occasion—the car may start and then it may not—but if we kept records over a long period of time, we should find that the proportion of "successes" is very close to 0.90.

In accordance with the frequency interpretation of probability, we estimate the probability of an event by observing what fraction of the time similar events have occurred in the past.

**EXAMPLE 5.23** A sample survey conducted in a recent year showed that among 8,319 women in their twenties who remarried after being divorced, 1,358 divorced again. What is the probability that a woman in her twenties who remarried after being divorced will divorce again?

**Solution**   In the past this happened $\frac{1,358}{8,319} \cdot 100 = 16.3\%$ of the time (rounded to the nearest tenth of a percent), so we can use 0.163 as an estimate of the desired probability. ■

EXAMPLE **5.24**   Records show that 34 of 956 persons who recently visited Central Africa came down with malaria. What is the probability that a person who visits Central Africa at about the same time of the year will *not* catch this disease?

**Solution**   Since $956 - 34 = 922$ of the 956 persons did not catch the disease, we estimate the desired probability to be roughly $\frac{922}{956} = 0.96$. ■

When probabilities are estimated in this way, it is only reasonable to ask whether the estimates are any good. In Chapter 14 we shall answer this question in some detail, but for now let us refer to an important theorem called the **law of large numbers**. Informally, this theorem may be stated as follows:

THE LAW OF LARGE NUMBERS

> *If a situation, trial, or experiment is repeated again and again, the proportion of successes will tend to approach the probability that any one outcome will be a success.*

This theorem is known informally as the "law of averages." It is a statement about the long-run proportion of successes and it has little to say about any single trial.

In the first six editions of this book we illustrated the law of large numbers by repeatedly flipping a coin and recording the accumulated proportion of heads after each fifth flip. Since then we have used the **computer simulation** shown in Figure 5.3, *where the 1s and 0s denote heads and tails*.

**Figure 5.3**
Computer simulation of 100 flips of a coin.

Reading across successive rows, we find that among the first five simulated flips there are 2 heads, among the first 10 there are 5 heads, among the first 15 there are 8 heads, among the first 20 there are 10 heads, among the first 25 there are 13 heads, ..., and among all hundred there are 51 heads. The corresponding proportions, plotted in Figure 5.4, are $\frac{2}{5} = 0.40$, $\frac{5}{10} = 0.50$, $\frac{8}{15} = 0.53$, $\frac{10}{20} = 0.50$, $\frac{13}{25} = 0.52, \ldots$, and $\frac{51}{100} = 0.51$. Observe that the proportion of heads fluctuates but comes closer and closer to 0.50, the probability of heads for each flip of the coin.

**Figure 5.4**
Graph illustrating law of large numbers.

In the frequency interpretation, the probability of an event is defined in terms of what happens to similar events in the long run, so let us examine briefly whether it is at all meaningful to talk about the probability of an event that can occur only once. For instance, can we assign a probability to the event that Ms. Bertha Jones will be able to leave the hospital within four days after having an appendectomy, or to the event that a certain major-party candidate will win an upcoming gubernatorial election. If we put ourselves in the position of Ms. Jones's doctor, we might check medical records, discover that patients left the hospital within four days after an appendectomy in, say 78% of hundreds of cases, and apply this figure to Ms. Jones. This may not be of much comfort to Ms. Jones, but it does provide a meaning for a probability statement about her leaving the hospital within four days—the probability is 0.78.

This illustrates that when we make a probability statement about a specific (nonrepeatable) event, the frequency interpretation of probability leaves us no choice but to refer to a set of similar events. As can well be imagined, however, this can easily lead to complications, since the choice of similar events is generally neither obvious nor straightforward. With reference to Ms. Jones's appendectomy, we might consider as similar only cases in which the patients were of the same sex, only cases in which the patients were also of the same age as Ms. Jones, or only cases in which the patients were also of the same height and weight as Ms. Jones. Ultimately, the choice of similar events is a matter of personal judgment, and it is by no means contradictory that we can arrive at different probability estimates, all valid, concerning the same event.

With regard to the question whether a certain major-party candidate will win an upcoming gubernatorial election, suppose that we ask the persons who have conducted a poll how sure they are that the candidate will win. If they say they are "95% sure" (that is, if they assign a probability of 0.95 to the candidate winning the election), this is not meant to imply that he would win 95% of the time if he ran for office a great number of times. Rather, it means that the pollsters' prediction is based on a method that works 95% of the time. It is in this way that we must interpret many of the probabilities attached to statistical results.

Finally, let us mention a third probability concept that is currently enjoying favor. According to this point of view, probabilities are interpreted as **personal** or **subjective** evaluations. They reflect one's belief with regard to the uncertainties that are involved, and they apply especially when there is little or no direct evidence, so that there really is no choice but to consider collateral (indirect) information, educated guesses, and perhaps intuition and other subjective factors. Subjective probabilities are sometimes determined by putting the issues in question on a money basis, as will be explained in Sections 6.3 and 7.1.

## EXERCISES

**5.43** When a card is drawn from a well-shuffled deck of 52 playing cards, what are the probabilities of getting
(a) the king of hearts;
(b) a red face card (jack, queen, or king);
(c) a 5, a 6, or a 7;
(d) a diamond?

**5.44** When two cards are drawn from a well-shuffled deck of playing cards, and the first card is not replaced before the second card is drawn, what are the probabilities of getting
(a) two queens;
(b) two clubs?

**5.45** When three cards are drawn without replacement from a well-shuffled deck of 52 playing cards, what is the probability of getting a jack, a queen, and a king?

**5.46** If we roll a balanced die, what is the probability of getting
(a) a 3;
(b) an even number;
(c) a number greater than 4?

**5.47** If we roll a pair of balanced dice, one red and one green, list the 36 possible outcomes and determine the probabilities of getting a total of
(a) 4;
(b) 9;
(c) 7 or 11.

**5.48** If H stands for heads and T for tails, the 16 possible outcomes for four flips of a coin are HHHH, HHHT, HHTH, HTHH, THHH, HHTT, HTHT, HTTH, THHT, THTH, TTHH, HTTT, THTT, TTHT, TTTH, and TTTT. Assuming that these outcomes are all equally likely, find the probabilities of getting 0, 1, 2, 3, or 4 heads.

**5.49** A bowl contains 15 red beads, 30 white beads, 20 blue beads, and 7 black beads. If one of the beads is drawn at random, what are the probabilities that it will be
(a) red;
(b) white or blue;
(c) black;
(d) neither white nor black?

**5.50** A leather bag contains 24 dimes dated 2000, 14 dimes dated 1999, and 10 dimes dated 2002. If one of the dimes is picked at random, what is the probability that it will be

(a) dated 2002;

(b) dated 1999 or 2002?

**5.51** The balls used in selecting numbers for BINGO carry the numbers 1, 2, 3, ..., 75. If one of the balls is selected at random, what are the probabilities that it will be

(a) an even number;

(b) a number that is 15 or below;

(c) a number that is 60 or above?

**5.52** If a game has $n$ equally likely outcomes, what is the probability of each individual outcome?

**5.53** Among the 12 applicants for managerial positions at a chain of movie theaters, eight have a degree from a community college. If three of the applicants are randomly selected, what are the probabilities that

(a) all three of them have a degree from a community college;

(b) only one of them has a degree from a community college?

**5.54** A carton of 24 light bulbs includes two that are defective. If two of the bulbs are chosen at random, what are the probabilities that

(a) neither bulb will be defective;

(b) one of the bulbs will be defective;

(c) both bulbs will be defective?

**5.55** A hoard of medieval coins discovered in what is now Belgium included 20 struck in Antwerp and 16 struck in Brussels. If a person chooses five of these coins at random, what are the probabilities that she will get

(a) two coins struck in Antwerp and three struck in Brussels;

(b) four coins struck in Antwerp and one coin struck in Brussels?

**5.56** According to the National Center for Health Statistics, there were 725,192 deaths from heart disease in a recent year from a total of 2,391,399 people who died of all causes. Find the probability that the cause of death of a person who died in that year was heart disease.

**5.57** In a poll conducted by a newspaper, 424 of 954 readers claimed that the coverage of local sports was inadequate. Based on these figures, estimate the probability that any one of its readers, selected at random, will support this claim.

**5.58** In a lengthy study of shoplifting at a department store, it was found that 816 of 4,800 shoplifters were not caught until his or her fifth try. Based on these figures, estimate the probability that a person shoplifting at this department store will not get caught until his or her fifth try.

**5.59** If 678 of 904 cars passed a state-operated emission test on the first try, estimate the probability that any one car will pass this test on the first try.

**5.60** If 1,558 of 2,050 persons visiting the Grand Canyon said that they expect to return within a few years, estimate the probability that any one person visiting the Grand Canyon expects to return within a few years.

**5.61** Weather bureau statistics show that in a city near Portland it has been overcast 28 times in the last 52 years on the first Sunday in May, when a service club holds its annual picnic. Based on these figures, estimate the probability that it will be overcast on next year's annual picnic of the service club.

**5.62**   To get a "feeling" for the law of large numbers, a student tossed a coin 150 times, getting

$$
\begin{array}{cccccccccc}
1 & 0 & 1 & 0 & 0 & 1 & 0 & 0 & 0 & 0 \\
1 & 1 & 0 & 1 & 0 & 1 & 0 & 1 & 1 & 0 \\
1 & 1 & 0 & 1 & 0 & 0 & 0 & 0 & 1 & 0 \\
0 & 1 & 1 & 0 & 0 & 1 & 1 & 1 & 0 & 1 \\
1 & 0 & 1 & 0 & 1 & 1 & 1 & 0 & 0 & 1 \\
1 & 1 & 0 & 0 & 0 & 1 & 0 & 0 & 1 & 0 \\
0 & 1 & 0 & 1 & 1 & 1 & 0 & 0 & 0 & 0 \\
1 & 1 & 1 & 1 & 1 & 0 & 1 & 0 & 0 & 0 \\
0 & 0 & 1 & 0 & 1 & 1 & 1 & 1 & 0 & 0 \\
0 & 1 & 0 & 1 & 0 & 1 & 0 & 1 & 0 & 0 \\
1 & 0 & 0 & 1 & 1 & 1 & 0 & 0 & 0 & 1 \\
0 & 0 & 0 & 1 & 0 & 1 & 1 & 1 & 0 & 1 \\
0 & 0 & 1 & 1 & 1 & 1 & 0 & 0 & 0 & 1 \\
1 & 0 & 0 & 1 & 1 & 1 & 1 & 0 & 0 & 1 \\
0 & 0 & 1 & 0 & 1 & 1 & 1 & 1 & 0 & 0 \\
\end{array}
$$

where 1 denotes *heads* and 0 denotes *tails*. Using these data, duplicate the work on page 117 that led to Figure 5.4.

**5.63**   Repeat Exercise 5.62, getting your data with a computer or a graphing calculator.

**5.64**   Record the last digit of the number on the license plate of 200 cars and plot the accumulated proportion of 5s after each 25 cars. Judge whether the resulting graph supports the law of large numbers.

*(What we mean by "chance," "randomness," and "probability" is subject to all sorts of myths and misconceptions. Some of these are illustrated by the exercises that follow.)*

**5.65**   Some philosophers have argued that if we have absolutely no information about the likelihood of the different possibilities, it is reasonable to regard them all as equally likely. This is sometimes referred to as the principle of *equal ignorance*. Discuss the argument that human life either does or does not exist elsewhere in the universe, and since we really have no information one way or the other, the probability that human life exists elsewhere in the universe is $\frac{1}{2}$.

**5.66**   The following illustrates how one's intuition can be misleading in connection with probabilities: A box contains 100 beads, some red and some white. One bead will be drawn, and you are asked to call beforehand whether it is going to be red or white. Would you be willing to bet even money (say, you will win $5 if you are right and lose $5 if you are wrong) if

(a)   you have no idea how many of the beads are red and how many are white;

(b)   you are told that 50 of the beads are red and 50 are white?

Strange as it may seem, most persons are more willing to gamble under condition (b) than under condition (a).

**5.67**   The following is a good example of the difficulties in which we may find ourselves if we use only "common sense," or intuition, in judgments concerning probabilities:

*"Among three indistinguishable boxes one contains 2 pennies, one contains a penny and a dime, and one contains 2 dimes. Selecting*

> *one of these boxes at random (each box has a probability of 1/3),*
> *one coin is taken out at random (each coin has a probability of 1/2)*
> *without looking at the other. The coin that is taken out of the box is*
> *a penny, and without giving the matter too much thought, we may*
> *well be inclined to say that there is a probability of 1/2 that the other*
> *coin in the box is also a penny. After all, the penny must have come*
> *either from the box with the penny and the dime or from the box with*
> *two pennies. In the first case the other coin is a dime, in the second*
> *case it is a penny, and it would seem reasonable to say that these two*
> *possibilities are equally likely.''*

Actually, the correct value of the probability that the other coin is also a penny is 2/3, and it will be left to the reader to verify this result by mentally labeling the two pennies in the first box $P_1$ and $P_2$, the two dimes in the third box $D_1$ and $D_2$, and drawing a tree diagram showing the six possible (and equally likely) outcomes of the experiment.

**5.68**  Discuss the following assertion: If a meteorologist says that the probability for rain on the next day is 0.30, whatever happens on that day cannot prove him right or wrong.

**5.69**  Discuss the following assertion: Since probabilities are measures of uncertainty, the probability we assign to a future event will always increase as we get more information.

**5.70**  The probability that a patient will survive minor surgery is 0.98 for hospital A and 0.86 for hospital B; the probability that a patient will survive major surgery is 0.73 for hospital A and 0.66 for hospital B. Can we conclude that any kind of surgery at hospital A is a better risk than any kind of surgery at hospital B?

**5.71**  No diagnostic tests are infallible, so imagine that the probability is 0.95 that a certain test will diagnose a diabetic correctly as being diabetic, and it is 0.05 that it will diagnose a person who is not diabetic as being diabetic. It is known that roughly 10% of the population is diabetic. Guess at the probability that a person diagnosed as being diabetic actually is diabetic. (This problem will be discussed further in Exercise 6.75.)

**5.72**  Some persons claim that if the probability of something happening is more than 0.50, it is confirmed if the event actually happens and refuted if the event does not happen. Correspondingly, if the probability of something happening is less than 0.50, it is confirmed if the event does not happen and it is refuted if the event happens. Comment on this method of confirming and refuting probabilities.

# CHECKLIST OF KEY TERMS (with page references to their definitions)

Binomial coefficients, 108
  Table XI, 516
Classical probability concept, 114
Combinations, 108
Computer simulation, 117
Factorial notation, 105
Frequency interpretation, 116
Law of large numbers, 117
Mathematics of chance, 100
Multiplication of choices, 102
  generalized, 104

Odds, 114
Permutations, 105
Personal probability, 119
Probability, 100, 114
Probability theory, 100
Random selection, 115
Subjective probability, 119
Tree diagram, 101

# REFERENCES

*Informal introductions to probability, written primarily for the layperson, may be found in*

GARVIN, A.D., *Probability in Your Life*. Portland, Maine: J. Weston Walch Publisher, 1978.

HUFF, D., and GEIS, I., *How to Take a Chance*. New York: W. W. Norton & Company, Inc., 1959.

KOTZ, S., and STROUP, D. E., *Educated Guessing: How to Cope in an Uncertain World*. New York: Marcel Dekker, Inc., 1983.

LEVINSON, H. C., *Chance, Luck, and Statistics*. New York: Dover Publications, Inc., 1963.

MOSTELLER, F., KRUSKAI, W. H., LINK, R. F., PIETERS, R. S., and RISING, G. R., *Statistics by Example: Weighing Chances*. Reading, Mass.: Addison-Wesley Publishing Company, Inc., 1973.

WEAVER, W., *Lady Luck: The Theory of Probability*. New York: Dover Publications, Inc., 1982.

*For fascinating reading on the history of probability, see*

DAVID, F. N., *Games, Gods and Gambling*. New York: Hafner Press, 1962.

*and the first three chapters of*

STIGLER, S. M., *The History of Statistics*. Cambridge, Mass.: Harvard University Press, 1986.

*To supplement Exercises 5.65 through 5.72, further examples of myths and misconceptions about probability, including some fascinating paradoxes, may be found in*

BENNETT, D. J., *Randomness*. Cambridge, Mass.: Harvard University Press, 1998.

# SOME RULES
# OF
# PROBABILITY

**6.1**  Sample Spaces and Events  125

**6.2**  The Postulates of Probability  132

**6.3**  Probabilities and Odds  135

**6.4**  Addition Rules  139

**6.5**  Conditional Probability  145

**6.6**  Multiplication Rules  147

***6.7**  Bayes' Theorem  152

Checklist of Key Terms  157

References  157

n the study of probability there are basically three kinds of questions:

> 1. **What do we mean when we say that the probability of an event is, say, 0.50, 0.78, or 0.24?**
> 2. **How are the numbers we refer to as probabilities determined, or measured in actual practice?**
> 3. **What are the mathematical rules that probabilities must obey?**

For the most part, we have already studied the first two kinds of questions in Chapter 5. In the classical probability concept we are concerned with equally likely possibilities, count the ones that are favorable, and use the formula $\frac{s}{n}$. In the frequency interpretation we are concerned with proportions of "successes" in the long run and base our estimates on what has happened in the past. When it comes to subjective probabilities we are concerned with a measure of a person's belief, and in Section 6.3 (and later in Section 7.1) we shall see how such probabilities can actually be determined.

After some preliminaries in Section 6.1, we shall concentrate in this chapter on the mathematical rules that probabilities must obey; namely, on what we call the **theory of probability**. This includes the basic postulates in Section 6.2, the relationship between probabilities and odds in Section 6.3, the addition rules in Section 6.4, conditional probabilities in Section 6.5, the multiplication rules in Section 6.6, and finally Bayes' theorem in Section 6.7.

# 6.1 SAMPLE SPACES AND EVENTS

In statistics, we use the word "experiment" in a very unconventional way. For lack of a better term, we use it for any process of observation or measurement. Thus, an **experiment** may consist of counting how many employees of a government agency are absent on a given day, checking whether a switch is "on" or "off," or finding out whether or not a person is married. An experiment may also consist of the complicated process of predicting trends in the economy, finding the source of a social unrest, or diagnosing the cause of a disease. A result of an experiment, being an instrument reading, a "yes" or "no" answer, or a value obtained through lengthy calculations, is called the **outcome** of the experiment.

For each experiment, the set of all possible outcomes is called the **sample space** and it is usually denoted by the letter $S$. For instance, if a zoologist must choose three of 24 guinea pigs for classroom demonstrations, the sample space consists of the $\binom{24}{3} = \frac{24 \cdot 23 \cdot 22}{3 \cdot 2 \cdot 1} = 2{,}024$ different ways, or combinations, in which the selection can be made. Also, if the dean of a college must assign two of his 84 faculty members as advisors to a political science club, the sample space consists of the $\binom{84}{2} = \frac{84 \cdot 83}{2 \cdot 1} 3{,}486$ ways in which this choice can be made.

When we study the outcomes of an experiment, we usually identify the various possibilities with numbers, points, or some other kinds of symbols, so that we can treat all questions about them mathematically, without having to go through long verbal descriptions of what has taken place, is taking place, or will take place. For instance, if there are eight candidates for a scholarship and we let $a$, $b$, $c$, $d$, $e$, $f$, $g$, and $h$ denote that it is awarded to Ms. Adam, Mr. Bean, Miss Clark, Mrs. Daly, Mr. Earl, Ms. Fuentes, Ms. Gardner, or Ms. Hall, then the sample space for this experiment (namely, for this selection) is the set $S = \{a, b, c, d, e, f, g, h\}$. The use of points rather than letters or numbers to denote the elements of a sample space has the advantage that it makes it easier to visualize the various possibilities and perhaps discover special patterns among the different outcomes.

Usually, we classify sample spaces according to the number of elements, or points, that they contain. The ones we have mentioned so far in this section contained 2,024, 3,486, and 8 elements. We refer to them all as **finite**. In this chapter we shall consider only sample spaces that are finite, but in later chapters we shall consider also samples spaces that are **infinite**. An infinite sample space arises, for example, when we deal with quantities such as temperatures, weights, or distances, which are measured on continuous scales. Even when we throw a dart at a target, there is a continuum of points we may hit.

Any **subset** of a sample space is referred to as an **event**. By subset we mean any part of a set, including the set as a whole and the **empty set**, denoted by Ø, which has no elements at all. For instance, for the example dealing with the eight candidates for the scholarship, $M = \{b, e\}$ denotes the event that one of the two men will get the scholarship.

**EXAMPLE 6.1**   A used-car dealer has three 2005 Dodge Rams on his lot, to be sold by either of two salespersons. He is interested in how many of these trucks each of the two salespersons will sell in a given week. Using two coordinates, so that $(1, 1)$, for example, denotes the outcome that each of the two salespersons will sell one of the three trucks and $(2, 0)$ denotes the outcome that the first of the two salespersons will sell two of the trucks and the second will sell none, list all the possible outcomes of this "experiment." Also, draw a diagram showing the corresponding points of the sample space.

**Figure 6.1**
Sample space for
Example 1.

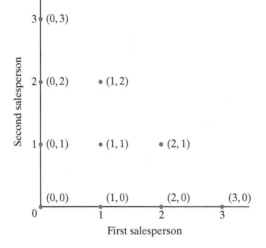

**Solution**   The 10 possible outcomes are $(0, 0)$, $(0, 1)$, $(0, 2)$, $(0, 3)$, $(1, 0)$, $(1, 1)$, $(1, 2)$, $(2, 0)$, $(2, 1)$, and $(3, 0)$, and the corresponding points are shown in Figure 6.1.

**EXAMPLE  6.2**   With reference to Example 6.1 and Figure 6.1, express in words what events are represented by $A = \{(2, 0), (1, 1), (0, 2)\}$, $B = \{(0, 0), (1, 0), (2, 0), (3, 0)\}$, and $C = \{(0, 2), (1, 2), (0, 3)\}$.

**Solution**   $A$ is the event that between them the two salespersons will sell two of the trucks; $B$ is the event that the second salesperson will not sell any of the trucks; and $C$ is the event that the second salesperson will sell at least two of the trucks.

In Example 6.2, events $B$ and $C$ have no elements (outcomes) in common and they are referred to as **mutually exclusive**; that is, the occurrence of either one precludes the occurrence of the other. Clearly, if the second salesperson will not sell any of the trucks, he or she cannot very well sell at least two. Observe also that events $A$ and $B$ are not mutually exclusive, and neither are events $A$ and $C$. The first pair shares the outcome $(2, 0)$ and the second pair shares the outcome $(0, 2)$.

In many probability problems we are interested in events that can be expressed by forming **unions**, **intersections**, and/or **complements**. The reader is probably familiar with these elementary set operations, but if not, the union of two sets $X$ and $Y$, denoted by $X \cup Y$, is the event that consists of all the elements (outcomes) contained in $X$, in $Y$, or in both. The intersection of two events $X$ and $Y$, denoted by $X \cap Y$, is the event that consists of all the elements contained in both $X$ and $Y$, and the complement of $X$, denoted by $X'$, is the event that consists of all the elements of the sample space that are not contained in $X$. We usually read $\cup$ as "or," $\cap$ as "and," and $X'$ as "not $X$."

**EXAMPLE  6.3**    With reference to Examples 6.1 and 6.2, list the outcomes comprising $B \cup C$, $A \cap C$, and $B'$. Also express each of these events in words.

**Solution**    Since $B \cup C$ contains all the elements (outcomes) that are in $B$, in $C$, or in both, we find that

$$B \cup C = \{(0, 0), (1, 0), (2, 0), (3, 0), (0, 2), (1, 2), (0, 3)\}$$

and this is the event that the second salesperson will sell 0, 2, or 3 of the trucks; namely, that the second salesperson will not sell just one of the trucks. As we already pointed out, events $A$ and $C$ share the outcome $(0, 2)$, and only $(0, 2)$, so that

$$A \cap C = \{(0, 2)\}$$

and this is the event that the first salesperson does not sell any of the trucks and the second salesperson sells two. Since $B'$ contains all the elements not contained in $B$, we can write

$$B' = \{(0, 1), (1, 1), (2, 1), (0, 2), (1, 2), (0, 3)\}$$

and this is the event that the second salesperson will sell at least one of the trucks. ∎

Sample spaces and events, particularly relationships among events, are often illustrated by **Venn diagrams** such as those of Figures 6.2 and 6.3. In each case, the sample space is represented by a rectangle, and events by circles or parts

**Figure 6.2**
Venn diagrams.

Figure 6.3
Venn diagram.

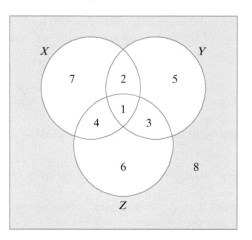

of circles within the rectangle. The tinted regions of the four Venn diagrams of Figure 6.2 represent event $X$, the complement of event $X$, the union of events $X$ and $Y$, and the intersection of events $X$ and $Y$.

**EXAMPLE  6.4**    If $X$ is the event that Mr. Jones is a graduate student at Stanford University and $Y$ is the event that he likes to go sailing in Monterey Bay, what events are represented by the tinted regions of the four Venn diagrams of Figure 6.2?

**Solution**    The tinted region of the first diagram represents the event that Mr. Jones is a graduate student at Stanford University. The tinted region of the second diagram represents the event Mr. Jones is not a graduate student at Stanford University. The tinted region of the third diagram represents the event Mr. Jones is a graduate student at Stanford University and/or likes to go sailing in Monterey Bay. The tinted region of the fourth diagram represents the event that Mr. Jones is a graduate student at Stanford University and likes to go sailing in Monterey Bay. ◼

When we deal with three events, we draw the circles as in Figure 6.3. In this diagram, the circles divide the sample space into eight regions numbered 1 through 8, and it is easy to determine whether the corresponding events are in $X$ or in $X'$, in $Y$ or in $Y'$, and in $Z$ or in $Z'$.

**EXAMPLE  6.5**    If $X$ is the event that employment will go up, $Y$ is the event that stock prices will go up, and $Z$ is the event that interest rates will go up, express in words what events are represented by region 4, regions 1 and 3 together, and regions 3, 5, 6, and 8 together in Figure 6.3.

**Solution**    Since region 4 is contained in $X$ and $Z$ but not in $Y$, it represents the event that employment and interest rates will go up but stock prices will not go up. Since regions 1 and 3 together constitute the region common to $Y$ and $Z$, they represent the event that stock prices and interest rates will go up. Since regions 3, 5, 6, and 8 together constitute the entire region outside $X$, they represent the event that employment will not go up. ◼

**6.1**  With reference to the illustration on page 125, the one concerning the scholarship, let $U = \{b, e, h\}$ and $V = \{e, f, g, h\}$, and list the outcomes that comprise $U'$, $U \cap V$, and $U \cup V'$. Also, express each of these events in words.

**6.2**  With reference to Exercise 6.1, are events $U$ and $V$ mutually exclusive?

**6.3**  With reference to the sample space of Figure 6.1, list the sets of points that constitute the following events:

(a)  One of the three trucks will remain unsold;

(b)  The two salespersons will sell equally many trucks;

(c)  Both salespersons will sell at least one truck.

**6.4**  At least one of two professors and two of five graduate assistants must be present when a chemistry lab is being used.

(a)  Using two coordinates as in Example 6.1, so that $(1, 4)$, for example, denotes the presence of one of the professors and four of the graduate assistants, list the eight possibilities.

(b)  Draw a diagram similar to that of Figure 6.1.

**6.5**  With reference to Exercise 6.4, list in words what events are represented by

(a)  $K = \{(1, 2), (2, 3)\}$;

(b)  $L = \{(1, 3), (2, 2)\}$;

(c)  $M = \{(1, 2), (2, 2)\}$.

Also, determine which of the three pairs of events, $K$ and $L$, $K$ and $M$, and $L$ and $M$, are mutually exclusive.

**6.6**  A literary critic has two days on which to take a look at some of the seven books that have recently been released. She wants to check out at least five of the books, but not more than four on either day.

(a)  Using two coordinates so that $(2, 3)$, for example, represents the event that she will take a look at two books on the first day and three books on the second day, list the nine possibilities and draw a diagram of the sample space similar to that of Figure 6.1.

(b)  List the points of the sample space that constitute event $T$ that she will take a look at five of the books, event $U$ that she will take a look at more books on the first day than on the second day, and event $V$ that she will take a look at three books on the second day.

**6.7**  With reference to Exercise 6.6, list the points of the sample space that constitute events $T \cap U$, $U \cap V$, and $V \cap T'$.

**6.8**  A small marina has three fishing boats that are sometimes in dry dock for repairs.

(a)  Using two coordinates, so that $(2, 1)$, for example, represents the event that two of the fishing boats are in dry dock and one is rented out for the day, and $(0, 2)$ represents the event that none of the boats is in dry dock and two are rented out for the day, draw a diagram similar to that of Figure 6.1 showing the 10 points of the corresponding sample space.

(b)  If $K$ is the event that at least two of the boats are rented out for the day, $L$ is the event that more boats are in dry dock than are rented out for the day, and $M$ is the event that all the boats that are not in dry dock are rented out for the day, list the outcomes that comprise each of these events.

(c)  Which of the three pairs of events, $K$ and $L$, $K$ and $M$, and $L$ and $M$, are mutually exclusive?

**6.9**  With reference to Exercise 6.8, list the points of the sample space that constitute events $K'$ and $L \cap M$. Also, express each of these events in words.

**6.10** To construct sample spaces for experiments where we deal with categorical data, we often code the various alternatives by assigning them numbers. For instance, if individuals are asked whether their favorite color is red, yellow, blue, green, brown, white, purple, or some other color, we might assign these alternatives the codes 1, 2, 3, 4, 5, 6, 7, and 8. If

$$A = \{3, 4\}, B = \{1, 2, 3, 4, 5, 6, 7\}, \quad \text{and} \quad C = \{6, 7, 8\}$$

list the outcomes comprising the events $B'$, $A \cap B$, $B \cap C'$, and $A \cup B'$. Also express each of these events in words.

**6.11** Among six applicants for an executive job, $A$ is a college graduate, foreign born, and single; $B$ is not a college graduate, foreign born, and married; $C$ is a college graduate, native born, and married; $D$ is not a college graduate, native born, and single; $E$ is a college graduate, native born, and married; and $F$ is not a college graduate, native born, and married. One of these applicants is to get the job, and the event that the job is given to a college graduate, for example, is denoted $\{A, C, E\}$. State in a similar manner the event that the job is given to

(a) a single person;

(b) a native-born college graduate;

(c) a married person who is foreign born.

**6.12** Which of the following pairs of events are mutually exclusive? Explain your answers.

(a) A driver getting a ticket for speeding and a ticket for going through a red light.

(b) Being foreign-born and being President of the United States.

(c) A baseball player getting a walk and hitting a home run in the same at bat.

(d) A baseball player getting a walk and hitting a home run in the same game.

**6.13** Which of the following pairs of events are mutually exclusive? Explain your answers.

(a) Having rain and sunshine on the 4th of July, 2005.

(b) A person wearing black shoes and green socks.

(c) A person leaving Los Angeles by jet at 11 P.M. and arriving in New York City on the same day.

(d) A person having a degree from U.C.L.A. and a degree from the University of Chicago.

**6.14** With reference to Figure 6.4, $D$ is the event that a graduate student speaks very good Dutch and $H$ is the event that he is going to attend a university in Hilversum. In words, what events are represented by regions 1 and 3 together, regions 3 and 4 together, and by regions 1, 2, 3, and 4?

**6.15** Venn diagrams are also useful in determining the numbers of possible outcomes associated with various events. Suppose that one of the 360 members of a golf club is to be chosen Player of the Year. If 224 of the members play at least once a week, 98 of them are lefthanded, and 50 of the lefthanded members play at least once a week, how many of the possible choices would be

(a) lefthanded members who do not play at least once a week;

(b) members who play at least once a week but are not lefthanded;

(c) members who do not play at least once a week and are not lefthanded?

Figure 6.4
Venn diagram for
Example 6.14.

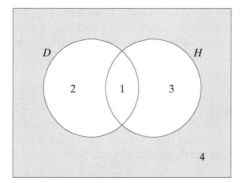

**6.16**  One of the 200 business majors at a college is to be chosen for the student senate. If 77 of these students are enrolled in a course in accounting, 64 are enrolled in a course in business law, and 92 are not enrolled in either course, how many of the outcomes correspond to the choice of a business major who is enrolled in both courses?

**6.17**  In Figure 6.5, $E$, $T$, and $N$ are the events that a car brought to a garage needs an engine overhaul, transmission repairs, or new tires. Express in words what events are represented by region 1, region 3, region 7, regions 1 and 4 together, regions 2 and 5 together, and regions 7, 4, 6, and 8 together.

**6.18**  With reference to Exercise 6.17 and Figure 6.5, list the regions or combinations of regions that represent the events that a car brought to the garage will need

(a)  transmission repairs, but neither an engine overhaul nor new tires;

(b)  an engine overhaul and transmission repairs;

(c)  transmission repairs or new tires, but not an engine overhaul;

(d)  new tires.

**6.19**  As we pointed out in Exercise 6.15, Venn diagrams are also useful in determining the numbers of outcomes associated with various circumstances. Among 60 houses advertised for sale there are 8 with swimming pools, three or more bedrooms, and wall-to-wall carpeting; 5 with swimming pools, three or more bedrooms, but no wall-to-wall carpeting; 3 with swimming pools, wall-to-wall carpeting, but fewer than three bedrooms; 8 with swimming pools but neither wall-to-wall carpeting nor three or more bedrooms; 24 with three or more bedrooms, but neither a swimming pool nor wall-to-wall carpeting; 2 with three or more bedrooms, wall-to-wall carpeting,

Figure 6.5
Venn diagram for
Exercise 6.17.

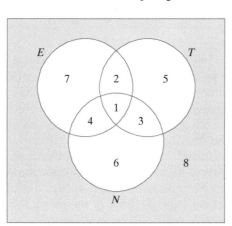

but no swimming pool; 3 with wall-to-wall carpeting, but neither a swimming pool nor three or more bedrooms; and 7 without any of these features. If one of these houses is to be chosen for a television commercial, how many outcomes correspond to the choice of

(a) a house with a swimming pool;

(b) a house with wall-to-wall carpeting?

**6.20** In a psychiatric study reported in the *New England Journal of Medicine*, psychiatrists reported on mental status evaluations. Letting *A* stand for the event that the patients have auditory delusions and *V* stand for the event that the patients have visual delusions, state in words the probabilities expressed by

(a) $P(A')$;

(b) $P(V')$;

(c) $P(A \cup V)$;

(d) $P(A \cap V)$;

(e) $P(A' \cap V')$;

(f) $P(A \cap V')$.

# 6.2 THE POSTULATES OF PROBABILITY

Probabilities always pertain to the occurrence of events, and now that we have learned how to deal mathematically with events, let us turn to the rules that probabilities must obey. To formulate these rules, we continue the practice of denoting events by capital letters and write the probability of event *A* as $P(A)$, the probability of event *B* as $P(B)$, and so on. As before, we denote the set of all possible outcomes, the sample space, by the letter *S*.

Most basic among all the rules of probability are the three **postulates**, which, as we shall state them here, apply when the sample space *S* is finite. Beginning with the first two, we write

**F**IRST TWO POSTULATES OF PROBABILITY

> 1. *Probabilities are positive real numbers or zero; symbolically, $P(A) \geq 0$ for any event A.*
> 2. *Every sample space has probability 1; symbolically, $P(S) = 1$ for any sample space S.*

To justify these two postulates, as well as the third one that follows, let us show that they are in agreement with the classical probability concept as well as the frequency interpretation. In Section 6.3 we shall see to what extent the postulates are compatible also with subjective probabilities.

The first two postulates are in agreement with the classical probability concept because the fraction $\frac{s}{n}$ is always positive or zero, and for the entire sample space (which includes all *n* outcomes) the probability is $\frac{s}{n} = \frac{n}{n} = 1$. When it comes to the frequency interpretation, the proportion of the time that an event will occur cannot be a negative number and since one of the outcomes in a sample space must always occur (that is, 100% of the time), the probability that one of the outcomes in a sample space will occur is 1.

Although a probability of 1 is thus identified with certainty, in actual practice we also assign a probability of 1 to events that are "practically certain" to occur. For instance, we would assign a probability of 1 to the event that at least one person will vote in the next presidential election, even though it is not logically certain. Similarly, we would assign a probability of 1 to the event that not every student entering college in the fall of 2005 will apply for admission to Princeton University.

The third postulate of probability is especially important, but it is not quite as obvious as the other two.

**T**HIRD POSTULATE OF PROBABILITY

> **3.** *If two events are mutually exclusive, the probability that one or the other will occur equals the sum of their probabilities. Symbolically,*
>
> $$P(A \cup B) = P(A) + P(B)$$
>
> *for any two mutually exclusive events A and B.*

For instance, if the probability that weather conditions will improve during a certain week is 0.62 and the probability that they will remain unchanged is 0.23, then the probability that they will either improve or remain unchanged is $0.62 + 0.23 = 0.85$. Similarly, if the probabilities that a student will get an $A$ or a $B$ in a course are 0.13 and 0.29, then the probability that he or she will get either an $A$ or a $B$ is $0.13 + 0.29 = 0.42$.

To show that the third postulate is also compatible with the classical probability concept, let $s_1$ and $s_2$ denote the number of equally likely possibilities that comprise events $A$ and $B$. Since $A$ and $B$ are mutually exclusive, no two of these possibilities are alike and all $s_1 + s_2$ of them comprise event $A \cup B$. Thus,

$$P(A) = \frac{s_1}{n} \quad P(B) = \frac{s_2}{n} \quad P(A \cup B) = \frac{s_1 + s_2}{n}$$

and $P(A) + P(B) = P(A \cup B)$.

Insofar as the frequency interpretation is concerned, if one event occurs, say, 36% of the time, another event occurs 41% of the time, and they cannot both occur at the same time (that is, they are mutually exclusive), then one or the other will occur $36 + 41 = 77\%$ of the time. This is in agreement with the third postulate.

By using the three postulates of probability, we can derive many further rules according to which probabilities must "behave"—some of them are easy to prove and some are not, but they all have important applications. Among the immediate consequences of the three postulates we find that probabilities can never be greater than 1, that an event that cannot occur has probability 0, and that the probabilities that an event will occur and that it will not occur always add up to 1. Symbolically,

$$P(A) \leq 1 \quad \textit{for any event A}$$

$$P(\emptyset) = 0$$

$$P(A) + P(A') = 1 \quad \textit{for any event A}$$

The first of these results simply expresses the fact that there cannot be more favorable outcomes than there are outcomes, or that an event cannot occur more than 100% of the time. The second result expresses the fact that when an event cannot occur there are $s = 0$ favorable outcomes, or that such an event occurs 0% of the time. In actual practice, we also assign 0 probability to events that are so unlikely that we are "practically certain" they will not occur. For instance, we would assign 0 probability to the event that a monkey set loose on a typewriter will by chance type Plato's *Republic* word for word without a single mistake. Or, when flipping a coin, we would assign 0 probability to the event that it will land on its edge.

The third result can also be derived from the postulates of probability, and it can easily be seen that it is compatible with the classical probability concept and the frequency interpretation. In the classical concept, if there are $s$ "successes" there are $n - s$ "failures," the corresponding probabilities are $\frac{s}{n}$ and $\frac{n-s}{n}$, and their sum is

$$\frac{s}{n} + \frac{n-s}{n} = \frac{n}{n} = 1$$

In accordance with the frequency interpretation, we can say that if some investments are successful 22% of the time, then they are not successful 78% of the time, the corresponding probabilities are 0.22 and 0.78, and their sum is 1.

The examples that follow show how the postulates and the rules we gave previously are put to use in actual practice.

**EXAMPLE 6.6** If $A$ and $B$ are the events that *Consumer Union* will rate a car stereo good or poor, $P(A) = 0.24$ and $P(B) = 0.35$, determine the following probabilities:

**(a)** $P(A')$;

**(b)** $P(A \cup B)$;

**(c)** $P(A \cap B)$.

**Solution**   **(a)** Using the third of the three rules on page 133, we find that $P(A')$, the probability that *Consumer Union* will not rate the stereo as being good, is $1 - 0.24 = 0.76$.
**(b)** Since events $A$ and $B$ are mutually exclusive, we can use the third postulate of probability and write $P(A \cup B) = P(A) + P(B) = 0.24 + 0.35 = 0.59$ for the probability that the car stereo will be rated good or poor.
**(c)** Since events $A$ and $B$ are mutually exclusive, they cannot both occur and $P(A \cap B) = P(\emptyset) = 0$.

In problems like this it usually helps to draw a Venn diagram, fill in the probabilities associated with the various regions, and then read the answers directly off the diagram.

**EXAMPLE 6.7** If $C$ and $D$ are the events that Dr. Miller will be in his office at 9 A.M. or in the hospital, $P(C) = 0.48$ and $P(D) = 0.27$, find $P(C' \cap D')$.

**Figure 6.6**
Venn diagram for
Example 6.7.

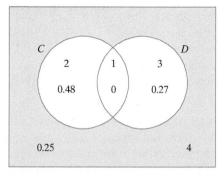

**Solution**    Drawing the Venn diagram of Figure 6.6, we first put a 0 into region 1 since $C$ and $D$ are mutually exclusive events. It follows that the 0.48 probability of event $C$ must go into region 2 and that the 0.27 probability of event $D$ must go into region 3. Hence, the remaining probability, $1 - (0.48 + 0.27) = 0.25$, must go into region 4. Since event $C' \cap D'$ is represented by the region outside both circles, namely, region 4, we find that the answer is $P(C' \cap D') = 0.25$. ◼

# 6.3 PROBABILITIES AND ODDS

If an event is twice as likely to occur than not to occur, we say that the **odds** are 2 to 1 that it will occur, if an event is three times as likely to occur than not to occur, we say that the odds are 3 to 1 that it will occur; if an event is ten times as likely to occur than not to occur, we say that the odds are 10 to 1 that it will occur; and so forth. In general,

> **The odds that an event will occur are given by the ratio of the probability that it will occur to the probability that it will not occur.**

Symbolically,

EXPRESSING ODDS
IN TERMS OF
PROBABILITIES

> *If the probability of an event is p, the odds for its occurrence are a to b, where a and b are positive values such that*
> $$\frac{a}{b} = \frac{p}{1 - p}$$

It is customary to express odds as ratios of two positive integers having no common factors. Also, if an event is more likely not to occur than to occur, it is customary to quote the odds that it will not occur rather than the odds that it will occur.

EXAMPLE **6.8**    What are the odds for the occurrence of an event if its probability is

(a) $\frac{5}{9}$;

(b) 0.85;

(c) 0.20?

**Solution**   (a) By definition, the odds are $\frac{5}{9}$ to $1 - \frac{5}{9} = \frac{4}{9}$, or 5 to 4.
(b) By definition, the odds are 0.85 to $1 - 0.85 = 0.15$, 85 to 15, or better 17 to 3.
(c) By definition, the odds are 0.20 to $1 - 0.20 = 0.80$, 20 to 80, or 1 to 4, and we would quote the odds as 4 to 1 *against* the occurrence of the event. In general, this is how we would quote the odds, when an event is more likely *not* to occur.  ◼

In betting, the term "odds" is also used to denote the ratio of the wager of one party to that of another. For instance, if a gambler says that he will give 3 to 1 odds on the occurrence of an event, he means that he is willing to bet $3 against $1 (or perhaps $30 against $10 or $1,500 against $500) that the event will occur. If such **betting odds** actually equal the odds that the event will occur, we say that the betting odds are **fair**. If a gambler really believes that a bet is fair, then he is, at least in principle, willing to bet on either side. The gambler in this situation would also be willing to bet $1 against $3 (or $10 against $30 or $500 against $1,500) that the event will not occur.

**EXAMPLE 6.9**   Records show that $\frac{1}{12}$ of the trucks weighed at a certain check point in Nevada carry too heavy a load. If someone offers to bet $40 against $4 that the next truck weighed at this check point will not carry too heavy a load, are these betting odds fair?

**Solution**   Since the probability is $1 - \frac{1}{12} = \frac{11}{12}$ that the truck will not carry too heavy a load, the odds are 11 to 1, and the bet would be fair if the person offered to bet $44 against $4 that the next truck weighed at the check point will not carry too heavy a load. Thus, the $40 against $4 bet is not fair; it favors the person offering the bet.  ◼

This discussion of odds and betting odds provides the groundwork for a way of measuring **subjective probabilities**. If a businessman "feels" that the odds for the success of a new clothing store are 3 to 2, this means that he is willing to bet (or considers it fair to bet) $300 against $200, or perhaps $3,000 against $2,000, that the new store will be a success. In this way he is expressing his belief regarding the uncertainties connected with the success of the store, and to convert it into a probability we take the equation

$$\frac{a}{b} = \frac{p}{1-p}$$

with $a = 3$ and $b = 2$ and solve it for $p$. In general, solving the equation $\frac{a}{b} = \frac{p}{1-p}$, for $p$, we get the following result, which the reader will be asked to verify in Exercise 6.39.

**E**XPRESSING PROBABILITIES IN TERMS OF ODDS

> *If the odds are a to b that an event will occur, the probability of its occurrence is*
>
> $$p = \frac{a}{a+b}$$

Now, if we substitute $a = 3$ and $b = 2$ into this formula for $p$, we find that, according to the businessman, the probability of the new clothing store's success is $\frac{3}{3+2} = \frac{3}{5}$ or 0.60.

**EXAMPLE 6.10**    If an applicant for a position as a football coach feels that the odds are 7 to 1 that he will get the job, what subjective probability is he thus assigning to his getting the job?

**Solution**    Substituting $a = 7$ and $b = 1$ into the preceding formula for $p$, we get $p = \frac{7}{7+1} = \frac{7}{8} = 0.875$.    ■

Let us now see whether subjective probabilities determined in this way "behave" in accordance with the postulates of probability. Since $a$ and $b$ are positive numbers, $\frac{a}{a+b}$ cannot be negative and this satisfies the first postulate. As for the second postulate, observe that the more certain we are that an event will occur, the "better" odds we should be willing to give—say, 99 to 1, 9,999 to 1, or perhaps even 999,999 to 1. The corresponding probabilities are 0.99, 0.9999, and 0.999999, and it can be seen that the more certain we are that an event will occur, the closer its probability will be to 1.

The third postulate does not necessarily apply to subjective probabilities, but proponents of the subjectivist point of view impose it as a **consistency criterion**. In other words, if a person's subjective probabilities "behave" in accordance with the third postulate, he or she is said to be consistent; otherwise, the person's probability judgments must be taken with a grain of salt.

**EXAMPLE 6.11**    A newspaper columnist feels that the odds are 2 to 1 that interest rates will go up before the end of the year, 1 to 5 that they will remain unchanged, and 8 to 3 that they will go up or remain unchanged. Are the corresponding probabilities consistent?

**Solution**    The corresponding probabilities that interest rates will go up before the end of the year, that they will remain unchanged, or that they will go up or remain unchanged are, respectively, $\frac{2}{2+1} = \frac{2}{3}$, $\frac{1}{1+5} = \frac{1}{6}$, and $\frac{8}{8+3} = \frac{8}{11}$. Since $\frac{2}{3} + \frac{1}{6} = \frac{5}{6}$ and not $\frac{8}{11}$, the probabilities are not consistent. Therefore, the newspaper columnist's judgment must be questioned.    ■

**EXERCISES**

**6.21**    In a study of the future needs of an airport, $C$ is the event that there will be enough capital for the planned expansion and $E$ is the event that the planned expansion will provide enough parking. State in words what probabilities are expressed by $P(C')$, $P(E')$, $P(C' \cap E)$, and $P(C \cap E')$.

**6.22**    With reference to Exercise 6.21, express symbolically the probabilities that there will be
(a)  enough capital but not enough parking;
(b)  neither enough capital nor enough parking.

**6.23**    With regard to a scheduled symphony concert, let $A$ denote the event that there will be a good attendance and $W$ denote the event that more than half the crowd will walk out during the intermission. State in words what probabilities are expressed by $P(A')$, $P(A' \cup W)$, and $P(A \cap W')$.

**6.24**   With reference to Exercise 6.23, let $F$ denote the event that the soloist will fail to show up and express symbolically the probabilities that

(a) the soloist will fail to show up and more than half the crowd will walk out during the intermission;

(b) the soloist will show up and there will be a good attendance.

**6.25**   Which of the postulates of probability are violated by the following statements?

(a) Since their car broke down, the probability that they will be late is $-0.40$.

(b) The probability that a mineral specimen will contain copper is 0.26 and the probability that it will not contain copper is 0.64.

(c) The probability that a lecture will be entertaining is 0.35, and the probability that it will not be entertaining is four times as large.

(d) The probabilities that a student will spend an evening studying or watching television are, respectively, 0.22 and 0.48, and the probability that it will be one or the other is 0.80.

**6.26**   Which of the three rules on page 132–133 are violated by the following statements?

(a) The probability that a surgical procedure will be successful is 0.73 and the probability that it will not be successful is 0.33.

(b) The probability that two mutually exclusive events will both occur is always equal to 1.

(c) The probability that a new vaccine will be effective is 1.09.

(d) The probability that an event will occur and that it will not occur is always equal to 1.

**6.27**   A decathlon is an athletic contest of 10 events to determine the best track and field athlete. If an athlete feels that the odds are 2 to 1 that he can complete the decathlon and 3 to 1 that he cannot complete this contest, can these odds be right? Explain.

**6.28**   Convert each of the following odds to probabilities.

(a) The odds are 5 to 3 that a realtor will sell a certain house within a month.

(b) The odds against getting zero heads in eight flips of a balanced coin are 15 to 1.

**6.29**   Explain in words why each of the following inequalities must be false:

(a) $P(A \cup B) < P(A)$;

(b) $P(A \cap B) > P(A)$.

**6.30**   Make up numerical examples in which two events $A$ and $B$ are mutually exclusive and events $A'$ and $B'$

(a) are not mutually exclusive;

(b) are also mutually exclusive.

**6.31**   Under what condition are events $A$ and $B$ as well as events $A$ and $B'$ mutually exclusive?

**6.32**   When entering data into a computer, the probability that a student will make at most three mistakes per 1,000 keystrokes is 0.64, and the probability of making anywhere from 4 to 6 mistakes per 1,000 keystrokes is 0.21. Find the probabilities that in 1,000 keystrokes the student will make

(a) at least 4 mistakes;

(b) at most 6 mistakes;

(c) at least 7 mistakes.

**6.33**   Convert each of the following probabilities to odds or odds to probabilities:

(a) The probability of getting at least two heads in four flips of a balanced coin is $\frac{11}{16}$.

(b) If three ceramic tiles are randomly chosen from a carton of twelve, of which three have blemishes, the odds are 34 to 21 that at least one of them will have a blemish.

(c) If a pollster randomly selects five of 24 households to be included in a survey, the probability is $\frac{5}{24}$ that any particular household will be included.

(d) If a secretary randomly puts six letters into six envelopes that are already addressed, the odds are 719 to 1 that not all the letters will end up in the right envelopes.

**6.34**   A soccer fan is offered a bet of $15 against his $5 that the United States will lose its first World Cup match. What does it tell us about the subjective probability the fan assigns to the United States winning this match if he is unwilling to accept the bet?

**6.35**   A television producer is willing to bet $1,200 against $1,000, but not $1,500 against $1,000 that a new game show will be a success. What does this tell us about the probability that the producer assigns to the show's success?

**6.36**   Asked about his political future, a party official replies that the odds are 2 to 1 that he will not run for the House of Representatives and 4 to 1 that he will not run for the Senate. Furthermore, he feels that the odds are 7 to 5 that he will run for one or the other. Are the corresponding probabilities consistent?

**6.37**   A high school principal feels that the odds are 7 to 5 against her getting a $1,000 raise and 11 to 1 against her getting a $2,000 raise. Furthermore, she feels that it is an even-money bet that she will get one of these raises or the other. Discuss the consistency of the corresponding subjective probabilities.

**6.38**   Some events are so unlikely that we choose to assign them probabilities of zero. Would you assign zero probabilities to the events that

(a) if you randomly strike the keys, your computer will correctly print Lincoln's Gettysburg Address;

(b) lightning will strike the same tree on four successive days;

(c) thirteen cards randomly dealt from an ordinary deck of 52 playing cards will all be spades?

**6.39**   Verify algebraically that the equation $\frac{a}{b} = \frac{p}{1-p}$, solved for $p$, yields $p = \frac{a}{a+b}$.

# 6.4 ADDITION RULES

The third postulate of probability applies only to two mutually exclusive events, but it can easily be generalized in two ways, so that it will apply to more than two mutually exclusive events and also to two events that need not be mutually exclusive. We say that $k$ events are mutually exclusive if no two of them have any elements in common. In that case, we can repeatedly use the third postulate and, thus, show that

**GENERALIZATION OF POSTULATE 3**

> *If $k$ events are mutually exclusive, the probability that one of them will occur equals the sum of their individual probabilities; symbolically,*
>
> $$P(A_1 \cup A_2 \cup \cdots \cup A_k) = P(A_1) + P(A_2) + \cdots + P(A_k)$$
>
> *for any mutually exclusive events $A_1, A_2, \ldots,$ and $A_k$.*

Here again, we read $\cup$ as "or."

EXAMPLE **6.12**   The probabilities that a certain person looking for a new car will end up buying a Chevrolet, a Ford, or a Honda are, respectively, 0.17, 0.22, and 0.08. Assuming that she will buy just one car, what is the probability that it will be one of the three kinds?

**Solution**   Since the three possibilities are mutually exclusive, direct substitution yields $0.17 + 0.22 + 0.08 = 0.47$.   ■

EXAMPLE **6.13**   The probability that a consumer testing service will rate a new camera poor, fair, good, very good, or excellent are 0.07, 0.16, 0.34, 0.32, and 0.11. What is the probability that it will rate the new camera good, very good, or excellent?

**Solution**   Since the five possibilities are mutually exclusive, the probability is $0.34 + 0.32 + 0.11 = 0.77$.   ■

The job of assigning probabilities to all possible events connected with a given situation can be very tedious. Indeed, it can be shown that if a sample space has 10 elements (points or outcomes) we can form more than 1,000 different events, and if a sample space has 20 elements we can form more than 1 million.[†] Fortunately, it is seldom necessary to assign probabilities to all possible events (that is, to all possible subsets of a sample space). The following rule, which is a direct application of the preceding generalization of the third postulate of probability, makes it easy to determine the probability of any event on the basis of the probabilities associated with the individual outcomes in a sample space:

**R**ULE FOR CALCULATING THE PROBABILITY OF AN EVENT

> *The probability of any event A is given by the sum of the probabilities of the individual outcomes comprising A.*

EXAMPLE **6.14**   Referring again to Example 6.1, which dealt with two car salespersons and three 1998 Dodge Ram trucks, suppose that the ten points of the sample space, shown originally in Figure 6.1, have the probabilities shown in Figure 6.7. Find the probabilities that

**(a)** between them, the two salespersons will sell two of the trucks;
**(b)** the second salesperson will not sell any of the trucks;
**(c)** the second salesperson will sell at least two of the trucks.

**Solution**   **(a)** Adding the probabilities associated with the points $(2, 0)$, $(1, 1)$, and $(0, 2)$, we get $0.08 + 0.06 + 0.04 = 0.18$.
**(b)** Adding the probabilities associated with the points $(0, 0)$, $(1, 0)$, $(2, 0)$, and $(3, 0)$, we get $0.44 + 0.10 + 0.08 + 0.05 = 0.67$.

---

[†] In general, if a sample space has $n$ elements, we can form $2^n$ different events. Each element is either included or excluded for a given event, so by the multiplication of choices there are $2 \cdot 2 \cdot 2 \cdot \ldots \cdot 2 = 2^n$ possibilities. Note that $2^{10} = 1,024$ and $2^{20} = 1,048,576$.

**Figure 6.7**
Sample space for
Example 6.14.

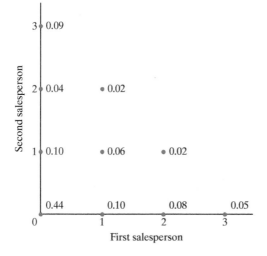

**(c)** Adding the probabilities associated with the points $(0, 2)$, $(1, 2)$, and $(0, 3)$, we get $0.04 + 0.02 + 0.09 = 0.15$.

In the special case where the outcomes are all equiprobable, the preceding rule leads to the formula $P(A) = \frac{s}{n}$, which we used earlier in connection with the classical probability concept. Here, $n$ is the total number of individual outcomes in the sample space and $s$ is the number of "successes," namely, the number of outcomes comprising event $A$.

**EXAMPLE  6.15**    Given that the 44 points (outcomes) in the sample space of Figure 6.8 are all equiprobable, find $P(A)$.

**Solution**    Since the 44 points (outcomes) are equiprobable, each one has the probability $\frac{1}{44}$, and since a count shows that there are 10 outcomes in $A$, $P(A) = \frac{1}{44} + \cdots + \frac{1}{44} = \frac{10}{44}$ or approximately 0.23.

Since the third postulate and its generalization apply only to mutually exclusive events, they cannot be used, for example, to determine the probability that a bird watcher, on a walk in the desert, will spot a roadrunner or a cactus wren. Also, they cannot be used to determine the probability that a person shopping in a department store will buy a shirt, a sweater, a belt, or a tie. In the first case, the bird watcher could spot both, a roadrunner and a cactus wren, and in the second case, the person doing the shopping could buy several of the items mentioned.

**Figure 6.8**
Sample space with
44 equiprobable
outcomes.

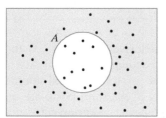

**Figure 6.9**
Venn diagram for
example.

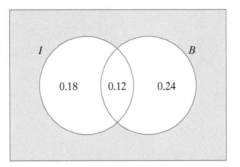

To find a formula for $P(A \cup B)$ that holds regardless of whether or not events $A$ and $B$ are mutually exclusive, consider the Venn diagram of Figure 6.9. It concerns the job applications of a recent college graduate, with $I$ and $B$ denoting her getting a job offer from an investment broker; and her getting a job offer from a bank. It can be seen from the Venn diagram of Figure 6.9 that

$$P(I) = 0.18 + 0.12 = 0.30$$
$$P(B) = 0.12 + 0.24 = 0.36$$

and

$$P(I \cup B) = 0.18 + 0.12 + 0.24 = 0.54$$

where we could add the respective probabilities because they pertain to mutually exclusive events (nonoverlapping regions of the Venn diagram).

Had we erroneously used the third postulate to calculate $P(I \cup B)$, we would have obtained $P(I) + P(B) = 0.30 + 0.36 = 0.66$, which exceeds the correct value by 0.12. This error results from adding $P(I \cap B) = 0.12$ in twice, once in $P(I) = 0.30$ and once in $P(B) = 0.36$, and we could correct for this by subtracting 0.12 from 0.66. Thus, we could write

$$P(I \cup B) = P(I) + P(B) - P(I \cap B)$$
$$= 0.30 + 0.36 - 0.12 = 0.54$$

and this agrees, as it should, with the result obtained before.

Since the argument used in this example holds for any two events $A$ and $B$, we can now state the following **general addition rule**, which applies regardless of whether $A$ and $B$ are mutually exclusive events:

$G$ENERAL
ADDITION RULE

$$P(A \cup B) = P(A) + P(B) - P(A \cap B)$$

When $A$ and $B$ are mutually exclusive, $P(A \cap B) = 0$ and the preceding formula reduces to that of the third postulate of probability. In this connection, the third postulate is also referred to as the **special addition rule**. To add to this terminology, the generalization of the third postulate on page 139 is sometimes referred to as the **generalized** (special) **addition rule**.

EXAMPLE **6.16**   The probabilities that it will rain in Tucson, Arizona, on a day in mid-August, that there will be a thunderstorm on that day, and that there will be rain as well as a thunderstorm are 0.27, 0.24, and 0.15. What is the probability that there will be rain and/or a thunderstorm in Tucson on such a day?

**Solution**   If $R$ denotes rain and $T$ denotes a thunderstorm, we have $P(R) = 0.27$, $P(T) = 0.24$, and $P(R \cap T) = 0.15$. Substituting these values into the formula for the general addition rule, we get

$$P(R \cup T) = P(R) + P(T) - P(R \cap T)$$
$$= 0.27 + 0.24 - 0.15$$
$$= 0.36$$

EXAMPLE **6.17**   In a sample survey conducted in a suburban community, the probabilities are 0.92, 0.53, and 0.48 that a family selected at random will own a family sedan, a sports utility vehicle, or both. What is the probability that such a family will own a family sedan, a sports utility vehicle, or both?

**Solution**   Substituting these values into the formula for the general addition rule, we get $0.92 + 0.53 - 0.48 = 0.97$. Had we *erroneously* used the third postulate of probability, we would have obtained the impossible result $0.92 + 0.53 = 1.45$.

The general addition rule can be generalized further so that it applies to three or more events that need not be mutually exclusive, but we shall not go into this here.

**EXERCISES**

**6.40**   A city's police department needs new tires for its patrol cars. The probabilities that it will buy Firestone, Goodyear, Michelin, Goodrich, or Pirelli tires are, respectively, 0.19, 0.26, 0.25, 0.20, and 0.07. Find the probabilities that this police department will buy
(a) Goodyear or Goodrich tires;
(b) Firestone or Michelin tires;
(c) Goodyear, Michelin, or Pirelli tires.

**6.41**   With reference to Exercise 6.40, what is the probability that the police department will buy some other kind of tires?

**6.42**   The probabilities of rolling a total of $2, 3, 4, \ldots, 11$, or 12 with a pair of balanced dice are, respectively, $\frac{1}{36}, \frac{2}{36}, \frac{3}{36}, \frac{4}{36}, \frac{5}{36}, \frac{6}{36}, \frac{5}{36}, \frac{4}{36}, \frac{3}{36}, \frac{2}{36}$, and $\frac{1}{36}$. What are the probabilities of rolling
(a) a total of 7 or 11;
(b) a total of 2, 3, or 12;
(c) a total that is an odd number, that is 3, 5, 7, 9, or 11?

**6.43**   Sometimes a laboratory assistant has lunch at the cafeteria where she works, sometimes she brings her own lunch, sometimes she has lunch at a nearby restaurant, sometimes she goes home for lunch, and sometimes she skips lunch to lose weight. If the corresponding probabilities are 0.23, 0.31, 0.15, 0.24, and 0.07, find the probabilities that she will
(a) have lunch at the cafeteria or the nearby restaurant;

(b)  bring her own lunch, go home for lunch, or skip lunch altogether;

(c)  have lunch at the cafeteria or go home for lunch;

(d)  not skip lunch to lose weight.

**6.44**   Figure 6.10 pertains to the number of persons that are invited to look over a new development of luxury homes and the number of persons that will actually attend. If the 45 outcomes (points of the sample space) are all equiprobable, what are the probabilities that

(a)  at most six of the persons will attend;

(b)  at least seven of the persons will attend;

(c)  only two of the invited persons will not attend?

**Figure 6.10**
Sample space for Exercise 6.44.

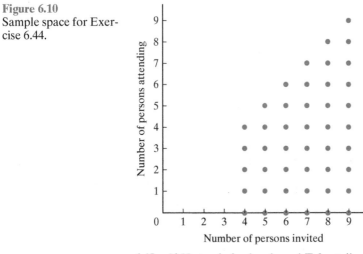

Number of persons invited

**6.45**   If H stands for heads and T for tails, the 32 possible outcomes for five flips of a coin are HHHHH, HHHHT, HHHTH, HHTHH, HTHHH, THHHH, HHHTT, HHTHT, HHTTH, HTHHT, HTHTH, HTTHH, THHHT, THHTH, THTHH, TTHHH, HHTTT, HTHTT, HTTHT, HTTTH, THHTT, THTHT, THTTH, TTHHT, TTHTH, TTTHH, HTTTT, THTTT, TTHTT, TTTHT, TTTTH, and TTTTT. If all these 32 possibilities are equally likely, what are the probabilities of getting 0, 1, 2, 3, 4, or 5 heads?

**6.46**   The probabilities that a person convicted of drunk driving will spend a night in jail, have his license revoked, or both are, respectively, 0.68, 0.51, and 0.22. What is the probability that a person convicted of drunk driving will spend a night in jail and/or have his license revoked?

**6.47**   An auction house has two appraisers of precious jewelry. The probability that the older of the two will not be available is 0.33, the probability that the other one will not be available is 0.27, and the probability that both of them will not be available is 0.19. What is the probability that either or both of them will not be available?

**6.48**   A professor feels that the odds are 3 to 2 against his getting a promotion, there is a fifty-fifty chance that he will get a raise, and the odds are 4 to 1 against his getting both. What are the odds that he will get a promotion and/or a raise?

**6.49**   For married couples living in a suburb, the probabilities that the husband, the wife, or both will vote in a gubernatorial election are, respectively, 0.39, 0.46, and 0.31. What is the probability that either or both will vote in the election?

**6.50**   Explain why there must be a mistake in each of the following statements:
   (a) The probabilities that the management of a professional basketball team will fire the coach, the general manager, or both are 0.85, 0.49, and 0.27.
   (b) The probabilities that a patient at a hospital will have a high temperature, high blood pressure, or both are 0.63, 0.29, and 0.45.

# 6.5 CONDITIONAL PROBABILITY

If we ask for the probability of an event but fail to specify the sample space, we may well get different answers and they can all be correct. For instance, if we ask for the probability that a lawyer will make more than $200,000 a year within 10 years after passing the bar, we may get one answer that applies to all persons practicing law in the United States, another one that applies to corporation lawyers, another one that applies to lawyers employed by the federal government, another one that applies to lawyers who specialize in divorce cases, and so forth. Since the choice of the sample space is by no means always self-evident, it helps to use the symbol $P(A|S)$ to denote the **conditional probability** of event $A$ relative to the sample space $S$, or as we often call it "the probability of $A$ given $S$." The symbol $P(A|S)$ makes it explicit that we are referring to a particular sample space $S$, and it is generally preferable to the abbreviated notation $P(A)$ unless the tacit choice of $S$ is clearly understood. It is also preferable when we have to refer to different sample spaces in the same problem.

To elaborate on the idea of a conditional probability, suppose that a consumer research organization has studied the service under warranty provided by the 200 tire dealers in a large city, and that their findings are summarized in the following table:

|  | Good service under warranty | Poor service under warranty | Total |
|---|---|---|---|
| Name-brand tire dealers | 64 | 16 | 80 |
| Off-brand tire dealers | 42 | 78 | 120 |
| Total | 106 | 94 | 200 |

If one of these tire dealers is randomly selected (that is, each one has the probability $\frac{1}{200}$ of being selected), we find that the probabilities of event $N$ (choosing a name-brand dealer), event $G$ (choosing a dealer who provides good service under warranty), and event $N \cap G$ (choosing a name-brand dealer who provides good service under warranty) are

$$P(N) = \frac{80}{200} = 0.40$$

$$P(G) = \frac{106}{200} = 0.53$$

and

$$P(N \cap G) = \frac{64}{200} = 0.32$$

All these probabilities were calculated by means of the formula $\frac{s}{n}$ for equally likely possibilities.

Since the second of these possibilities is particularly disconcerting—there is almost a fifty–fifty chance of choosing a dealer who does not provide good service under warranty—let us see what will happen if we limit the choice to name-brand dealers. This reduces the sample space to the 80 choices corresponding to the first row of the table, and we find that the probability of choosing a name-brand dealer who will provide good service under warranty is

$$P(G|N) = \frac{64}{80} = 0.80$$

This is quite an improvement over $P(G) = 0.53$, as might have been expected. Note that the conditional probability that we have obtained here,

$$P(G|N) = 0.80$$

can also be written as

$$P(G|N) = \frac{\frac{64}{200}}{\frac{80}{200}} = \frac{P(N \cap G)}{P(N)}$$

namely, as the ratio of the probability of choosing a name-brand dealer who provides good service under warranty to the probability of choosing a name-brand dealer.

Generalizing from this example, let us now make the following definition of conditional probability, which applies to any two events $A$ and $B$ belonging to a given sample space $S$:

**DEFINITION OF CONDITIONAL PROBABILITY**

> *If $P(B)$ is not equal to zero, then the conditional probability of A relative to B, namely, the probability of A given B, is*
>
> $$P(A|B) = \frac{P(A \cap B)}{P(B)}$$

When $P(B)$ is equal to zero, the conditional probability of $A$ relative to $B$ is undefined.

**EXAMPLE 6.18**   With reference to the tire dealers of the illustration on page 145, what is the probability that an off-brand dealer will provide good service under warranty; namely, what is the probability $P(G|N')$?

**Solution**   As can be seen from the table,

$$P(G \cap N') = \frac{42}{200} = 0.21 \quad \text{and} \quad P(N') = \frac{120}{200} = 0.60$$

so that substitution into the formula yields

$$P(G|N') = \frac{P(G \cap N')}{P(N')} = \frac{0.21}{0.60} = 0.35$$

Of course, we could have obtained this result directly from the second row of the table on page 145 by writing

$$P(G|N') = \frac{42}{120} = 0.35$$

**EXAMPLE 6.19**    At a certain elementary school, the probability that a student selected at random will come from a one-parent home is 0.36 and the probability that he or she will come from a one-parent home and also be a low achiever (get mostly $D$'s and $F$'s) is 0.27. What is the probability that such a randomly selected student will be a low achiever given that he or she comes from a one-parent home?

**Solution**    If we let $L$ denote a low achiever and $O$ a student from a one-parent home, we have $P(O) = 0.36$ and $P(O \cap L) = 0.27$, and we get

$$P(L|O) = \frac{P(O \cap L)}{P(O)} = \frac{0.27}{0.36} = 0.75$$

**EXAMPLE 6.20**    The probability that Henry will like a new movie is 0.70 and the probability that Jean, his girlfriend, will like it is 0.60. If the probability is 0.28 that he will like it and she will dislike it, what is the probability that he will like it given that she is not going to like it?

**Solution**    If $H$ and $J$ are the events that Henry will like the new movie and that Jean will like it, we have $P(J') = 1 - 0.60 = 0.40$ and $P(H \cap J') = 0.28$, and we get

$$P(H|J') = \frac{P(H \cap J')}{P(J')} = \frac{0.28}{0.40} = 0.70$$

What is special and interesting about this result is that $P(H)$ and $P(H|J')$ both equal 0.70 and, as the reader will be asked to verify in Exercise 6.66, it follows from the given information that $P(H|J)$ is also equal to 0.70. Thus, the probability of event $H$ is the same regardless of whether or not event $J$ has occurred, occurs, or will occur, and we say that event $H$ is independent of event $J$. In general, it can easily be verified that when one event is independent of another, the second event is also independent of the first, and we say that the two events are **independent**. When two events are not independent, we say that they are **dependent**.

When giving these definitions, we should have specified that neither $H$ nor $J$ can have zero probability, for in that case some of the conditional probabilities would not even have existed. This is why an alternative definition of independence, which we shall give on page 149, is often preferred.

# 6.6 MULTIPLICATION RULES

In Section 6.5 we used the formula $P(A|B) = \frac{P(A \cap B)}{P(B)}$ only to define and calculate conditional probabilities, but if we multiply by $P(B)$ on both sides of the equation, we get the following formula, which enables us to calculate the probability that two events will both occur:

GENERAL
MULTIPLICATION
RULE

$$P(A \cap B) = P(B) \cdot P(A|B)$$

As we have indicated in the margin, this formula is called the **general multiplication rule**, and it states that the probability that two events will both occur is the product of the probability that one of the events will occur and the conditional probability that the other event will occur given that the first event has occurred, occurs, or will occur. As it does not matter which event is referred to as $A$ and which is referred to as $B$, the formula can also be written as

$$P(A \cap B) = P(A) \cdot P(B|A)$$

EXAMPLE  **6.21**   A panel of jurors consists of 15 persons who have had no education beyond high school and 9 who have had some college education. If a lawyer randomly chooses two of them to ask some questions, what is the probability that neither of them will have had any college education?

**Solution**   If $A$ is the event that the first person selected has not had any college education, then $P(A) = \frac{15}{24}$. Also, if $B$ is the event that the second person picked has not had any college education, it follows that $P(B|A) = \frac{14}{23}$, since there are only 14 persons without any college education among the 23 who are left after one person without any college education has been picked. Hence, the general multiplication rule yields

$$P(A \cap B) = P(A) \cdot P(B|A) = \frac{15}{24} \cdot \frac{14}{23} = \frac{105}{276}$$

or approximately 0.38.

EXAMPLE  **6.22**   Suppose that the probability is 0.45 that a rare tropical disease is diagnosed correctly and, if diagnosed correctly, the probability is 0.60 that the patient will be cured. What is the probability that a person who has the disease will be diagnosed correctly and cured?

**Solution**   Using the general multiplication rule, we get $(0.45)(0.60) = 0.27$.

When $A$ and $B$ are independent events, we can substitute $P(A)$ for $P(A|B)$ in the first of the two formulas for $P(A \cap B)$, or $P(B)$ for $P(B|A)$ in the second, and we obtain

SPECIAL
MULTIPLICATION
RULE (INDEPENDENT
EVENTS)

*If A and B are independent events, then*

$$P(A \cap B) = P(A) \cdot P(B)$$

In words, the probability that two independent events will both occur is simply the product of their respective probabilities.

As can easily be shown, it is also true that if $P(A \cap B) = P(A) \cdot P(B)$, then $A$ and $B$ are independent events. Dividing by $P(B)$, we get

$$\frac{P(A \cap B)}{P(B)} = P(A)$$

and then replacing $\frac{P(A \cap B)}{P(B)}$ with $P(A|B)$ in accordance with the definition of a conditional probability, we arrive at the result that $P(A|B) = P(A)$, namely, that $A$ and $B$ are independent. Therefore, we can use the special multiplication rule as a *definition* of independence that makes it very easy to check whether two events $A$ and $B$ are independent.

**EXAMPLE 6.23**   Check for each of the following pairs of events whether they are independent:

(a) Events $A$ and $B$ for which $P(A) = 0.40$, $P(B) = 0.90$, and $P(A \cap B) = 0.36$.

(b) Events $C$ and $D$ for which $P(C) = 0.75$, $P(D) = 0.80$, and $P(C \cap D') = 0.15$.

(c) Events $E$ and $F$ for which $P(E) = 0.30$, $P(F) = 0.35$, and $P(E' \cap F') = 0.40$.

**Solution**   (a) Since $(0.40)(0.90) = 0.36$, the two events are independent.

(b) Since $P(D') = 1 - 0.80 = 0.20$ and $(0.75)(0.20) = 0.15$, the events $C$ and $D'$, and hence also the events $C$ and $D$, are independent.

(c) Since $P(E') = 1 - 0.30 = 0.70$, $P(F') = 1 - 0.35 = 0.65$, and $(0.70)(0.65) = 0.455$ and not 0.40, the events $E'$ and $F'$, and hence also the events $E$ and $F$, are not independent. ∎

The special multiplication rule can easily be generalized so that it applies to the occurrence of three or more independent events. Again, we multiply together all the individual probabilities.

**EXAMPLE 6.24**   If the probability is 0.70 that any one person interviewed at a shopping mall will be against an increase in the sales tax to finance a new football stadium, what is the probability that among four persons interviewed at the mall the first three will be against the increase in the sales tax, but the fourth will not be against it?

**Solution**   Assuming independence, we multiply all the probabilities, getting $(0.70)(0.70)(0.70)(0.30) = 0.1029$. ∎

**EXAMPLE 6.25**   With reference to Example 6.21, what is the probability that if three of the members of the panel are randomly picked by the lawyer, none of them will have had a college education?

**Solution**   Since the probability is $\frac{15}{24}$ that the first person picked will not have had a college education, it is $\frac{14}{23}$ that the second person picked will not have had a college education given that the first person picked had no college education, and $\frac{13}{22}$ that the third person picked will not have had a college education given that the first two persons picked had no college education, the probability asked for is $\frac{15}{24} \cdot \frac{14}{23} \cdot \frac{13}{22} = \frac{455}{2,024}$, or approximately 0.225.    ▪

**6.51**   If $A$ is the event that an astronaut is a member of the armed services, $T$ is the event that he was once a test pilot, and $W$ is the event that he is a well-trained scientist, express each of the following probabilities in symbolic form:
   (a)   the probability that an astronaut who was once a test pilot is a member of the armed services;
   (b)   the probability that an astronaut who is a member of the armed services is a well-trained scientist;
   (c)   the probability that an astronaut who is not a well-trained scientist was once a test pilot;
   (d)   the probability that an astronaut who is not a member of the armed services and was never a test pilot is a well-trained scientist.

**6.52**   With reference to Exercise 6.51, express each of the following probabilities in words: $P(A|W)$, $P(A'|T')$, $P(A' \cap W|T)$, and $P(A|W \cap T)$.

**6.53**   A guidance department tests students in various ways. If $I$ is the event that a student scores high in intelligence, $A$ is the event that a student rates high on a social adjustment scale, and $N$ is the event that a student displays neurotic tendencies, express symbolically the probabilities that
   (a)   a student who scores high in intelligence will display neurotic tendencies;
   (b)   a student who does not rate high on the social adjustment scale will not score high in intelligence;
   (c)   a student who displays neurotic tendencies will neither score high in intelligence nor rate high on the social adjustment scale.

**6.54**   Among the 30 applicants for a position at a credit union, some are married and some are not, some have had experience in banking and some have not, with the exact breakdown being

|  | Married | Single |  |
|---|:---:|:---:|:---:|
| Some experience | 6 | 3 | 9 |
| No experience | 12 | 9 | 21 |
|  | 18 | 21 | 30 |

If the branch manager randomly chooses the applicant to be interviewed first, $M$ denotes the event that the first applicant to be interviewed is married, and $E$ denotes the event that the first applicant to be interviewed has had some experience in banking, express in words and also evaluate the following probabilities: $P(M)$, $P(M \cap E)$, and $P(E|M)$.

**6.55**   Use the results obtained in Exercise 6.54 to verify that

$$P(E|M) = \frac{P(M \cap E)}{P(M)}$$

**6.56**   With reference to Exercise 6.54, express in words and evaluate the following probabilities: $P(E')$, $P(M' \cap E')$, and $P(M'|E')$.

**6.57**   Use the results obtained in Exercise 6.56 to verify that

$$P(M'|E') = \frac{P(M' \cap E')}{P(E')}$$

**6.58**   The probability that a bus from Seattle to Vancouver will leave on time is 0.80, and the probability that it will leave on time and also arrive on time is 0.72.

(a)  What is the probability that a bus that leaves on time will also arrive on time?

(b)  If the probability that such a bus will arrive on time is 0.75, what is the probability that a bus that arrives on time also left on time?

**6.59**   A survey of women in executive positions showed that the probability is 0.80 that such a woman will enjoy making financial decisions, and 0.44 that such a woman will enjoy making financial decisions and also be willing to assume substantial risks. What is the probability that a woman in an executive position who enjoys making financial decisions will be willing to assume substantial risks?

**6.60**   The probability that a woman attending a junior college will buy a portable computer is 0.75; if she buys such a computer, the odds are 4 to 1 that her grades will go up. What is the probability that such a student will buy a portable computer and have her grades go up?

**6.61**   Among 40 pieces of luggage loaded on a bus from Heathrow airport to downtown London, 30 are destined for the Dorchester and 10 are destined for the Savoy. If two of them were stolen from the bus when it stopped at a red light, what is the probability that both of them should have gone to the Dorchester?

**6.62**   If two cards are drawn at random from an ordinary deck of 52 playing cards, what are the probabilities that they will both be hearts if

(a)  the first card is replaced before the second card is drawn;

(b)  the first card is not replaced before the second card is drawn?

The distinction between the two parts of this exercise is important in statistics. What we do in part (a) is called **sampling with replacement** and what we do in part (b) is called **sampling without replacement**.

**6.63**   If $P(A) = 0.80$, $P(C) = 0.95$, and $P(A \cap C) = 0.76$, are events $A$ and $C$ independent?

**6.64**   If $P(M) = 0.15$, $P(N) = 0.82$, and $P(M \cap N) = 0.12$, are events $M$ and $N$ independent?

**6.65**   In Example 6.20 we gave $P(H \cap J') = 0.28$, and accordingly we wrote 0.28 in the region corresponding to $H \cap J'$ in the Venn diagram of Figure 6.11. Making use of the fact that we gave $P(H) = 0.70$ and $P(J) = 0.60$ in Example 6.20, fill in the probabilities associated with the other three regions of the Venn diagram of Figure 6.11.

**6.66**   Using the probabilities associated with the four regions of the Venn diagram of Figure 6.11 (see Exercise 6.65), show that

(a)  $P(H|J) = 0.70$, which verifies what we said on page 147 about event $H$ being independent of event $J$.

**Figure 6.11**
Venn diagram for
Exercise 6.65.

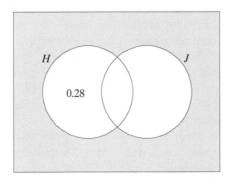

(b)  $P(J) = P(J|H) = P(J|H') = 0.60$, which verifies what we said on page 147 about event $J$ being independent of event $H$.

**6.67**   In a Western community, the probability of passing the road test for a driver's license on the first try is 0.75. After that, the probability of passing becomes 0.60 regardless of how many times a person has failed. What is the probability of getting one's license on the fourth try?

**6.68**   Strange as it may seem, it is possible for an event $A$ to be independent of two events $B$ and $C$ taken individually, but not when they are taken together. Verify that this is the case in the situation pictured in Figure 6.12 by showing that $P(A|B) = P(A)$ and $P(A|C) = P(A)$, but $P(A|B \cup C) \neq P(A)$.

**Figure 6.12**
Venn diagram for
Exercise 6.68.

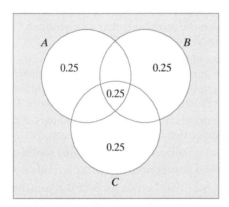

**6.69**   During the month of September, the probability that a rainy day will be followed by another rainy day in a given city is 0.70, and the probability that a sunny day will be followed by a rainy day is 0.40. Assuming that each day is classified as being either rainy or sunny and that the weather on any one day depends only on what happened on the day before, what is the probability that a rainy day will be followed by two more rainy days, then two sunny days, and finally another rainy day?

# *6.7 BAYES' THEOREM

Although $P(A|B)$ and $P(B|A)$ look quite a bit alike, there is a great difference between the probabilities they represent. For instance, on page 146 we calculated the probability $P(G|N)$ that a name-brand tire dealer will provide good service under warranty, but what do we mean when we write $P(N|G)$? This is the probability that a tire dealer who provides good service under warranty is a name-brand dealer. To give another example, suppose that $B$ represents the event that a person committed a burglary and $G$ represents the event that he or she is found guilty of the crime. Then $P(G|B)$ is the probability that the person who committed the burglary will be found guilty of the crime, and $P(B|G)$ is the probability that the person who is found guilty of the burglary actually committed it. Thus, in both of these examples we turned things around—cause, so to speak, became effect and effect became cause.

   Since there are many problems in statistics that involve such pairs of conditional probabilities, let us find a formula that expresses $P(B|A)$ in terms of $P(A|B)$ for any two events $A$ and $B$. To this end we equate the expressions for $P(A \cap B)$ in the two forms of the general multiplication rule on page 148, and we get

$$P(A) \cdot P(B|A) = P(B) \cdot P(A|B)$$

and, hence,

$$P(B|A) = \frac{P(B) \cdot P(A|B)}{P(A)}$$

after we divide by $P(A)$.

**EXAMPLE 6.26**    In a state where cars have to be tested for the emission of pollutants, 25% of all cars emit excessive amounts of pollutants. When tested, 99% of all cars that emit excessive amounts of pollutants will fail, but 17% of the cars that do not emit excessive amounts of pollutants will also fail. What is the probability that a car that fails the test actually emits excessive amounts of pollutants?

**Solution**    Letting $A$ denote the event that a car fails the test and $B$ the event that it emits excessive amounts of pollutants, we first translate the given percentages into probabilities and write $P(B) = 0.25$, $P(A|B) = 0.99$, and $P(A|B') = 0.17$. To calculate $P(B|A)$ by means of the formula

$$P(B|A) = \frac{P(B) \cdot P(A|B)}{P(A)}$$

We shall first have to determine $P(A)$, and to this end let us look at the tree diagram of Figure 6.13. Here $A$ is reached either along the branch that passes through $B$ or along the branch that passes through $B'$, and the probabilities of this happening are, respectively, $(0.25)(0.99) = 0.2475$ and $(0.75)(0.17) = 0.1275$.

Since the two alternatives represented by the two branches are mutually exclusive, we find that $P(A) = 0.2475 + 0.1275 = 0.3750$. Thus, substitution into the formula for $P(B|A)$ yields

$$P(B|A) = \frac{P(B) \cdot P(A|B)}{P(A)} = \frac{0.2475}{0.3750} = 0.66$$

This is the probability that a car that fails the test actually emits excessive amounts of pollutants. ■

With reference to the tree diagram of Figure 6.13, we can say that $P(B|A)$ is the probability that event $A$ was reached via the upper branch of the tree, and we showed that it equals the ratio of the probability associated with that branch

**Figure 6.13**
Tree diagram for emission testing example.

of the tree to the sum of the probabilities associated with both branches. This argument can be generalized to the case where there are more than two possible "causes," namely, more than two branches leading to event A. With reference to Figure 6.14, we can say that $P(B_i|A)$ is the probability that event A was reached via the $i$th branch of the tree (for $i = 1, 2, 3, \ldots,$ or $k$), and it can be shown that it equals the ratio of the probability associated with the $i$th branch to the sum of the probabilities associated with all $k$ branches leading to A. Formally, we write

**BAYES' THEOREM**

If $B_1$, $B_2$, $\ldots$, and $B_k$ are mutually exclusive events of which one must occur, then

$$P(B_i|A) = \frac{P(B_i) \cdot P(A|B_i)}{P(B_1) \cdot P(A|B_1) + P(B_2) \cdot P(A|B_2) + \cdots + P(B_k) \cdot P(A|B_k)}$$

for $i = 1, 2, \ldots,$ or $k$.

Note that the expression in the denominator actually equals $P(A)$. This formula for calculating $P(A)$ when A is reached via one of several intermediate steps is called the **rule of elimination** or the **rule of total probability**.

**Figure 6.14**
Tree diagram for
Bayes' theorem.

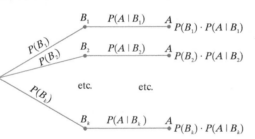

**EXAMPLE 6.27**   In a cannery, assembly lines, I, II, and III account for 50, 30, and 20% of the total output. If 0.4% of the cans from assembly line I are improperly sealed, and the corresponding percentages for assembly lines II and III are 0.6% and 1.2%, what is the probability that an improperly sealed can (discovered at the final inspection of outgoing products) will have come from assembly line I?

**Solution**   Letting A denote the event that a can is improperly sealed, and $B_1$, $B_2$, and $B_3$ denote the events that a can comes from assembly lines I, II, or III, we can translate the given percentages into probabilities and write $P(B_1) = 0.50$, $P(B_2) = 0.30$, $P(B_3) = 0.20$, $P(A|B_1) = 0.004$, $P(A|B_2) = 0.006$, and $P(A|B_3) = 0.012$. Thus, the probabilities associated with the three branches of the tree diagram of Figure 6.15 are $(0.50)(0.004) = 0.0020$, $(0.30)(0.006) = 0.0018$, and $(0.20)(0.012) = 0.0024$, and the rule of elimination yields $P(A) = 0.0020 + 0.0018 + 0.0024 = 0.0062$. Then, substituting this result together with the probability associated with the first branch of the tree diagram, we get

$$P(B_1|A) = \frac{0.0020}{0.0062} = 0.32$$

rounded to two decimal places.

**Figure 6.15**
Tree diagram for
Example 6.27.

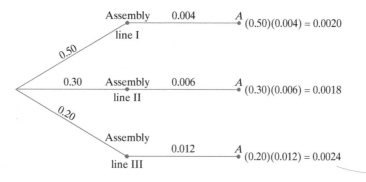

As can be seen from the two examples of this section, Bayes' formula is a relatively simple mathematical rule. There can be no question about its validity, but criticism has frequently been raised about its applicability. This is because it involves a "backward" or "inverse" sort of reasoning, namely, reasoning from effect to cause. For instance, in Example 6.26 we wondered whether a car's failing the test was brought about, or caused, by its emitting excessive amounts of pollutants. Similarly, in the example immediately preceding we wondered whether an improperly sealed can was produced, or caused, by assembly line I. It is precisely this aspect of Bayes' theorem that makes it play an important role in statistical inference, where our reasoning goes from sample data that are observed to the populations from which they came.

EXERCISES

**\*6.70**   The probability is 0.60 that a famous Nigerian distance runner will enter the Boston marathon. If he does not enter, the probability that last year's winner will repeat is 0.66, but if he enters, the probability that last year's winner will repeat is only 0.18. What is the probability that last year's winner will repeat?

**\*6.71**   With reference to Exercise 6.70, suppose that we have been out of touch, but hear on the radio that last year's winner won again. What is the probability that the Nigerian distance runner did not enter the race?

**\*6.72**   In a T-maze, a rat is given food if it turns left and an electric shock if it turns right. On the first trial there is a fifty-fifty chance that a rat will turn either way; then, if it receives food on the first trial, the probability that it will turn left on the second trial is 0.68, and if it receives a shock on the first trial, the probability that it will turn left on the second trial is 0.84. What is the probability that a rat will turn left on the second trial?

**\*6.73**   With reference to Exercise 6.72, what is the probability that a rat that turns left on the second trial will also have turned left on the first trial?

**\*6.74**   At an electronics plant, it is known from past experience that the probability is 0.86 that a new worker who has attended the company's training program will meet his production quota, and that the corresponding probability is 0.35 for a new worker who has not attended the company's training program. If 80% of all new workers attend the training program, what is the probability that
   (a) a new worker will not meet his production quota;
   (b) a new worker who does not meet his production quota will not have attended the company's training program?

**\*6.75**  In Exercise 5.71 we asked the reader to imagine that the probability is 0.95 that a certain test will correctly diagnose a person with diabetes as being diabetic, and that it is 0.05 that the test will incorrectly diagnose a person without diabetes as being diabetic. Given that roughly 10% of the population is diabetic, the reader was then asked to guess at the probability that a person diagnosed as being diabetic actually has diabetes. Now use Bayes' theorem to answer this question correctly.

**\*6.76**  (From Hans Reichenbach's *The Theory of Probability*, University of California Press, 1949.) Mr. Smith's gardener is not dependable; the probability that he will forget to water the rosebush during Smith's absence is $\frac{2}{3}$. The rosebush is in questionable condition anyhow; if watered, the probability of its withering is $\frac{1}{2}$, but if it is not watered, the probability of its withering is $\frac{3}{4}$. Upon returning, Smith finds that the rosebush has withered. What is the probability that the gardener did not water the rosebush?

**\*6.77**  Two firms $V$ and $W$ consider bidding on a road-building job that may or may not be awarded depending on the amount of the bids. Firm $V$ submits a bid and the probability is $\frac{3}{4}$ that it will get the job provided firm $W$ does not bid. The odds are 3 to 1 that $W$ will bid, and if it does, the probability that $V$ will get the job is only $\frac{1}{3}$.

    (a)  What is the probability that $V$ will get the job?

    (b)  If $V$ gets the job, what is the probability that $W$ did not bid?

**\*6.78**  A computer software firm maintains a telephone hotline service for its customers. The firm finds that 48% of the calls involve questions about the application of the software, 38% involve issues of incompatibility with the hardware, and 14% involve the inability to install the software on the user's machine. These three categories of problems can be resolved with probabilities 0.90, 0.15, and 0.80, respectively.

    (a)  Find the probability that a call to the hotline involves a problem that cannot be resolved.

    (b)  If a call involves a problem that cannot be resolved, what is the probability that it concerned incompatibility with the hardware?

**\*6.79**  An explosion in a liquefied natural gas tank undergoing repair could have occurred as the result of static electricity, malfunctioning electrical equipment, an open flame in contact with the liner, or purposeful action (industrial sabotage). Interviews with engineers who were analyzing the risks led to estimates that such an explosion would occur with probability 0.25 as a result of static electricity, 0.20 as a result of malfunctioning electric equipment, 0.40 as a result of an open flame, and 0.75 as a result of purposeful action. These interviews also yielded subjective estimates of the probabilities of the four causes of 0.30, 0.40, 0.15, and 0.15, respectively. Based on all this information, what is the most likely cause of the explosion? (From I. Miller, and J. E. Freund, *Probability and Statistics for Engineers*, 3rd ed. Upper Saddle River, N.J.: Prentice Hall, Inc., 1985.)

**\*6.80**  To get answers to sensitive questions, we sometimes use a method called the **randomized response technique**. Suppose, for instance, that we want to determine what percentage of the students at a large university smoke marijuana. We construct 20 flash cards, write "I smoke marijuana at least once a week" on 12 of the cards, where 12 is an arbitrary choice, and "I do not smoke marijuana at least once a week" on the others. Then we let each student (in the sample interviewed) select one of the cards at random, and respond "yes" or "no" without divulging the question.

    (a)  Establish a relationship between $P(Y)$, the probability that a student will give a "yes" response, and $P(M)$, the probability that a student randomly selected at that university smokes marijuana at least once a week.

(b) If 106 of 250 students answered "yes" under these conditions, use the result of part (a) and $\frac{160}{250}$ as an estimate of $P(Y)$ to estimate $P(M)$.

**\*6.81** "Among three indistinguishable boxes one contains two pennies, one contains a penny and a dime, and one contains two dimes. We randomly choose one of the three boxes, and randomly pick one of the two coins in that box without looking at the other. If the coin we pick is a penny, what is the probability that the other coin in the box is also a penny? Without giving the matter too much thought, we might argue that there is a fifty–fifty chance that the other coin is also a penny. After all, the coin we picked must have come from the box with the penny and the dime or from the box with the two pennies. In the first case the other coin is a dime, in the second case the other coin is a penny, and it would seem reasonable to say that these two possibilities are equally likely."

Use Bayes' theorem to show that the probability is actually $\frac{2}{3}$ that the other coin is also a penny.

# CHECKLIST OF KEY TERMS (with page references to their definitions)

Addition rules, 142
*Bayes' theorem, 154
Betting odds, 136
Complement, 126
Conditional probability, 145
Consistency criterion, 137
Dependent events, 147
Empty set, 125
Event, 125
Experiment, 125
Fair odds, 136
Finite sample space, 125
General addition rule, 142
General multiplication rule, 148
Generalized addition rule, 142
Independent events, 147
Infinite sample space, 125
Intersection, 126

Multiplication rules, 148
Mutually exclusive events, 126
Odds, 135
Outcome, 125
Postulates of probability, 132
*Randomized response technique, 156
*Rule of elimination, 154
*Rule of total probability, 154
Sample space, 125
Sampling with replacement, 151
Sampling without replacement, 151
Special addition rule, 142
Special multiplication rule, 148
Subjective probability, 136
Subset, 125
Theory of probability, 124
Union, 126
Venn diagram, 127

# REFERENCES

*More detailed, though still elementary, treatments of probability may be found in*

BARR, D. R., and ZEHNA, P. W., *Probability: Modeling Uncertainty*. Reading, Mass.: Addison-Wesley Publishing Company, Inc., 1983.

DRAPER, N. R., and LAWRENCE, W. E., *Probability: An Introductory Course*. Chicago: Markham Publishing Co., 1970.

FREUND, J. E., *Introduction to Probability*. New York: Dover Publications, Inc., 1993 Reprint.

GOLDBERG, S., *Probability—An Introduction*. Englewood Cliffs, N.J.: Prentice Hall, 1960.

HODGES, J. L., and LEHMANN, E. L., *Elements of Finite Probability*, 2nd ed. San Francisco: Holden-Day, Inc., 1970.

MOSTELLER, F., ROURKE, R. E. K., and THOMAS, G. B., *Probability with Statistical Applications*, 2nd ed. Reading, Mass.: Addison-Wesley Publishing Company, Inc., 1970.

SCHEAFFER, R. L. and MENDENHALL, W., *Introduction to Probability: Theory and Applications*. North Scituate, Mass.: Duxbury Press, 1975.

*The following is an introduction to the mathematics of gambling and various games of chance:*

PACKEL, E. W., *The Mathematics of Games and Gambling*. Washington, D.C.: Mathematics Association of America, 1981.

# EXPECTATIONS AND DECISIONS

**7.1**   Mathematical Expectation  160

*7.2   Decision Making  164

*7.3   Statistical Decision Problems  168

Checklist of Key Terms  171

References  171

S ection 7.1 is a discussion of mathematical expectation. Section 7.2 deals with the use of mathematical expectation in the decision-making process. Section 7.3 deals with the science of decision making under uncertainty, which is useful in work in management science. These topics are included here to familiarize the reader with the basic role that decision theory plays in the foundations of statistics. A conceptually more informal approach to decision making is adequate for the remainder of this book.

When decisions are made in the face of uncertainties, they are seldom based on probabilities alone. Usually, we must also know something about the potential consequences (profits, losses, penalties, or rewards). If we must decide whether or not to buy a new car, knowing that our old car will soon require repairs is not enough—to make an intelligent decision we must know, among other things, the cost of the repairs and the trade-in value of the old car. Also, suppose that a building contractor has to decide whether to bid on a construction job that promises a profit of $120,000 with probability 0.20 or a loss of $27,000 (perhaps, due to a strike) with probability 0.80. The probability that the contractor will make a profit is not very high, but on the other hand, the amount he stands to gain is much greater than the amount he stands to lose. Both of these examples demonstrate the need for a method of combining probabilities and consequences, and this is why we introduce the concept of a **mathematical expectation** in Section 7.1.

In Chapters 11 through 18 we deal with many different problems of inference. We estimate unknown quantities, test hypotheses (assumptions or claims), and make predictions, and in problems like these it is essential that, directly or indirectly, we pay attention to the consequences of what we do. After all, if there is nothing at stake, no penalties or rewards, and nobody cares, why not estimate the average weight of gray squirrels as 315.2 pounds, why not accept the claim that by adding water to gasoline we can get 150 miles per gallon with the old family car, and why not predict that by the year 2250 the average person will live to be 200 years old? *WHY NOT, if there is nothing at stake, no penalties or rewards, and nobody cares?* So, in Section 7.2 we give some examples that show how mathematical expectations based on penalties, rewards, and other kinds of payoffs can be used in making decisions, and in Section 7.3 we show how such factors may have to be considered in choosing appropriate statistical techniques.

# 7.1 MATHEMATICAL EXPECTATION

If a mortality table tells us that in the United States a 50-year-old woman can expect to live 32 more years, this does not mean that anyone really expects a 50-year-old woman to live until her 82nd birthday and then die the next day. Similarly, if we read that in the United States a person can expect to eat 104.4 pounds of beef and drink 39.6 gallons of soft drinks per year, or that a child in the age group from 6 to 16 can expect to visit a dentist 2.2 times a year, it must be apparent that the word "expect" is not being used in its colloquial sense. A child cannot go to the dentist 2.2 times, and it would be surprising, indeed, if we found somebody who actually ate 104.4 pounds of beef and drank 39.6 gallons of soft drinks in any given year. So far as 50-year-old women are concerned, some will live another 12 years, some will live another 20 years, some will live another 40 years, . . ., and the life expectancy of "32 more years" will have to be interpreted as an average, or as we call it here, a **mathematical expectation**

Originally, the concept of a mathematical expectation arose in connection with games of chance, and in its simplest form it is the product of the amount that a player stands to win and the probability that the player will win.

**EXAMPLE 7.1**   What is the mathematical expectation of a player who stands to win $50 if and only if a balanced coin comes up tails?

**Solution**   If the coin is balanced and randomly tossed—namely, if the probability of tails is $\frac{1}{2}$—the mathematical expectation is $50 \cdot \frac{1}{2} = \$25$.   ◼

**EXAMPLE 7.2**   What is the mathematical expectation of someone who buys one of 2,000 raffle tickets for a trip to Aruba worth $1,960?

**Solution**   Since the probability of winning the trip is $\frac{1}{2,000} = 0.0005$, the mathematical expectation is $1,960(0.0005) = \$0.98$. Thus, financially speaking, it would be foolish to pay more than 98 cents for one of these tickets, unless the proceeds of the raffle went to a worthy cause or unless the person buying the raffle ticket derived some pleasure from placing the bet.   ◼

In both of these examples there is only one prize, yet two possible payoffs (outcomes). In Example 7.1 the prize was $50 and the two outcomes were $50 and nothing at all; in Example 7.2 the prize was the trip worth $1,960 and the two outcomes were the trip worth $1,960 and nothing at all. In Example 7.2 we could also argue that one of the raffle tickets will pay the equivalent of $1,960, each of the other 1,999 tickets will not pay anything at all, so that, altogether, the 2,000 raffle tickets will pay the equivalent of $1,960, or on the average $\frac{1,960}{2,000} = \$0.98$ per ticket. This average is the mathematical expectation. To generalize the concept of a mathematical expectation, consider the following change in the raffle of Example 7.2:

**EXAMPLE  7.3**  What is the mathematical expectation per ticket if the raffle also awards a second prize consisting of dinner for two worth $200 at a famous restaurant and a third prize of two movie tickets worth $16?

**Solution**  Now we can argue that one of the raffle tickets will pay the equivalent of $1,960, another ticket will pay the equivalent of $200, a third ticket will pay the equivalent of $16, and each of the other 1,997 tickets will not pay anything at all. Altogether, the 2,000 tickets will, thus, pay the equivalent of $1,960 + 200 + 16 = \$2,176$, or on the average the equivalent of $\frac{2,176}{2,000} = \$1.088$ per ticket. Looking at the example in a different way, we could argue that if the raffle were repeated many times, a person holding one of the tickets would get nothing $\frac{1,997}{2,000} \cdot 100 = 99.85\%$ of the time (or with probability 0.9985) and each of the three prizes $\frac{1}{2,000} \cdot 100 = 0.05\%$ of the time (or with probability 0.0005). On the average, a person holding one of the tickets would thus win the equivalent of $0(0.9985) + 1,960(0.0005) + 200(0.0005) + 16(0.0005) = \$1.088$, which is the sum of the products obtained by multiplying each payoff by the corresponding probability. ■

Generalizing from this example leads to the following definition:

MATHEMATICAL
EXPECTATION

> *If the probabilities of obtaining the amounts $a_1, a_2, \ldots,$ or $a_k$ are $p_1, p_2, \ldots,$ and $p_k$, where $p_1 + p_2 + \cdots + p_k = 1$, then the mathematical expectation is*
>
> $$E = a_1 p_1 + a_2 p_2 + \cdots + a_k p_k$$

Each amount is multiplied by the corresponding probability, and the mathematical expectation, $E$, is given by the sum of all these products. In the $\sum$ notation, $E = \sum a \cdot p$.

Insofar as the $a$'s are concerned, it is important to keep in mind that they are positive when they represent profits, winnings, or gains (namely, amounts that we receive), and that they are negative when they represent losses, penalties, or deficits (namely, amounts that we have to pay).

**EXAMPLE  7.4**  What is our mathematical expectation if we win $25 when a die comes up 1 or 6 and lose $12.50 when it comes up 2, 3, 4, or 5?

**Solution**  The amounts are $a_1 = 25$ and $a_2 = -12.5$, and the probabilities are $p_1 = \frac{2}{6} = \frac{1}{3}$ and $p_2 = \frac{4}{6} = \frac{2}{3}$ (if the die is balanced and randomly tossed). Thus, our mathematical expectation is

$$E = 25 \cdot \frac{1}{3} + (-12.5) \cdot \frac{2}{3} = 0$$

■

This example illustrates what we mean by an **equitable** or **fair game**. It is a game that does not favor either player; namely, a game in which each player's mathematical expectation is zero.

**EXAMPLE 7.5**   The probabilities are 0.22, 0.36, 0.28, and 0.14 that an investor will be able to sell a piece of land for a mountain cabin at a profit of $2,500, at a profit of $1,500, at a profit of $500, or at a loss of $500. What is the investor's expected profit?

**Solution**   Substituting $a_1 = 2,500$, $a_2 = 1,500$, $a_3 = 500$, $a_4 = -500$, $p_1 = 0.22$, $p_2 = 0.36$, $p_3 = 0.28$, and $p_4 = 0.14$ into the formula for $E$, we get

$$E = 2,500(0.22) + 1,500(0.36) + 500(0.28) - 500(0.14)$$

$$= \$1,160$$

Although we referred to the quantities $a_1, a_2, \ldots,$ and $a_k$ as "amounts," they need not be amounts of money. On page 161 we said that a child in the age group from 6 to 16 can expect to visit a dentist 2.2 times a year. This value is a mathematical expectation, namely, the sum of the products obtained by multiplying $0, 1, 2, 3, 4, \ldots,$ by the corresponding probabilities that a child in that age group will visit a dentist that many times a year.

**EXAMPLE 7.6**   At a certain airport, the probabilities are 0.06, 0.21, 0.24, 0.18, 0.14, 0.10, 0.04, 0.02, and 0.01 that an airline office will receive 0, 1, 2, 3, 4, 5, 6, 7, or 8 complaints per day about luggage handling. How many such complaints can this airline office expect per day?

**Solution**   Substituting into the formula for a mathematical expectation, we get

$$E = 0(0.06) + 1(0.21) + 2(0.24) + 3(0.18) + 4(0.14)$$

$$+ 5(0.10) + 6(0.04) + 7(0.02) + 8(0.01)$$

$$= 2.75$$

In all of the examples of this section we were given the values of $a$ and $p$ (or the values of the $a$'s and $p$'s) and calculated $E$. Now let us consider an example in which we are given values of $a$ and $E$ to arrive at some result about $p$, and also an example in which we are given values of $p$ and $E$ to arrive at some result about $a$.

**EXAMPLE 7.7**   To defend a client in a liability suit resulting from a car accident, a lawyer must decide whether to charge a straight fee of $7,500 or a contingent fee, which she will get only if her client wins. How does she feel about her client's chances if

(a) she prefers the straight fee of $7,500 to a contingent fee of $25,000;

(b) she prefers a contingent fee of $60,000 to the straight fee of $7,500?

**Solution**   (a) If she feels that the probability is $p$ that her client will win, the lawyer associates a mathematical expectation of $25,000p$ with the contingent fee of $25,000. Since she feels that $7,500 is preferable to this expectation, we can write $7{,}500 > 25{,}000p$ and, hence,

$$p < \frac{7{,}500}{25{,}000} = 0.30$$

**(b)** Now the mathematical expectation associated with the contingent fee is $60{,}000p$, and since she feels that this is preferable to \$7,500, we can write $60{,}000p > 7{,}500$ and, hence,

$$p > \frac{7{,}500}{60{,}000} = 0.125$$ ■

Combining the results of parts (a) and (b) of this example we have shown here that $0.125 < p < 0.30$, where $p$ is the lawyer's subjective probability about her client's success. To narrow it down further, we might vary the contingent fee as in Exercises 7.9 and 7.10.

EXAMPLE **7.8**  A friend says that he would "give his right arm" for our two tickets to an NBA playoff game. To put this on a cash basis, we propose that he pay us \$220 (the actual price of the two tickets), but he will get the tickets only if he draws a jack, queen, king, or ace from an ordinary deck of 52 playing cards; otherwise, we keep the tickets and his \$220. What are the two tickets worth to our friend if he feels that this arrangement is fair?

**Solution**  Since there are four jacks, four queens, four kings, and four aces, the probability that our friend will get the two tickets is $\frac{16}{52}$. Hence, the probability that he will not get the tickets is $1 - \frac{16}{52} = \frac{36}{52}$, and the mathematical expectation associated with the gamble is

$$E = a \cdot \frac{16}{52} + 0 \cdot \frac{36}{52} = a \cdot \frac{16}{52}$$

where $a$ is the amount, he feels, the tickets are worth. Putting this mathematical expectation equal to \$220, which he considers a fair price to pay for taking the risk, we get

$$a \cdot \frac{16}{52} = 220 \quad \text{and} \quad a = \frac{52 \cdot 220}{16} = \$715$$

This is what the two tickets are worth to our friend. ■

EXERCISES

**7.1**  A service club has printed and sold 3,000 raffle tickets for a painting worth \$750. What is the mathematical expectation of a person who holds one of these raffle tickets?

**7.2**  With reference to Exercise 7.1, what would have been the mathematical expectation of a person holding one of the tickets, if the service club had sold only 1,875 of the tickets?

**7.3**  As part of a promotional scheme, a soap manufacturer offers a first prize of \$3,000 and a second prize of \$1,000 to persons chosen at random from among 15,000 persons willing to try a new product and send in their name and address on the label. What is the mathematical expectation of a person taking part in this promotion?

**7.4**   A jeweler wants to unload 45 men's watches that cost her $12 each. She wraps these 45 watches and also five men's watches that cost her $600 each in identically shaped unmarked boxes and lets each customer take his or her pick.

(a)   Find each customer's mathematical expectation.

(b)   What is the jeweler's expected profit per customer if she charges $100 for the privilege of taking a pick?

**7.5**   At the end of a golf tournament paying the winner $300,000 and the runner-up $120,000, two golfers are tied for first place. In the play-off, what are the two golfers' mathematical expectations if

(a)   they are evenly matched;

(b)   the younger one is favored by odds of 3 to 2?

**7.6**   If the two teams in a "best of seven" play-off are evenly matched, the probabilities that the series will last 4, 5, 6, or 7 games are, respectively, 1/8, 1/4, 5/16, and 5/16. Under these conditions, how many games can such a series be expected to last?

**7.7**   An importer pays $12,000 for a shipment of bananas, and the probabilities that he will be able to sell them for $16,000, $13,000, $12,000, or only $10,000 are, respectively, 0.25, 0.46, 0.19, and 0.10. What is his expected gross profit?

**7.8**   A security service knows from experience that the probabilities of 2, 3, 4, 5, or 6 false alarms on any given evening are 0.12, 0.26, 0.37, 0.18, and 0.07. How many false alarms can they expect on any given evening?

**7.9**   With reference to Example 7.7, suppose that the lawyer prefers a straight fee of $7,500 to a contingent fee of $30,000. How does she feel about the chances that her client will win?

**7.10**   With reference to Example 7.7, suppose that the lawyer prefers a contingent fee of $37,500 to a straight fee of $7,500. How does she feel about the chances that her client will win?

**7.11**   One contractor offers to do a road repair job for $45,000, while another contractor offers to do the job for $50,000 with a penalty of $12,500 if the job is not finished on time. If the person who lets out the contract for the job prefers the second offer, what does this tell us about her assessment of the probability that the second contractor will not finish the job on time?

**7.12**   Mr. Smith feels that it is just about a toss-up whether to accept a cash prize of $26 or to gamble on two flips of a coin, where he is to receive an electric drill if the coin comes up heads both times, while otherwise he is to receive $5. What cash value does he attach to owning the drill?

**7.13**   Mr. Jones would like to beat Mr. Brown in an upcoming golf tournament, but his chances are nil unless he takes $400 worth of lessons, which (according to the pro at his club) will give him a fifty–fifty chance. If Mr. Jones can expect to break even if he takes the lessons and bets Mr. Brown $1,000 against $x$ dollars that he will win, find $x$.

# *7.2 DECISION MAKING

In the face of uncertainty, mathematical expectations can often be used to great advantage in making decisions. In general, if we must choose between two or more alternatives, it is considered rational to select the one with the "most promising" mathematical expectation—the one that maximizes expected profits, minimizes expected costs, maximizes expected tax advantages, minimizes expected losses, and so on. This approach to decision making has intuitive appeal, but it is not without complications. In many problems it is difficult, if not impossible, to assign numerical values to all of the $a$'s (amounts) and $p$'s (probabilities) in the formula for $E$. Some of these will be illustrated in the examples that follow.

**EXAMPLE 7.9**    The research division of a pharmaceutical company has already spent $400,000 to determine whether a new medication for the prevention of seasickness is effective. Now the director of the division must decide whether to spend an additional $200,000 to complete the tests, knowing that the probability of success is only $\frac{1}{3}$. He also knows that if the tests are continued and the medication proves to be effective, this would result in a profit of $1,500,000 to his company. Of course, if the tests are continued and the medication turns out to be ineffective, this would mean $600,000 "down the drain." He also knows that if the tests are not continued and the medication is successfully produced by a competitor, this would entail an additional loss of $100,000 due to his company's being put at a competitive disadvantage. What should he decide to do in order to maximize his company's expected profit?

**Solution**    In a problem like this it usually helps to present the given information in a table such as the following, called a **payoff table**:

|  | Continue tests | Discontinue tests |
|---|---|---|
| Medication is effective | 1,500,000 | −500,000 |
| Medication is not effective | −600,000 | −400,000 |

Using all this information and the $\frac{1}{3}$ and $\frac{2}{3}$ probabilities for the effectiveness and ineffectiveness of the medication, the director of the research division can argue that the expected profit, if the tests are continued, is

$$1,500,000 \cdot \frac{1}{3} + (-600,000) \cdot \frac{2}{3} = \$100,000$$

If the tests are discontinued, there will be a loss of

$$(-500,000) \cdot \frac{1}{3} + (-400,000)\frac{2}{3} \approx -\$433,333.$$

Since an expected profit of $100,000 is preferable to an expected loss of $433,333, the director's decision is to continue the tests.    ■

The way in which we have studied this problem is called a **Bayesian analysis**. In this kind of analysis, probabilities are assigned to the alternatives about which uncertainties exist (the **states of nature**, which in our example were the effectiveness and the ineffectiveness of the medication); then we choose whichever alternative promises the greatest expected profit or the smallest expected loss. As we have said, this approach to decision making is not without complications. If mathematical expectations are to be used for making decisions, it is essential that our appraisals of all relevant probabilities and payoffs are fairly close.

EXAMPLE **7.10**   With reference to Example 7.9, suppose that the director of the research division has an assistant who strongly feels that he greatly overestimated the probability of success; namely, that the probability of the medication's effectiveness should be $\frac{1}{15}$ instead of $\frac{1}{3}$. How does this change in the probability of success affect the result?

**Solution**   With this change, the expected profit becomes

$$1,500,000 \cdot \frac{1}{15} + (-600,000) \cdot \frac{14}{15} = -\$460,000$$

if the tests are continued, and

$$(-500,000) \cdot \frac{1}{15} + (-400,000) \cdot \frac{14}{15} \approx -\$406,667$$

if the tests are discontinued. Neither of these alternatives looks very promising, but since an expected loss of \$406,667 is preferable to an expected loss of \$460,000, the decision reached in Example 7.9 should be reversed.   ◼

EXAMPLE **7.11**   Now suppose that the same assistant tells the director of the research division that the anticipated profit of \$1,500,000 was also wrong; he feels that it should have been \$2,300,000. How will this change affect the result?

**Solution**   With this change and the $\frac{1}{15}$ probability of success as in Example 7.10, the expected profit becomes

$$2,300,000 \cdot \frac{1}{15} + (-600,000) \cdot \frac{14}{15} = -\$406,667$$

if the tests are continued, and it is also $-\$406,667$ if the tests are discontinued, exactly as in Example 7.10. Again, the result has changed; now it seems that the decision might be left to the toss of a coin.   ◼

**EXERCISES**

*7.14   A grab-bag contains 5 packages worth \$1 apiece, 5 packages worth \$3 apiece, and 10 packages worth \$5 apiece. Is it rational to pay \$4 for the privilege of selecting one of these packages at random?

*7.15   A contractor must choose between two jobs. The first promises a profit of \$120,000 with a probability of $\frac{3}{4}$ or a loss of \$30,000 (due to strikes and other delays) with a probability of $\frac{1}{4}$; the second job promises a profit of \$180,000 with a probability of $\frac{1}{2}$ or a loss of \$45,000 with a probability of $\frac{1}{2}$. Which job should the contractor choose so as to maximize his expected profit?

*7.16   A landscape architect must decide whether to bid on the landscaping of a public building. What should she do if she figures that the job promises a profit of \$10,800 with probability of 0.40 or a loss of \$7,000 (due to a lack of rain or perhaps an early frost) with probability 0.60, and it is not worth her time unless the expected profit is at least $1,000?

**\*7.17** A truck driver has to deliver a load of building materials to one of two construction sites, a barn that is 18 miles from the lumberyard or a shopping center that is 22 miles from the lumberyard. He has misplaced the order indicating where the load should go. Also, he must return to the lumberyard after the delivery. The barn and shopping center are 8 miles apart. To complicate matters, the telephone at the lumberyard is not working. If the driver feels that the probability is $\frac{1}{6}$ that the load should go to the barn and $\frac{5}{6}$ that the load should go to the shopping center, where should he go first so as to minimize the distance that he will have to drive?

**\*7.18** With reference to Exercise 7.17, where should the driver go first so as to minimize the expected driving distance if the probabilities are $\frac{1}{3}$ and $\frac{2}{3}$ instead of $\frac{1}{6}$ and $\frac{5}{6}$?

**\*7.19** With reference to Exercise 7.17, where should the driver go first so as to minimize the expected driving distance if the probabilities are $\frac{1}{4}$ and $\frac{3}{4}$ instead of $\frac{1}{6}$ and $\frac{5}{6}$?

**\*7.20** The management of a mining company must decide whether to continue an operation at a certain location. If they continue and are successful, they will make a profit of $4,500,000; if they continue and are not successful, they will lose $2,700,000; if they do not continue but would have been successful if they had continued, they will lose $1,800,000 (for competitive reasons); and if they do not continue and would not have been successful if they had continued, they will make a profit of $450,000 (because funds allocated to the operation remain unspent). What decision would maximize the company's expected profit if it is felt that there is a fifty–fifty chance for success?

**\*7.21** With reference to Exercise 7.20, show that it does not matter what they decide to do if it is felt that the probabilities for and against success are $\frac{1}{3}$ and $\frac{2}{3}$.

**\*7.22** A group of investors must decide whether to arrange for the financing to build a new arena or to continue holding its sports promotions at the gymnasium of a community college. They figure that if the new arena is built and they can get a professional basketball franchise, there will be a profit of $2,050,000 over the next five years; if the new arena is built and they cannot get a professional basketball franchise, there will be a deficit of $500,000; if the new arena is not built and they get a professional basketball franchise, there will be a profit of $1,000,000; and if the new arena is not built and they cannot get a professional basketball franchise, there will be a profit of only $100,000 from their other promotions.

(a) Present all this information in a table like that on page 165.

(b) If the investors believe an official of the professional basketball league who tells them that the odds are 2 to 1 against their getting the franchise, what should they decide to do so as to maximize the expected profit over the next five years?

(c) If the investors believe the sports editor of a local newspaper who tells them that the odds are really only 3 to 2 against their getting the franchise, what should they decide to do so as to maximize the expected profit over the next five years?

**\*7.23** In the absence of any information about relevant probabilities, a pessimist may well try to minimize the maximum loss or maximize the minimum profit, that is, use the **minimax** or **maximin criterion**.

(a) With reference to Example 7.9, suppose that the director of the research division of the pharmaceutical company has no idea about the probability that the medication will be effective. What decision would minimize the maximum loss?

(b) With reference to Exercise 7.17, suppose that the truck driver has no idea about the chances that the building materials should go to either of the two construction sites. Where should he go first so as to minimize the maximum distance he has to drive?

**\*7.24**  In the absence of any information about relevant probabilities, an optimist may well try to minimize the minimum loss or maximize the maximum profit, that is, use the **minimin** or **maximax criterion**.

(a) With reference to Example 7.10, suppose that the assistant of the director of research also has no idea about the probability that the medication will be effective. What should he recommend so as to maximize the maximum profit?

(b) With reference to Exercise 7.17, suppose that the truck driver has no idea about the probabilities that the building material should go to one of the construction sites or the other. Where should he go first so as to minimize the minimum distance he has to drive?

**\*7.25**  With reference to Exercise 7.20, suppose that there is no information about the potential success of the mining operation. What should the CEO recommend to the board of directors, if he is always

(a) very optimistic;

(b) very pessimistic?

**\*7.26**  With reference to Exercise 7.22, suppose that there is no information about the group's chances of getting the franchise. Would one of the investors vote for or against building the new arena if he is

(a) a confirmed optimist;

(b) a confirmed pessimist?

# **\*7.3** STATISTICAL DECISION PROBLEMS

Modern statistics, with its emphasis on inference, may be looked upon as the art, or science, of decision making under uncertainty. This approach to statistics, called **decision theory**, dates back only to the middle of the twentieth century and the publication of John von Neumann and Oscar Morgenstern's *Theory of Games and Economic Behavior* in 1944 and Abraham Wald's *Statistical Decision Functions* in 1950. Since the study of decision theory is quite complicated mathematically, we limit our discussion here to an example in which the method of Section 7.2 is applied to a problem that is of a statistical nature.

**EXAMPLE 7.12**  On the five teams appointed by the government to study gender discrimination in business, 1, 2, 5, 1, and 6 of the members are women. The teams are randomly assigned to various cities, and the mayor of one city hires a consultant to predict how many of the members on the team sent to her city will be women. If the consultant is paid $300 plus a bonus of $600, which he will receive only if his prediction is correct (that is, exactly on target), what prediction maximizes the amount of money he can expect to get?

**Solution**  If the consultant's prediction is 1, which is the mode of the five numbers, he will make only $300 with probability $\frac{3}{5}$ and $900 with probability $\frac{2}{5}$. So he can expect to make

$$300 \cdot \frac{3}{5} + 900 \cdot \frac{2}{5} = \$540$$

As can easily be verified, that is the best he can do. If his prediction is 2, 5, or 6, he can expect to make

$$300 \cdot \frac{4}{5} + 900 \cdot \frac{1}{5} = \$420$$

and for any other prediction his expectation is only $300.

This example illustrates the (perhaps obvious) fact that if one has to pick the exact value on the nose and there is no reward for being close, the best prediction is the mode. To illustrate further how the consequences of one's decisions may dictate the choice of a statistical method of decision or prediction, let us consider the following variation of Example 7.12.

**EXAMPLE 7.13** Suppose that the consultant is paid $600 minus an amount of money equal in dollars to 40 times the magnitude of the error. What prediction will maximize the amount of money he can expect to get?

**Solution** Now it is the median that yields the best predictions. If the consultant's prediction is 2—that is, the median of 1, 1, 2, 5, and 6—the magnitude of the error will be 1, 0, 3, or 4, depending on whether 1, 2, 5, or 6 of the members of the team sent to the city will be women. Consequently, he will get $560, $600, $480, or $440 with probabilities $\frac{2}{5}$, $\frac{1}{5}$, $\frac{1}{5}$, and $\frac{1}{5}$, and he can expect to make

$$560 \cdot \frac{2}{5} + 600 \cdot \frac{1}{5} + 480 \cdot \frac{1}{5} + 440 \cdot \frac{1}{5} = \$528$$

It can be shown that the consultant's expectation is less than $528 for any value other than 2, but we shall verify this only for the mean of the five numbers, which is 3. In that case, the magnitude of the error will be 2, 1, 2, or 3, depending on whether 1, 2, 5, or 6 of the members of the team sent to her city will be women. Thus, the consultant will get $520, $560, $520, or $480 with probabilities $\frac{2}{5}$, $\frac{1}{5}$, $\frac{1}{5}$, and $\frac{1}{5}$, and he can expect to make

$$520 \cdot \frac{2}{5} + 560 \cdot \frac{1}{5} + 520 \cdot \frac{1}{5} + 480 \cdot \frac{1}{5} = \$520$$

The mean comes into its own right when the penalty, the amount subtracted, increases more rapidly with the size of the error; namely, when it is proportional to its square.

**EXAMPLE 7.14** Suppose that the consultant is paid $600 minus an amount of money equal in dollars to 20 times the square of the error. What prediction will maximize the amount of money that he can expect to get?

**Solution** If the consultant's prediction is $\frac{1+2+1+5+6}{5} = 3$, the squares of the errors will be 4, 1, 4, or 9 depending on whether 1, 2, 5, or 6 of the members of the team sent to the city will be women. Correspondingly, the consultant will get $520, $580, $520, or $420 with probabilities $\frac{2}{5}$, $\frac{1}{5}$, $\frac{1}{5}$, and $\frac{1}{5}$, and he can expect to make

$$520 \cdot \frac{2}{5} + 580 \cdot \frac{1}{5} + 520 \cdot \frac{1}{5} + 420 \cdot \frac{1}{5} = \$512$$

As can be verified, the consultant's expectation is less than $512 for any other prediction (see Exercise 7.27). ◼

This third case is of special importance in statistics, as it ties in closely with the *method of least squares*. We shall study this method in Chapter 16, where it is used in fitting curves to observed data, but besides this it has other important applications in the theory of statistics. The idea of working with the squares of the errors is justified on the grounds that in actual practice the seriousness of an error often increases rapidly with the size of the error, more rapidly than the magnitude of the error itself.

The greatest difficulty in applying the methods of this chapter to realistic problems in statistics is that we seldom know the exact values of all the risks that are involved; that is, we seldom know the exact values of the "payoffs" corresponding to the various eventualities. For instance, if the Food and Drug Administration must decide whether or not to release a new drug for general use, how can it put a cash value on the damage that might be done by not waiting for a more thorough analysis of possible side effects or on the lives that might be lost by not making the drug available to the public right away? Similarly, if a faculty committee must decide which of several applicants should be admitted to a medical school or, perhaps, receive a scholarship, how can they possibly foresee all the consequences that might be involved?

The fact that we seldom have adequate information about relevant probabilities also provides obstacles to finding suitable decision criteria; without them, is it reasonable to base decisions, say, on optimism or pessimism, as we did in Exercises 7.23 and 7.24? Questions like these are difficult to answer, but their analysis serves the important purpose of revealing the logic that underlies statistical thinking.

## EXERCISES

**\*7.27** With reference to Example 7.14, where the consultant is paid $600 minus an amount of money equal in dollars to 20 times the square of the error, what can the consultant expect to make if
(a) his prediction is 1, the mode;
(b) his prediction is 2, the median?
(In general, it can be shown that for any set of numbers $x_1, x_2, \ldots,$ and $x_n$, the quantity $\sum(x - k)^2$ is smallest when $k = \bar{x}$. In this case, the amount subtracted from $600 is smallest when the prediction is $\bar{x} = 3$.)

**\*7.28** The ages of the seven entries in an essay contest are 17, 17, 17, 18, 20, 21, and 23, and their chances of winning are all equal. If we want to predict the age of the winner and there is a reward for being right, but none for being close, what prediction maximizes the expected reward?

**\*7.29** With reference to Exercise 7.28, what prediction maximizes the expected reward if
(a) there is a penalty proportional to the size of the error;
(b) there is a penalty proportional to the square of the error?

**\*7.30** Some of the used cars on a lot are priced at $1,895, some are priced at $2,395, some are priced at $2,795, and some are priced at $3,495. If we want to predict the price of the car that will be sold first, what prediction would minimize the maximum size of the error? What is the name of this statistic, which is mentioned in one of the exercises in Chapter 3?

# CHECKLIST OF KEY TERMS (with page references to their definitions)

*Bayesian analysis, 165
*Decision theory, 168
*Equitable game, 162
*Fair game, 162
  Mathematical expectation, 159, 160
*Maximax criterion, 168

*Maximin criterion, 167
*Minimax criterion, 167
*Minimin criterion, 168
*Payoff table, 165
*States of nature, 165

# REFERENCES

*More detailed treatments of the subject matter of this chapter may be found in*

BROSS, I. D. J., *Design for Decision*. New York: Macmillan Publishing Co., Inc., 1953.

JEFFREY, R. C., *The Logic of Decision*. New York: McGraw-Hill Book Company, 1965.

*and in some textbooks on business statistics. Some fairly elementary material on decision theory can be found in*

CHERNOFF, H., and MOSES, L. E., *Elementary Decision Theory*. New York: Dover Publications, Inc., 1987 reprint.

*"The Dowry Problem" and "A Tie Is Like Kissing Your Sister" are two amusing examples of decision making given in*

HOLLANDER, M., and PROSCHAN, F., *The Statistical Exorcist: Dispelling Statistics Anxiety*. New York: Marcel Dekker, Inc., 1984.

**R.48** With reference to Figure R.1, express symbolically the events that are represented by regions 1, 2, 3, and 4 of the Venn diagram.

**R.49** With reference to Figure R.1, assign the probabilities 0.48, 0.12, 0.32, and 0.08 to the events represented by regions 1, 2, 3, and 4 of the Venn diagram, and find $P(C)$, $P(D')$, and $P(C \cap D')$.

**R.50** Use the results of Exercise R.49 to check whether events $C$ and $D'$ are independent.

**R.51** The probabilities that a newspaper will receive $0, 1, 2, \ldots, 7$, or at least 8 letters to the editor about an unpopular decision of the school board are 0.01, 0.02, 0.05, 0.14, 0.16, 0.20, 0.18, 0.15, and 0.09, respectively. What are the probabilities that the newspaper will receive

(a) at most 4 letters to the editor about the school board decision;

(b) at least 6;

(c) from 3 to 5?

**R.52** Determine whether each of the following is true or false:

(a) $\frac{1}{4!} + \frac{1}{6!} = \frac{31}{6!}$;

(b) $\frac{20!}{17!} = 20 \cdot 19 \cdot 18 \cdot 17$;

(c) $5! + 6! = 7 \cdot 5!$;

(d) $3! + 2! + 1! = 6$.

**R.53** A small real estate office has five part-time salespersons. Using two coordinates, so that (3, 1), for example, represents the event that three of the salespersons are at work and one of them is busy with a customer and (2, 0) represents the event that two of the salespersons are at work but neither of them is busy with a customer, draw a diagram similar to that of Figure 6.1, showing the 21 points of the corresponding sample space.

**R.54** With reference to Exercise R.53, assume that each of the 21 points of the sample space has the probability $\frac{1}{21}$, and find the probabilities that

(a) at least three salespersons are at work;

(b) at least three salespersons are busy with a customer;

(c) none of the salespersons is busy with a customer;

(d) only one person at work is not busy with a customer.

**R.55** One substitute quarterback is willing to give odds of 3 to 1, but not odds of 4 to 1, that he will be able to beat another substitute quarterback in the 40-yard dash. What does this tell us about the probability he assigns to his being faster than the other substitute quarterback?

**Figure R.1**
Venn diagram for Exercises R.48, R.49, and R.50.

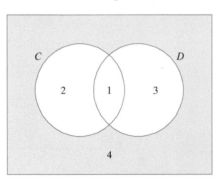

**\*R.56**  The manufacturer of a new battery additive has to decide whether to sell his product for $1.00 a can or for $1.25 with a "double-your-money-back-if-not-satisfied guarantee." How does he feel about the chances that a person will actually ask for double his or her money back if

(a)  he decides to sell the product for $1.00;

(b)  he decides to sell the product for $1.25 with the guarantee;

(c)  he cannot make up his mind?

**R.57**  In how many different ways can a person buy half a pound each of four of the 15 kinds of coffee carried by a gourmet food shop?

**R.58**  Suppose that someone flips a coin 100 times and gets 34 heads, which is far short of the number of heads he might expect. Then he flips the coin another 100 times and gets 46 heads, which is again short of the number of heads he might expect. Comment on his claim that the law of large numbers is letting him down.

**R.59**  If 1,134 of the 1,800 students attending a small college are residents of the community in which the college is located, estimate the probability that any one student attending the college, chosen at random, will be a resident of the community in which the college is located.

**R.60**  On a low-cholesterol diet, a person is allowed to eat four eggs in three weeks, with no more than two eggs in any one week. Draw a tree diagram to show the various ways in which he or she can plan to distribute the four eggs among three weeks.

**\*R.61**  The mortgage manager for a bank figures that if an applicant for a $150,000 home mortgage is a good risk and the bank accepts him, the bank's profits will be $8,000. If the applicant is a bad risk and the bank accepts him, the bank will lose $20,000. If the manager turns down the applicant, there will be no profit or loss either way. What should the manager do if

(a)  he wishes to maximize the expected profit and feels that the probability is 0.10 that the applicant is a bad risk;

(b)  he wishes to maximize the expected profit and feels that the probability is 0.30 that the applicant is a bad risk;

(c)  he wishes to minimize the maximum loss and has no idea about the probability that the applicant is a bad risk?

**R.62**  As part of a promotional scheme in Arizona and New Mexico, a company distributing frozen foods will award a grand prize of $100,000 to some person sending in his or her name on an entry blank, with the option of including a label from one of the company's products. A breakdown of the 225,000 entries received is shown in the following table:

|  | With label | Without label |
|---|---|---|
| Arizona | 120,000 | 42,000 |
| New Mexico | 30,000 | 33,000 |

If the winner of the grand prize is chosen by lot, but the drawing is rigged so that by including a label the probability of winning the grand prize is tripled, what are the probabilities that the grand prize will be won by someone who

(a)  included a label;

(b)  is from New Mexico?

**R.63**   A test consists of eight true–false questions and four multiple choice questions, with each having four different answers. In how many different ways can a student check off one answer to each question?

**R.64**   Among 1,200 women interviewed, 972 said that they prefer being called "Mrs. or Miss" rather than "Ms." Estimate the probability that a woman prefers being called "Mrs. or Miss" rather than "Ms."

**R.65**   A donut shop has 12 jelly donuts on hand and 15 donuts with chocolate icing. If the first customer buys one donut and the second customer buys one of each kind, how many choices does the second customer have if

(a)   the first customer buys a jelly donut;

(b)   the first customer buys a donut with chocolate icing?

**R.66**   A hotel gets cars for its guests from three rental agencies, 20% from agency $X$, 40% from agency $Y$, and 40% from agency $Z$. If 14% of the cars from $X$, 4% from $Y$, and 8% from $Z$ need tune-ups, what is the probability that

(a)   a car needing a tune-up will be delivered to one of the guests;

(b)   if a car needing a tune-up is delivered to one of the guests, it came from agency $X$?

**R.67**   If the probability is 0.24 that any one sportswriter will give the University of Nebraska's football team a preseason ranking of No. 1, what is the probability that three sportswriters, chosen at random, will all rank the University of Nebraska football team as No. 1?

**R.68**   If $Q$ is the event that a person is qualified for a civil service job and $A$ is the event that he will get appointed to the job, express in words what probabilities are represented by $P(Q')$, $P(A|Q)$, $P(A'|Q')$, and $P(Q'|A)$.

**R.69**   Convert each of the following odds to probabilities.

(a)   The odds are 21 to 3 that a certain driver will not win the Indianapolis 500.

(b)   If four cards are drawn with replacement from an ordinary deck of 52 playing cards, the odds are 11 to 5 that at most two of them will be black.

**R.70**   The probabilities that a towing service will receive 0, 1, 2, 3, 4, 5, or 6 calls for help during the evening rush hour are, respectively 0.05, 0.12, 0.31, 0.34, 0.12, 0.05, and 0.01. How many calls for help can the towing service expect during the evening rush hour?

**R.71**   If $P(A) = 0.37$, $P(B) = 0.25$, and $P(A \cup B) = 0.62$, are events $A$ and $B$

(a)   mutually exclusive;

(b)   independent?

**R.72**   Explain why there must be a mistake in each of the following:

(a)   $P(A) = 0.53$ and $P(A \cap B) = 0.59$.

(b)   $P(C) = 0.83$ and $P(C') = 0.27$.

(c)   For the independent events $E$ and $F$, $P(E) = 0.60$, $P(F) = 0.15$, and $P(E \cap F) = 0.075$.

**R.73**   Mrs. Jones feels that it is a toss-up whether to accept $30 in cash or to gamble on drawing a bead from an urn containing 15 red beads and 45 blue beads, with the provision that she is to receive $3 if she draws a red bead or a bottle of fancy perfume if she draws a blue bead. What value, or utility, does she assign to the bottle of perfume?

**R.74**   An artist feels that she can get $5,000 for one of her paintings on display at an art show if it wins a prize, but only $2,000 if it does not win a prize. How does she rate her chances of its winning a prize if she decides to sell the painting for $3,000 before the prize winners are announced?

**R.75** The kinds of cars considered for the fleet of a cab company average 32, 30, 30, 33, and 30 miles per gallon. Assuming that each of the five kinds of cars has an equal chance of being selected, the cab company's accountant wants to prepare the company's budget. What figure would he use for the cars' mileage per gallon if

(a) it is much more important to him to be right rather than to be close;

(b) he wants to minimize the square of his error?

**R.76** The operating room of a hospital has six humidity settings and eight temperature settings. In how many different ways can these two variables be set?

**R.77** How many different permutations are there of the letters in the word

(a) house;

(b) murder;

(c) runner;

(d) paddled?

**R.78** A racing car driver feels that the odds are 5 to 1 that he will not win a NASCAR race, 8 to 1 that he will not come in second, and 2 to 1 that he will come in neither first nor second. Are the corresponding probabilities consistent?

**R.79** The faculty of an industrial design department of an engineering school includes three with Ph.D.'s, two with M.A.'s, and five with M.F.A.'s. If three of them are randomly selected to serve on a curriculum committee, find the probabilities that the committee will include

(a) one with each kind of degree;

(b) only those with M.F.A.'s;

(c) one with a Ph.D. and two with M.F.A.'s.

**R.80** In Figure R.2, event $B$ is the event that a person traveling in the South Pacific will visit Bora Bora, $M$ is the event that the person will visit Moorea, and $T$ is the event that the person will visit Tahiti. Explain in words what events are represented by region 4, regions 1 and 3 together, regions 3 and 6 together, and regions 2, 5, 7, and 8 together.

**Figure R.2**
Venn diagram for
Exercise R.80.

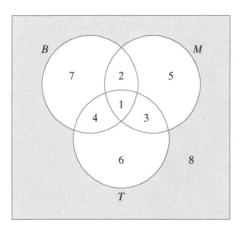

**R.81** Lie detectors have been used during wartime to uncover security risks. As is well known, lie detectors are not infallible. Let us suppose that the probability is 0.10 that the lie detector will fail to detect a person who is a security risk and that the probability is 0.08 that the lie detector will incorrectly label a person who is not a security risk. If 2% of the persons who are given the test are actually security risks, what is the probability that

(a) a person labeled a security risk by a lie detector is in fact a security risk;

(b) a person cleared by a lie detector is in fact not a security risk?

**R.82** One of two partners in a salvage operation feels that the odds are at most 3 to 1 that they have located the right shipwreck, and the other partner feels that the odds are at least 13 to 7. Find betting odds that would be agreeable to both partners. (Note that the answer is not unique.)

**R.83** Sometimes we may prefer an option that has an inferior mathematical expectation. Suppose that you have the choice of investing $5,000 in a federally insured certificate of deposit paying 4.5% or in a mining stock that pays no dividend but has averaged a growth rate of 6.2%. Why may a person conceivably prefer the certificate of deposit?

**R.84** The probabilities that a research worker's success will lead to a raise in salary, a promotion, or both, are, respectively, 0.33, 0.40, and 0.25. What is the probability that it will lead to either or both kinds of rewards?

**R.85** In Figure R.3, each outcome in event $A$ is twice as likely as each outcome in event $A'$. What is the probability of event $A$?

**R.86** Draw a tree diagram to determine the number of ways in which we can get a total of 6 in three rolls of a die.

**R.87** Suppose that we answer a true–false test consisting of 18 questions by tossing a coin. What is the probability of getting ten right and eight wrong?

**R.88** Four persons are getting ready for a game of bridge. In how many different ways can they choose partners?

**R.89** In how many different ways can the members of a family put their four cars in

(a) a four-car garage;

(b) in four of five parking spaces?

**R.90** Two friends are betting on repeated flips of a balanced coin. One has $7 at the start and the other has $3, and after each flip the loser pays the winner $1. If $p$ is the probability that the one who starts with $7 will win his friend's $3 before he loses his own $7, explain why $3p - 7(1 - p)$ should equal 0, and then solve the equation

$$3p - 7(1 - p) = 0$$

for $p$. Generalize this result to the case where the two players start with $a$ dollars and $b$ dollars, respectively.

**Figure R.3**
Venn diagram for
Exercise R.85.

# 8 PROBABILITY DISTRIBUTIONS†

**8.1**   Random Variables 178

**8.2**   Probability Distributions 179

**8.3**   The Binomial Distribution 181

**8.4**   The Hypergeometric Distribution 189

**8.5**   The Poisson Distribution 193

*__8.6__   The Multinomial Distribution 197

**8.7**   The Mean of a Probability Distribution 198

**8.8**   The Standard Deviation of a Probability Distribution 201

Checklist of Key Terms 205

References 205

ncertainties face us no matter where we turn. An investor can never be certain about the next day's price of a share of General Motors stock, nor is the traveler sure about the exact arrival time of a flight from Dallas to New York. The president of a technology manufacturing company cannot be sure about the potential demand for a new software package, nor can an automotive engineer be sure of the useful life of a car battery. We cannot tell exactly how many leaves will be left on a tree after the first frost, nor how many misprints there will be in a company's annual report. In each case we are concerned with a number about which we cannot be sure, namely, the value of a **random variable**.

Since random variables are neither random nor variables, why do they go by this name? This is hard to say, but a mathematics professor with a good sense of humor likened them to alligator pears, or avocados, which are neither alligators nor pears.

In the study of random variables we are usually interested in the probabilities with which they take on the various values within their range,

---

† Since there are quite a few computer printouts and reproductions from the display screen of a graphing calculator in this chapter and in subsequent chapters, let us repeat from the Preface and the footnote to page 12 that the purpose of the printouts and the graphing calculator reproductions is to make the reader aware of the existence of these technologies for work in statistics. Let us make it clear, however, that neither computers nor graphing calculators are required for the use of our text. Indeed, the book can be used effectively by readers who do not possess or have easy access to computers and statistical software, or to graphing calculators. Some of the exercises are labeled with special icons for the use of a computer or a graphing calculator, but these exercises are optional.

namely, in their **probability distributions**. The general introduction of random variables and probability distributions in Sections 8.1 and 8.2 will be followed by the discussion of some of the most important probability distributions in Sections 8.3 through 8.6. Then, we discuss some ways of describing the most relevant features of probability distributions in Sections 8.7 and 8.8.

# 8.1 RANDOM VARIABLES

To be more explicit about the concept of a random variable, let us return to Examples 6.1 and 6.14, dealing with the used-car dealer, his two salespersons, and the three 2004 Dodge Ram trucks. This time, however, suppose that the used-car dealer is interested in how many of the trucks the two salespersons *together* will sell during the given week. This total is a random variable, and its values are shown in blue in Figure 8.1, added to each point of the sample space given originally in Figure 6.7.

In more advanced work in mathematics, associating numbers with the points of a sample space is just a way of defining a function. This means that, strictly speaking, random variables are functions and not variables, but most beginners find it easier to think of random variables simply as quantities that can take on different values depending on chance. For instance, the number of speeding tickets issued each day on the freeway between Indio and Blyth in California is a random variable, and so is the annual production of coffee in Brazil, the number of persons visiting Disneyland each week, the wind velocity at Kennedy airport, the size of the audience at a baseball game, and the number of mistakes a person makes when typing a report.

Random variables are usually classified according to the number of values they can assume. When we talk about the number of points we roll with a pair of dice, the rise or decline of the prime (interest) rate, or the date of birth of

**Figure 8.1**
Sample space with values of random variable in blue.

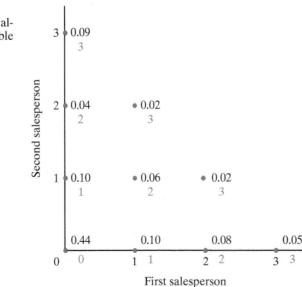

the next president of the United States, there are in each case finitely many possibilities, and we refer to them as values of a finite random variable. On the other hand, if we have a robot slam a car door until it begins to show signs of wear, if we count flashes of lightning on a stormy evening, or if an inspector counts the number of imperfections in 100 yards of fabric, there is (logically speaking, at least) no limit to the number of values which these random variables can assume, and we refer to them as *countably infinite random variables*. Often we treat random variables as if they were countably infinite even that we know that, practically speaking, there must be an upper limit. The trouble is that often we do not know the value of this limit, so we play it safe by allowing for a countable infinity. Taken together, finite and countably infinite values are referred to as discrete: namely, random variables that can take on a finite number of values or a countable infinity (as many values are there are whole numbers). For instance, the number of trucks that, between them, the two salespersons will sell is a discrete random variable that can take on a finite set of values, the four numbers 0, 1, 2, and 3. In contrast, the roll on which a die comes up 6 for the first time is a **discrete random variable** that can take on the countable infinity of values 1, 2, 3, 4, .... It is possible, though highly unlikely, that it will take a thousand rolls of the die, a million rolls, or even more, until we finally get a 6. There are also continuous random variables when we deal with quantities measured on a continuous scale, say, time, weight, or distance. These will be taken up in Chapter 9.

# 8.2 PROBABILITY DISTRIBUTIONS

To get the probability that a random variable will take on any particular value, we simply add the probabilities associated with all the points of the sample space for which the random variable takes on that value. For instance, to find the probability that the two salespersons together will sell two of the trucks in the given week, we add the probabilities associated with the points $(0, 2)$, $(1, 1)$, and $(2, 0)$ in Figure 8.1, getting $0.04 + 0.06 + 0.08 = 0.18$. Similarly, the probability that the random variable will take on the value 1 is $0.10 + 0.10 = 0.20$, and it will be left to the reader to verify that for 0 and 3 the probabilities are 0.44 and 0.18. All this is summarized in the following table:

| Number of trucks sold | Probability |
|---|---|
| 0 | 0.44 |
| 1 | 0.20 |
| 2 | 0.18 |
| 3 | 0.18 |

This table and the two that follow illustrate what we mean by a **probability distribution**—it is a correspondence that assigns probabilities to the values of a random variable.

Another correspondence like this is given by the following table, which pertains to the number of points we roll with a balanced die:

| Number of points we roll with a die | Probability |
|:---:|:---:|
| 1 | $\frac{1}{6}$ |
| 2 | $\frac{1}{6}$ |
| 3 | $\frac{1}{6}$ |
| 4 | $\frac{1}{6}$ |
| 5 | $\frac{1}{6}$ |
| 6 | $\frac{1}{6}$ |

Finally, for four flips of a balanced coin there are the sixteen equally likely possibilities HHHH, HHHT, HHTH, HTHH, THHH, HHTT, HTHT, HTTH, THHT, THTH, TTHH, HTTT, THTT, TTHT, TTTH, and TTTT, where H stands for heads and T for tails. Counting the number of heads in each case and using the formula $\frac{s}{n}$ for equiprobable outcomes, we get the following probability distribution for the total number of heads:

| Number of heads | Probability |
|:---:|:---:|
| 0 | $\frac{1}{16}$ |
| 1 | $\frac{4}{16}$ |
| 2 | $\frac{6}{16}$ |
| 3 | $\frac{4}{16}$ |
| 4 | $\frac{1}{16}$ |

When possible, we try to express probability distributions by means of formulas that enable us to calculate the probabilities associated with the various values of a random variable. For instance, for the number of points we roll with a balanced die we can write

$$f(x) = \frac{1}{6} \quad \text{for} \quad x = 1, 2, 3, 4, 5, \text{ and } 6$$

where $f(1)$ denotes the probability of rolling a 1, $f(2)$ denotes the probability of rolling a 2, and so on, in the usual functional notation. Here we wrote the probability that the random variable will take on the value $x$ as $f(x)$, but we could just as well write it as $g(x), h(x), m(x)$, etc.

**EXAMPLE 8.1**   Verify that for the number of heads obtained in four flips of a balanced coin the probability distribution is given by

$$f(x) = \frac{\dbinom{4}{x}}{16} \quad \text{for} \quad x = 0, 1, 2, 3, \text{ and } 4$$

**Solution**   By direct calculation, or by using Table XI at the end of the book, we find that
$\binom{4}{0} = 1, \binom{4}{1} = 4, \binom{4}{2} = 6, \binom{4}{3} = 4$, and $\binom{4}{4} = 1$. Thus, the probabilities
for $x = 0, 1, 2, 3$, and 4 are $\frac{1}{16}, \frac{4}{16}, \frac{6}{16}, \frac{4}{16}$, and $\frac{1}{16}$, which agrees with the values
given in the table on page 516.   ■

Since the values of probability distributions are probabilities, and since
random variables have to take on one of their values, we have the following two
rules that apply to any probability distribution:

---

**The values of a probability distribution must be numbers on the interval
from 0 to 1.**

**The sum of all the values of a probability distribution must be equal to 1.**

---

These rules enable us to determine whether or not a function (given by an equation
or by a table) can serve as the probability distribution of some random variable.

**EXAMPLE 8.2**   Check whether the correspondence given by

$$f(x) = \frac{x+3}{15} \quad \text{for} \quad x = 1, 2, \text{ and } 3$$

can serve as the probability distribution of some random variable.

**Solution**   Substituting $x = 1, 2$, and 3 into $\frac{x+3}{15}$, we get $f(1) = \frac{4}{15}, f(2) = \frac{5}{15}$, and $f(3) = \frac{6}{15}$.
Since none of these values is negative or greater than 1, and since their sum is

$$\frac{4}{15} + \frac{5}{15} + \frac{6}{15} = 1$$

the given function can serve as the probability distribution of some random
variable.   ■

# 8.3 THE BINOMIAL DISTRIBUTION

There are many applied problems in which we are interested in the probability
that an event will occur $x$ times out of $n$. For instance, we may be interested in
the probability of getting 45 responses to 400 questionnaires sent out as part of a
sociological study, the probability that 5 of 12 mice will survive for a given length
of time after the injection of a cancer-inducing substance, the probability that
45 of 300 drivers stopped at a road block will be wearing their seat belts, or the
probability that 66 of 200 television viewers (interviewed by a rating service) will
recall what products were advertised on a given program. To borrow from the
language of games of chance, we could say that in each of these examples we are
interested in the probability of getting "$x$ successes in $n$ trials," or in other words,
"$x$ successes and $n - x$ failures in $n$ attempts."

In the problems we shall study in this section, we always make the following assumptions:

> **There is a fixed number of trials.**
> **The probability of a success is the same for each trial.**
> **The trials are all independent.**

Thus, the theory we develop does not apply, for example, if we are interested in the number of dresses that a woman may try on before she buys one (where the number of trials is not fixed), if we check every hour whether traffic is congested at a certain intersection (where the probability of "success" is not constant), or if we are interested in the number of times that a person voted for the Republican candidate in the last five presidential elections (where the trials are not independent).

In what follows, we will be able to obtain a formula to solve problems that meet the conditions listed in the preceding paragraph. If $p$ and $1 - p$ are the probabilities of a success and a failure on any given trial, then the probability of getting $x$ successes and $n - x$ failures *in some specific order* is $p^x(1 - p)^{n-x}$. Clearly, in this product of $p$'s and $(1 - p)$'s there is one factor $p$ for each success, one factor $1 - p$ for each failure, and the $x$ factors $p$ and $n - x$ factors $1 - p$ are all multiplied together by virtue of the generalization of the special multiplication rule for more than two independent events. Since this probability applies to any point of the sample space that represents $x$ successes and $n - x$ failures (in some specific order), we have only to count how many points of this kind there are, and then multiply $p^x(1 - p)^{n-x}$ by this number. Clearly, the number of ways in which we can choose the $x$ trials on which the successes are to occur is $\binom{n}{x}$, and we have thus arrived at the following result:

---

**B**INOMIAL
DISTRIBUTION

> *The probability of getting $x$ successes in $n$ independent trials is*
>
> $$f(x) = \binom{n}{x} p^x(1 - p)^{n-x} \quad for \ x = 0, 1, 2, \ldots, or \ n$$
>
> *where $p$ is the constant probability of a success for each trial.*

---

It is customary to say here that the number of successes in $n$ trials is a random variable having the **binomial probability distribution**, or simply the **binomial distribution**. The binomial distribution is called by this name because for $x = 0, 1, 2, \ldots,$ and $n$, the values of the probabilities are the successive terms of the binomial expansion of $[(1 - p) + p]^n$.

**EXAMPLE  8.3**   Verify that the formula we gave in Example 8.1 for the probability of getting $x$ heads in four flips of a balanced coin is, in fact, the one for the binomial distribution with $n = 4$ and $p = \frac{1}{2}$.

**Solution**   Substituting $n = 4$ and $p = \frac{1}{2}$ into the formula for the binomial distribution, we get

$$f(x) = \binom{4}{x} \left(\frac{1}{2}\right)^x \left(1 - \frac{1}{2}\right)^{4-x} = \binom{4}{x} \left(\frac{1}{2}\right)^4 = \frac{\binom{4}{x}}{16}$$

for $x = 0, 1, 2, 3$, and 4. This is exactly the formula given in Example 8.1.   ▪

**EXAMPLE 8.4**   If the probability is 0.70 that any one registered voter (randomly selected from official rolls) will vote in a given election, what is the probability that two of five registered voters will vote in the election?

**Solution**   Substituting $x = 2$, $n = 5$, $p = 0.70$, and $\binom{5}{2} = 10$ into the formula for the binomial distribution, we get

$$f(2) = \binom{5}{2} (0.70)^2 (1 - 0.70)^{5-2} = 10(0.70)^2(0.30)^3 = 0.132$$   ▪

Following is an example where we calculate all the probabilities of a binomial distribution.

**EXAMPLE 8.5**   The probability is 0.30 that a person shopping at a certain supermarket will take advantage of its special promotion of ice cream. Find the probabilities that among six persons shopping at this market there will be 0, 1, 2, 3, 4, 5, or 6 who will take advantage of the promotion. Also, draw a histogram of this probability distribution.

**Solution**   Assuming that the selection is random, we substitute $n = 6$, $p = 0.30$, and, respectively, $x = 0, 1, 2, 3, 4, 5$, and 6 into the formula for the binomial distribution, and we get

$$f(0) = \binom{6}{0} (0.30)^0(0.70)^6 = 0.118$$

$$f(1) = \binom{6}{1} (0.30)^1(0.70)^5 = 0.303$$

$$f(2) = \binom{6}{2} (0.30)^2(0.70)^4 = 0.324$$

$$f(3) = \binom{6}{3} (0.30)^3(0.70)^3 = 0.185$$

$$f(4) = \binom{6}{4} (0.30)^4(0.70)^2 = 0.060$$

$$f(5) = \binom{6}{5} (0.30)^5(0.70)^1 = 0.010$$

$$f(6) = \binom{6}{6} (0.30)^6(0.70)^0 = 0.001$$

all rounded to three decimals. A histogram of this distribution is shown in Figure 8.2.   ▪

**Figure 8.2**
Histogram of binomial
distribution with $n = 6$
and $p = 0.30$.

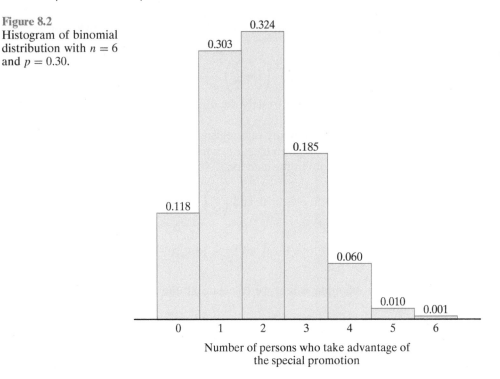

Number of persons who take advantage of
the special promotion

   In case the reader does not care much for ice cream and is not particularly interested in personal buying habits, let us stress the importance of the binomial distribution as a **statistical model**. The results of the preceding example apply also if the probability is 0.30 that the energy cell of a watch will last two years under normal usage, and we want to know the probabilities that, among six of these cells, 0, 1, 2, 3, 4, 5, or 6 will last two years under normal usage; if the probability is 0.30 that an embezzler will be caught and brought to trial, and we want to know the probabilities that, among six embezzlers, 0, 1, 2, 3, 4, 5, or 6 will be caught and brought to trial; if the probability is 0.30 that the head of a household owns at least one life insurance policy, and we want to know the probabilities that, among six heads of households, 0, 1, 2, 3, 4, 5, or 6 will own at least one life insurance policy; or if the probability that a person having a certain disease will live for another ten years is 0.30, and we want to know the probabilities that, among six persons having the disease, 0, 1, 2, 3, 4, 5, or 6 will live another ten years. The argument we have presented here is precisely like the one we used in Section 1.2, where we tried to impress upon the reader the generality of statistical techniques.

   In actual practice, binomial probabilities are seldom found by direct substitution into the formula. Sometimes we use approximations such as those discussed later in this chapter and in Chapter 9, and sometimes we use special tables such as Table V at the end of this book. Computer software is now the most common source for binomial probabilities. At the end of this chapter are listed some references for detailed binomial tables in book form (although they are falling into disuse).

Table V is limited to the binomial probabilities for $n = 2$ to $n = 20$ and $p = 0.05, 0.1, 0.2, 0.3, 0.4, 0.5, 0.6, 0.7, 0.8, 0.9,$ and $0.95$, all rounded to three decimals. Where values are omitted in this table, they are less than 0.0005 and, hence, rounded to 0.000 to three decimals.

**EXAMPLE 8.6**    The probability that a lunar eclipse will be obscured by clouds at an observatory near Buffalo, New York, is 0.60. Use Table V to find the probabilities that

(a) at most three of ten lunar eclipses will be obscured by clouds at that location;
(b) at least seven of ten lunar eclipses will be obscured by clouds at that location.

**Solution**    (a) For $n = 10$ and $p = 0.60$, the entries in Table V corresponding to $x = 0$, 1, 2, and 3 are 0.000, 0.002, 0.011, and 0.042. Thus, the probability that at most three of ten lunar eclipses will be obscured by clouds at that location is approximately

$$0.000 + 0.002 + 0.011 + 0.042 = 0.055$$

It is only approximate since the entries in Table V are all rounded to three decimals.

(b) For $n = 10$ and $p = 0.60$, the entries in Table V corresponding to $x = 7$, 8, 9, and 10 are 0.215, 0.121, 0.040, and 0.006. Thus, the probability that at least seven of ten lunar eclipses will be obscured by clouds at that location is approximately

$$0.215 + 0.121 + 0.040 + 0.006 = 0.382$$

■

We were able to use Table V in this example, because $p = 0.60$ is one of the few values for which Table V provides binomial probabilities. Had the probability been, say, 0.51 or 0.63 that a lunar eclipse will be obscured by clouds at the given location, we would have had to use one of the more detailed tables listed on page 205, a computer as in the example that follows, or as a last resort, the formula for the binomial distribution.

**EXAMPLE 8.7**    With reference to Example 8.6, suppose that the probability had been 0.63 that a lunar eclipse will be obscured by clouds at the given observatory. Use the computer printout of Figure 8.3 to rework Example 8.6 with $p = 0.63$ substituted for $p = 0.60$.

**Solution**    (a) For $n = 10$ and $p = 0.63$, the entries in the computer printout of Figure 8.3 corresponding to $x = 0$, 1, 2, and 3 are 0.0000, 0.0008, 0.0063, and 0.0285. Thus, the probability that at most three of the ten lunar eclipses will be obscured by clouds at that location is approximately

$$0.0000 + 0.0008 + 0.0063 + 0.0285 = 0.0356$$

Note that this answer is also shown in the lower part of the printout of Figure 8.3. It is the cumulative probability under the $P(X< = x)$ caption corresponding to $x = 3.00$.

(b) For $n = 10$ and $p = 0.63$, the entries in the computer printout of Figure 8.3 corresponding to $x = 7$, 8, 9, and 10 are 0.2394, 0.1529, 0.0578, and 0.0098.

**Figure 8.3**
Computer printout for
Example 8.7.

---

## Probability Density Function

```
Binomial with n = 10 and p = 0.630000

            x          P( X = x)
         0.00           0.0000
         1.00           0.0008
         2.00           0.0063
         3.00           0.0285
         4.00           0.0849
         5.00           0.1734
         6.00           0.2461
         7.00           0.2394
         8.00           0.1529
         9.00           0.0578
        10.00           0.0098
```

## Cumulative Distribution Function

```
Binomial with n = 10 and p = 0.630000

            x          P( X <= x)
         0.00           0.0000
         1.00           0.0008
         2.00           0.0071
         3.00           0.0356
         4.00           0.1205
         5.00           0.2939
         6.00           0.5400
         7.00           0.7794
         8.00           0.9323
         9.00           0.9902
        10.00           1.0000
```

---

Thus, the probability that at least seven of lunar eclipses will be obscured by clouds at the given location is

$$0.2394 + 0.1529 + 0.0578 + 0.0098 = 0.4599$$

Again, the answer can be obtained from the lower part of the printout. It is 1 minus the entry corresponding to $x = 6$ in the $P(X <= x)$ column, namely $1 - 0.5400 = 0.4600$. Of course, the small difference between 0.4599 and 0.4600 is due to rounding. ∎

When we observe a value of a random variable having the binomial distribution—for instance, when we observe the number of heads in 25 flips of a coin, the number of seeds (in a package of 24 seeds) that germinate, the number of students (among 200 interviewed) who are opposed to a change in student activity fees, or the number of automobile accidents (among 20 investigated) that are due to drunk driving—we say that we are **sampling a binomial population**. This terminology is widely used in statistics.

EXERCISES

**8.1**   In each case determine whether the given values can serve as the values of the probability distribution of some random variable that can take on the values 1, 2, and 3, and explain your answers:

(a)  $f(1) = 0.52$, $f(2) = 0.26$, and $f(3) = 0.32$;

(b)  $f(1) = 0.18$, $f(2) = 0.02$, and $f(3) = 1.00$;

(c)  $f(1) = \frac{10}{33}$, $f(2) = \frac{1}{3}$, and $f(3) = \frac{12}{33}$.

**8.2**   In each case determine whether the given values can serve as the values of the probability distribution of some random variable that can take on the values 1, 2, 3, and 4, and explain your answers:

(a)  $f(1) = 0.20$, $f(2) = 0.80$, $f(3) = 0.20$, and $f(4) = -0.20$;

(b)  $f(1) = 0.25$, $f(2) = 0.17$, $f(3) = 0.39$, and $f(4) = 0.19$;

(c)  $f(1) = \frac{1}{17}$, $f(2) = \frac{7}{17}$, $f(3) = \frac{6}{17}$, and $f(4) = \frac{2}{17}$.

**8.3**   For each of the following, determine whether it can serve as the probability distribution of some random variable:

(a)  $f(x) = \frac{1}{7}$   for $x = 1, 2, 3, 4, 5, 6, 7$;

(b)  $g(y) = \frac{1}{9}$   for $y = 0, 1, 2, 3, 4, 5, 6, 7, 8, 9$;

(c)  $f(x) = \frac{x+2}{18}$   for $x = 1, 2, 3, 4$.

**8.4**   For each of the following, determine whether it can serve as the probability distribution of some random variable:

(a)  $f(x) = \frac{x-1}{10}$   for $x = 0, 1, 2, 3, 4, 5$;

(b)  $h(z) = \frac{z^2}{30}$   for $z = 0, 1, 2, 3, 4$;

(c)  $f(y) = \frac{y+4}{y-4}$   for $y = 1, 2, 3, 4, 5$.

**8.5**   In a certain city, medical expenses are given as the reason for 75% of all personal bankruptcies. Use the formula for the binomial distribution to calculate the probability that medical expenses will be given as the reason for two of the next three personal bankruptcies filed in that city.

**8.6**   Use the formula for the binomial distribution to calculate the probability that four of six hibiscus plants will fail to survive a frost if the probability is 0.30 that any such plant will survive a frost. Also, check the answer in Table V.

**8.7**   A doctor knows from experience that 10% of the patients to whom he prescribes a certain blood pressure medication will have undesirable side effects. Use the formula for the binomial distribution to calculate the probability that none of four patients to whom he prescribes the medication will have undesirable side effects. Also, check the answer in Table V.

**8.8**   It has been claimed that 80% of all industrial accidents can be prevented by paying strict attention to safety regulations. If this is so, find the probability that four of six industrial accidents can thus be prevented, using

(a)  the formula for the binomial distribution;

(b)  Table V.

**8.9**   Experience has shown that 30% of the rocket launchings at a NASA base have to be delayed due to weather conditions. Use Table V to determine the probabilities that among ten rocket launchings at that base

(a)  at most three will have to be delayed due to weather conditions;

(b)  at least six will have to be delayed due to weather conditions.

**8.10**   A study shows that 95% of the watermelons shipped out by an agricultural cooperative are ripe and ready to eat. Find the probabilities that among 20 watermelons shipped out by the cooperative

(a) at most 16 are ripe and ready to eat;

(b) all of them are ripe and ready to eat.

**8.11**   A study shows that 60% of the divorce cases filed in a certain county give incompatibility as the legal reason. Find the probabilities that among 15 divorce cases filed in that county

(a) at most five will give incompatibility as the legal reason;

(b) anywhere from eight to eleven will give incompatibility as the legal reason;

(c) at least eleven will give incompatibility as the legal reason.

**8.12**   A frozen food distributor claims that 80% of its frozen chicken dinners contain at least three ounces of chicken. To check on this claim, a consumer testing service decides to check ten of these frozen chicken dinners and reject the claim unless at least seven of them contain at least three ounces of chicken. Find the probabilities that the testing service will make the error of

(a) rejecting the claim even though it is true;

(b) not rejecting the claim when in reality only 70% of the frozen chicken dinners contain at least three ounces of chicken.

**8.13**   A quality control engineer wants to check whether, in accordance with specifications, 90% of the products shipped are in perfect working condition. To this end, she randomly selects 12 items from each lot ready to be shipped and passes the lot only if all 12 are in perfect working condition. If one or more items are not in perfect working condition, she holds the lot for a complete inspection. Find the probabilities that she will commit the error of

(a) holding a lot for a complete inspection even though 90% of the items are in perfect working condition;

(b) letting a lot pass through even though only 80% of the items are in perfect working condition;

(c) letting a lot pass through even though only 70% of the items are in perfect working condition.

**8.14**   A study shows that 70% of all patients coming to a certain medical clinic have to wait at least 15 minutes to see their doctor. Find the probabilities that among ten patients coming to this clinic $0, 1, 2, 3, \ldots$, or 10 have to wait at least 15 minutes to see their doctor, and draw a histogram of this probability distribution.

**8.15**   Use Figure 8.3 to determine the probability that a random variable having the binomial distribution with $n = 10$ and $p = 0.63$ will take on a value less than five using

(a) the binomial probabilities;

(b) the cumulative probabilities.

**8.16**   Use Figure 8.3 to determine the probability that a random variable having the binomial distribution with $n = 10$ and $p = 0.63$ will take on a value greater than eight using

(a) the binomial probabilities;

(b) the cumulative probabilities.

**8.17**   In some situations where otherwise the binomial distribution applies, we are interested in the probability that the first success will occur on a given trial. For this to happen on the $x$th trial, it must be preceded by $x - 1$ failures for which the probability is $(1 - p)^{x-1}$, and it follows that the probability that the first success will occur on the $x$th trial is

$$f(x) = p(1 - p)^{x-1} \quad \text{for} \quad x = 1, 2, 3, 4, \ldots$$

This distribution is called the **geometric distribution** (because its successive values constitute a geometric progression) and it should be observed that there is a countable infinity of possibilities.[†] Using the formula, we find, for example, that for repeated rolls of a balanced die the probability that the first 6 will occur on the fifth roll is

$$\frac{1}{6}\left(\frac{5}{6}\right)^{5-1} = \frac{625}{7{,}776} \approx 0.080$$

(a) When taping a television commercial, the probability that a child actor will get his lines straight on any one take is 0.40. What is the probability that this child actor will finally get his lines straight on the fourth take?

(b) Suppose the probability is 0.25 that any given person will believe a rumor about the private life of a certain politician. What is the probability that the fifth person to hear the rumor will be the first one to believe it?

(c) The probability is 0.70 that a child exposed to a certain contagious disease will catch it. What is the probability that the third child exposed to the disease will be the first one to catch it?

# 8.4 THE HYPERGEOMETRIC DISTRIBUTION

In Exercise 6.62 we introduced the terms "sampling with replacement" and "sampling without replacement" in connection with drawings from a deck of cards. To carry this distinction further, let us point out that the binomial distribution applies when we sample with replacement and the trials are all independent, but that it does not apply when we sample without replacement. To introduce a probability distribution that applies when we sample without replacement, let us consider the following example: A nursery ships three-year-old citrus trees in batches of 24, and when they arrive at their destination, an inspector randomly chooses three from each batch. If these three trees are all healthy, the entire batch is accepted; otherwise, the other 21 trees in the batch are also inspected. Since a batch can be accepted without further inspection even though quite a few of the trees are not in good condition, this inspection procedure involves quite a risk. To illustrate the size of this risk, let us suppose that in reality 6 of the 24 trees are in poor condition, and let us determine the probability that the whole batch will, nevertheless, be accepted without further inspection. This means that we must find the probability of three successes (healthy trees) in three trials (trees inspected), and we might be tempted to argue that since 18 of the 24 trees in the batch are healthy, the probability is $\frac{18}{24} = \frac{3}{4}$ that any one of them is healthy, and hence the desired probability is

$$f(3) = \binom{3}{3}\left(\frac{3}{4}\right)^3\left(1 - \frac{3}{4}\right)^{3-3} = 0.42$$

This result, obtained with the formula for the binomial distribution, would be correct if sampling is with replacement, but that is not what we do in realistic

[†] As formulated in Chapter 6, the postulates of probability apply only when the sample space is finite. When the sample space is countably infinite, as is the case here, the third postulate must be modified accordingly. This will be explained on page 193.

problems of sampling inspection. To get the correct answer for our problem when sampling is without replacement, we might argue as follows: There are altogether $\binom{24}{3} = 2{,}024$ ways of choosing three of the 24 trees, and they are all equiprobable by virtue of the assumption that the selection is random. Among these, there are $\binom{18}{3} = 816$ ways of selecting three of the 18 healthy trees, and it follows that the desired probability is $\frac{816}{2{,}024} = 0.40$.

To generalize the method we used here, suppose that $n$ objects are to be chosen from a set of $a$ objects of one kind (successes) and $b$ objects of another kind (failures), the selection is without replacement, and we are interested in the probability of getting $x$ successes and $n - x$ failures. Arguing as before, we find that the $n$ objects can be chosen from the whole set of $a + b$ objects in $\binom{a+b}{n}$ ways, and that $x$ of the $a$ successes and $n - x$ of the $b$ failures can be chosen in $\binom{a}{x} \cdot \binom{b}{n-x}$ ways. It follows that for sampling without replacement the probability of "$x$ successes in $n$ trials" is

**HYPERGEOMETRIC DISTRIBUTION**

$$f(x) = \frac{\binom{a}{x} \cdot \binom{b}{n-x}}{\binom{a+b}{n}} \quad for \ x = 0, 1, 2, \ldots, or \ n$$

where $x$ cannot exceed $a$ and $n - x$ cannot exceed $b$. This is the formula for the **hypergeometric distribution**.

The next two examples illustrate use of the hypergeometric distribution in problems where we sample without replacement.

**EXAMPLE 8.8**   A mailroom clerk is supposed to send 6 of 15 packages to Europe by airmail, but he gets them all mixed up and randomly puts airmail postage on 6 of the packages. What is the probability that only three of the packages that are supposed to go by air get airmail postage?

**Solution**   Substituting $a = 6$, $b = 9$, and $x = 3$ into the formula for the hypergeometric distribution, we get

$$f(3) = \frac{\binom{6}{3} \cdot \binom{9}{6-3}}{\binom{15}{6}} = \frac{20 \cdot 84}{5{,}005} \approx 0.336$$

**EXAMPLE 8.9**   Among an ambulance service's 16 ambulances, five emit excessive amounts of pollutants. If eight of the ambulances are randomly picked for inspection, what is the probability that this sample will include at least three of the ambulances that emit excessive amounts of pollutants?

**Solution**  The probability we must find is $f(3) + f(4) + f(5)$, where each term in this sum is a value of the hypergeometric distribution with $a = 5$, $b = 11$, and $n = 8$. Substituting these quantities together with $x = 3$, 4, and 5 into the formula for the hypergeometric distribution, we get

$$f(3) = \frac{\binom{5}{3} \cdot \binom{11}{5}}{\binom{16}{8}} = \frac{10 \cdot 462}{12{,}870} = 0.359$$

$$f(4) = \frac{\binom{5}{4} \cdot \binom{11}{4}}{\binom{16}{8}} = \frac{5 \cdot 330}{12{,}870} = 0.128$$

$$f(5) = \frac{\binom{5}{5} \cdot \binom{11}{3}}{\binom{16}{8}} = \frac{1 \cdot 165}{12{,}870} \approx 0.013$$

and the probability that the sample will include at least three of the ambulances that emit excessive amounts of pollutants is

$$0.359 + 0.128 + 0.013 = 0.500$$

This result suggests that the inspection should, perhaps, have included more than eight of the ambulances, and it will be left to the reader to show in Exercise 8.18 that the probability of catching at least three of the ambulances that emit excessive amounts of pollutants would have been 0.76 if the inspection had included ten of the ambulances. ■

In the beginning of this section we gave an example where we erroneously used the binomial distribution instead of the hypergeometric distribution. The error was quite small, however—we got 0.42 instead of 0.40—and in actual practice the binomial distribution is often used to approximate the hypergeometric distribution. It is generally agreed that this approximation is satisfactory if $n$ does not exceed 5 percent of $a + b$, namely, if

$$n \leq (0.05)(a + b)$$

The main advantages of the approximation are that the binomial distribution has been tabulated much more extensively than the hypergeometric distribution, and that, between them, the formula for the binomial distribution is easier to use; that is, the binomial calculations are usually less complicated. Observe also that the binomial distribution is described by two parameters ($n$ and $p$), while the hypergeometric distribution requires three ($a$, $b$, and $n$).

**EXAMPLE 8.10**  In a federal prison, 120 of the 300 inmates are serving time for drug-related offenses. If eight of them are to be chosen at random to appear before a legislative committee, what is the probability that three of the eight will be serving time for drug-related offenses?

**Solution**   Since $n = 8$ and $a + b = 300$ and 8 is less than $0.05(300) = 15$, we can use the binomial approximation to the hypergeometric distribution. From Table V we find that for $n = 8$, $p = \frac{120}{300} = 0.40$, and $x = 3$, the probability asked for is 0.279. Fairly extensive calculations would show that the error of this approximation is only 0.003.   ▪

EXERCISES

8.18   In Example 8.9 we indicated that if ten of the ambulances had been inspected, the probability of including at least three of the ones that emit excessive amounts of pollutants would have been 0.76. Verify this probability.

8.19   Among the 14 auto mechanics at a dealership, ten are factory trained. If three of the mechanics are chosen at random for a special job, find the probabilities that
   (a)   all three of them will be factory trained;
   (b)   only two of them will be factory trained.

8.20   Among the 20 solar collectors on display at a trade show, 12 are flat-plate collectors and the others are concentrating collectors. If a person visiting the show randomly selects six of the solar collectors to check out, what is the probability that three of them will be flat-plate collectors?

8.21   Among the 12 male applicants for a job with the postal service, nine have working wives. If two of the applicants are randomly chosen for further consideration, what are the probabilities that
   (a)   neither has a working wife;
   (b)   only one has a working wife;
   (c)   both have working wives?

8.22   A customs inspector decides to inspect 3 of 16 shipments that arrive from Caracas by plane. If the selection is random and five of the shipments contain contraband, find the probabilities that the customs inspector will catch
   (a)   none of the shipments with contraband;
   (b)   only one of the shipments with contraband;
   (c)   two of the shipments with contraband;
   (d)   three of the shipments with contraband.

8.23   Check in each case whether the condition for the binomial approximation to the hypergeometric distribution is satisfied:
   (a)   $a = 140, b = 60$, and $n = 12$;
   (b)   $a = 220, b = 280$, and $n = 20$;
   (c)   $a = 250, b = 390$, and $n = 30$;
   (d)   $a = 220, b = 220$, and $n = 25$.

8.24   A shipment of 250 swimming pool pumps contains four with minor imperfections. Use the binomial approximation to the hypergeometric distribution to find the probability that if five of these pumps are randomly selected to be shipped to a retail outlet of pool supplies, they will include one with minor imperfections.

8.25   With reference to Exercise 8.24, find the error of this binomial approximation to the hypergeometric distribution.

8.26   Among the 200 employees of a company, 120 are union members while the others are not. If six of the employees are to be chosen by lot to serve on a committee that administers the pension fund, find the probability that three of them will be union members while the others are not, using
   (a)   the formula for the hypergeometric distribution;
   (b)   the binomial distribution with $p = \frac{120}{200} = 0.60$ and $n = 6$ as an approximation.

# 8.5 THE POISSON DISTRIBUTION

When $n$ is large and $p$ is small, binomial probabilities are often approximated by means of the formula

POISSON
APPROXIMATION
TO BINOMIAL
DISTRIBUTION

$$f(x) = \frac{(np)^x \cdot e^{-np}}{x!} \quad for \quad x = 0, 1, 2, 3, \ldots$$

which is a special form of the **Poisson distribution**, named after the French mathematician and physicist S. D. Poisson (1781–1840). In this formula, the irrational number $e = 2.71828\ldots$ is the base of the system of natural logarithms, and the necessary values of $e^{-np}$ may be obtained from Table XII at the end of the book. Note that, as in Exercise 8.17, we are faced here with a random variable that can take on a countable infinity of values (namely, as many values as there are whole numbers). Correspondingly, the third postulate of probability must be modified so that for any sequence of mutually exclusive events $A_1, A_2, A_3, \ldots$, the probability that one of them will occur is

$$P(A_1 \cup A_2 \cup A_3 \cup \cdots) = P(A_1) + P(A_2) + P(A_3) + \cdots$$

It is difficult to give precise conditions under which the Poisson approximation to the binomial distribution may be used; that is, explain precisely what we mean here by "when $n$ is large and $p$ is small." Although other books may give less stringent rules of thumb, we shall play it relatively safe and use the Poisson approximation to the binomial distribution only when

$$n \geq 100 \quad and \quad np < 10$$

To get some idea about the closeness of the Poisson approximation to the binomial distribution, consider the computer printouts of Figure 8.4, which show, one next to the other, the binomial probabilities with $n = 150$ and $p = 0.05$ and the Poisson probabilities with $np = 150(0.05) = 7.5$. [Comparing the probabilities in the columns headed $P(X = x)$, we find that the greatest difference, corresponding to $x = 8$, is $0.1410 - 0.1373 = 0.0037$.]

In Figure 8.4, MINITAB refers to probability distributions as probability density functions. Also, the parameter of this special form of the Poisson distribution is the product $np$, referred to here as "mu" and denoted by $\mu$, as will be explained in Section 8.7.

The two examples that follow illustrate the Poisson approximation to the binomial distribution.

EXAMPLE  8.11    It is known from experience that 2% of the books bound at a certain bindery have defective bindings. Use the Poisson approximation to the binomial distribution to find the probability that in a lot of 400 books bound at this bindery, five will have defective bindings.

**Figure 8.4**
Poisson approximation to binomial distribution.

| Probability Density Function | | Probability Density Function | |
|---|---|---|---|
| Binomial with n = 150 and p = 0.0500000 | | Poisson with mu = 7.50000 | |
| x | P( X = x) | x | P( X = x) |
| 0.00 | 0.0005 | 0.00 | 0.0006 |
| 1.00 | 0.0036 | 1.00 | 0.0041 |
| 2.00 | 0.0141 | 2.00 | 0.0156 |
| 3.00 | 0.0366 | 3.00 | 0.0389 |
| 4.00 | 0.0708 | 4.00 | 0.0729 |
| 5.00 | 0.1088 | 5.00 | 0.1094 |
| 6.00 | 0.1384 | 6.00 | 0.1367 |
| 7.00 | 0.1499 | 7.00 | 0.1465 |
| 8.00 | 0.1410 | 8.00 | 0.1373 |
| 9.00 | 0.1171 | 9.00 | 0.1144 |
| 10.00 | 0.0869 | 10.00 | 0.0858 |
| 11.00 | 0.0582 | 11.00 | 0.0585 |
| 12.00 | 0.0355 | 12.00 | 0.0366 |
| 13.00 | 0.0198 | 13.00 | 0.0211 |
| 14.00 | 0.0102 | 14.00 | 0.0113 |
| 15.00 | 0.0049 | 15.00 | 0.0057 |
| 16.00 | 0.0022 | 16.00 | 0.0026 |
| 17.00 | 0.0009 | 17.00 | 0.0012 |
| 18.00 | 0.0003 | 18.00 | 0.0005 |
| 19.00 | 0.0001 | 19.00 | 0.0002 |
| 20.00 | 0.0000 | 20.00 | 0.0001 |
| 21.00 | 0.0000 | 21.00 | 0.0000 |

**Solution** Since $n = 400 \geq 100$ and $np = 400(0.02) = 8 < 10$, the conditions for the approximation are satisfied. Thus, substitution of $np = 8$, $e^{-8} = 0.00033546$ (from Table XII), and $x = 5$ into the formula for the Poisson distribution yields

$$f(5) = \frac{8^5 \cdot e^{-8}}{5!} = \frac{(32,768)(0.00033546)}{120} \approx 0.0916$$

**EXAMPLE 8.12** Records show that the probability is 0.00006 that a car will have a flat tire while being driven through a certain tunnel. Use the Poisson approximation to the binomial distribution to find the probability that at least two of 10,000 cars will have a flat tire while being driven through that tunnel.

**Solution** Since $n = 10,000 \geq 100$ and $np = 10,000(0.00006) = 0.6 < 10$, the conditions for the approximation are satisfied. Rather than add the probabilities for $x = 2, 3, 4, \ldots$, we shall subtract from 1 the sum of the probabilities for $x = 0$ and $x = 1$. Thus, substituting $np = 0.6$, $e^{-0.6} = 0.5488$ (from Table XII), and respectively, $x = 0$ and $x = 1$ into the formula for the Poisson distribution, we get

$$f(0) = \frac{(0.6)^0 \cdot e^{-0.6}}{0!} = \frac{1(0.5488)}{1} = 0.5488$$

$$f(1) = \frac{(0.6)^1 \cdot e^{-0.6}}{1!} \approx \frac{(0.6)(0.5488)}{1} = 0.3293$$

and, finally, $1 - (0.5488 + 0.3293) = 0.1219$.

**Figure 8.5**
Computer printout of the Poisson distribution with $np = 0.60$.

| **Probability Density Function** | | **Cumulative Distribution Function** | |
| --- | --- | --- | --- |
| Poisson with mu = 0.600000 | | Poisson with mu = 0.600000 | |
| x | P( X = x) | x | P( X <= x) |
| 0.00 | 0.5488 | 0.00 | 0.5488 |
| 1.00 | 0.3293 | 1.00 | 0.8781 |
| 2.00 | 0.0988 | 2.00 | 0.9769 |
| 3.00 | 0.0198 | 3.00 | 0.9966 |
| 4.00 | 0.0030 | 4.00 | 0.9996 |
| 5.00 | 0.0004 | 5.00 | 1.0000 |
| 6.00 | 0.0000 | 6.00 | 1.0000 |

In actual practice, Poisson probabilities (values of Poisson distributions) are rarely calculated by direct substitution into the formula for the Poisson distribution or by using a special table. Such tasks can be greatly simplified by using a computer. For instance, had we used the computer printout of Figure 8.5 in Example 8.12, we would have obtained directly $1 - 0.8781 = 0.1219$, where 0.8781 is the value corresponding to $x = 1.00$ in the column headed $P(X < = x)$.

In some instances, hypergeometric distributions can be approximated by binomial distributions, which in turn can be approximated by Poisson distributions. Consider, for example, the following:

**EXAMPLE 8.13**   A 2002 audit of 4,000 sales of a department store included 28 with mistakes in billing. Now, a CPA wants to investigate the audit by rechecking a random subsample of 150 of the 4,000 sales. In particular, he would like to know the probability of finding two of these 150 sales with mistakes in billing.

**Solution**   This calls for the hypergeometric probability with $x = 2$, $a = 28$, $b = 4,000 - 28 = 3,972$, and $n = 150$, but since $150 \leq 0.05(4,000) = 200$, we can use the binomial approximation to the hypergeometric distribution. Furthermore, since this is the binomial distribution with $n = 150$ and $p = 28/4,000 = 0.007$, for which $n \geq 100$ and $np = 150(0.007) = 1.05 < 10$, we can approximate it with the Poisson distribution with $np = 1.05$. Thus, we can approximate the original hypergeometric probability with the Poisson probability with $x = 2$ and $np = 1.05$. That is,

$$f(2) = \frac{1.05^2 \cdot e^{-1.05}}{2!} = \frac{1.1025(0.349938)}{2} = 0.1929$$

where the value of $e^{-1.05}$ was obtained with a statistical calculator.

Since there are many situations where $n$ or $a + b$ are large, the Poisson distribution provides very useful approximations. Note also that the Poisson distribution involves only one parameter, $np$, whereas the binomial distribution involves two, $n$ and $p$, and the hypergeometric distribution involves three, $a$, $b$, and $n$.

The Poisson distribution has many important applications that have no direct connection with the binomial distribution. In that case $np$ is replaced by the parameter $\lambda$ (Greek lowercase *lambda*) and we calculate the probability of getting $x$ successes by means of the formula

## POISSON DISTRIBUTION

$$f(x) = \frac{\lambda^x \cdot e^{-\lambda}}{x!} \quad for\ x = 0, 1, 2, 3, \ldots$$

where $\lambda$ is interpreted as the expected, or average, number of successes, as is explained in the last paragraph of Section 8.7.

This formula applies to many situations where we can expect a fixed number of "successes" per unit time (or for some other kind of unit), say, when a bank can expect to receive six bad checks per day, when 1.6 accidents can be expected per day at a busy intersection, when eight small pieces of meat can be expected in a frozen meat pie, when 5.6 imperfections can be expected per roll of cloth, when 0.03 complaint per passenger can be expected by an airline, and so on.

**EXAMPLE 8.14**   Given that a bank receives on the average $\lambda = 6$ bad checks per day, what is the probability that it will receive four bad checks on any one day?

**Solution**   Substituting $x = 4$ and $\lambda = 6$ into the preceding formula for the Poisson distribution, we get

$$f(4) = \frac{6^4 \cdot e^{-6}}{4!} = \frac{1{,}296(0.002479)}{24} = 0.1339$$

where the value of $e^{-6}$ was obtained from Table XII.

**EXAMPLE 8.15**   If $\lambda = 5.6$ imperfections can be expected per roll of a given kind of curtain material, what is the probability that a roll will have $x = 3$ imperfections?

**Solution**   Substituting $x = 3$ and $\lambda = 5.6$ into the preceding formula for the Poisson distribution, we get

$$f(3) = \frac{5.6^3 \cdot e^{-5.6}}{3!} = \frac{175.616(0.003698)}{6} \approx 0.1082$$

where the value of $e^{-5.6}$ was obtained from Table XII.

## EXERCISES

**8.27**   Check in each case whether the values of $n$ and $p$ satisfy the rule of thumb that we gave on page 193 for using the Poisson approximation to the binomial distribution:

(a)   $n = 250$ and $p = \frac{1}{20}$;

(b)   $n = 400$ and $p = \frac{1}{50}$;

(c)   $n = 90$ and $p = \frac{1}{10}$.

**8.28**   Check in each case whether the values of $n$ and $p$ satisfy the rule of thumb that we gave on page 193 for using the Poisson approximation to the binomial distribution:

(a)   $n = 300$ and $p = 0.01$;

(b)   $n = 600$ and $p = 0.02$;

(c)   $n = 75$ and $p = 0.1$.

**8.29**   Based on her experience, a hospital administrator feels that 5% of all newly admitted patients must be put immediately into intensive care. Use this figure to estimate the probability that among 120 newly admitted patients, three will have to be put immediately into intensive care.

**8.30**   If 3% of the persons who look at a model home are seriously interested in buying one like it, what is the Poisson probability that among 180 persons who look at a model home, five will be seriously interested in buying one like it?

**8.31**   Suppose that in a certain city, 5% of all licensed drivers will be involved in at least one accident in any given year. Use the formula for the Poisson distribution to approximate the probability that among 150 licensed drivers in that city, at most two will be involved in at least one accident in a given year.

**8.32**   Use Figure 8.4 to
   (a)   verify the result obtained for Exercise 8.31;
   (b)   find the error of this Poisson approximation to the binomial distribution.

**8.33**   Can the hypergeometric distribution with $n = 120, a = 50$, and $b = 3,150$ be approximated with a Poisson distribution?

**8.34**   The number of complaints that a dry cleaning establishment receives per day is a random variable having the Poisson distribution with $\lambda = 3.2$. Find the probabilities that on any given day it will receive
   (a)   only two complaints;
   (b)   at most two complaints.

**8.35**   The number of monthly breakdowns of the kind of computer used by an office is a random variable having the Poisson distribution with $\lambda = 1.6$. Find the probabilities that this kind of computer will function for a month
   (a)   without a breakdown;
   (b)   with one breakdown;
   (c)   with two breakdowns.

**8.36**   With reference to Exercise 8.35, suppose that the office has four of the computers. What is the probability that all four of them will function for a month without a breakdown? What assumption has to made to determine this probability?

# *8.6 THE MULTINOMIAL DISTRIBUTION

An important generalization of the binomial distribution arises when there are more than two possible outcomes for each trial, the probabilities of the various outcomes remain the same for each trial, and the trials are all independent. This is the case, for example, when we repeatedly roll a die, where each trial has six possible outcomes; when students are asked whether they like a certain new recording, dislike it, or don't care; or when a U.S. Department of Agriculture inspector grades beef as prime, choice, good, commercial, or utility.

If there are $k$ possible outcomes for each trial and their probabilities are $p_1, p_2, \ldots$, and $p_k$, it can be shown that the probability of $x_1$ outcomes of the first kind, $x_2$ outcomes of the second kind, $\ldots$, and $x_k$ outcomes of the $k$th kind in $n$ trials is given by

**MULTINOMIAL DISTRIBUTION**

$$\frac{n!}{x_1! x_2! \cdots x_k!} p_1^{x_1} \cdot p_2^{x_2} \cdots p_k^{x_k}$$

This distribution is called the **multinomial distribution**.

**EXAMPLE 8.16** In a large city, network TV has 30% of the viewing audience on Friday nights, a local channel has 20%, cable TV has 40%, and 10% of the viewing audience is watching videocassettes. What is the probability that among seven television viewers randomly selected in that city on a Friday night three will be viewing network TV, one will be watching the local channel, two will be watching cable TV, and one will be watching a videocassette?

**Solution** Substituting $n = 7$, $x_1 = 3$, $x_2 = 1$, $x_3 = 2$, $x_4 = 1$, $p_1 = 0.30$, $p_2 = 0.20$, $p_3 = 0.40$, and $p_4 = 0.10$ into the formula for the multinomial distribution, we get

$$\frac{7!}{3! \cdot 1! \cdot 2! \cdot 1!} \cdot (0.30)^3 (0.20)^1 (0.40)^2 (0.10)^1 = 0.036$$

rounded to three decimals.

**EXERCISES**

**\*8.37**  For a car being tested at a state inspection station, the probability that it will pass on the first try is 0.70, the probability that it will pass on the second try is 0.20, and the probability that it will pass on the third try is 0.10. What is the probability that among ten cars being tested, six will pass on the first try, three will pass on the second try, and one will pass on the third try?

**\*8.38**  According to the Mendelian theory of heredity, if plants with round yellow seeds are crossbred with plants with wrinkled green seeds, the probabilities of getting a plant that produces round yellow seeds, wrinkled yellow seeds, round green seeds, or wrinkled green seeds are, respectively, $\frac{9}{16}$, $\frac{3}{16}$, $\frac{3}{16}$, and $\frac{1}{16}$. What is the probability that among nine plants thus obtained there will be four that produce round yellow seeds, two that produce wrinkled yellow seeds, three that produce round green seeds, and none that produce wrinkled green seeds?

**\*8.39**  The probabilities are 0.60, 0.20, 0.10, and 0.10, respectively, that a state income tax form will be filled out correctly, that it will contain only errors favoring the taxpayer, that it will contain only errors favoring the government, and that it will contain both kinds of errors. What is the probability that among ten such tax forms (randomly selected for audit) seven will be filled out correctly, one will contain only errors favoring the taxpayer, one will contain only errors favoring the government, and one will contain both kinds of errors?

## 8.7 THE MEAN OF A PROBABILITY DISTRIBUTION

When we showed on page 162 that an airline office at a certain airport can expect 2.75 complaints per day about its luggage handling, we arrived at this result by using the formula for a mathematical expectation, namely, by adding the products obtained by multiplying 0, 1, 2, 3, ... by the corresponding probabilities that the office will receive 0, 1, 2, 3, ... complaints about its luggage handling on any given day. Here the number of complaints is a random variable and 2.75 is its **expected value**.

If we apply the same argument to the first illustration of Section 8.2, we find that, between them, the two salespersons can expect to sell

$$0(0.44) + 1(0.20) + 2(0.18) + 3(0.18) = 1.10$$

of the trucks. In this case, the number of trucks is a random variable and 1.1 is its expected value.

As we explained in Chapter 7, mathematical expectations must be interpreted as averages, or means, and it is customary to refer to the expected value of a random variable as its **mean**, or as the **mean of its probability distribution**. In general, if a random variable takes on the values $x_1, x_2, x_3, \ldots$, or $x_k$, with the probabilities $f(x_1)$, $f(x_2)$, $f(x_3)$, $\ldots$, and $f(x_k)$, its expected value is

$$x_1 \cdot f(x_1) + x_2 \cdot f(x_2) + x_3 \cdot f(x_3) + \cdots + x_k \cdot f(x_k)$$

and in the $\sum$ notation we write

**M**EAN OF A
PROBABILITY
DISTRIBUTION

$$\mu = \sum x \cdot f(x)$$

Like the mean of a population, the mean of a probability distribution is denoted by the Greek lowercase $\mu$ (*mu*). The notation is the same, for as we pointed out in connection with the binomial distribution, when we observe a value of a random variable, we refer to its distribution as the population we are sampling. For instance, the histogram of Figure 8.2 on page 184 may be looked upon as the population we are sampling when we observe a value of a random variable having the binomial distribution with $n = 6$ and $p = 0.30$.

**EXAMPLE  8.17**    Find the mean of the second probability distribution of Section 8.2; the one that pertained to the number of points we roll with a balanced die.

**Solution**    Since the probabilities of rolling a 1, 2, 3, 4, 5, or 6 are all $\frac{1}{6}$, we get

$$\mu = 1 \cdot \tfrac{1}{6} + 2 \cdot \tfrac{1}{6} + 3 \cdot \tfrac{1}{6} + 4 \cdot \tfrac{1}{6} + 5 \cdot \tfrac{1}{6} + 6 \cdot \tfrac{1}{6} = 3\tfrac{1}{2}$$

**EXAMPLE  8.18**    With reference to Example 8.5, find the mean number of persons, among six shopping at the supermarket, who will take advantage of the special promotion.

**Solution**    Substituting $x = 0, 1, 2, 3, 4, 5$, and 6, and the probabilities on page 183 into the formula for $\mu$, we get

$$\mu = 0(0.118) + 1(0.303) + 2(0.324) + 3(0.185)$$

$$+ 4(0.060) + 5(0.010) + 6(0.001)$$

$$= 1.802$$

When a random variable can take on many different values, the calculation of $\mu$ may become very laborious. For instance, if we want to know how many persons can be expected to contribute to a charity, when 2,000 are solicited for funds and the probability is 0.40 that any one of them will make a contribution, we might consider calculating the 2,001 probabilities corresponding to 0, 1, 2, 3, $\ldots$, 1,999, or 2,000 of them making a contribution and then substitute into the formula for $\mu$. Not seriously, though, and we might argue instead that in the long run 40% of the persons will make a contribution, 40% of 2,000 is 800, and hence we can expect that 800 of the 2,000 persons will make a contribution. Similarly, if a balanced coin is flipped 1,000 times, we might argue that in the long run heads will come up 50% of the time, and hence that we can expect $1,000(0.50) = 500$ heads. These two results are correct; both problems deal with random variables having binomial distributions, and it can be shown that in general

## MEAN OF A BINOMIAL DISTRIBUTION

$$\mu = n \cdot p$$

In words, the mean of a binomial distribution is simply the product of the number of trials and the probability of success on an individual trial.

**EXAMPLE 8.19**   With reference to Example 8.5, use the formula for the mean of a binomial distribution to find the mean number of persons, among six shopping at the supermarket, who can be expected to take advantage of the special promotion.

**Solution**   Since we are dealing with the binomial distribution having $n = 6$ and $p = 0.30$, we get $\mu = 6(0.30) = 1.80$. The small difference between the values obtained here and in Example 8.18 is due to rounding the probabilities used in Example 8.18 to three decimals.   ■

It is important to remember that the formula $\mu = n \cdot p$ applies only to binomial distributions. There are other formulas for other distributions; for instance, for the hypergeometric distribution the formula for the mean is

## MEAN OF A HYPERGEOMETRIC DISTRIBUTION

$$\mu = \frac{n \cdot a}{a + b}$$

**EXAMPLE 8.20**   Among twelve school buses, five have worn brakes. If six of these buses are randomly picked for inspection, how many of them can be expected to have worn brakes?

**Solution**   Since we are sampling without replacement, we have here a hypergeometric situation with $a = 5, b = 7$, and $n = 6$. Substituting these values into the special formula for the mean of a hypergeometric distribution, we get

$$\mu = \frac{6 \cdot 5}{5 + 7} = 2.5$$

This should not come as a surprise—half of the school buses are chosen for inspection and, hence, half of the ones with faulty brakes can be expected to be included in the sample.   ■

Also, the mean of the Poisson distribution with the parameter $\lambda$ is $\mu = \lambda$, and this agrees with what we suggested earlier, namely, that $\lambda$ is to be interpreted as an average. Derivations of all these special formulas may be found in textbooks on mathematical statistics.

# 8.8 THE STANDARD DEVIATION OF A PROBABILITY DISTRIBUTION

In Chapter 4 we saw that the most widely used measures of variation are the variance and its square root, the standard deviation, which measure variability by averaging the squared deviations from the mean. For probability distributions we measure variability in nearly the same way, but instead of averaging the squared deviations from the mean, we find their expected value. In general, if a random variable takes on the values $x_1, x_2, x_3, \ldots$, or $x_k$, with the probabilities $f(x_1), f(x_2), f(x_3), \ldots$, and $f(x_k)$, and the mean of this probability distribution is $\mu$, then the deviations from the mean are $x_1 - \mu, x_2 - \mu, x_3 - \mu, \ldots$, and $x_k - \mu$, and the expected value of their squares is

$$(x_1 - \mu)^2 \cdot f(x_1) + (x_2 - \mu)^2 \cdot f(x_2) + \cdots + (x_k - \mu)^2 \cdot f(x_k)$$

Thus, in the $\sum$ notation we write

**V**ARIANCE OF
A PROBABILITY
DISTRIBUTION

$$\sigma^2 = \sum (x - \mu)^2 \cdot f(x)$$

which we refer to as the **variance of the random variable** or **the variance of its probability distribution**. As in the preceding section, and for the same reason, we denote this description of a probability distribution with the same symbol as the corresponding description of a population. Since the square root of the variance of a population, like that of the variance of a sample, is an important measure of variation, we refer to it as the **population standard variation**.

**EXAMPLE 8.21**    With reference to Example 8.5, find the **standard deviation** of the number of persons among six shoppers at the supermarket, who will take advantage of the special promotion.

**Solution**    In Example 8.19 we showed that $\mu = 1.80$ for this random variable, so that we can arrange the calculations as follows:

| Number of persons | Probability | Deviation from mean | Squared deviation from mean | $(x - \mu)^2 f(x)$ |
|---|---|---|---|---|
| 0 | 0.118 | −1.8 | 3.24 | 0.38232 |
| 1 | 0.303 | −0.8 | 0.64 | 0.19392 |
| 2 | 0.324 | 0.2 | 0.04 | 0.01296 |
| 3 | 0.185 | 1.2 | 1.44 | 0.26640 |
| 4 | 0.060 | 2.2 | 4.84 | 0.29040 |
| 5 | 0.010 | 3.2 | 10.24 | 0.10240 |
| 6 | 0.001 | 4.2 | 17.64 | 0.01764 |
| | | | | $\sigma^2 = 1.26604$ |

The values in the right-hand column were obtained by multiplying the squared deviations from the mean by their probabilities, and the total of this column is the variance of the distribution. Thus, the standard deviation is

$$\sigma = \sqrt{1.26604} = 1.13$$

rounded to two decimals.   ∎

The calculations were easy in this example, although we could have simplified them by using the shortcut formula

$$\sigma^2 = \sum x^2 \cdot f(x) - \mu^2$$

Easier yet would have been to use the special formula for the variance of a binomial distribution, namely,

**VARIANCE OF BINOMIAL DISTRIBUTION**

$$\sigma^2 = np(1 - p)$$

For Example 8.21, where we had $n = 6$ and $p = 0.30$, this would have yielded $\sigma^2 = 6(0.30)(0.70) = 1.26$ and $\sigma = 1.12$. The difference between the results obtained here and previously is due to rounding. Had we carried the square roots an extra decimal, the difference between the results would have been only 0.003.

Intuitively speaking, the standard deviation of a probability distribution measures the expected size of the chance fluctuations of a corresponding random variable. When $\sigma$ is small, there is a high probability that we will get a value close to the mean; when $\sigma$ is large, we are more likely to get a value far away from the mean. This important idea is expressed formally by Chebyshev's theorem, which we introduced in Section 4.3, as it pertains to numerical data. For probability distributions, Chebyshev's theorem may be stated as follows:

**CHEBYSHEV'S THEOREM**

*The probability that a random variable will take on a value within $k$ standard deviations of the mean is at least $1 - \frac{1}{k^2}$.*

Thus, the probability of getting a value within two standard deviations of the mean (a value between $\mu - 2\sigma$ and $\mu + 2\sigma$) is at least $1 - \frac{1}{2^2} = \frac{3}{4}$, the probability of getting a value within five standard deviations of the mean (a value between $\mu - 5\sigma$ and $\mu + 5\sigma$) is at least $1 - \frac{1}{5^2} = \frac{24}{25}$, and so forth. Note that in the formulation of this theorem, the phrase "within $k$ standard deviations of the mean" does not include the endpoints $\mu - k\sigma$ and $\mu + k\sigma$.

**EXAMPLE 8.22**   The number of telephone calls that an answering service receives between 9 A.M. and 10 A.M. is a random variable whose distribution has the mean $\mu = 27.5$ and the standard deviation $\sigma = 3.2$. What does Chebyshev's theorem with $k = 3$ tell us about the number of telephone calls that the answering service can be expected to receive between 9 A.M. and 10 A.M.?

**Solution**   Since $\mu - 3\sigma = 27.5 - 3(3.2) = 17.9$ and $\mu + 3\sigma = 27.5 + 3(3.2) = 37.1$, we can assert with a probability of at least $1 - \frac{1}{3^2} = \frac{8}{9}$, or approximately 0.89, that the answering service will receive between 17.9 and 37.1 calls, namely, anywhere from 18 to 37 calls.   ∎

**EXAMPLE 8.23**   What does Chebyshev's theorem with $k = 5$ tell us about the number of heads, and hence the proportion of heads, we might get in 400 flips of a balanced coin?

**Solution**   Here we are dealing with a random variable having the binomial distribution with $n = 400$ and $p = 0.50$, so that $\mu = 400(0.50) = 200$ and $\sigma = \sqrt{400(0.50)(0.50)} = 10$. Since $\mu - 5\sigma = 200 - 5 \cdot 10 = 150$ and $\mu + 5\sigma = 200 + 5 \cdot 10 = 250$, we can assert with a probability of at least $1 - \frac{1}{5^2} = \frac{24}{25} = 0.96$ that we will get between 150 and 250 heads or that the proportion of heads will be between $\frac{150}{400} = 0.375$ and $\frac{250}{400} = 0.625$.   ■

To continue with this example, the reader will be asked to show in Exercise 8.53 that for $n = 10,000$ flips of a balanced coin the probability is at least 0.96 that the proportion of heads will be between 0.475 and 0.525, and that for 1,000,000 flips of a balanced coin the probability is at least 0.96 that the proportion of heads will be between 0.4975 and 0.5025. All this provides support for the law of large numbers, introduced in Chapter 5 in connection with the frequency interpretation of probability.

**EXERCISES**

**8.40**   Suppose that the probabilities are 0.40, 0.30, 0.20, and 0.10, respectively, that 1, 2, 3, or 4 new cholesterol-lowering drugs will be approved by the FDA in the next year.
  (a)   Use the formula that defines the mean of a probability distribution to find the mean of this distribution.
  (b)   Use the formula that defines the variance of a probability distribution to find the variance of this distribution.

**8.41**   It is usually worthwhile to simplify the calculation of the population variance or the population standard deviation by using the computing formula for $\sigma^2$. Like the one for the sample variance, it has the advantage that we do not have to work with the deviations from the mean. Use this computing formula to rework part (b) of Exercise 8.40.

**8.42**   Under fairly normal conditions, the probabilities are 0.20, 0.50, 0.10, 0.10, and 0.10, respectively, that 0, 1, 2, 3, or 4 hurricanes will strike the Caribbean island of Martinique in any one year. Find the mean and the variance of this probability distribution.

**8.43**   A study shows that 27% of all patients coming to a certain medical clinic have to wait at least half an hour to see their doctor. Use the probabilities in Figure 8.6 to calculate $\mu$ and $\sigma$ for the number of patients, among 12 coming to this clinic, who have to wait at least half an hour to see their doctor.

**8.44**   In Section 8.2 we showed that the probabilities of getting 0, 1, 2, 3, or 4 heads in four flips of a balanced coin are, respectively, $\frac{1}{16}$, $\frac{4}{16}$, $\frac{6}{16}$, $\frac{4}{16}$, and $\frac{1}{16}$. Use the formulas that define $\mu$ and $\sigma^2$ to find the mean and the standard deviation of this probability distribution.

**8.45**   Rework Exercise 8.44, using the special formulas for the mean and the standard deviation of a binomial distribution.

**8.46**   If 80% of certain videocassette recorders will function successfully through the 90-day warranty period, find the mean and standard deviation of the number of these videocassette recorders, among 10 randomly selected, that will function successfully through the 90-day warranty period, using
  (a)   Table V, the formula that defines $\mu$, and the computing formula for $\sigma^2$;

Figure 8.6
Computer printout for
Exercise 8.43.

**Probability Density Function**

Binomial with n = 18 and p = 0.270000

```
      x        P( X = x)
   0.00         0.0035
   1.00         0.0231
   2.00         0.0725
   3.00         0.1431
   4.00         0.1985
   5.00         0.2055
   6.00         0.1647
   7.00         0.1044
   8.00         0.0531
   9.00         0.0218
  10.00         0.0073
  11.00         0.0020
  12.00         0.0004
  13.00         0.0001
  14.00         0.0000
```

(b) the special formulas for the mean and the standard deviation of the binomial distribution.

**8.47**   Find the mean and the standard deviation of each of the following binomial random variables:

(a) the number of heads obtained in 484 flips of a balanced coin;

(b) the number of 3s obtained in 720 rolls of a balanced die;

(c) the number of persons, among 600 invited, who can be expected to attend the opening of a new branch bank, when the probability is 0.30 that any one of them will attend;

(d) the number of defectives in a sample of 600 parts made by a machine, when the probability is 0.04 that any one of the parts is defective;

(e) the number of students, among 800 interviewed, who do not like the food served at the university cafeteria, when the probability is 0.65 that any one of them does not like the food.

**8.48**   A study shows that 60% of all first-class letters between any two cities in Ohio are delivered within 48 hours. Find the mean and the variance of the number of such letters, among eight, that are delivered within 48 hours, using

(a) Table V, the formula that defines $\mu$, and the computing formula for $\sigma^2$;

(b) the special formulas for the mean and the standard deviation of a binomial distribution.

**8.49**   In Example 8.9, which deals with a random variable having a hypergeometric distribution with $a = 5, b = 11$, and $n = 8$, we showed that the probabilities are, respectively, 0.359, 0.128, and 0.013 that it will take on the values 3, 4, and 5. As can easily be verified, the corresponding probabilities for 0, 1, or 2 successes are 0.013, 0.128, and 0.359. Use all these probabilities to calculate the mean of this hypergeometric distribution. Also, use the special formula for the mean of a hypergeometric distribution to verify the result.

**8.50**   In a Southwestern city, the annual number of days with above 100 degree temperatures is a random variable with $\mu = 138$ and $\sigma = 9$. What does Chebyshev's theorem with $k = 4$ tell us about the number of days with above-100-degree temperature there will be in that city in any one year?

**8.51**   If the number of gamma rays emitted per second by a certain radioactive substance is a random variable having the Poisson distribution with $\lambda = 2.5$, the probabilities that it will emit 0, 1, 2, 3, 4, 5, 6, 7, 8, or 9 gamma rays in any one second are, respectively, 0.082, 0.205, 0.256, 0.214, 0.134, 0.067, 0.028, 0.010, 0.003, and 0.001. Calculate the mean and use the result to verify the special formula $\mu = \lambda$ for a random variable having the Poisson distribution with the parameter $\lambda$.

**8.52**  Among eight faculty members considered for promotions, four have Ph.D. degrees and four do not.

(a)  If four of them are chosen at random, find the probabilities that 0, 1, 2, 3, or all 4 of them have Ph.D.'s.

(b)  Use the results of part (a) to determine the mean of this probability distribution.

(c)  Use the special formula for the mean of a hypergeometric distribution to verify the result of part (b).

**8.53**  Use Chebyshev's theorem to show that the probability is at least 0.96 that

(a)  for 10,000 flips of a balanced coin the proportion of heads will be between 0.475 and 0.525;

(b)  for 1,000,000 flips of a balanced coin the proportion of heads will be between 0.4975 and 0.5025.

# CHECKLIST OF KEY TERMS (with page references to their definitions)

Binomial distribution, 182
Discrete random variable, 179
Expected value of a random variable, 198
Geometric distribution, 189
Hypergeometric distribution, 190
Mean of a probability distribution, 199
*Multinomial distribution, 197
Poisson distribution, 193
Population standard deviation 201

Probability distribution, 178, 179
Random variable, 177
Sampling a binomial population, 186
Standard deviation of probability
    distribution, 201
Statistical model, 184
Variance of a probability
    distribution, 201
Variance of a random variable, 201

# REFERENCES

*A great deal of information about various probability distributions may be found in*

HASTINGS, N. A. J., and PEACOCK, J. B., *Statistical Distributions*. London: Butterworth & Company (Publishers) Ltd., 1975.

*More detailed tables of binomial probabilities may be found in*

ROMIG, H. G., *50–100 Binomial Tables*. New York: John Wiley & Sons, Inc., 1953. *Tables of the Binomial Probability Distribution, National Bureau of Standards Applied Mathematics Series No. 6*. Washington, D.C.: U.S. Government Printing Office, 1950.

*and a detailed table of Poisson probabilities is given in*

MOLINA, E. C., *Poisson's Exponential Binomial Limit*. Princeton, N.J.: D. Van Nostrand Company, Inc., 1947.

*The wide availability of computer programs for binomial and Poisson probabilities makes it unlikely that any of the aforementioned tables will be extended or updated.*

# THE NORMAL DISTRIBUTION

**9.1**   Continuous Distributions  207

**9.2**   The Normal Distribution  209

**\*9.3**   A Check for Normality  218

**9.4**   Applications of the Normal Distribution  219

**9.5**   The Normal Approximation to the Binomial
        Distribution  222

        Checklist of Key Terms  228

        References  228

**C**ontinuous random variables arise when we deal with quantities that are measured on a continuous scale. Some examples are the weight of a jar of instant coffee, the tar content of a cigarette, the floor space of an office, and the temperature inside a produce-hauling truck. It is true that there exist infinitely many possibilities in situations like this, but in practice we always round to the nearest integer or to one or more decimals. If we did not round, we would find that the probabilities associated with individual values of continuous random variables are all equal to zero. Surely, we would be willing to give any odds that a car will not be traveling, at exactly $20\pi$ miles per hour, where $\pi$ is the irrational number 3.1415926... which arises in connection with the area of a circle. We should be willing to give any odds that a jar of instant coffee does not contain exactly $\sqrt{36.5} = 6.0415229...$ ounces of coffee. More realistically, we might be interested in the probability that a car will be traveling anywhere from 60 to 65 miles per hour rounded to the nearest mile, which really means that it will be traveling anywhere from 59.5 to 65.5 miles per hour. Similarly, we might be interested in the probability that a jar of instant coffee contains anywhere from 5.9 to 6.1 ounces of coffee rounded to the nearest tenth of an ounce, which really means that it contains anywhere from 5.85 to 6.15 ounces of coffee. Thus, when dealing with continuous random variables we are never really interested in probabilities associated with individual outcomes, but in probabilities associated with intervals or regions.

In this chapter we shall learn how to determine, and work with, probabilities relating to continuous random variables. The place of histograms will be taken by continuous curves, as in Figure 9.1, picturing them mentally as being

**Figure 9.1**
Continuous distribution
curve.

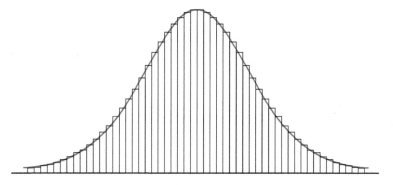

approximated by histograms with narrower and narrower classes. After a general introduction to **continuous distributions** in Section 9.1, we shall devote the remainder of this chapter to the **normal distribution**, which is basic to most of the "bread and butter" techniques of modern statistics. Various applications of the normal distribution will be discussed in Sections 9.4 and 9.5, following the optional material in Section 9.3 that concerns a method of deciding whether observed data follow the general pattern of a normal distribution.

# 9.1 CONTINUOUS DISTRIBUTIONS

In the histograms we have seen thus far, the frequencies, percentages, proportions, or probabilities were represented by the heights of the rectangles, or by their areas. In the continuous case, we also represent probabilities by areas—not by areas of rectangles, but by areas under continuous curves. This is illustrated by Figure 9.2, where the diagram on the left shows a histogram of the probability distribution of a discrete random variable that takes on only the values 0, 1, 2, ... and 10. The probability that it will take on the value 3, for example, is given by the area of the lightly tinted rectangle, and the probability that it will take on a value greater than or equal to 8 is given by the sum of the areas of the three darker tinted rectangles. The diagram on the right pertains to a continuous random variable that can take on any value on the interval from 0 to 10. The probability that it will take on a value on the interval from 2.5 to 3.5, for example, is given by the area of the lightly tinted region under the curve, and the probability that it will take on a value greater than or equal to 8 is given by the area of the darker tinted region under the curve.

Continuous curves such as the one shown on the right in Figure 9.2 are the graphs of functions called **probability densities**, or informally, **continuous**

**Figure 9.2**
Histogram of a prob-
ability distribution and
graph of a continuous
distribution.

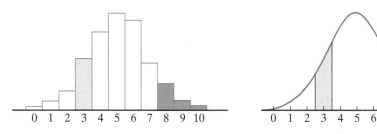

**distributions.** The term "probability density" comes from physics, where the terms "weight" and "density" are used in just about the same way in which we use the terms "probability" and "probability density" in statistics. As is illustrated by Figure 9.3, probability densities are characterized by the fact that *the area under the curve between any two values **a** and **b** gives the probability that a random variable having this continuous distribution will take on a value on the interval from a to b.*

**Figure 9.3**
Continuous
distribution.

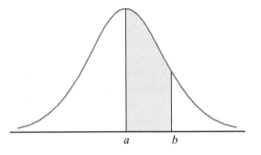

Observe from Figure 9.1 that when *a* and *b* are very close together, the height of the curve and the height of the rectangle with the interval from *a* to *b* as its base are nearly equal. Thus, the area under the curve from *a* to *b* nearly equals that of the corresponding rectangle.

It follows that the values of a continuous distribution must be nonnegative, and that the total area under the curve, representing the certainty that a random variable must take on one of its values, is always equal to 1.

**EXAMPLE  9.1**    Verify that $f(x) = \frac{x}{8}$ can serve as the probability density of a random variable defined over the interval from $x = 0$ to $x = 4$.

**Solution**    The first condition is satisfied since $\frac{x}{8}$ is nonnegative (positive or zero) for all values of $x$ on the interval from 0 to 4. Insofar as the second condition is concerned, it can be seen from Figure 9.4 that the total area under the

**Figure 9.4**
Diagram for
Examples 9.1
and 9.2.

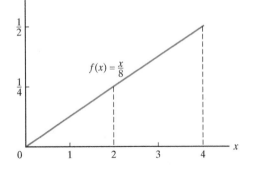

curve from $x = 0$ to $x = 4$ is that of a triangle, whose base is 4 and whose height is $\frac{4}{8} = \frac{1}{2}$. The usual formula for the area of a triangle yields the required $\frac{1}{2} \cdot 4 \cdot \frac{1}{2} = 1$.

**EXAMPLE 9.2**   With reference to Example 9.1, find the probabilities that a random variable having the given probability density will take on a value

(a) less than 2;

(b) less than or equal to 2.

**Solution**   (a) The probability is given by the area of the triangle bounded by the dashed line $x = 2$. Its base is 2, its height is $\frac{2}{8} = \frac{1}{4}$, and its area is $\frac{1}{2} \cdot 2 \cdot \frac{1}{4} = \frac{1}{4}$.

(b) The probability is the same as that of part (a), namely, $\frac{1}{4}$.

This example illustrates the important fact that, in the continuous case, the probability is zero that a random variable will take on any particular value. In our example, the probability is zero that the random variable will take on the value 2, and by 2 we mean *exactly* 2, and we do not include nearby values such as 1.9999998 or 2.0000001.

A consequence of measuring (rather than counting) is that we must assign probability zero to any particular outcome. We assert that the probability is zero that an individual will have a weight of *exactly* 145.27 pounds or that a horse will run a race in *exactly* 58.442 seconds. Observe, however, that even though every particular outcome has probability zero, the process will still produce a value (whether or not we can measure it with extra-fine precision); thus, events of probability zero not only can occur, but must occur when we deal with measured (continuous) random variables.

Statistical descriptions of continuous distributions are as important as descriptions of probability distributions or distributions of observed data, but most of them, including the mean and the standard deviation, cannot be defined without using calculus. Informally, though, we can always picture continuous distributions as being approximated by histograms of probability distributions (see Figure 9.1), whose mean and standard deviation we can calculate. Then, if we choose histograms with narrower and narrower classes, the means and the standard deviations of the corresponding probability distributions will approach the mean and the standard deviation of the continuous distribution. Actually, the mean and the standard deviation of a continuous distribution measure the same properties as the mean and the standard deviation of a probability distribution—the expected value of a random variable having the given distribution, and the square root of the expected value of its squared deviations from the mean. More intuitively, the mean $\mu$ of a continuous distribution is a measure of its center, or middle, and the standard deviation $\sigma$ of a continuous distribution is a measure of its dispersion, or spread.

# 9.2 THE NORMAL DISTRIBUTION

Among many different continuous distributions used in statistics, the most important is the **normal distribution**, whose study dates back to eighteenth-century investigations concerning the nature of errors of measurement. It was observed that discrepancies among repeated measurement of the same physical quantity displayed a surprising degree of regularity. The distribution of the discrepancies could be closely approximated by a certain continuous curve, referred to as the "normal curve of errors" and attributed to the laws of chance. The mathematical equation for this type of curve is

$$f(x) = \frac{1}{\sigma\sqrt{2\pi}}e^{-\frac{1}{2}\left(\frac{x-\mu}{\sigma}\right)^2}$$

for $-\infty < x < \infty$, where $e$ is the irrational number $2.71828\ldots$, which we met on page 193 in connection with the Poisson distribution. We have given this equation only to point out some of the key features of normal distributions; it is not used in any of our calculations.

The graph of a normal distribution is a bell-shaped curve that extends indefinitely in both directions. Although this may not be apparent from a small drawing like that of Figure 9.5, the curve comes closer and closer to the horizontal axis without ever reaching it, no matter how far we go in either direction away from the mean. Fortunately, it is seldom necessary to extend the tails of a normal distribution very far because the area under the curve more than four or five standard deviations away from the mean is negligible for most practical purposes.

**Figure 9.5**
Normal distribution curve.

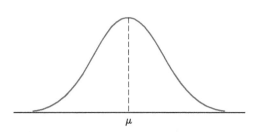

$\mu$

An important feature of normal distributions, apparent from the preceding equation, is that they depend only on the two quantities $\mu$ and $\sigma$, which are, indeed, the mean and the standard deviation. In other words, there is one and only one normal distribution with a given mean $\mu$ and a given standard deviation $\sigma$. The fact that we will get different curves depending on the values of $\mu$ and $\sigma$ is illustrated by Figure 9.6. At the top there are two normal curves with unequal means but equal standard deviations; the curve to the right has the higher mean. In the middle are two normal curves with equal means but unequal standard deviations; the lower and flatter curve has the higher standard deviation. At the bottom are two normal curves with unequal means and unequal standard deviations.

In all our work with normal distributions, we shall be concerned only with areas under their curves—so-called **normal-curve areas**—and such areas are found in practice from tables such as Table I at the end of the book. As it is physically impossible, but also unnecessary, to construct separate tables of normal-curve areas for all conceivable pairs of values of $\mu$ and $\sigma$, we tabulate these areas only for the normal distribution with $\mu = 0$ and $\sigma = 1$, called the **standard normal distribution**. Then, we obtain areas under any normal curve by performing the change of scale (see Figure 9.7) that converts the units of measurement from the original scale, or $x$-scale, into **standard units, standard scores**, or **z-scores**, by means of the formula

**Figure 9.6**
Three pairs of normal distributions.

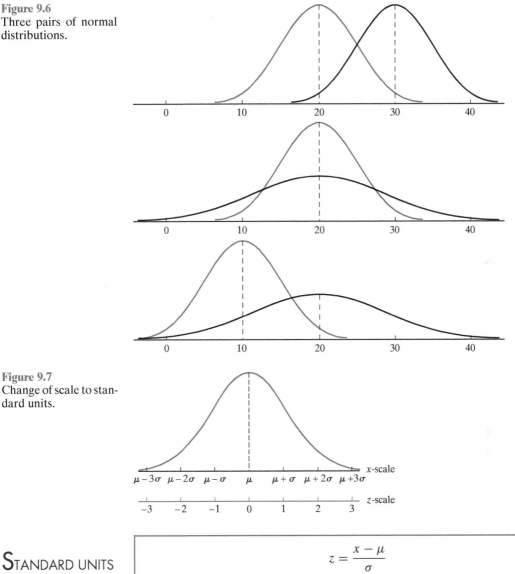

**Figure 9.7**
Change of scale to standard units.

STANDARD UNITS

$$z = \frac{x - \mu}{\sigma}$$

In this new scale, the z-scale, a value of z simply tells us how many standard deviations the corresponding value of x lies above or below the mean of its distribution.

The entries in Table I are the areas under the standard normal curve between the mean $z = 0$ and $z = 0.00, 0.01, 0.02, \ldots, 3.08$, and 3.09, and also $z = 4.00, z = 5.00$, and $z = 6.00$. In other words, the entries in Table I are normal-curve areas like that of the tinted region of Figure 9.8.

Table I has no entries corresponding to negative values of z, for these are not needed by virtue of the symmetry of any normal curve about its mean. This follows from the equation on page 210, which remains unchanged when we substitute

**Figure 9.8**
Tabulated normal
curve area.

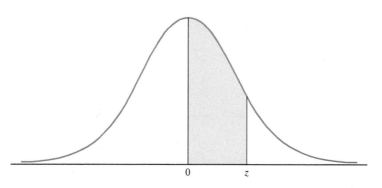

$-(x - \mu)$ for $x - \mu$. Specifically, $f(\mu - a) = f(\mu + a)$, meaning that we get the same value for $f(x)$ when we go the distance $a$ to the left or to the right of $\mu$.

**EXAMPLE 9.3**  Find the standard-normal-curve area between $z = -1.20$ and $z = 0$.

**Solution**  As can be seen from Figure 9.9, the area under the curve between $z = -1.20$ and $z = 0$ equals the area under the curve between $z = 0$ and $z = 1.20$. So we look up the entry corresponding to $z = 1.20$ in Table I and we get 0.3849.  ■

Questions concerning areas under normal distributions arise in various ways, and the ability to find any desired area quickly can be a big help. Although the table gives only areas between $z = 0$ and selected positive values of $z$, we often have to find areas to the left or to the right of given positive or negative values of $z$, or areas between two given values of $z$. This is easy, provided that we remember exactly what areas are represented by the entries in Table I, and also that the standard normal distribution is symmetrical about $z = 0$, so that the area under the curve to the left of $z = 0$ and that to the right of $z = 0$ are both equal to 0.5000.

**EXAMPLE 9.4**  Find the standard-normal-curve area

**(a)** to the left of $z = 0.94$;
**(b)** to the right of $z = -0.65$;
**(c)** to the right of $z = 1.76$;

**Figure 9.9**
Diagram for
Example 9.3.

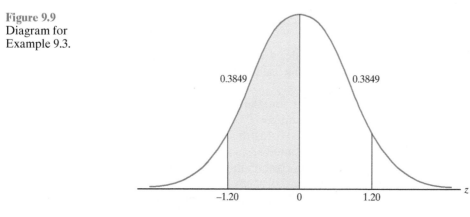

Figure 9.10
Diagram for
Example 9.4.

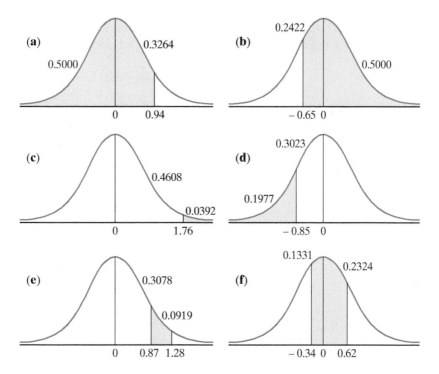

**(d)** to the left of $z = -0.85$;

**(e)** between $z = 0.87$ and $z = 1.28$;

**(f)** between $z = -0.34$ and $z = 0.62$.

**Solution** For each of the six parts, refer to the corresponding diagram in Figure 9.10.

**(a)** The area to the left of $z = 0.94$ is 0.5000 plus the entry in Table I corresponding to $z = 0.94$, or $0.5000 + 0.3264 = 0.8264$.

**(b)** The area to the right of $z = -0.65$ is 0.5000 plus the entry in Table I corresponding to $z = 0.65$, or $0.5000 + 0.2422 = 0.7422$.

**(c)** The area to the right of $z = 1.76$ is 0.5000 minus the entry in Table I corresponding to $z = 1.76$, or $0.5000 - 0.4608 = 0.0392$.

**(d)** The area to the left of $z = -0.85$ is 0.5000 minus the entry in Table I corresponding to $z = 0.85$, or $0.5000 - 0.3023 = 0.1977$.

**(e)** The area between $z = 0.87$ and $z = 1.28$ is the difference between the entries in Table I corresponding to $z = 0.87$ and $z = 1.28$, or $0.3997 - 0.3078 = 0.0919$.

**(f)** The area between $z = -0.34$ and $z = 0.62$ is the sum of the entries in Table I corresponding to $z = 0.34$ and $z = 0.62$, or $0.1331 + 0.2324 = 0.3655$. ◼

In both of the preceding examples we dealt directly with the standard normal distribution. Now let us consider an example where $\mu$ and $\sigma$ are not 0 and 1, so that we must first convert to standard units.

**EXAMPLE 9.5** If a random variable has the normal distribution with $\mu = 10$ and $\sigma = 5$, what is the probability that it will take on a value on the interval from 12 to 15?

**Solution**   The probability is given by the area of the tinted region of Figure 9.11. Converting $x = 12$ and $x = 15$ to standard units, we get

$$z = \frac{12 - 10}{5} = 0.40 \quad \text{and} \quad z = \frac{15 - 10}{5} = 1.00$$

and since the corresponding entries in Table I are 0.1554 and 0.3413, we find that the probability asked for in this example is $0.3413 - 0.1554 = 0.1859$.

**Figure 9.11**
Diagram for
Example 9.5.

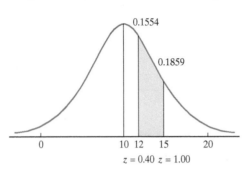

**EXAMPLE 9.6**   If $z_\alpha$ denotes the value of $z$ for which the standard normal-curve area to its right is equal to $\alpha$ (lowercase Greek *alpha*), find

**(a)** $z_{0.01}$;       **(b)** $z_{0.05}$.

**Solution**   For both parts refer to Figure 9.12.

**(a)** Since $z_{0.01}$ corresponds to an entry of $0.5000 - 0.0100 = 0.4900$ in Table I, we look for the entry closest to 0.4900 and find 0.4901 corresponding to $z = 2.33$; thus we let $z_{0.01} = 2.33$.

**(b)** Since $z_{0.05}$ corresponds to an entry of $0.5000 - 0.0500 = 0.4500$ in Table I, we look for the entry closest to 0.4500 and find 0.4495 and 0.4505 corresponding to $z = 1.64$ and $z = 1.65$; thus, we let $z_{0.05} = 1.645$.

Table I also enables us to verify the remark on page 80 that for frequency distributions having the general shape of the cross section of a bell, about 68% of the values will lie within one standard deviation of the mean, about 95% will lie within two standard deviations of the mean, and about 99.7% will lie within three standard deviations of the mean. These percentages apply to frequency distributions having the general shape of a normal distribution, and the reader will be asked to verify them in the first three parts of Exercise 9.15. The other

**Figure 9.12**
Diagram for
Example 9.6.

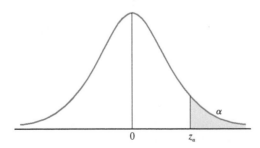

two parts of that exercise show that, although the "tails" extend indefinitely in both directions, the standard-normal-curve area to the right of $z = 4$ or $z = 5$, or to the left of $z = -4$ or $z = -5$, is negligible.

Although this chapter is devoted to the normal distribution, and its importance cannot be denied, it would be a serious mistake to think that the normal distribution is the only continuous distribution that matters in the study of statistics. In Chapter 11 and in subsequent chapters we shall meet other continuous distributions that play important roles, in problems of statistical inference.

**EXERCISES**

**9.1**   For each case involving areas under the standard normal curve, decide whether the first or second area is bigger or the two areas are equal:
  (a)   the area to the right of $z = 1.5$ or the area to the right of $z = 2$;
  (b)   the area to the left of $z = -1.5$ or the area to the left of $z = -2$;
  (c)   the area to the right of $z = 2$ or the area to the left of $z = -2$.

**9.2**   For each case involving areas under the standard normal curve, decide whether the first or second area is bigger or the two areas are equal:
  (a)   the area to the left of $z = 0$ or the area to the right of $z = -0.1$;
  (b)   the area to the right of $z = 0$ or the area to the left of $z = 0$;
  (c)   the area to the right of $z = -1.4$ or the area to the left of $z = -1.4$.

**9.3**   For each case involving areas under the standard normal curve, decide whether the first or second area is bigger or the two areas are equal:
  (a)   the area between $z = 0$ and $z = 1.3$ or the area between $z = 0$ and $z = 1$;
  (b)   the area between $z = -0.2$ and $z = 0.2$ or the area between $z = -0.4$ and $z = 0.4$;
  (c)   the area between $z = -1$ and $z = 1.5$ or the area between $z = -1.5$ and $z = 1$.

**9.4**   Find the standard normal curve area that lies
  (a)   between $z = 0$ and $z = 0.87$;
  (b)   between $z = -1.66$ and $z = 0$;
  (c)   to the right of $z = 0.48$.

**9.5**   Find the standard normal curve area that lies
  (a)   to the right of $z = -0.27$;
  (b)   to the left of $z = 1.30$;
  (c)   to the left of $z = -0.79$.

**9.6**   Find the area under the standard normal curve that lies
  (a)   between $z = 0.45$ and $z = 1.23$;
  (b)   between $z = -1.15$ and $z = 1.85$;
  (c)   between $z = -1.35$ and $z = 0.48$.

**9.7**   Find the standard normal curve area that lies
  (a)   between $z = -0.77$ and $z = 0.77$;
  (b)   to the right of $z = -1.39$.

**9.8**   Find the standard normal curve area that lies
  (a)   to the left of $z = 0.27$;
  (b)   between $z = 1.69$ and $z = 2.33$.

**9.9**   Find $z$ if
  (a)   the normal curve area between 0 and $z$ is 0.4484;
  (b)   the normal curve area to the left of $z$ is 0.9868.

**9.10**  Find $z$ if
  (a)  the normal curve area to the right of $z$ is 0.8413;
  (b)  the normal curve area to the left of $z$ is 0.8051.

**9.11**  Find the area under the graph of the standard normal distribution which lies
  (a)  between $z = 0.85$ and $z = 1.85$;
  (b)  between $z = -2.00$ and $z = -0.57$.

**9.12**  For each case involving random variables with normal distributions, decide whether the first probability is bigger, the second probability is bigger, or the two probabilities are equal:
  (a)  the probability that a random variable having the normal distribution with $\mu = 50$ and $\sigma = 10$ takes on a value less than 60 or the probability that a random variable having the normal distribution with $\mu = 500$ and $\sigma = 100$ takes on a value less than 600;
  (b)  the probability that a random variable having the normal distribution with $\mu = 40$ and $\sigma = 5$ takes on a value greater than 40 or the probability that a random variable having the normal distribution with $\mu = 50$ and $\sigma = 5$ takes on a value greater than 40;
  (c)  the probability that a random variable having the normal distribution with $\mu = 50$ and $\sigma = 10$ takes on a value less than 60 or the probability that a random variable having the normal distribution with $\mu = 50$ and $\sigma = 20$ takes on a value less than 60;
  (d)  the probability that a random variable having the normal distribution with $\mu = 100$ and $\sigma = 5$ takes on a value greater than 110 or the probability that a random variable having the normal distribution with $\mu = 108$ and $\sigma = 5$ takes on a value greater than 110.

**9.13**  Find $z$ if the standard-normal-curve area
  (a)  between 0 and $z$ is 0.4788;
  (b)  to the left of $z$ is 0.8365;
  (c)  between $-z$ and $z$ is 0.8584;
  (d)  to the left of $z$ is 0.3409.

**9.14**  Find $z$ if the standard-normal-curve area
  (a)  between 0 and $z$ is 0.1480;
  (b)  to the right of $z$ is 0.7324;
  (c)  between $-z$ and $z$ is 0.9328.

**9.15**  Find the area under the standard normal curve between $-z$ and $z$ if
  (a)  $z = 1$;
  (b)  $z = 2$;
  (c)  $z = 3$;
  (d)  $z = 4$;
  (e)  $z = 5$.

**9.16**  With $z_\alpha$ defined as in Example 9.6, verify that
  (a)  $z_{0.025} = 1.96$;
  (b)  $z_{0.005} = 2.575$.

**9.17**  If a random variable has the normal distribution with $\mu = 82.0$ and $\sigma = 4.8$, find the probabilities that it will take on a value
  (a)  less than 89.2;
  (b)  greater than 78.4;

(c)  between 83.2 and 88.0;

(d)  between 73.6 and 90.4.

**9.18**  Given a normal curve with $\mu = 36.3$ and $\sigma = 8.1$, find the area under the curve between 31.6 and 40.1.

**9.19**  A normal distribution has the mean $\mu = 62.4$. Find its standard deviation if 20% of the area under the curve lies to the right of 79.2.

**9.20**  A random variable has a normal distribution with $\sigma = 10$. If the probability that the random variable will take on a value less than 82.5 is 0.8212, what is the probability that it will take on a value greater than 58.3?

**9.21**  Another continuous distribution, called the **exponential distribution**, has many important applications. If a random variable has an exponential distribution with the mean $\mu$, the probability that it will take on a value between 0 and any given nonnegative value of $x$ is $1 - e^{-x/\mu}$ (see Figure 9.13). Here $e$ is the irrational number that also appears in the formula for the normal distribution. Many calculators have keys for computing expressions of the form $e^{-x/\mu}$, and selected values may be obtained from Table XII. Find the probability that a random variable having the exponential distribution with $\mu = 10$ will take on a value

(a)  less than 4;

(b)  between 5 and 9;

**9.22**  The lifetime of a certain electronic component is a random variable that has the exponential distribution with the mean $\mu = 2,000$ hours. Use the formula of Exercise 9.21 to find the probabilities that such a component will last

(a)  at most 2,400 hours;

(b)  at least 1,600 hours;

(c)  between 1,800 and 2,200 hours.

**9.23**  According to medical research, the time between successive reports of a rare tropical disease is a random variable having the exponential distribution with the mean $\mu = 120$ days. Find the probabilities that the time between successive reports of the disease will

(a)  exceed 240 days;

(b)  exceed 360 days;

(c)  be less than 60 days.

**9.24**  In the region around a geological fault line, the time between aftershocks that follow a major earthquake is a random variable with the exponential distribution with the mean $\mu = 36$ hours. Find the probabilities that the time between successive aftershocks will

(a)  be less than 18 hours;

(b)  exceed 72 hours;

(c)  be between 36 and 108 hours.

**Figure 9.13**
Exponential
distribution.

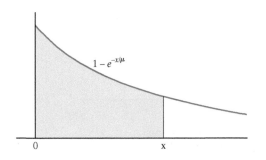

# *9.3 A CHECK FOR NORMALITY

In many of the procedures we shall discuss in subsequent chapters, it will be assumed that our data come from normal populations; namely, that we are dealing with values of random variables having normal distributions. For many years, this assumption was checked with the use of a special kind of graph paper, called **normal probability paper** or **arithmetic probability paper**. Such graph paper could be obtained at most college or university bookstores, stationary stores, or at businesses offering technical supplies. Nowadays, normal probability paper has become a curio, difficult to find, since the job of checking for normality has been taken over by computers (with special statistical software) and other technology. As we have pointed out repeatedly, computers with appropriate software or some other technology can be very helpful in statistics, but neither of these tools is essential for the use of this text. That is why this section is marked optional with an asterisk.

Actually, the plots shown in Figures 9.14 and 19.15 share the characteristic scales of normal probability paper. The horizontal axes, used for the values of a random variable, the same one for both diagrams, have ordinary scales with equal subdivisions. On the other hand, the vertical axes are scaled in such a way that the graph of any cumulative normal distribution becomes a straight line. This is precisely how we judge the "normality" of any set of data. *If the pattern we get for our data is nearly that of a straight line, we are justified in concluding that they come from a **normal population**.*

Figure 9.14 shows an example of a **normal probability plot** obtained with the use of a computer and MINITAB software. Figure 9.15 does the same

**Figure 9.14**
Normal probability plot obtained with computer and MINITAB software.

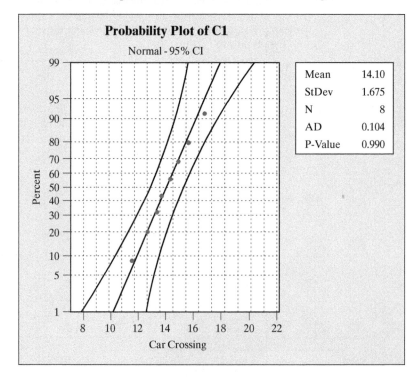

**Figure 9.15**
Normal probability
plot obtained with
a TI-83Plus graphing
calculator.

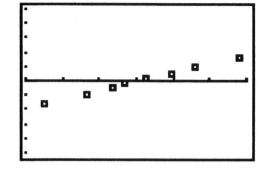

with the use of a graphing calculator. Either of these normal probability plots could serve as a preliminary procedure for evaluating, say, the durability of a new paint for highway center lines. It so happens that the statistical procedure used by a Department of Motor Vehicles to judge the durability of such paint works only when its data constitute a sample from a normal population. So the Department of Motor Vehicles painted test strips across heavily traveled roads in eight different locations and found that the paint deteriorated after having been crossed by 14.26, 16.78, 13.65, 11.53, 12.64, 13.37, 15.60, and 14.94 million cars. Then, entering these data into their computer and following the instructions of their MINITAB program for a normal probability plot, they obtained the normal probability plot shown in Figure 9.14. As can be seen by inspection, the eight points are all very close to a straight line; most certainly, close enough to conclude that the data constitute a sample from a normal population. On a less subjective level, one can continue with one of several more formal testing procedures of the normality of a set of sample data. Information for several such procedures, not needed at this time, was deleted from the MINITAB display in Figure 9.14.

Graphing calculators generate probability plots in a way that is somewhat different from the MINITAB-generated plot of Figure 9.14. The one shown in Figure 9.15 pertains to the same data, but the theory behind it is different. Again, the linearity of the plot is a sign of normality; that is, it indicates that the data are values of a random variable having a normal or a near-normal distribution.

# 9.4 APPLICATIONS OF THE NORMAL DISTRIBUTION

Let us now consider some applications where it can be assumed in each case that the random variables under consideration have normal distributions or can be approximated closely with normal distributions. The most straightforward applications of normal distributions arise when we are given the values of both of its parameters, $\mu$ and $\sigma$. As we pointed out on page 211 in connection with the equation of the normal curve, these two parameters completely specify the normal distribution; that is, any area under the normal curve can be determined by converting to standard units and then referring to Table I at the end of the book. What follows illustrates this kind of application.

EXAMPLE **9.7**    If the amount of cosmic radiation to which a person is exposed while flying by jet across the United States is a random variable having a normal distribution with

$\mu = 4.35$ mrem and $\sigma = 0.59$ mrem, find the probabilities that a person on such a flight will be exposed to

(a) more than 5.00 mrem of cosmic radiation;

(b) anywhere from 3.00 to 4.00 mrem of cosmic radiation.

**Solution**   (a) This probability is given by the area of the tinted region under the curve of the diagram at the top of Figure 9.16, namely, the area under the curve to the right of

$$z = \frac{5.00 - 4.35}{0.59} \approx 1.10$$

Since the entry in Table I corresponding to $z = 1.10$ is 0.3643, we find that the probability is $0.5000 - 0.3643 = 0.1357$, or approximately 0.14, that a person will be exposed to more than 5.00 mrem of cosmic radiation on such a flight.

(b) This probability is given by the area of the tinted region under the curve of the diagram at the bottom of Figure 9.16, namely, the area under the curve between

$$z = \frac{3.00 - 4.35}{0.59} \approx -2.29 \quad \text{and} \quad z = \frac{4.00 - 4.35}{0.59} \approx -0.59$$

Since the entries in Table I corresponding to $z = 2.29$ and $z = 0.59$ are, respectively, 0.4890 and 0.2224, we find that the probability is $0.4890 - 0.2224 = 0.2666$, or approximately 0.27, that a person will be exposed to anywhere from 3.00 to 4.00 mrem of cosmic radiation on such a flight.  ▪

Although the normal distribution is a continuous distribution that applies to continuous random variables, it is often used to approximate distributions of discrete random variables, which can take on only a finite number of values

**Figure 9.16**
Diagram for
Example 9.7.

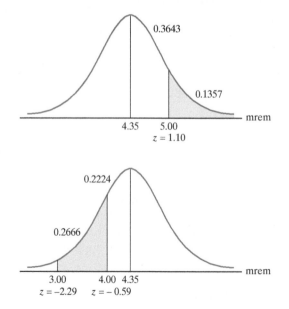

or as many values as there are positive integers. To do this, we must use the **continuity correction** illustrated in the following example. Otherwise, it is again an application with given (actually estimated) values of the mean and the standard deviation.

EXAMPLE  **9.8**   In a study of aggressive behavior, male white mice, returned to the group in which they live after four weeks of isolation, averaged 18.6 fights in the first five minutes with a standard deviation of 3.3 fights. If it can be assumed that the distribution of this random variable (the number of fights which such a mouse gets into under the stated conditions) can be approximated closely with a normal distribution, what is the probability that such a mouse will get into at least 15 fights in the first five minutes?

**Solution**   The answer is given by the area of the tinted region under the curve in Figure 9.17, namely, by the area under the curve to the right of 14.5, and not 15. The reason for this is that the number of fights is a whole number. Hence, if we want to approximate the distribution of this random variable with a normal distribution, we must "spread" its values over a continuous scale, and we do this by representing each whole number $k$ by the interval from $k - \frac{1}{2}$ to $k + \frac{1}{2}$. For instance, 5 is represented by the interval from 4.5 to 5.5, 10 is represented by the interval from 9.5 to 10.5, 20 is represented by the interval from 19.5 to 20.5, and the probability of 15 or more is given by the area under the curve to the right of 14.5. Accordingly, we get

$$z = \frac{14.5 - 18.6}{3.3} \approx -1.24$$

and it follows from Table I that the area of the tinted region under the curve in Figure 9.17, the probability that such a mouse will get into at least 15 fights in the first five minutes is $0.5000 + 0.3925 = 0.8925$, or approximately 0.89.   ◼

**Figure 9.17**
Diagram for
Example 9.8.

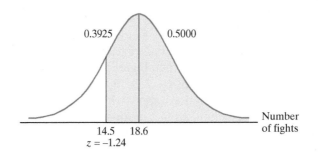

Somewhat more difficult are the applications where only one of the parameters, $\mu$ or $\sigma$, is given, supplemented by a value of $x$ and the area under the curve to its left or to its right. What follows illustrates this kind of application with given values of $\sigma$ and $x$, and the area under the curve to the left of this value of $x$.

EXAMPLE  **9.9**   The actual amount of instant coffee that a filling machine deposits into "6-ounce" jars varies from jar to jar and may be looked upon as a random variable having

**Figure 9.18**
Diagram for
Example 9.9.

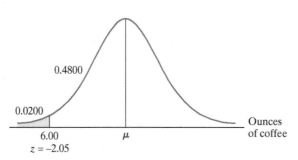

a normal distribution with a standard deviation of 0.04 ounce. If only 2% of the jars are to contain less than 6 ounces of coffee, what must be the mean fill of these jars?

**Solution**   Here we are given $\sigma = 0.04$, $x = 6.00$, a normal-curve area (that of the tinted region under the curve in Figure 9.18), and we are asked to find $\mu$. The value of $z$ for which the entry in Table I is closest to $0.5000 - 0.0200 = 0.4800$ is $z = 2.05$ corresponding to 0.4798, so that

$$-2.05 = \frac{6.00 - \mu}{0.04}$$

Then, solving for $\mu$, we get $6.00 - \mu = -2.05(0.04) = -0.082$ and $\mu = 6.00 + 0.082 = 6.082$, or approximately 6.08. The mean fill must be 6.08 ounces.   ■

All the examples of this section dealt with random variables having normal distributions, or distributions that can be approximated closely with normal curves. When we observe a value (or values) of a random variable having a normal distribution, we may say that we are **sampling a normal population**; this is consistent with the terminology introduced at the end of Section 8.3.

# **9.5** THE NORMAL APPROXIMATION TO THE BINOMIAL DISTRIBUTION

The normal distribution provides a close approximation to the binomial distribution when $n$, the number of trials, is large and $p$, the probability of a success on an individual trial, is close to $\frac{1}{2}$. Figure 9.19 shows the histograms of binomial distributions with $p = \frac{1}{2}$ and $n = 2, 5, 10$, and 25, and it can be seen that with increasing $n$ these distributions approach the symmetrical bell-shaped pattern of the normal distribution. In fact, normal distributions with the mean $\mu = np$ and the standard deviation $\sigma = \sqrt{np(1 - p)}$ can often be used to approximate

**Figure 9.19**
Binomial distributions
with $p = \frac{1}{2}$.

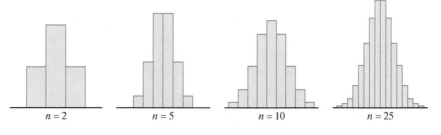

binomial probabilities when $n$ is not all that large and $p$ differs quite a bit from $\frac{1}{2}$. Since "not all that large" and "differs quite a bit" are not very precise terms, let us state the following rule of thumb:

---

**It is considered sound practice to use the normal approximation to the binomial distribution only when $np$ and $n(1 - p)$ are both greater than 5; symbolically, when**

$$np > 5 \quad \text{and} \quad n(1 - p) > 5$$

---

Figure 9.19 shows clearly how the shape of a binomial distribution approaches that of a normal distribution and it reminds us that we are approximating the distribution of a discrete random variable (indeed, a finite random variable) with the distribution of one that is continuous. We are prepared for this from the solution of Example 9.8, where we introduced the continuity correction of spreading each whole number $k$ over the continuous interval from $k - \frac{1}{2}$ to $k + \frac{1}{2}$.

**EXAMPLE  9.10**   Use the normal distribution to approximate the binomial probability of getting 6 heads and 10 tails in 16 flips of a balanced coin, and compare the result with the corresponding value in Table V.

**Figure 9.20**
Diagram for
Example 9.10.

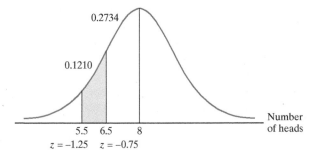

**Solution**   With $n = 16$ and $p = \frac{1}{2}$, we find that $np = 16 \cdot \frac{1}{2} = 8$ and $n(1 - p) = 16(1 - \frac{1}{2}) = 8$ are both greater than 5 and, hence, that the normal approximation is justified. The normal approximation to the probability of 6 heads and 10 tails is given by the tinted area under the curve in Figure 9.20 with 6 heads represented by the interval from 5.5 to 6.5. Since $\mu = 16 \cdot \frac{1}{2} = 8$ and $\sigma = \sqrt{16 \cdot \frac{1}{2} \cdot \frac{1}{2}} = 2$, we get

$$z = \frac{5.5 - 8}{2} = -1.25 \quad \text{and} \quad z = \frac{6.5 - 8}{2} = -0.75$$

in standard units for $x = 5.5$ and $x = 6.5$. The entries corresponding to $z = 1.25$ and $z = 0.75$ in Table I are 0.3944 and 0.2734, and hence we get $0.3944 - 0.2734 = 0.1210$ for the normal approximation to the binomial probability of getting 6 heads and 10 tails in 16 flips of a balanced coin. Since the corresponding entry in Table V is 0.122, it follows that the **percentage error** is $\frac{0.001}{0.122} \cdot 100\% = 0.82\%$.

The normal approximation to the binomial distribution is of special help if, without it, the individual binomial probabilities would have to be calculated for numerous values of $x$. For instance, in the example that follows we would have to determine and add the individual probabilities for 9, 10, 11, ..., 148, 149, and 150 successes in 150 trials, unless we used some form of approximation to get the probability of at least 9 successes in 150 trials directly, say, in terms of one normal curve area.

**EXAMPLE  9.11**    Suppose that 5% of the adobe bricks shipped by a manufacturer have minor blemishes. Use the normal approximation to the binomial distribution to approximate the probability that among 150 adobe bricks shipped by the manufacturer at least nine will have minor blemishes. Also use the computer printout of Figure 8.4 on page 194 to calculate the error and the percentage error of this approximation.

**Solution**    Since $150(0.05) = 7.5$ and $150(1 - 0.05) = 142.5$ are both greater than 5, the rule of thumb for using the normal approximation to the binomial distribution is satisfied. In accordance with the continuity correction explained on page 221, we represent 9 of the bricks by the interval from 8.5 to 9.5; that is, we shall have to determine the tinted region under the curve in Figure 9.21. Since $\mu = 150(0.05) = 7.5$ and $\sigma = \sqrt{150(0.05)(0.95)} \approx 2.67$, we get

$$z = \frac{8.5 - 7.5}{2.67} \approx 0.37$$

in standard units for $x = 8.5$. The corresponding entry in Table I is 0.1443, and hence we get $0.5000 - 0.1443 = 0.3557$ for the normal approximation to the probability that at least 9 of the adobe bricks will have minor blemishes. Since the printout of Figure 8.4 shows that the corresponding binomial probability is $0.1171 + 0.0869 + 0.0582 + \cdots + 0.0001 = 0.3361$, we find that the error of our approximation is $0.3557 - 0.3361 = 0.0196$, or approximately 0.02. The corresponding percentage error is $\frac{0.0196}{0.3361} \cdot 100 \approx 5.8\%$. ∎

The percentage error obtained in this example is fairly substantial, and it serves to illustrate that satisfying the rule of thumb on page 223 does not necessarily guarantee that we will get a good approximation. For instance, when $p$ is quite small and we want to approximate the probability of a value that is not too close to the mean, the approximation may be quite poor even though the rule of thumb is satisfied. If we had wanted to approximate the probability that there will be only one brick with minor blemishes among the 150 shipped by the manufacturer, the percentage error would have been more than 100%. Since we do not expect the reader to become an instant expert at using the normal approximation to the binomial distribution, let us add that we have presented it

**Figure 9.21**
Diagram for
Example 9.11.

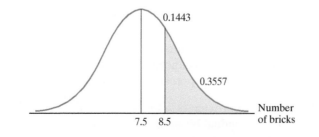

0.1443

0.3557

Number
of bricks

7.5    8.5

here mainly because it will be needed in later chapters for large-sample inferences concerning proportions.

EXERCISES

**9.25** The time it takes to assemble an "easy to assemble" bookcase from a Danish import house is a random variable having a normal distribution with $\mu = 17.40$ minutes and $\sigma = 2.20$ minutes. What is the probability that this kind of bookcase can be assembled by a person in

(a) less than 19.0 minutes;

(b) anywhere from 12.0 to 15.0 minutes?

**9.26** With reference to Exercise 9.25, for what length of time or less is the probability 0.20 that such an assembly can be finished?

**9.27** The reduction of a person's oxygen consumption during periods of transcendental meditation may be looked upon as a random variable having the normal distribution with $\mu = 38.6$ cc per minute and $\sigma = 6.5$ cc per minute. Find the probabilities that during a period of transcendental meditation a person's oxygen consumption will be reduced by

(a) at least 33.4 cc per minute;

(b) at most 34.7 cc per minute.

**9.28** The grapefruits grown in a large orchard have a mean weight of 18.2 ounces with a standard deviation of 1.2 ounces. Assuming that the distribution of the weights of these grapefruits has roughly the shape of a normal distribution, what percentage of the grapefruits weigh

(a) less than 16.1 ounces;

(b) more than 17.3 ounces;

(c) anywhere from 16.7 to 18.8 ounces?

**9.29** With reference to Exercise 9.28, find the weight above which one can expect the heaviest 80% of these grapefruits.

**9.30** A manufacturer needs coil springs that can stand a load of at least 20.0 pounds. Among two suppliers, Supplier A can provide coil springs that, on the average, can stand a load of 24.5 pounds with a standard deviation of 2.1 pounds, and Supplier B can provide coil springs that, on the average, can stand a load of 23.3 pounds with a standard deviation of 1.6 pounds. If we can assume that the distributions of these loads can be approximated with normal distributions, determine which of the two suppliers can provide the manufacturer with the smaller percentage of unsatisfactory coil springs.

**9.31** The lengths of full-grown scorpions of a certain variety have a mean of 1.96 inches and a standard deviation of 0.08 inch. Assuming that the distribution of these lengths has roughly the shape of a normal distribution, find what percentage of these scorpions will be at least 2.0 inches long.

**9.32** With reference to Exercise 9.31, what value would one find the longest 5% of these scorpions?

**9.33** In 1945, after World War II, all service personnel were given point scores based on length of service, number of purple hearts, number of other decorations, number of campaigns, etc. Assuming that the distribution of these point scores can be approximated closely with a normal distribution with $\mu = 63$ and $\sigma = 20$, how many from an army of 8,000,000 would be discharged if the army discharged all those with more than 79.0 points? (Courtesy Department of Mathematics, U.S. Military Academy.)

**9.34**   The distribution of the IQs of the 4,000 employees of a large company has a mean of 104.5, a standard deviation of 13.9, and its shape is roughly that of a normal distribution. Given that a certain job requires a minimum IQ of 95 and bores those with an IQ over 110, how many of the company's employees are suitable for this job on the basis of IQ alone? (Use the continuity correction.)

**9.35**   The daily number of complaints at a department store has close to a normal distribution with $\mu = 25.8$. Also, the odds are 4 to 1 that on any one day there will be at least 18 complaints.

(a)   What is the standard deviation of this normal distribution?

(b)   What is the probability that there will be at least 30 complaints on one day?

**9.36**   The annual number of earthquakes, the world over, is a random variable having approximately a normal distribution with $\mu = 20.8$. What is the standard deviation of this distribution if the probability is 0.70 that there will be at least 18 earthquakes?

**9.37**   It is known from experience that the number of outgoing calls from an office building between 3 and 5 P.M. is a normal random variable (or very close to it) with a mean of 338. If the probability is 0.06 that there will be more than 400 calls, what is the probability that there will be fewer than 300 calls?

**9.38**   A random variable has a normal distribution with $\sigma = 4.0$. If the probability is 0.9713 that it will take on a value less than 82.6, what is the probability that it will take on a value between 70.0 and 80.0?

**9.39**   A student plans to answer a true–false test by flipping a coin. What is the probability that the student will get at least 12 correct answers out of $n = 20$, using

(a)   the normal approximation to the binomial distribution with $n = 20$ and $p = 0.50$;

(b)   Table V?

**9.40**   With reference to Exercise 9.39, find the percentage error of the approximation.

**9.41**   Check in each case whether the conditions for the normal approximation to the binomial distribution are satisfied:

(a)   $n = 32$ and $p = \frac{1}{7}$;

(b)   $n = 75$ and $p = 0.10$;

(c)   $n = 50$ and $p = 0.08$.

**9.42**   Check in each case whether the conditions for the normal approximation to the binomial distribution are satisfied:

(a)   $n = 150$ and $p = 0.05$;

(b)   $n = 60$ and $p = 0.92$;

(c)   $n = 120$ and $p = \frac{1}{20}$.

**9.43**   Records show that 80% of the customers at a restaurant pay with a credit card. Use the normal approximation to the binomial distribution to find the probability that at least 170 out of 200 customers of the restaurant will pay by credit card.

**9.44**   The probability is 0.26 that a cloud seeded with silver iodide will fail to show spectacular growth. Use the normal approximation to the binomial distribution to find the probability that among 30 clouds seeded with silver iodide

(a)   at most 8 will fail to show spectacular growth;

(b)   only 8 will fail to show spectacular growth.

**9.45**   With reference to Exercise 9.44, use the MINITAB printout shown in Figure 9.22 to find the percentage error of both parts of that exercise.

**Figure 9.22**
Binomial distribution with $n = 30$ and $p = 0.26$.

**Probability Density Function**

Binomial with n = 30 and p = 0.260000

| x | P(X = x) |
|---|---|
| 0.00 | 0.0001 |
| 1.00 | 0.0013 |
| 2.00 | 0.0064 |
| 3.00 | 0.0210 |
| 4.00 | 0.0499 |
| 5.00 | 0.0911 |
| 6.00 | 0.1334 |
| 7.00 | 0.1606 |
| 8.00 | 0.1623 |
| 9.00 | 0.1394 |
| 10.00 | 0.1028 |
| 11.00 | 0.0657 |
| 12.00 | 0.0365 |
| 13.00 | 0.0178 |
| 14.00 | 0.0076 |
| 15.00 | 0.0028 |
| 16.00 | 0.0009 |

**Figure 9.23**
Binomial distribution with $n = 50$ and $p = 0.22$.

**Cumulative Distribution Function**

Binomial with n = 50 and p = 0.220000

| x | P(X <= x) |
|---|---|
| 5.00 | 0.0233 |
| 6.00 | 0.0555 |
| 7.00 | 0.1126 |
| 8.00 | 0.1991 |
| 9.00 | 0.3130 |
| 10.00 | 0.4448 |
| 11.00 | 0.5799 |
| 12.00 | 0.7037 |
| 13.00 | 0.8058 |
| 14.00 | 0.8819 |
| 15.00 | 0.9335 |
| 16.00 | 0.9653 |
| 17.00 | 0.9832 |
| 18.00 | 0.9925 |
| 19.00 | 0.9969 |
| 20.00 | 0.9988 |

**9.46**  Studies have shown that 22% of all patients taking a certain antibiotic will get a headache. Use the normal approximation to the binomial distribution to find the probability that among 50 patients taking this antibiotic

(a)  at least 10 will get a headache;

(b)  at most 15 will get a headache.

**9.47**  What is the probability of getting at least 12 replies to questionnaires mailed to 100 persons when the probability is 0.18 that any one of them will reply?

**9.48**  A bank manager has determined from experience that the time required for a security guard to make his rounds in a bank building is a random variable having an approximately normal distribution with $\mu = 18.0$ minutes and $\sigma = 3.2$ minutes. What are the probabilities that a security guard will complete his rounds of the bank building in

(a)  less than 15 minutes;

(b)  15 to 20 minutes;

(c)  more than 20 minutes.

# CHECKLIST OF KEY TERMS (with page references to their definitions)

*Arithmetic probability paper, 218
Continuity correction, 221
Continuous distribution, 207
Continuous random variable, 206
Exponential distribution, 217
Normal approximation to binomial
  distribution, 222
Normal distribution, 207, 209
Normal population, 218
*Normal probability paper, 218

*Normal probability plot, 218
Normal-curve area, 210
Percentage error, 223
Probability density, 207
Sampling a normal population, 222
Standard normal distribution, 210
Standard scores, 210
Standard units, 210
z-scores, 210

# REFERENCES

*Detailed information about various continuous distributions may be found in*

HASTINGS, N. A. J., and PEACOCK, J. B., *Statistical Distributions*. London: Butterworth & Company (Publishers) Ltd., 1975.

*and further information about the normal approximation to the binomial distribution in*

GREEN, J., and ROUND-TURNER, J., "The Error in Approximating Cumulative Binomial and Poisson Probabilities," *Teaching Statistics*, May 1986.

*More extensive tables of normal-curve areas, as well as tables for other continuous distributions, may be found in*

FISHER, R. A., and YATES, F., *Statistical Tables for Biological, Agricultural and Medical Research*. Cambridge: The University Press, 1954.

PEARSON, E. S., and HARTLEY, H. O., *Biometrika Tables for Statisticians*, Vol. I. New York: John Wiley & Sons, Inc., 1968.

# 10

## SAMPLING AND SAMPLING DISTRIBUTIONS

**10.1**   Random Sampling  230

***10.2**   Sample Designs  236

***10.3**   Systematic Sampling  237

***10.4**   Stratified Sampling  237

***10.5**   Cluster Sampling  239

**10.6**   Sampling Distributions  242

**10.7**   The Standard Error of the Mean  245

**10.8**   The Central Limit Theorem  248

**10.9**   Some Further Considerations  249

***10.10**  Technical Note (Simulation)  252

Checklist of Key Terms  254

References  255

Most everyone knows, at least intuitively, what the word "sample" means, and if we check it out, we find that a sample is "a part to show what the rest is like" according to one dictionary, and "a portion, part or piece, taken or shown as representative of the whole" according to another. This may well make sense to some readers, but what about "the rest" and "the whole"?

The whole question, "What is a sample a sample of?" is of critical importance in statistics. For instance, if a buyer of a chain of food stores inspects contents of a few baskets of fresh strawberries at a certain farm in California, are these strawberries to be looked upon as a sample of all the strawberries for sale at that farm on that particular day, all the strawberries which are regularly for sale at that farm, all the strawberries for sale at all farms in that locality, or perhaps all the strawberries for sale at all farms in the state of California? Similarly, if we observe the hair styles of some of the women who visit an art museum on a given day, do our observations reflect the hair styles of all women who visit that museum, all women who enjoy art, or all women in the United States?

As may be apparent, both of these questions can be answered by "Take your pick!", but it should be understood that in some instances we would have good samples which lend themselves to meaningful generalizations, while in other instances it would be utter folly to make any generalization whatsoever. Since this is only one of the problems that arise in connection with sampling, we shall treat this subject here in some detail.

In most of the methods we shall study in the remainder of this book, it will be assumed that we are dealing with so-called **random samples**. We pay this much attention to random samples, which are defined and discussed in Section 10.1, because they permit valid, or logical, generalizations. As we shall see, however, random sampling is not always feasible, or even desirable, and some alternative sampling procedures are mentioned in optional sections of this chapter.

In Section 10.6 we introduce the related concept of a **sampling distribution**, which tells us how quantities determined from samples may vary from sample to sample. Then, in Sections 10.7 through 10.9 we learn how such chance variations can be measured, predicted, and perhaps even controlled.

# 10.1 RANDOM SAMPLING

In Section 3.1 we distinguished between populations and samples, stating that a population consists of all conceivably possible (or hypothetically possible) observations of a given phenomenon, while a sample is simply part of a population. In what follows, we shall distinguish further between two kinds of populations—**finite populations** and **infinite populations**.

A population is finite if it consists of a finite, or fixed, number of elements, measurements, or observations. Examples of finite populations are the net weights of the 3,000 cans of paint in a certain production lot, the SAT scores of all the freshmen admitted to a given college in the fall of the year 2006, the 52 different cards in an ordinary deck of playing cards, and the daily high temperatures recorded at a weather station during the years 2001 through 2006.

In contrast, a population is infinite if, hypothetically at least, it contains infinitely many elements. This would be the case, for example, when we observe values of a continuous random variable, say, when we repeatedly measure the boiling point of a silicon compound. It would also be the case when we observe the totals obtained in repeated rolls of a pair of dice and when we sample with replacement from a finite population. There is no limit to the number of times that we can measure the boiling point of the silicon compound, no limit to the number of times that we can roll a pair of dice, and no limit to the number of times that we can draw an element from a finite population and replace it before the next one is drawn.

To present the idea of **random sampling from a finite population**, let us first see how many different samples of size $n$ can be drawn from a finite population of size $N$. Referring to the rule for the number of combinations of $n$ objects taken $r$ at a time on page 108, we find that, with a change of letters, the answer is $\binom{N}{n}$.

**EXAMPLE 10.1**   How many different samples of size $n$ can be drawn from a finite population of size $N$, when

(a) $n = 2$ and $N = 12$;

(b) $n = 3$ and $N = 50$?

**Solution**   (a) There are $\binom{12}{2} = \dfrac{12 \cdot 11}{2} = 66$ different samples.

(b) There are $\binom{50}{3} = \dfrac{50 \cdot 49 \cdot 48}{3!} = 19{,}600$ different samples.

Based on the result that there are $\binom{N}{n}$ different samples of size $n$ from a finite population of size $N$, let us now give the following definition of a **random sample** (sometimes referred to also as a **simple random sample**) from a finite population:

---

**A sample of size $n$ from a finite population of size $N$ is random if it is chosen in such a way that each of the $\binom{N}{n}$ possible samples has the same probability, $\frac{1}{\binom{N}{n}}$, of being selected.**

---

For instance, if a population consists of the $N = 5$ elements $a$, $b$, $c$, $d$, and $e$ (which might be the annual incomes of five persons, the weights of five cows, or five different models of airplanes), there are $\binom{5}{3} = 10$ possible samples of size $n = 3$. They consist of the elements $abc$, $abd$, $abe$, $acd$, $ace$, $ade$, $bcd$, $bce$, $bde$, and $cde$. If we choose one of these samples in such a way that each one has the probability $\frac{1}{10}$ of being selected, we call this sample a random sample.

Next comes the question of how random samples are drawn in actual practice. In a simple situation like the one described immediately above, we could write each of the ten samples on a slip of paper, put them in a hat, shuffle them thoroughly, and then draw one without looking. Obviously, though, this would be impractical in a more realistically complex situation where $n$ and $N$, or only $N$, are large. For instance, for $n = 4$ and $N = 100$ we would have to label and draw one of $\binom{100}{4} = 3{,}921{,}225$ slips of paper.

Fortunately, we can draw a random sample from a finite population without listing all possible samples, which we mentioned here only to stress the point that the selection of a random sample must depend entirely on chance. Instead of listing all possible samples, we can write each of the $N$ elements of the finite population on a slip of paper, and draw $n$ of them one at a time without replacement, making sure in each of the successive drawings that all of the remaining elements of the population have the same chance of being selected. It is easy to show mathematically that this also leads to the probability $\frac{1}{\binom{N}{n}}$ for each possible sample.

For instance, to take a random sample of $n = 12$ of $N = 138$ archaeological sites for supplementary funding of excavations, we could write the numbers $1, 2, 3, \ldots$, 137, and 138 on 138 slips of paper, mix them up thoroughly (say, in a proverbial hat), and then draw 12, one at a time without replacement, without looking.

Even an easy procedure like this can be simplified further. Nowadays, the easiest way of taking a random sample from a finite population is to base the selection on **random numbers** that are generated by means of statistical calculators or computers.

EXAMPLE **10.2**  With reference to the aforementioned archaeological sites, which presumably have been numbered 001 to 138, use a statistical calculator to generate a random sample of $n = 12$ of the sites.

**Solution**   Using only the first three digits of the four-digit random numbers generated, omitting 000 and those exceeding 138, and also omitting any number that has already been selected, we get 041, 021, 079, 084, 016, 108, 029, 003, 100, 046, 136, and 075. The archaeological sites associated with these numbers constitute the sample.

Had we wanted to use a computer in Example 10.2 instead of a statistical calculator, **MINITAB** or some other software might have yielded a printout similar to the one shown in Figure 10.1. The one shown here consists of the archaeological sites associated with the numbers 24, 131, 69, 113, 127, 5, 57, 52, 7, 13, 11, and 64. (The other numbers in the printout are just the rows and the columns of the worksheet where the computer put the data.)

**Figure 10.1**
Computer-generated sample for Example 10.2.

**Random Integers 1–138**

|   | C1 | C2 | C3 | C4 | C5 | C6 |
|---|----|----|----|----|----|----|
| 1 | 24 | 131 | 69 | 113 | 127 | 5 |
| 2 | 57 | 52 | 7 | 13 | 11 | 64 |

Only a few decades ago, random samples were generated almost exclusively with the use of published tables of random numbers. Such tables consist of page after page of nothing but the digits 0, 1, 2, 3, ..., 8, and 9 arranged in rows and columns. Such tables were constructed with the use of census data, decimal expansions of irrational numbers, tables of logarithms, Selective Service lotteries, and assorted electronic gadgetry. There were even some arithmetical procedures, such as the one where we begin with a three- or four-digit number, square it, and use the middle three or four digits to begin the sequence of random digits. Then we square this number and repeat this process over and over again. Eventually, these methods were replaced by computer-generated tables of random numbers, and even these have been replaced by the methods that we used in Example 10.2 to generate our own random numbers.

Nevertheless, since the necessary technology may not always be available, we shall illustrate here the use of a published table of random numbers. For this purpose, we shall refer to Figure 10.2, which consists of a page reproduced from *Tables of 105,000 Random Decimal Digits*, published by the Interstate Commerce Commission, Bureau of Transport Economics and Statistics.

**EXAMPLE 10.3**   Repeat Example 10.2, reading three-digit numbers off Figure 10.2, with the same restrictions as in the solution of Example 10.2. Use the 11th, 12th, and 13th columns, starting with the 6th row and going down the page. Figure 10.2 is comprised of 35 columns and 50 rows.

**Solution**   Again omitting the number 000 and all numbers exceeding 138, and making sure that no value was chosen more than once, it can easily be verified that the random

**Figure 10.2**
Sample page
of random digits.

| | | | | | | |
|---|---|---|---|---|---|---|
| 94620 | 27963 | 96478 | 21559 | 19246 | 88097 | 44926 |
| 60947 | 60775 | 73181 | 43264 | 56895 | 04232 | 59604 |
| 27499 | 53523 | 63110 | 57106 | 20865 | 91683 | 80688 |
| 01603 | 23156 | 89223 | 43429 | 95353 | 44662 | 59433 |
| 00815 | 01552 | 06392 | 31437 | 70385 | 45863 | 75971 |
| 83844 | 90942 | 74857 | 52419 | 68723 | 47830 | 63010 |
| 06626 | 10042 | 93629 | 37609 | 57215 | 08409 | 81906 |
| 56760 | 63348 | 24949 | 11859 | 29793 | 37457 | 59377 |
| 64416 | 29934 | 00755 | 09418 | 14230 | 62887 | 92683 |
| 63569 | 17906 | 38076 | 32135 | 19096 | 96970 | 75917 |
| 22693 | 35089 | 72994 | 04252 | 23791 | 60249 | 83010 |
| 43413 | 59744 | 01275 | 71326 | 91382 | 45114 | 20245 |
| 09224 | 78530 | 50566 | 49965 | 04851 | 18280 | 14039 |
| 67625 | 34683 | 03142 | 74733 | 63558 | 09665 | 22610 |
| 86874 | 12549 | 98699 | 54952 | 91579 | 26023 | 81076 |
| 54548 | 49505 | 62515 | 63903 | 13193 | 33905 | 66936 |
| 73236 | 66167 | 49728 | 03581 | 40699 | 10396 | 81827 |
| 15220 | 66319 | 13543 | 14071 | 59148 | 95154 | 72852 |
| 16151 | 08029 | 36954 | 03891 | 38313 | 34016 | 18671 |
| 43635 | 84249 | 88984 | 80993 | 55431 | 90793 | 62603 |
| 30193 | 42776 | 85611 | 57635 | 51362 | 79907 | 77364 |
| 37430 | 45246 | 11400 | 20986 | 43996 | 73122 | 88474 |
| 88312 | 93047 | 12088 | 86937 | 70794 | 01041 | 74867 |
| 98995 | 58159 | 04700 | 90443 | 13168 | 31553 | 67891 |
| 51734 | 20849 | 70198 | 67906 | 00880 | 82899 | 66065 |
| 88698 | 41755 | 56216 | 66852 | 17748 | 04963 | 54859 |
| 51865 | 09836 | 73966 | 65711 | 41699 | 11732 | 17173 |
| 40300 | 08852 | 27528 | 84648 | 79589 | 95295 | 72895 |
| 02760 | 28625 | 70476 | 76410 | 32988 | 10194 | 94917 |
| 78450 | 26245 | 91763 | 73117 | 33047 | 03577 | 62599 |
| 50252 | 56911 | 62693 | 73817 | 98693 | 18728 | 94741 |
| 07929 | 66728 | 47761 | 81472 | 44806 | 15592 | 71357 |
| 09030 | 39605 | 87507 | 85446 | 51257 | 89555 | 75520 |
| 56670 | 88445 | 85799 | 76200 | 21795 | 38894 | 58070 |
| 48140 | 13583 | 94911 | 13318 | 64741 | 64336 | 95103 |
| 36764 | 86132 | 12463 | 28385 | 94242 | 32063 | 45233 |
| 14351 | 71381 | 28133 | 68269 | 65145 | 28152 | 39087 |
| 81276 | 00835 | 63835 | 87174 | 42446 | 08882 | 27067 |
| 55524 | 86088 | 00069 | 59254 | 24654 | 77371 | 26409 |
| 78852 | 65889 | 32719 | 13758 | 23937 | 90740 | 16866 |
| 11861 | 69032 | 51915 | 23510 | 32050 | 52052 | 24004 |
| 67699 | 01009 | 07050 | 73324 | 06732 | 27510 | 33761 |
| 50064 | 39500 | 17450 | 18030 | 63124 | 48061 | 59412 |
| 93126 | 17700 | 94400 | 76075 | 08317 | 27324 | 72723 |
| 01657 | 92602 | 41043 | 05686 | 15650 | 29970 | 95877 |
| 13800 | 76690 | 75133 | 60456 | 28491 | 03845 | 11507 |
| 98135 | 42870 | 48578 | 29036 | 69876 | 86563 | 61729 |
| 08313 | 99293 | 00990 | 13595 | 77457 | 79969 | 11339 |
| 90974 | 83965 | 62732 | 85161 | 54330 | 22406 | 86253 |
| 33273 | 61993 | 88407 | 69399 | 17301 | 70975 | 99129 |

numbers we get for the sample are 007, 012, 031, 135, 114, 120, 047, 124, 070, 009, 118, and 094. Thus, the sample consists of the archaeological sites associated with the numbers 7, 12, 31, 135, 114, 120, 47, 124, 70, 9, 118, and 94.

When lists are available so that items can readily be numbered, it is easy to draw random samples with the use of calculators, computers, or random number tables. Unfortunately, however, there are many situations where it is impossible to proceed in the ways we have just described. For instance, if we want to use a sample to estimate the mean outside diameter of thousands of ball bearings packed in a large crate, or if we want to estimate the mean height of the trees in a forest, it would be impossible to number the ball bearings or the trees, choose random numbers, and then locate and measure the corresponding ball bearings or trees. In these and in many similar situations, all we can do is proceed according to the dictionary definition of the word "random," namely, "haphazardly, without aim or purpose." That is, we must not select or reject any element of a population because of its seeming typicalness or lack of it, nor must we favor or ignore any part of a population because of its accessibility or lack of it, and so forth. With some reservations, such samples can often be treated as if they were, in fact, random samples.

So far we have discussed random sampling only in connection with finite populations. For infinite populations we say that

> **A sample of size $n$ from an infinite population is random if it consists of values of independent random variables having the same distribution.**

As we pointed out in connection with the binomial and normal distributions, it is this "same" distribution that we refer to as the population being sampled. Also, by "independent" we mean that the probabilities relating to any one of the random variables are the same regardless of what values may have been observed for the other random variables.

For instance, if we get 2, 5, 1, 3, 6, 4, 4, 5, 2, 4, 1, and 2 in twelve rolls of a die, these numbers constitute a random sample if they are values of independent random variables having the same probability distribution.

$$f(x) = \frac{1}{6} \quad \text{for} \quad x = 1, 2, 3, 4, 5, \text{ or } 6$$

To give another example of a random sample from an infinite population, suppose that eight students obtained the following measurements of the boiling point of a silicon compound: 136, 153, 170, 148, 157, 152, 143, and 150 degrees Celsius. According to the definition, these values constitute a random sample if they are values of independent random variables having the same distribution, say, the normal distribution with $\mu = 152$ and $\sigma = 10$. To judge whether this is actually the case, we would have to ascertain, among other things, that the eight students' measuring techniques are equally precise (so that $\sigma$ is the same for each of the random variables) and that there was no collaboration (which might make the random variables dependent). In practice, it is not an easy task to judge whether a set of data may be looked upon as a random sample, and we shall go into this further in Chapter 18.

**10.1**   How many different samples of size $n = 2$ can be drawn from a finite population of size

(a) $N = 6$;        (c) $N = 32$;

(b) $N = 20$;      (d) $N = 75$?

**10.2**   How many different samples of size $n = 3$ can be drawn from a finite population of size

(a) $N = 8$;        (c) $N = 30$;

(b) $N = 26$;      (d) $N = 40$?

**10.3**   What is the probability of each possible sample if a random sample of size $n = 4$ is drawn from a finite population of size

(a) $N = 12$;

(b) $N = 20$?

**10.4**   What is the probability of each possible sample if a random sample of size $n = 6$ is drawn from a finite population of size

(a) $N = 10$;

(b) $N = 15$?

**10.5**   List the $\binom{6}{3} = 20$ possible samples of size $n = 3$ that can be drawn from a finite population whose elements are denoted by $u$, $v$, $w$, $x$, $y$, and $z$.

**10.6**   With reference to Exercise 10.5, what is the probability that one of the random samples will include the element denoted by $u$?

**10.7**   With reference to Exercise 10.5, what is the probability that one of the random samples will include the elements denoted by $u$ and $v$?

**10.8**   A college bookstore stocks seven different books on the history of art, offering a 10% discount on any three. How many different choices does a customer have?

**10.9**   A person planning a trip to Southern California has friends in Pasadena, Long Beach, San Diego, Oxnard, and Santa Monica. If he randomly chooses three of these places, what are the probabilities

(a) of each possible selection;

(b) that the selection will include Long Beach;

(c) that the selection will include Long Beach and San Diego?

**10.10**   A research organization wants to include 6 of the 50 states of the United States in a marketing survey. If the states are numbered 01, 02, 03, ..., 49, and 50 in alphabetic order and the research organization uses the 6th and 7th columns of the table in Figure 10.2, going down the page starting with the third row, which states will be included in the survey? A list that associates numbers with the elements of a population for obtaining a sample is called a **sampling frame**. For this exercise, such a list may be obtained from a listing of area codes in a telephone directory.

**10.11**   A hematologist wants to recheck a sample of $n = 10$ of the 653 blood specimens analyzed by her laboratory in a given month. In her records, these blood specimens are numbered 3250, 3251, 3252, ..., 3901, and 3902. Which specimens would she select if she used columns 21, 22, and 23 of the table in Figure 10.2, going down the page starting with row 16? (Since all the numbers begin with a 3, this digit can be omitted in the selection of the sample.)

**10.12**  Use a statistical calculator, a graphing calculator, or a computer to rework Exercise 10.11.

**10.13**  Three hundred sales have been recorded in the shoe department of a department store over a period of two weeks, with the corresponding invoices numbered from 251 through 550. If an auditor wants to verify $n = 12$ of the invoices selected at random, which ones would he check if he used the 18th, 19th, and 20th columns of the table in Figure 10.2, going down the page starting with the top row?

**10.14**  Use a statistical calculator, a graphing calculator, or a computer to rework Exercise 10.13.

**10.15**  A county assessor wants to reassess a random sample of 50 of the county's 7,964 one-family homes, and he asks you to help him with the selection of the sample. First you create a sampling frame by assigning these homes the numbers 0001, 0002, 0003, ..., 7963, and 7964, and then you use the first four columns of the table in Figure 10.2, going down the page starting with the first row, and continuing with the next four columns, also going down the page starting with the first row. By numbers, which of the one-family homes would thus be selected?

**10.16**  Use a computer to rework Exercise 10.15.

**10.17**  On page 231 we said that a random sample can be drawn from a finite population without listing all possible samples; instead, we merely number (or label) the $N$ elements of the finite population, and then draw $n$ of them one at a time without replacement, making sure in each of the successive drawings that all of the remaining elements have the same probability of being selected. Verify this for the example on page 231, which dealt with random samples of size $n = 3$ drawn from the finite population that consists of the elements $a$, $b$, $c$, $d$, and $e$, by showing that the probability of any particular sample drawn by this method (say, $bce$) is again $\frac{1}{10}$.

**10.18**  Use the same kind of argument as in Exercise 10.17 to verify that each possible random sample of size $n = 3$, drawn one at a time from a finite population of size $N = 100$, has the probability $1 \left/ \binom{100}{3} \right. = \frac{1}{161,700}$.

**10.19**  Use the same kind of argument as in Exercise 10.17 to verify that each possible random sample of size $n$, drawn one at a time from a finite population of size $N$, has the probability $1 \left/ \binom{N}{n} \right.$.

# *10.2 SAMPLE DESIGNS

The only kind of samples we have discussed so far are random samples, and we did not even consider the possibility that under certain conditions there may be samples that are better (say, easier to obtain, cheaper, or more informative) than random samples, and we did not go into any details about the question of what might be done when random sampling is impossible. Indeed, there are many other ways of selecting a sample from a population, and there is an extensive literature devoted to the subject of designing sampling procedures.

In statistics, a **sample design** is a definite plan, determined completely before any data are actually collected, for obtaining a sample from a given population. Thus, the plan to take a simple random sample of 12 of a city's 247 drugstores by using a table of random numbers in a prescribed way constitutes a sample design.

In the next three sections we shall discuss some sample designs that apply mainly to large-scale operations, surveys and the like, and we might have referred to this material as **survey sampling**. We marked it optional, but not because it is unimportant. The whole subject of survey sampling is rarely covered in a general introductory course in statistics, simply because there is not enough time in the kind of course for which this book is designed.

# *10.3 SYSTEMATIC SAMPLING

In some instances, the most practical way of sampling is to select, say, every 20th name on a list, every 12th house on one side of a street, every 50th piece coming off an assembly line, and so on. This is called **systematic sampling**, and an element of randomness can be introduced into this kind of sampling by using random numbers to pick the unit with which to start. Although a systematic sample may not be a random sample in accordance with the definition, it is often reasonable to treat systematic samples as if they were random samples; indeed, in some instances, systematic samples actually provide an improvement over simple random samples in as much as the samples are spread more evenly over the entire populations.

The real danger in systematic sampling lies in the possible presence of hidden periodicities. For instance, if we inspect every 40th piece made by a particular machine, the results would be very misleading if, because of a regularly recurring failure, every 10th piece produced by the machine has blemishes. Also, a systematic sample might yield biased results if we interview the residents of every 12th house along a certain route and it so happens that each 12th house along the route is a corner house on a double lot.

# *10.4 STRATIFIED SAMPLING

If we have information about the makeup of a population (that is, its composition) and this is of relevance to our investigation, we may be able to improve on random sampling by **stratification**. This is a procedure that consists of stratifying (or dividing) the population into a number of nonoverlapping subpopulations, or **strata**, and then taking a sample from each stratum. If the items selected from each stratum constitute simple random samples, the entire procedure—first stratification and then random sampling—is called **stratified (simple) random sampling**.

Suppose, for instance, that we want to estimate the mean weight of four persons on the basis of a sample of size 2, and that the (unknown) weights of the four persons are 115, 135, 185, and 205 pounds. Thus, the mean weight we want to estimate is

$$\mu = \frac{115 + 135 + 185 + 205}{4} = 160 \text{ pounds}$$

If we take an ordinary random sample of size 2 from this population, the $\binom{4}{2} = 6$ possible samples are 115 and 135, 115 and 185, 115 and 205, 135 and 185, 135 and 205, and 185 and 205, and the corresponding means are 125, 150, 160, 160, 170, and 195. Observe that since each of these samples has the probability $\frac{1}{6}$, the probabilities are $\frac{1}{3}, \frac{1}{3}$, and $\frac{1}{3}$ that our error (the difference between the sample mean and $\mu = 160$) will be 0, 10, or 35.

Now suppose that we know that two of these persons are men and two are women, and suppose that the (unknown) weights of the men are 185 and 205 pounds, while the (unknown) weights of the women are 115 and 135 pounds. Stratifying our sample (by sex) and randomly choosing one of the two men and one of the two women, we find that there are only the four stratified samples 115 and 185, 115 and 205, 135 and 185, and 135 and 205. The means of these samples are 150, 160, 160, and 170, and now the probabilities are $\frac{1}{2}$ and $\frac{1}{2}$ that our error will be 0 or 10. Clearly, stratification has greatly improved our chances of getting a good (close) estimate of the weight of the four persons. See, however, Exercise 10.24.

Essentially, the goal of stratification is to form strata in such a way that there is some relationship between being in a particular stratum and the answer sought in the statistical study, and that within the separate strata there is as much homogeneity (uniformity) as possible. In our example there is such a connection between sex and weight and there is much less variability in weight within each of the two groups than there is within the entire population.

In the preceding example, we used **proportional allocation**, which means that the sizes of the samples from the different strata are proportional to the sizes of the strata. In general, if we divide a population of size $N$ into $k$ strata of size $N_1, N_2, \ldots$, and $N_k$, and take a sample of size $n_1$ from the first stratum, a sample of size $n_2$ from the second stratum, $\ldots$, and a sample of size $n_k$ from the $k$th stratum, we say that the allocation is proportional if

$$\frac{n_1}{N_1} = \frac{n_2}{N_2} = \cdots = \frac{n_k}{N_k}$$

or if these ratios are as nearly equal as possible. In the example dealing with the weights we had $N_1 = 2, N_2 = 2, n_1 = 1$, and $n_2 = 1$, so that

$$\frac{n_1}{N_1} = \frac{n_2}{N_2} = \frac{1}{2}$$

and the allocation was, indeed, proportional.

As can easily be verified, allocation is proportional if

**SAMPLE SIZES FOR PROPORTIONAL ALLOCATION**

$$n_i = \frac{N_i}{N} \cdot n \quad for\ i = 1, 2, \ldots, and\ k$$

where $n = n_1 + n_2 + \cdots + n_k$ is the total size of the sample. When necessary, we use the integers closest to the values given by this formula.

**EXAMPLE  10.4**   A stratified sample of size $n = 60$ is to be taken from a population of size $N = 4,000$, which consists of three strata of size $N_1 = 2,000$, $N_2 = 1,200$, and $N_3 = 800$. If the allocation is to be proportional, how large a sample must be taken from each stratum?

**Solution**   Substituting into the formula, we get

$$n_1 = \frac{2,000}{4,000} \cdot 60 = 30 \qquad n_2 = \frac{1,200}{4,000} \cdot 60 = 18$$

and

$$n_3 = \frac{800}{4,000} \cdot 60 = 12$$

This example illustrates proportional allocation, but we should add that there exist other ways of allocating portions of a sample to the different strata. One of these, called **optimum allocation**, is described in Exercise 10.29. It accounts not only for the size of the strata, as in proportional allocation, but also for the variability (of whatever characteristic is of concern) within the strata.

Also, stratification is not limited to a single variable of classification, or characteristic, and populations are often stratified according to several characteristics. For instance, in a systemwide survey designed to determine the attitude of its students, say, toward a new tuition plan, a state college system with 17 colleges might stratify its sample not only with respect to the colleges, but also with respect to students' class standing, sex, and major. So, part of the sample would be allocated to junior women in college A majoring in engineering, another part to sophomore men in college L majoring in English, and so on. Up to a point, stratification like this, called **cross stratification**, will increase the precision (reliability) of estimates and other generalizations, and it is widely used, particularly in opinion sampling and market research.

In stratified sampling, the cost of taking random samples from the individual strata is often so high that interviewers are simply given quotas to be filled from the different strata, with few (if any) restrictions on how they are to be filled. For instance, in determining voters' attitudes toward increased medical coverage for elderly persons, an interviewer working a certain area might be told to interview 6 male self-employed homeowners under 30 years of age, 10 female wage earners in the 45–60 age bracket who live in apartments, 3 retired males over 60 who live in trailers, and so on, with the actual selection of the individuals being left to the interviewer's discretion. This is called **quota sampling**, and it is a convenient, relatively inexpensive, and sometimes necessary procedure, but as it is often executed, the resulting samples do not have the essential features of random samples. In the absence of any controls on their choice, interviewers naturally tend to select individuals who are most readily available—persons who work in the same building, shop in the same store, or perhaps reside in the same general area. Quota samples are thus essentially **judgment samples**, and inferences based on such samples generally do not lend themselves to any sort of formal statistical evaluation.

# *10.5 CLUSTER SAMPLING

To illustrate another important kind of sampling, suppose that a large foundation wants to study the changing patterns of family expenditures in the San Diego area. In attempting to complete schedules for 1,200 families, the foundation finds that simple random sampling is practically impossible, since suitable lists are not available and the cost of contacting families scattered over a wide area (with possibly two or three callbacks for the not-at-homes) is very high. One way in which a sample can be taken in this situation is to divide the total area of interest into a number of smaller, nonoverlapping areas, say, city blocks. A number of these blocks are then randomly selected, with the ultimate sample consisting of all (or samples of) the families residing in these blocks.

In this kind of sampling, called **cluster sampling**, the total population is divided into a number of relatively small subdivisions, and some of these subdivisions, or clusters, are randomly selected for inclusion in the overall sample. If the clusters are geographic subdivisions, as in the preceding example, this kind of sampling is also called **area sampling**. To give another example of cluster sampling, suppose that the dean of students of a university wants to know how fraternity men at the school feel about a certain new regulation. He can take a cluster sample by interviewing some or all of the members of several randomly selected fraternities.

Although estimates based on cluster samples are usually not as reliable as estimates based on simple random samples of the same size (see Exercise 10.29), they are often more reliable per unit cost. Referring again to the survey of family expenditures in the San Diego area, it is easy to see that it may well be possible to take a cluster sample several times the size of a simple random sample for the same cost. It is much cheaper to visit and interview families living close together in clusters than families selected at random over a wide area.

In practice, several of the methods of sampling we have discussed may well be used in the same study. For instance, if government statisticians want to study the attitude of American elementary school teachers toward certain federal programs, they might first stratify the country by states or some other geographic subdivisions. To take a sample from each stratum, they might then use cluster sampling, subdividing each stratum into a number of smaller geographic subdivisions (say, school districts), and finally, they might use simple random sampling or systematic sampling to select a sample of elementary school teachers within each cluster.

EXERCISES

*10.20  Following are the percentages of persons 25 years old and over with some college education, but no degree, in the 50 states, listed in alphabetic order, as reported in a recent census.

| 16.8 | 27.6 | 25.4 | 16.6 | 22.6 | 24.0 | 15.9 | 16.9 | 19.4 | 17.0 |
| 20.1 | 24.2 | 19.4 | 16.6 | 17.0 | 21.9 | 15.2 | 17.2 | 16.1 | 18.6 |
| 15.8 | 20.4 | 19.0 | 16.9 | 18.4 | 22.1 | 21.1 | 25.8 | 18.0 | 15.5 |
| 20.9 | 15.7 | 16.8 | 20.5 | 17.0 | 21.3 | 25.0 | 12.9 | 15.0 | 15.8 |
| 18.8 | 16.9 | 21.1 | 27.9 | 14.7 | 18.5 | 25.0 | 13.2 | 16.7 | 24.2 |

List the ten possible systematic samples of size $n = 5$ that can be taken from this list by starting with one of the first ten numbers and then taking each tenth number.

*10.21  With reference to Exercise 10.20, list the five possible systematic samples of size $n = 10$ that can be taken from this list by starting with one of the first five numbers and then taking each fifth number.

**\*10.22** The following are consecutive monthly figures on the volume of mail (in millions of ton-miles) carried by domestic air operations over a four-year period.

| | | | | | | | | | | | |
|---|---|---|---|---|---|---|---|---|---|---|---|
| 67 | 62 | 75 | 67 | 70 | 68 | 64 | 70 | 66 | 73 | 73 | 97 |
| 76 | 73 | 80 | 78 | 78 | 72 | 75 | 75 | 73 | 83 | 76 | 108 |
| 84 | 78 | 86 | 85 | 81 | 78 | 78 | 75 | 78 | 86 | 76 | 111 |
| 79 | 77 | 87 | 84 | 82 | 77 | 79 | 77 | 80 | 84 | 78 | 117 |

List the six possible systematic samples of size $n = 8$ that can be taken from this list by starting with one of the first six numbers and then taking each sixth number.

**\*10.23** If one of the six systematic samples of Exercise 10.22 is randomly chosen to estimate the average monthly volume of mail, explain why there is a serious risk of getting a very misleading result.

**\*10.24** To generalize the example on page 240, suppose that we want to estimate the mean weight of six persons, whose (unknown) weights are 115, 125, 135, 185, 195, and 205 pounds.

(a) List all possible random samples of size 2 that can be taken from this population, calculate their means, and determine the probability that the mean of such a sample will differ by more than 5 from 160, the actual mean weight of the six persons.

(b) Suppose that the first three weights are those of women and the other three are those of men. List all possible stratified samples of size 2 that can be taken by randomly choosing one of the three women and one of the three men, calculate their means, and determine the probability that the mean of such a sample will differ by more than 5 from 160, the actual mean weight of the six persons.

(c) Suppose that three of the persons, those with weights 125, 135, and 185 pounds, are under 25 years of age, while those remaining are over 25 years of age. List all possible stratified samples of size 2 that can be taken by randomly choosing one of the three younger persons and one of the three older persons, calculate their means, and determine the probability that the mean of such a sample will differ by more than 5 from 160, the actual mean weight of the six persons.

(d) Compare the results of parts (a), (b), and (c).

**\*10.25** Based on their volume of sales, 9 of the 12 new car dealers in a city are classified as being small, and the other three are classified as being large. How many different stratified samples of four of these new car dealers can we choose if

(a) half of the sample is to be allocated to each of the strata;

(b) the allocation is to be proportional?

**\*10.26** Among 36 persons nominated for a city council, 18 are lawyers, 12 are business executives, and 6 are teachers. How many different stratified samples of six of these persons can we choose if

(a) a third of the sample is to be allocated to each of the strata;

(b) the allocation is to be proportional?

**\*10.27** A stratified sample of size $n = 40$ is to be taken from a population of size $N = 1,000$, which consists of four strata of size $N_1 = 250$, $N_2 = 600$, $N_3 = 100$, and $N_4 = 50$. If the allocation is to be proportional, how large a sample must be taken from each of the four strata?

**\*10.28** With reference to part (b) of Exercise 10.24, list all possible cluster samples of size $n = 2$ that can be taken by randomly choosing either two of the three women or two of the three men, calculate their means, and determine the probability that the mean of such a sample will differ by more than 5 from 160, the actual mean weight of the six persons. Compare this probability with those obtained in parts (a) and (b) of Exercise 10.24. What does this show about the relative merits of simple random sampling, stratified sampling, and cluster sampling in the given situation?

*10.29  In stratified sampling with proportional allocation, the importance of differences in stratum size is accounted for by letting the larger strata contribute relatively more items to the sample. However, strata differ not only in size but also in variability, and it would seem reasonable to take larger samples from the more variable strata and smaller samples from the less variable strata. If we let $\sigma_1, \sigma_2, \ldots,$ and $\sigma_k$ denote the standard deviations of the $k$ strata, we can account for both differences in stratum size and differences in stratum variability by requiring that

$$\frac{n_1}{N_1 \sigma_1} = \frac{n_2}{N_2 \sigma_2} = \cdots = \frac{n_k}{N_k \sigma_k}$$

The sample sizes for this kind of allocation, called **optimum allocation**, are given by the formula

$$n_i = \frac{n \cdot N_i \sigma_i}{N_1 \sigma_1 + N_2 \sigma_2 + \cdots + N_k \sigma_k}$$

for $i = 1, 2, \ldots,$ and $k$, where, if necessary, we round to the nearest integer.

(a)  A sample of size $n = 100$ is to be taken from a population consisting of two strata for which $N_1 = 10,000$, $N_2 = 30,000$, $\sigma_1 = 45$, and $\sigma_2 = 60$. To attain optimum allocation, how large a sample must be taken from each of the two strata?

(b)  A sample of size $n = 84$ is to be taken from a population consisting of three strata for which $N_1 = 5,000$, $N_2 = 2,000$, $N_3 = 3,000$, $\sigma_1 = 15$, $\sigma_2 = 18$, and $\sigma_3 = 5$. To attain optimum allocation, how large a sample must be taken from each of the three strata?

# 10.6 SAMPLING DISTRIBUTIONS

The sample mean, the sample median, and the sample standard deviation are examples of random variables, whose values will vary from sample to sample. Their distributions, which reflect such chance variations, play a fundamental role in statistical inference, and they are referred to as **sampling distributions**. In this chapter, we shall concentrate primarily on the sample mean and its sampling distribution, but in some of the exercises on page 252, and in later chapters we shall also consider the sampling distributions of other statistics.

To give an example of a sampling distribution, let us construct the one for the mean of a random sample of size $n = 2$ drawn without replacement from the finite population of size $N = 5$, whose elements are the numbers 3, 5, 7, 9, and 11. The mean of this population is

$$\mu = \frac{3 + 5 + 7 + 9 + 11}{5} = 7$$

and its standard deviation is

$$\sigma = \sqrt{\frac{(3 - 7)^2 + (5 - 7)^2 + (7 - 7)^2 + (9 - 7)^2 + (11 - 7)^2}{5}} = \sqrt{8}$$

Now, if we take a random sample of size $n = 2$ from this population, there are the $\binom{5}{2} = 10$ possibilities 3 and 5, 3 and 7, 3 and 9, 3 and 11, 5 and 7, 5 and 9, 5 and 11, 7 and 9, 7 and 11, and 9 and 11. Their means are 4, 5, 6, 7, 6, 7, 8, 8, 9, and 10, and since sampling is random, each of these ten values has the probability $\frac{1}{10}$. Thus, we arrive at the following sampling distribution of the mean:

| $\bar{x}$ | Probability |
|-----------|-------------|
| 4 | $\frac{1}{10}$ |
| 5 | $\frac{1}{10}$ |
| 6 | $\frac{2}{10}$ |
| 7 | $\frac{2}{10}$ |
| 8 | $\frac{2}{10}$ |
| 9 | $\frac{1}{10}$ |
| 10 | $\frac{1}{10}$ |

A histogram of this probability distribution is shown in Figure 10.3.

**Figure 10.3**
Sampling distribution of the mean.

An examination of this sampling distribution reveals some pertinent information about the chance variations of the mean of a random sample of size $n = 2$ from the given finite population. For instance, we find that the probability is $\frac{6}{10}$ that a sample mean will differ from the population mean $\mu = 7$ by 1 or less, and that the probability is $\frac{8}{10}$ that a sample mean will differ from the population mean $\mu = 7$ by 2 or less. The first case corresponds to $\bar{x} = 6, 7$, or 8, and the second case corresponds to $\bar{x} = 5, 6, 7, 8$, or 9. So, if we did not know the mean of the given population and wanted to estimate it with a random sample of size $n = 2$, this would give us some idea about the potential size of our error.

Further useful information about this sampling distribution of the mean can be obtained by calculating its mean and its standard deviation, denoted, respectively, by $\mu_{\bar{x}}$ and $\sigma_{\bar{x}}$. Here, the subscript $\bar{x}$ is used to distinguish between the parameters of this sampling distribution and those of the original population. Using again the definitions of the mean and the standard deviation of a probability distribution, we get

$$\mu_{\bar{x}} = 4 \cdot \frac{1}{10} + 5 \cdot \frac{1}{10} + 6 \cdot \frac{2}{10} + 7 \cdot \frac{2}{10} + 8 \cdot \frac{2}{10} + 9 \cdot \frac{1}{10} + 10 \cdot \frac{1}{10}$$

$$= 7$$

and

$$\sigma_{\bar{x}}^2 = (4-7)^2 \cdot \frac{1}{10} + (5-7)^2 \cdot \frac{1}{10} + (6-7)^2 \cdot \frac{2}{10} + (7-7)^2 \cdot \frac{2}{10}$$

$$+ (8-7)^2 \cdot \frac{2}{10} + (9-7)^2 \cdot \frac{1}{10} + (10-7)^2 \cdot \frac{1}{10}$$

$$= 3$$

so that $\sigma_{\bar{x}} = \sqrt{3}$. Observe that, at least for this example,

---

$\mu_{\bar{x}}$, **the mean of the sampling distribution of** $\bar{x}$, **equals** $\mu$, **the mean of the population;**

$\sigma_{\bar{x}}$, **the standard deviation of the sampling distribution of** $\bar{x}$, **is smaller than** $\sigma$, **the standard deviation of the population.**

---

These relationships are of fundamental importance, and we shall return to them in Section 10.7.

To illustrate the concept of a sampling distribution, we took a very small sample of size $n = 2$ from a very small finite population of size $N = 5$, but it would be difficult to duplicate this method to construct a sampling distribution of the mean of a large random sample from a large population. For instance, for $n = 10$ and $N = 100$, we would have had to list more than 17 trillion samples.

To get some idea about the sampling distribution of the mean of a somewhat larger sample from a large finite population, we shall use a **computer simulation**. In other words, we shall leave it to a computer to take repeated random samples from a given population, determine their means, and describe the distribution of these means in various ways. This will give us some idea about the overall shape and some of the key features of the real sampling distribution of the mean for random samples of the given size from the given population.

Without a computer, we can picture the simulation as follows: First, the numbers from 1 to 1,000 are written on 1,000 slips of paper (poker chips, small balls, or whatever may lend itself to drawing random samples). Then, a random sample of size $n = 15$ is drawn without replacement from this population and its values are recorded. We replace the sample before the next one is drawn, and we repeat this process until 100 random samples have been obtained.

The population we are dealing with here is said to have an **integer distribution** (also called a **discrete uniform distribution**), where each integer from 1 through $N$ has the probability $\frac{1}{N}$. Making use of the formulas for the sum and the sum of the squares of the integers from 1 through $N$, it can easily be shown that the mean and the standard deviation of this distribution are $\mu = \frac{N+1}{2}$ and $\sigma = \sqrt{\frac{N^2-1}{12}}$. For $N = 1,000$, their values are $\mu = 500.5$ and $\sigma = 288.67$ rounded to two decimals.

Actually using an appropriate computer package, MINITAB 13 in this case, we obtained the printout shown in Figure 10.4. The computer followed the

**Figure 10.4**
Computer simulation of a sampling distribution of the mean.

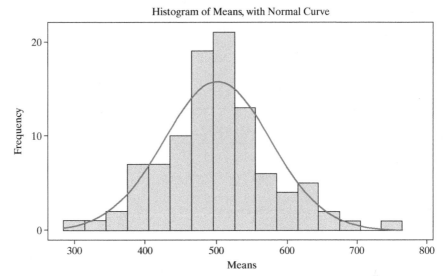

Histogram of Means, with Normal Curve

instruction of generating 100 random samples of size $n = 15$, each one without replacement, but replacing each sample before the next one was drawn.

As can be seen from Figure 10.4, the distribution of the 100 sample means is fairly symmetrical and bell shaped. In fact, the overall pattern follows quite

**Figure 10.5**
Description of the sampling distribution.

**Descriptive Statistics: Means**

| Variable | N | Mean | Median | TrMean | StDev | SE Mean |
|----------|------|--------|--------|--------|-------|---------|
| Means    | 100  | 502.40 | 497.47 | 501.15 | 75.64 | 7.56    |

| Variable | Minimum | Maximum | Q1 | Q3 |
|----------|---------|---------|--------|--------|
| Means    | 307.93  | 750.93  | 461.68 | 540.75 |

closely that of a normal curve. The computer also tells us (see Figure 10.5) that the mean of the 100 sample means is 502.40 and that their standard deviation is 75.64. In accordance with the relationships pointed out on page 244, the mean of the 100 sample means is very close to the mean of the population, and their standard deviation is (much) smaller than that of the population.

# 10.7 THE STANDARD ERROR OF THE MEAN

In most practical situations we cannot proceed as in the two illustrations of Section 10.6. That is, we cannot enumerate all possible samples or simulate a sampling distribution in order to judge how close a sample mean might be to the mean of the population from which the sample came. Fortunately, though, we can usually get the information we need from two theorems, which express essential facts about sampling distributions of the mean. One of these is discussed in this section and the other in Section 10.8.

The first of these two theorems expresses formally what we discovered from both of the examples of the preceding section—the mean of the sampling distribution of $\bar{x}$ equals the mean of the population sampled, and the standard deviation of the sampling distribution of $\bar{x}$ is smaller than the standard deviation of the population sampled. It may be phrased as follows: For random samples of size $n$ taken from a population with the mean $\mu$ and the standard deviation $\sigma$, the sampling distribution of $\bar{x}$ has the mean

MEAN OF
SAMPLING
DISTRIBUTION OF $\bar{x}$

$$\mu_{\bar{x}} = \mu$$

and the standard deviation

STANDARD ERROR
OF THE MEAN

$$\sigma_{\bar{x}} = \frac{\sigma}{\sqrt{n}} \quad or \quad \sigma_{\bar{x}} = \frac{\sigma}{\sqrt{n}} \cdot \sqrt{\frac{N-n}{N-1}}$$

depending on whether the population is infinite or finite of size $N$.

It is customary to refer to $\sigma_{\bar{x}}$ as the **standard error of the mean**, where "standard" is used in the sense of an average, as in "standard deviation." Its role in statistics is fundamental, as it measures the extent to which sample means can be expected to fluctuate, or vary, due to chance. If $\sigma_{\bar{x}}$ is small, the chances are good that the mean of a sample will be close to the mean of the population; if $\sigma_{\bar{x}}$ is large, we are more likely to get a sample mean which differs considerably from the mean of the population.

What determines the size of $\sigma_{\bar{x}}$ can be seen from the two preceding formulas. Both formulas (for infinite and finite populations) show that $\sigma_{\bar{x}}$ increases as the variability of the population increases, and that it decreases as the sample size increases. In fact, it is directly proportional to $\sigma$ and inversely proportional to the square root of $n$. (For finite populations it decreases even faster due to the $n$ appearing in $\sqrt{\frac{N-n}{N-1}}$.)

**EXAMPLE 10.5**   When we take a random sample from an infinite population, what happens to the standard error of the mean, and hence to the error we might expect when we use the mean of the sample to estimate the mean of the population, if the sample size is

(**a**) increased from 50 to 200;

(**b**) decreased from 360 to 40?

**Solution**   (**a**) The ratio of the two standard errors is

$$\frac{\frac{\sigma}{\sqrt{200}}}{\frac{\sigma}{\sqrt{50}}} = \frac{\sqrt{50}}{\sqrt{200}} = \sqrt{\frac{50}{200}} = \sqrt{\frac{1}{4}} = \frac{1}{2}$$

and where $n$ is quadrupled, the standard error of the mean is reduced, but only divided by 2.

**(b)** The ratio of the two standard errors is

$$\frac{\frac{\sigma}{\sqrt{40}}}{\frac{\sigma}{\sqrt{360}}} = \frac{\sqrt{360}}{\sqrt{40}} = \sqrt{9} = 3$$

and where $n$ is divided by 9, the standard error of the mean is increased, but only multiplied by 3.    ■

The factor $\sqrt{\frac{N-n}{N-1}}$ in the second formula for $\sigma_{\bar{x}}$ is called the **finite population correction factor**, for without it the two formulas for $\sigma_{\bar{x}}$ (for infinite and finite populations) are the same. In practice, it is omitted unless the sample constitutes at least 5% of the population, for otherwise it is so close to 1 that it has little effect on the value of $\sigma_{\bar{x}}$.

**EXAMPLE 10.6**   Find the value of the finite population correction factor for $N = 10,000$ and $n = 100$.

**Solution**   Substituting $N = 10,000$ and $n = 100$ into the formula for the finite population correction factor, we get

$$\sqrt{\frac{N-n}{N-1}} = \sqrt{\frac{10,000 - 100}{10,000 - 1}} = 0.995$$

and this is so close to 1 that the correction factor can be omitted for all practical purposes.    ■

Since we stated the formulas for the standard error of the mean without proof, let us verify the one for finite populations by referring to the results of the two illustrations of Section 10.6.

**EXAMPLE 10.7**   With reference to the illustration on page 242, where we had $N = 5, n = 2$, and $\sigma = \sqrt{8}$, verify that the second formula for $\sigma_{\bar{x}}$ will yield $\sqrt{3}$, namely, the value that we obtained on page 244.

**Solution**   Substituting $N = 5, n = 2$, and $\sigma = \sqrt{8}$ into the second of the two formulas for $\sigma_{\bar{x}}$, we get

$$\sigma_{\bar{x}} = \frac{\sqrt{8}}{\sqrt{2}} \cdot \sqrt{\frac{5-2}{5-1}} = \frac{\sqrt{8}}{\sqrt{2}} \cdot \sqrt{\frac{3}{4}} = \sqrt{\frac{8}{2} \cdot \frac{3}{4}} = \sqrt{3}$$    ■

**EXAMPLE 10.8**   With reference to the computer simulation of Figure 10.4, where we had $N = 1,000, n = 15$, and $\sigma = 288.67$, what value could we have expected for the standard deviation of the 100 sample means?

**Solution**   Substituting $N = 1,000, n = 15$, and $\sigma = 288.67$ into the second of the two formulas for $\sigma_{\bar{x}}$, we get

$$\sigma_{\bar{x}} = \frac{288.67}{\sqrt{15}} \cdot \sqrt{\frac{1,000 - 15}{1,000 - 1}} = 74.01$$

and this is quite close to 75.64, the value actually obtained for the computer simulation in Figure 10.5.    ■

# 10.8 THE CENTRAL LIMIT THEOREM

When a sample mean is used to estimate the mean of a population, the uncertainties about its potential error can be expressed in various ways. If we knew the exact sampling distribution of the mean, which, of course, we never do, we could proceed as in the first illustration of Section 10.6 and calculate the probabilities associated with errors of various size. Something else we rarely, if ever, do, is to use Chebyshev's theorem and assert with a probability of at least $1 - \frac{1}{k^2}$ that the mean of a random sample will differ from the mean of the population sampled by less than $k \cdot \sigma_{\bar{x}}$.

**EXAMPLE 10.9**  Based on Chebyshev's theorem with $k = 2$, what can we say about the potential size of our error if we are going to use a random sample of size $n = 64$ to estimate the mean of an infinite population with $\sigma = 20$?

**Solution**  Substituting $n = 64$ and $\sigma = 20$ into the appropriate formula for the standard error of the mean, we get

$$\sigma_{\bar{x}} = \frac{20}{\sqrt{64}} = 2.5$$

and it follows that we can assert with a probability of at least $1 - \frac{1}{2^2} = 0.75$ that the error will be less than $k \cdot \sigma_{\bar{x}} = 2(2.5) = 5$.  ■

The importance of this example is that it shows *how* we can make exact probability statements about the potential error when we estimate the mean of a population. The trouble with the use of Chebyshev's theorem is that "at least 0.75" does not tell us enough when in reality that probability may be, say, 0.998 or even 0.999. Whereas Chebyshev's theorem provides a logical link between errors and the probabilities that they may be committed, there exists another mathematical theorem that, in many instances, enables us to make much stronger probability statements about such potential errors.

This theorem, which is the second of the two theorems mentioned on page 245, is called the **central limit theorem** and, informally, it states that for large samples the sampling distribution of the mean can be approximated closely with a normal distribution. Recalling from Section 10.7 that

$$\mu_{\bar{x}} = \mu \quad \text{and} \quad \sigma_{\bar{x}} = \frac{\sigma}{\sqrt{n}}$$

for random samples from infinite populations, we can now say formally that

**CENTRAL LIMIT THEOREM**

> *If $\bar{x}$ is the mean of a random sample of size n from an infinite population with the mean $\mu$ and the standard deviation $\sigma$ and n is large, then*
>
> $$z = \frac{\bar{x} - \mu}{\sigma/\sqrt{n}}$$
>
> *has approximately the standard normal distribution.*

This theorem is of fundamental importance in statistics, as it justifies the use of normal-curve methods in a wide range of problems; it applies to infinite populations, and also to finite populations when $n$, though large, constitutes but a small portion of the population. We cannot say precisely how large $n$ must be so that the central limit theorem can be applied, but unless the population distribution has a very unusual shape, $n = 30$ is usually regarded as sufficiently large. When the population we are sampling has, itself, roughly the shape of a normal curve, the sampling distribution of the mean can be approximated closely with a normal distribution regardless of the size of $n$.

The central limit theorem can also be used for finite populations, but a precise description of the situations under which this can be done is rather complicated. The most common proper use is the case in which $n$ is large while $\frac{n}{N}$ is small.

Let us now see what probability will take the place of "at least 0.75" if we use the central limit theorem instead of Chebyshev's theorem in Example 10.9.

**Figure 10.6**
Diagram for
Example 10.10.

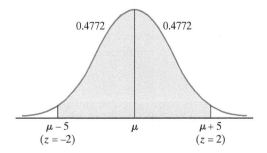

$$0.4772 \qquad 0.4772$$

$$\mu - 5 \qquad \mu \qquad \mu + 5$$
$$(z = -2) \qquad\qquad (z = 2)$$

**EXAMPLE  10.10**   Based on the central limit theorem, what is the probability that the error will be less than 5 when the mean of a random sample of size $n = 64$ is used to estimate the mean of an infinite population with $\sigma = 20$?

**Solution**   The probability is given by the tinted area under the curve in Figure 10.6, namely, by the standard-normal-curve area between

$$z = \frac{-5}{20/\sqrt{64}} = -2 \quad \text{and} \quad z = \frac{5}{20/\sqrt{64}} = 2$$

Since the entry in Table I corresponding to $z = 2.00$ is 0.4772, the probability asked for is $0.4772 + 0.4772 = 0.9544$. Thus, the statement that the probability is "at least 0.75" is replaced by the much stronger statement that the probability is about 0.95.    ▪

# 10.9 SOME FURTHER CONSIDERATIONS

In Sections 10.6 through 10.8, our main goal was to introduce the concept of a sampling distribution, and the one that we chose as an illustration was the sampling distribution of the mean. It should be clear, though, that instead of the mean we could have studied the median, the standard deviation, or some other statistic, and investigated its chance fluctuations. So far as the corresponding theory is concerned, this would have required a different formula for the standard error and theory analogous to, yet different from, the central limit theorem.

For instance, for large samples from continuous populations, the **standard error of the median** is approximately

$$\sigma_{\tilde{x}} = 1.25 \cdot \frac{\sigma}{\sqrt{n}}$$

where $n$ is the size of the sample and $\sigma$ is the population standard deviation. Note that comparison of the two formulas

$$\sigma_{\bar{x}} = \frac{\sigma}{\sqrt{n}} \quad \text{and} \quad \sigma_{\tilde{x}} = 1.25 \cdot \frac{\sigma}{\sqrt{n}}$$

reflects the fact that the mean is generally more reliable than the median (that is, it tends to expose us to smaller errors) when we estimate the mean of a symmetrical population. For symmetrical populations, the means of the sampling distributions of $\bar{x}$ and $\tilde{x}$ are both equal to the population mean $\mu$.

**EXAMPLE 10.11** How large a random sample do we need so that its mean is as reliable an estimate of the mean of a symmetrical continuous population as the median of a random sample of size $n = 200$?

**Solution** Equating the two standard error formulas and substituting $n = 200$ into the one for the standard error of the median, we get

$$\frac{\sigma}{\sqrt{n}} = 1.25 \cdot \frac{\sigma}{\sqrt{200}}$$

which, solved for $n$, yields $n = 128$. Thus, for the stated purpose, the mean of a random sample of size $n = 128$ is as "good" as the median of a random sample of size $n = 200$. ■

Also, a point worth emphasizing is that the illustrations of Section 10.6 were used as teaching aids, designed to convey the idea of a sampling distribution, but they do not reflect what we do in actual practice. In practice, we can seldom list all possible samples, and ordinarily we base an inference on one sample and not on 100 samples. In Chapter 11 and in subsequent chapters we shall go further into the problem of translating theory about sampling distributions into methods of evaluating the merits and shortcomings of statistical procedures.

Another point worth repeating concerns the $\sqrt{n}$ that appears in the denominator of the formulas for the standard error of the mean. It makes sense that when $n$ becomes larger and larger, our generalizations will be subject to smaller errors, but the $\sqrt{n}$ in the formulas for the standard error of the mean tells us that the gain in reliability is not proportional to the increase in the size of the sample. As we saw, quadrupling the size of the sample will only double the reliability of a sample mean as an estimate of the mean of a population. Indeed, to quadruple the reliability, we would have to multiply the sample size by 16. This relationship between reliability and sample size indicates that there are, to use a phrase from economics, diminishing returns to increasing the size of a sample. It seldom pays to take samples that are massively large.

**10.30** Suppose that in the illustration on page 242, where random samples of size $n = 2$ were drawn without replacement from the finite population that consists of the numbers 3, 5, 7, 9, and 11, sampling had been with replacement.

(a) List the 25 ordered samples that can be drawn with replacement from the given population and calculate their means. (By "ordered" we mean that 3 and 7, for example, is a different sample than 7 and 3.)

(b) Assuming that sampling is random, namely, that each of the ordered samples of part (a) has the probability $\frac{1}{25}$, construct the sampling distribution of the mean for random samples of size $n = 2$ drawn with replacement from the given population.

**10.31** With reference to Exercise 10.30 find the probabilities that the mean of a random sample of size $n = 2$ drawn with replacement from the given population will differ from $\mu = 7$ by

(a) 1 or less;   (b) at most 2.

**10.32** Calculate the standard deviation of the sampling distribution obtained in part (b) of Exercise 10.30 and verify the result by substituting $n = 2$ and $\sigma = \sqrt{8}$ into the first of the two standard error formulas on page 246.

**10.33** In each of the following examples, state whether we are sampling from a finite population or a hypothetically infinite population, and describe the population.

(a) A personnel manager selects 5 of 20 job applicants for an interview.

(b) We weigh a gold nugget three times and use the average we obtain as its weight.

(c) We observe how many heads we get in 100 flips of a balanced coin.

(d) We select 5 of 25 picture postcards displayed in a store to mail to friends.

(e) We observe the gasoline mileage obtained by our car for a period of time to estimate the miles per gallon for the car.

**10.34** Obtain the probability, by counting, of each possible sample if a random sample of size 2 is taken from

(a) a finite population of size 3;

(b) a finite population of size 4.

**10.35** When we sample from an infinite population, what happens to the standard error of the mean when the sample size is

(a) increased from 30 to 120;

(b) decreased from 245 to 5?

**10.36** When we sample from an infinite population, what happens to the standard error of the mean when the sample size is

(a) decreased from 1,000 to 10;

(b) increased from 80 to 500?

**10.37** What is the value of the finite population correction factor when

(a) $N = 100$ and $n = 10$;

(b) $N = 300$ and $n = 25$;

(c) $N = 5,000$ and $n = 100$?

**10.38** If the mean of a large random sample of size $n$ is used to estimate the mean of an infinite population with the standard deviation $\sigma$, there is a fifty–fifty chance that the error is less than

$$0.6745 \cdot \frac{\sigma}{\sqrt{n}}.$$

It has been the custom to refer to this quantity as the **probable error of the mean**.

(a) If a random sample of size 64 is drawn from an infinite population with $\sigma = 24.8$, what is the probable error of the mean?

(b) If a random sample of size $n = 144$ is drawn from a very large population of the fines paid for various traffic violations in a certain county in 2005 with $\sigma = \$219.12$, what is the probable error of the mean?

**10.39** If a random sample of size $n = 60$ is taken from a very large population (consisting of the IQ's of army inductees) which has the standard deviation $\sigma = 12.8$, determine the probable error of the mean.

**10.40** The mean of a random sample of size $n = 36$ is used to estimate the mean of an infinite population with the standard deviation $\sigma = 9$. What can we assert about the probability that the error of this estimate will be less than 4.5 if we use

(a) Chebyshev's theorem;

(b) the central limit theorem?

**10.41** The mean of a random sample of size $n = 25$ is used to estimate the mean attention span of persons over 65. Given that the standard deviation of the population sampled is $\sigma = 2.4$ minutes, what can we assert about the probability that the error of the estimate is less than 1.2 minutes if we use

(a) Chebyshev's theorem;

(b) the central limit theorem?

**10.42** The mean of a random sample of size $n = 100$ is going to be used to estimate the mean daily milk production of a very large herd of dairy cows. Given that the standard deviation of the population to be sampled is $\sigma = 3.6$ quarts, what can we assert about the probabilities that the error of this estimate will be

(a) more than 0.72 quart;

(b) less than 0.45 quart?

**10.43** If measurements of the specific gravity of a metal can be looked upon as a random sample from a normal population with the standard deviation $\sigma = 0.025$ ounce, what is the probability that the mean of a random sample of size $n = 16$ will be off by at most 0.01 ounce?

**10.44** Verify that the mean of a random sample of size $n = 256$ is as reliable an estimate of the mean of a symmetrical continuous population as the median of a random sample of size $n = 400$.

**10.45** How large a random sample do we have to take so that its median is as reliable an estimate of the mean of a symmetrical continuous population as the mean of a random sample of size $n = 144$?

# *10.10 TECHNICAL NOTE (SIMULATION)

Simulation provides one of the most effective ways of illustrating, and thus teaching, some of the basic concepts of statistics. It serves to demonstrate the validity of theory, where rigorous mathematical proofs are beyond the prerequisites for this book.

Also, as we shall see in the chapters that follow, the evaluation and interpretation of statistical techniques will often require that we imagine what would happen if experiments were repeated over and over again. Since, most of the time, such repetitions are neither practical nor feasible, we can resort instead to simulations, preferably with the use of computers. Simulation also plays a role in

Figure 10.7
Computer simulation
of Poisson data.

| Simulated Poisson Samples n = 5 | | | | |
|---|---|---|---|---|
| MTB > Random 40 c1-c5; | | | | |
| SUBC> Poisson 16. | | | | |
| MTB > Print c1-c5. | | | | |

Row

| Row | | | | | |
|---|---|---|---|---|---|
| 1 | 9 | 15 | 6 | 19 | 11 |
| 2 | 16 | 15 | 19 | 14 | 14 |
| 3 | 14 | 20 | 11 | 22 | 19 |
| 4 | 14 | 20 | 17 | 22 | 14 |
| 5 | 21 | 11 | 13 | 18 | 14 |
| 6 | 13 | 11 | 13 | 15 | 12 |
| 7 | 14 | 12 | 14 | 17 | 10 |
| 8 | 17 | 13 | 25 | 17 | 20 |
| 9 | 21 | 16 | 16 | 18 | 21 |
| 10 | 15 | 12 | 16 | 11 | 14 |
| 11 | 21 | 12 | 19 | 14 | 14 |
| 12 | 20 | 22 | 16 | 19 | 17 |
| 13 | 25 | 15 | 8 | 16 | 21 |
| 14 | 15 | 19 | 18 | 12 | 18 |
| 15 | 17 | 23 | 20 | 11 | 13 |
| 16 | 18 | 16 | 16 | 21 | 22 |
| 17 | 20 | 19 | 21 | 17 | 9 |
| 18 | 19 | 17 | 11 | 14 | 19 |
| 19 | 12 | 18 | 16 | 10 | 14 |
| 20 | 11 | 14 | 11 | 12 | 26 |
| 21 | 17 | 16 | 11 | 11 | 9 |
| 22 | 15 | 16 | 16 | 19 | 18 |
| 23 | 16 | 12 | 18 | 16 | 15 |
| 24 | 20 | 19 | 23 | 14 | 14 |
| 25 | 19 | 18 | 16 | 24 | 13 |
| 26 | 13 | 18 | 14 | 17 | 25 |
| 27 | 16 | 17 | 18 | 14 | 22 |
| 28 | 15 | 17 | 11 | 15 | 13 |
| 29 | 23 | 12 | 13 | 13 | 16 |
| 30 | 16 | 28 | 11 | 14 | 11 |
| 31 | 14 | 15 | 18 | 7 | 16 |
| 32 | 19 | 17 | 11 | 16 | 13 |
| 33 | 13 | 14 | 16 | 12 | 17 |
| 34 | 25 | 14 | 8 | 15 | 16 |
| 35 | 12 | 17 | 12 | 12 | 13 |
| 36 | 12 | 16 | 17 | 15 | 25 |
| 37 | 20 | 14 | 14 | 16 | 17 |
| 38 | 13 | 19 | 19 | 16 | 17 |
| 39 | 12 | 14 | 11 | 19 | 14 |
| 40 | 14 | 9 | 17 | 24 | 19 |

the development of statistical theory, for there are situations where simulation is easier than a detailed mathematical analysis.

Simulations of random samples can also be made with the use of a table of random numbers, but for use in the exercises that follow, we present in Figure 10.7 40 computer-simulated random samples, each consisting of $n = 5$ values of a random variable having the Poisson distribution with $\lambda = 16$, and hence $\mu = 16$ and $\sigma = 4$. (The reader may picture these figures as data on the number of emergency calls that an ambulance service receives in an afternoon, the number of calls that a switchboard receives during a ten-minute interval, or the number of pieces of junk mail that a person receives in Monday's mail.)

**\*10.46**  In Figure 10.7, each row constitutes a random sample of size $n = 5$ from a Poisson population with $\lambda = 16$, and hence with $\mu = 16$ and $\sigma = 4$.

(a)  Calculate the means of the 40 samples in Figure 10.7.

(b)  Calculate the mean and the standard deviation of the 40 means obtained in part (a), and compare the results with the corresponding values expected in accordance with the theory of Section 10.7.

**\*10.47**  Determine the medians of the 40 samples shown in Figure 10.7, calculate their standard deviation, and compare the result with the corresponding value expected in accordance with the theory of Section 10.9.

**\*10.48**  In Exercise 11.35 we shall present a way of estimating the population standard deviation in terms of the range (largest value minus the smallest). For this purpose, the range is divided by a constant depending on the size of the sample; for instance, by 2.33 for $n = 5$. Another way of saying this is that for samples of size $n = 5$, the mean of the sampling distribution of the range is $2.33\sigma$. To verify this, determine the ranges of the 40 samples shown in Figure 10.7 and then calculate their mean. This is an estimate of the mean of the sampling distribution of the range, and since the sample size is $n = 5$, divided by 2.33 it provides an estimate of $\sigma$, which is known to equal 4. Find the percentage error of this estimate.

**\*10.49**  On page 76 we explained that division by $n - 1$ in the formulas for the sample standard deviation and the sample variance serves to make $s^2$ an unbiased estimator of $\sigma^2$; namely, to make the mean of the sampling distribution of $s^2$ equal to $\sigma^2$. To verify this, find the mean of the variances of the 40 samples in Figure 10.7, which are 26.0, 4.3, 20.7, 12.8, 16.3, 2.2, 6.8, 19.8, 6.3, 4.3, 14.5, 5.7, 41.5, 8.3, 24.2, 7.8, 23.2, 12.0, 10.0, 40.7, 12.2, 2.7, 4.8, 15.5, 16.5, 22.3, 8.8, 5.2, 20.3, 49.5, 17.5, 10.2, 4.3, 37.3, 4.7, 23.5, 6.2, 6.2, 9.5, and 31.3. Since $\sigma^2$ is known to equal 16, calculate the percentage error of this estimate.

# CHECKLIST OF KEY TERMS (with page references to their definitions)

\*Area sampling, 240
Central limit theorem, 248
\*Cluster sampling, 240
Computer simulation, 244
\*Cross stratification, 239
Discrete uniform distribution, 244
Finite population, 230
Finite population correction factor, 247
Infinite population, 230
Integer distribution, 244
\*Judgment sample, 239
\*Optimum allocation, 239, 242
\*Probable error of the mean, 252
Proportional allocation, 238
\*Quota sampling, 239

Random numbers, 231
Random sample, 229, 231
Random sampling from a finite population, 230
\*Sample design, 236
Sampling distribution, 230, 242
Sampling frame, 235
Simple random sample, 231
Standard error of the mean, 246
Standard error of the median, 250
\*Strata, 237
\*Stratification, 237
\*Stratified sampling, 237
Stratified (simple) random sampling, 237
Survey sampling, 237
\*Systematic sampling, 237

# REFERENCES

*Among the many published tables of random numbers, one of the most widely used is*

RAND CORPORATION, *A Million Random Digits with 100,000 Normal Deviates.* New York: Macmillan Publishing Co., Inc., third printing 1966.

*There also exist calculators that are preprogrammed to generate random numbers, and it is fairly easy to program a computer so that a person can generate his or her own random numbers. The following is one of many articles on this subject.*

KIMBERLING, C., "Generate Your Own Random Numbers." *Mathematics Teacher,* February 1984.

*Interesting material on the early development of tables of random numbers may be found in*

BENNETT, D. J., *Randomness.* Cambridge, Mass.: Harvard University Press, 1998.

*Derivations of the various standard error formulas and more general formulations (and proof) of the central limit theorem may be found in most textbooks on mathematical statistics. All sorts of information about sampling is given in*

COCHRAN, W. G., *Sampling Techniques,* 3rd ed. New York: John Wiley & Sons, Inc., 1977.

SCHAEFFER, R. L., MENDENHALL, W., and OTT, L., *Elementary Survey Sampling,* 4th ed. Boston: PWS-Kent Publishing Co., 1990.

SLONIN, M. J., *Sampling in a Nutshell.* New York: Simon and Schuster, 1973.

WILLIAMS, W. H., *A Sampler on Sampling.* New York: John Wiley & Sons, Inc., 1978.

**R.91** Among 18 workers on a picket line, ten are men and eight are women. If a television crew randomly chooses four of them to be shown on camera, what is the probability that this will include

(a) only men;

(b) two men and two women?

**R.92** Find the standard-normal-curve area that lies

(a) to the left of $z = 1.65$;

(b) to the left of $z = -0.44$;

(c) between $z = 1.15$ and $z = 1.85$;

(d) between $z = -0.66$ and $z = 0.66$.

**R.93** An automobile manufacturer has produced a new model car with significant changes. Calculate the probability that two of five persons examining the car will like it, given the probability that any one of them will like it is 0.70.

**R.94** A customs official wants to check 12 of 875 shipments listed on a ship's manifest. Using the 28th, 29th, and 30th columns of the table in Figure 10.2, starting with the 6th row and going down the page, which ones (by number) will the customs official inspect?

**R.95** Check in each case whether the condition for the binomial approximation to the hypergeometric distribution is satisfied:

(a) $a = 40, b = 160$, and $n = 8$;

(b) $a = 100, b = 60$, and $n = 10$;

(c) $a = 68, b = 82$, and $n = 12$.

**R.96** If the amount of time a tourist spends in a cathedral is a random variable having the normal distribution with $\mu = 23.4$ minutes and $\sigma = 6.8$ minutes, find the probability that a tourist will spend

(a) at least 15 minutes;

(b) anywhere from 20 to 30 minutes.

**R.97** If a random sample of size $n = 6$ is to be chosen from a finite population of size $N = 45$, what is the probability of each possible sample?

**\*R.98** The probability that a baseball player will strike out on any given at bat is 0.36. What is the probability that he will strike out for the first time at his

(a) second at bat;

(b) fifth at bat?

(*Hint*: Use the formula for the geometric distribution.)

**R.99** Find the mean of the binomial distribution with $n = 8$ and $p = 0.40$, using

(a) Table V and the formula that defines $\mu$;

(b) the special formula for the mean of a binomial distribution.

**R.100** Use the normal distribution to approximate the binomial probability that at least 25 of 60 bee stings will cause some discomfort, if the probability is 0.48 that any one of them will cause some discomfort.

**R.101** Check in each case whether the conditions for the Poisson approximation to the binomial distribution are satisfied:

(a) $n = 180$ and $p = \frac{1}{9}$;

(b) $n = 480$ and $p = \frac{1}{60}$;

(c) $n = 575$ and $p = \frac{1}{100}$.

**R.102** It is known that 6% of all rats carry a certain disease. If we examine a random sample of 120 rats, will this satisfy the condition for using the Poisson approximation to the binomial distribution? If so, use the Poisson distribution to approximate the probability that only 5 of the rats will carry the disease.

**R.103** A random variable has a normal distribution with $\sigma = 4.0$. If the probability is 0.9713 that it will take on a value less than 82.6, what is the probability that it will take on a value between 70.0 and 80.0?

**R.104** A random sample of size $n = 3$ is drawn from the finite population that consists of the elements a, b, c, d, and e. What is the probability that any specific element, say, element b will be contained in the sample?

**R.105** A small cruise ship has deluxe outside cabins, standard outside cabins, and inside cabins, and the probabilities are 0.30, 0.60, and 0.10 that a travel agent will receive a reservation for the first, second, or third of these categories. If a travel agent receives nine reservations, what is the probability that four of them will be for a deluxe outside cabin, four will be for a standard outside cabin, and one will be for an inside cabin?

**R.106** In a certain community, the response time of an ambulance may be regarded as a random variable having a normal distribution with $\mu = 5.8$ minutes and a standard deviation of $\sigma = 1.2$ minutes. What is the probability that the ambulance will take at most 8.0 minutes to respond to a call?

**R.107** A zoo has a large collection of anteaters, including five males and four females. If a veterinarian randomly picks three of them for examination, what are the probabilities that

(a) none of them will be females;

(b) two of them will be females?

**R.108** It has been claimed that

(a) if the sample size is increased by 44%, the standard error of the mean is reduced by 20%;

(b) if the standard error of the mean is to be reduced by 20%, the sample size must be increased by 56.25%.

Which of these two statements is correct and which one is false?

**R.109** Figure R.4 shows the probability density of a continuous random variable that takes on values on the interval from 0 to 3.

(a) Verify that the total area under the curve is equal to 1.

(b) Find the probability that the random variable will take on a value greater than 1.5.

**Figure R.4**
Diagram for Exercise
R. 109.

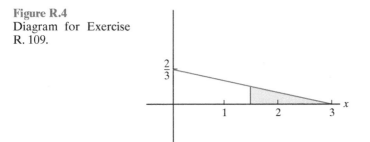

**\*R.110** Suppose that we want to find the probability that a random variable having the hypergeometric distribution with $n = 14, a = 180$, and $b = 120$ will take on the value $x = 5$.

(a) Verify that the binomial distribution with $n = 14$ and $p = \frac{180}{180+120} = \frac{3}{5}$ can be used to approximate this hypergeometric distribution.

(b) Verify that the normal distribution with $\mu = np = 14 \cdot \frac{3}{5} = 8.4$ and $\sigma = \sqrt{np(1-p)} = \sqrt{14(0.6)(0.4)} \approx 1.83$ can be used to approximate the binomial distribution of part (a).

(c) Use the normal distribution with $\mu = 8.4$ and $\sigma = 1.83$ to approximate the hypergeometric probability with $x = 5, n = 14, a = 180$, and $b = 120$.

**R.111** The amount of time that it takes an electrician to repair a ceiling fan can be treated as a random variable having the normal distribution with $\mu = 24.55$ minutes and $\sigma = 3.16$ minutes. Find the probability that it will take an electrician anywhere from 20.00 to 30.00 minutes to repair a ceiling fan.

**R.112** How many different samples of size

(a) $n = 3$ can be chosen from among $N = 14$ different magazines at a doctor's office;

(b) $n = 5$ can be chosen for a potential customer from among $N = 24$ houses listed for sale in Scottsdale, Arizona?

**R.113** A panel of 300 persons chosen for jury duty included only 30 persons under 25 years of age. For a particular narcotics case, the actual jury of 12 selected from this panel did not contain anyone under the age of 25. The youthful defendant's attorney complained that this jury of 12 is not representative. He argued that the probability of having one of the 12 jurors under 25 years of age should be *many times* the probability of having none of them under 25 years of age.

(a) Find the ratio of these two probabilities, using the hypergeometric distribution.

(b) Find the ratio of these two probabilities, using the binomial distribution as an approximation.

**R.114** During a busy weekend, 50 delivery vans obtained

| | | | | | | | | | |
|---|---|---|---|---|---|---|---|---|---|
| 23.2 | 26.7 | 21.5 | 23.8 | 19.1 | 22.3 | 27.4 | 22.4 | 20.6 | 23.5 |
| 16.5 | 22.2 | 21.9 | 14.4 | 25.6 | 23.0 | 25.4 | 21.2 | 16.8 | 28.4 |
| 20.5 | 21.5 | 22.6 | 19.8 | 20.5 | 21.7 | 16.3 | 18.9 | 24.0 | 21.3 |
| 22.2 | 24.8 | 17.5 | 18.0 | 21.4 | 22.5 | 20.6 | 17.7 | 15.9 | 22.5 |
| 26.7 | 21.3 | 24.5 | 19.3 | 25.4 | 20.0 | 16.5 | 21.1 | 23.8 | 20.5 |

miles per gallon. Use a computer or a graphing calculator to check whether these data can be looked upon as values of a random variable having a normal distribution.

**R.115** Refer to the upper part of the printout of Figure R.5 to find the probabilities that a random variable having the Poisson distribution with $\lambda = 1.6$ will take on

(a) a value less than 3;

(b) the value 3, 4, or 5;

(c) a value greater than 4.

**R.116** Repeat Exercise R.115, using the lower part of the printout of Figure R.5, namely, the cumulative probabilities.

**R.117** Use the upper part of Figure R.5 to calculate the mean of the Poisson distribution with $\lambda = 1.6$ and, thus, verify the formula $\mu = \lambda$.

**Figure R.5**
Poisson    distribution
with $\lambda = 1.6$.

**Probability Density Function**

Poisson with mu = 1.60000

|   x  | P( X = x ) |
|------|------------|
| 0.00 | 0.2019 |
| 1.00 | 0.3230 |
| 2.00 | 0.2584 |
| 3.00 | 0.1378 |
| 4.00 | 0.0551 |
| 5.00 | 0.0176 |
| 6.00 | 0.0047 |
| 7.00 | 0.0011 |
| 8.00 | 0.0002 |
| 9.00 | 0.0000 |

**Cumulative Distribution Function**

Poisson with mu = 1.60000

|   x  | P( X <= x ) |
|------|-------------|
| 0.00 | 0.2019 |
| 1.00 | 0.5249 |
| 2.00 | 0.7834 |
| 3.00 | 0.9212 |
| 4.00 | 0.9763 |
| 5.00 | 0.9940 |
| 6.00 | 0.9987 |
| 7.00 | 0.9997 |
| 8.00 | 1.0000 |
| 9.00 | 1.0000 |

**R.118** Use the upper part of Figure R.5 and the computing formula on page 193 to calculate the variance of the Poisson distribution with $\lambda = 1.6$ and, thus, verify the formula $\sigma^2 = \lambda$.

**R.119** What is the finite population correction factor if

(a) $N = 120$ and $n = 30$;

(b) $N = 400$ and $n = 50$?

**R.120** Determine in each case whether the following can be probability distributions (defined in each case for the given values of $x$) and explain your answers:

(a) $f(x) = \frac{1}{8}$ for $x = 0, 1, 2, 3, 4, 5, 6,$ and 7;

(b) $f(x) = \frac{x+1}{16}$ for $x = 1, 2, 3,$ and 4;

(c) $f(x) = \frac{(x-1)(x-2)}{20}$ for $x = 2, 3, 4,$ and 5.

**R.121** The number of blossoms on a rare cactus is a random variable having the Poisson distribution with $\lambda = 2.3$. What is the probability that such a cactus will have

(a) no blossoms;

(b) one blossom.

**R.122** The probabilities are 0.22, 0.34, 0.24, 0.13, 0.06, and 0.01 that 0, 1, 2, 3, 4, or 5 of a doctor's patients will come down with the flu while the doctor is out of town during the week after Christmas Day.

(a) Find the mean of this probability distribution.

(b) Use the computing formula to determine the variance of this probability distribution.

*R.123 Among 80 persons interviewed for certain jobs by a government agency, 40 are married, 20 are single, 10 are divorced, and 10 are widowed. In how many ways can a 10 percent stratified sample be chosen from among the persons interviewed if
  (a)  one-fourth of the sample is to be allocated to each group;
  (b)  the allocation is proportional?

R.124 Use the normal approximation to the binomial distribution to approximate the probabilities that a random variable having the binomial distribution with $n = 18$ and $p = 0.27$ will take on a value
  (a)  less than 6;
  (b)  anywhere from 4 to 8;
  (c)  greater than 6.

R.125 Use the upper part of Figure R.6 to determine the probabilities that a random variable having the binomial distribution with $n = 12$ and $p = 0.46$ will take on
  (a)  the value 6, 7, or 8;
  (b)  a value greater or equal to 9.

**Figure R.6**
Binomial   distribution
with  $n = 12$  and  $p = 0.46$.

**Probability Density Function**

Binomial with n = 12 and p = 0.460000

| x | P ( X = x) |
|---|---|
| 0.00 | 0.0006 |
| 1.00 | 0.0063 |
| 2.00 | 0.0294 |
| 3.00 | 0.0836 |
| 4.00 | 0.1602 |
| 5.00 | 0.2184 |
| 6.00 | 0.2171 |
| 7.00 | 0.1585 |
| 8.00 | 0.0844 |
| 9.00 | 0.0319 |
| 10.00 | 0.0082 |
| 11.00 | 0.0013 |
| 12.00 | 0.0001 |

**Cumulative Distribution Function**

Binomial with n = 12 and p = 0.460000

| x | P ( X <= x) |
|---|---|
| 0.00 | 0.0006 |
| 1.00 | 0.0069 |
| 2.00 | 0.0363 |
| 3.00 | 0.1199 |
| 4.00 | 0.2802 |
| 5.00 | 0.4986 |
| 6.00 | 0.7157 |
| 7.00 | 0.8742 |
| 8.00 | 0.9585 |
| 9.00 | 0.9905 |
| 10.00 | 0.9986 |
| 11.00 | 0.9999 |
| 12.00 | 1.0000 |

**R.126** Use the lower part of Figure R.6 to repeat Exercise R.125.

**R.127** Check in each case whether the conditions for the normal approximation to the binomial distribution are satisfied:

(a)  $n = 55$ and $p = \frac{1}{5}$;

(b)  $n = 105$ and $p = \frac{1}{35}$;

(c)  $n = 210$ and $p = \frac{1}{30}$;

(d)  $n = 40$ and $p = 0.95$.

**R.128** If a random sample of size $n = 3$ is to be chosen from a finite population of size $N = 70$, what is the probability of each possible sample?

**R.129** Among 12 Krugerrands, seven are genuine, three are gold-plated counterfeits, and the other two are pure brass counterfeits. If a not-very-knowledgeable buyer randomly picks three of these coins, what is the probability that he will get one of each kind?

**R.130** Find the mean of the binomial distribution with $n = 9$ and $p = 0.40$, using

(a)  Table V and the formula that defines $\mu$;

(b)  the special formula for the mean of a binomial distribution.

# 11.1

# PROBLEMS OF ESTIMATION

**11.1**  The Estimation of Means  263

**11.2**  The Estimation of Means ($\sigma$ Unknown)  268

**11.3**  The Estimation of Standard Deviations  274

**11.4**  The Estimation of Proportions  279

Checklist of Key Terms  286

References  286

Traditionally, statistical inference has been divided into problems of **estimation**, where we determine the values of population parameters; **tests of hypotheses**, where we must accept or reject assertions about populations and/or their parameters (or, perhaps, reserve judgment); and **problems of prediction**, where we forecast future values of random variables. In this chapter we shall concentrate on problems of estimation. Tests of hypotheses are treated in subsequent chapters, and problems of prediction are taken up in Chapters 16 and 17.

Problems of estimation can be found everywhere: in business, in science, as well as in everyday life. In business, a chamber of commerce may want to know the average income of the families in its community, and a real estate developer may want to know how many cars can be expected to drive by a certain location per day; in science, a mineralogist may wish to determine the average iron content of a given ore, and a biologist may want to know how many mutations will be produced in mice by a certain radiation; finally, in everyday life, a commuter may want to know how long on the average it will take her to drive to work, and a serious gardener may want to know what proportion of certain tulips can be expected to bloom.

In each of the examples of the preceding paragraph somebody was interested in determining the true value of some quantity, so that they were all problems of estimation. They would have been tests of hypotheses, however, if the chamber of commerce had wanted to decide on the basis of sample data whether the average family income in its community is really $43,000, if the commuter had wanted to see whether she can really expect (in the sense of an average) that it will take her 12.4 minutes to drive to work, or if the gardener had wanted to check whether it is true that 80% of his tulip bulbs can be expected to bloom. Now it must be decided in each case whether to accept or reject a hypothesis (namely, an assertion or claim) about the parameter of a population.

In this chapter, our focus is on **problems of estimation**, and we shall illustrate them in connection with the estimation of means.

Primarily, we shall distinguish here between point estimates and interval estimates. A point estimate consists of a single number, and the most widely used point estimate of a population mean is the mean of a suitable sample. Two other point estimates of population means are the sample median and average of the two extremes, namely, the midrange.

As its name implies, an interval estimate consists of an interval which, we hope, will contain the probability it is supposed to estimate. In fact, when we determine interval estimates, we always specify the probability that they will "do their job." Some interval estimates are concerned with measures of central tendency, some with measures of variation, and some with percentages. Conceptually, all such problems are treated in the same way, but there are differences in the particular methods that are employed.

Methods of estimating population means are taken up in Sections 11.1 and 11.2; those relating to measures of variability are treated in Section 11.3; and those relating to the estimation of percentages (also, proportions and probabilities) are discussed in Section 11.4.

# 11.1 THE ESTIMATION OF MEANS

To illustrate some of the problems we face when we estimate the mean of a population from sample data, let us refer to a study in which industrial designers want to determine the average (mean) time it takes an adult to assemble an "easy to assemble" toy. Using a random sample, they obtain the following data (in minutes) for 36 persons who assembled the toy:

$$
\begin{array}{cccccccccc}
17 & 13 & 18 & 19 & 17 & 21 & 29 & 22 & 16 & 28 \\
21 & 15 & 26 & 23 & 24 & 20 & 8 & 17 & 17 & 21 \\
32 & 18 & 25 & 22 & 16 & 10 & 20 & 22 & 19 & 14 \\
30 & 22 & 12 & 24 & 28 & 11 & & & & \\
\end{array}
$$

The mean of this sample is $\bar{x} = 19.9$ minutes, and in the absence of any other information, this figure can be used as an estimate of $\mu$, the "true" average time it takes an adult to assemble the given toy.

This kind of estimate is called a **point estimate**, since it consists of a single number, or a single point on the real number scale. Although this is the most common way in which estimates are expressed, it leaves room for many questions. By itself, it does not tell us on how much information the estimate is based, and it does not tell us anything about the possible size of the error. And, of course, we must expect an error. This should be clear from our discussion of the sampling distribution of the mean in Chapter 10, where we saw that the chance fluctuations of the mean (and, hence, its reliability as an estimate of $\mu$) depend on two things—the size of the sample and the size of the population standard deviation $\sigma$. Thus, we might supplement the estimate, $\bar{x} = 19.9$ minutes, with the information that it is the mean of a random sample of size $n = 36$, whose standard deviation is $s = 5.73$ minutes. Although this does not tell us the actual value of $\sigma$, the sample standard deviation can serve as an estimate of this quantity.

Scientific reports often present sample means in this way, together with the values of $n$ and $s$, but this does not supply readers of the report with a coherent picture unless they have had some formal training in statistics. To take care of this, we refer to the theory of Sections 10.7 and 10.8, and the definition in Example 9.6, according to which $z_\alpha$ is such that the area to its right under the standard normal curve is equal to $\alpha$, and hence the area under the standard normal curve between $-z_{\alpha/2}$ and $z_{\alpha/2}$ is equal to $1 - \alpha$. Making use of the fact that for large random samples from infinite populations, the sampling distribution of the mean is approximately a normal distribution with $\mu_{\bar{x}} = \mu$ and $\sigma_{\bar{x}} = \frac{\sigma}{\sqrt{n}}$, we find from Figure 11.1 that the probability is $1 - \alpha$ that the mean of a large random sample from an infinite population will differ from the mean of the population by at most $z_{\alpha/2} \cdot \frac{\sigma}{\sqrt{n}}$. In other words,

> **When we use $\bar{x}$ as an estimate of $\mu$, the probability is $1 - \alpha$ that this estimate will be "off" either way by at most**

**Figure 11.1**
Sampling distribution of the mean.

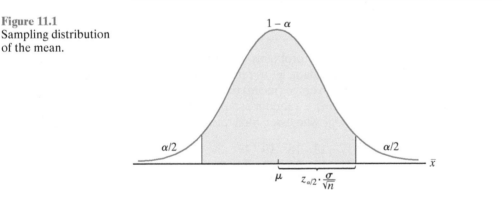

**MAXIMUM ERROR OF ESTIMATE**

$$E = z_{\alpha/2} \cdot \frac{\sigma}{\sqrt{n}}$$

This result applies when $n$ is large, $n \geq 30$, and the population is infinite (or large enough so that the finite population factor need not be used). The two values that are most commonly, though not necessarily, used for $1 - \alpha$ are 0.95 and 0.99, and the corresponding values of $\alpha/2$ are 0.025 and 0.005. As the reader was asked to verify in Exercise 9.16, $z_{0.025} = 1.96$ and $z_{0.005} = 2.575$.

**EXAMPLE 11.1**   A team of efficiency experts intends to use the mean of a random sample of size $n = 150$ to estimate the average mechanical aptitude of assembly-line workers in a large industry (as measured by a certain standardized test). If, based on experience, the efficiency experts can assume that $\sigma = 6.2$ for such data, what can they assert with probability 0.99 about the maximum error of their estimate?

**Solution**    Substituting $n = 150$, $\sigma = 6.2$, and $z_{0.005} = 2.575$ into the formula for $E$, we have

$$E = 2.575 \cdot \frac{6.2}{\sqrt{150}} \approx 1.30$$

Thus, the efficiency experts can assert with probability 0.99 that their error will be at most 1.30.    ■

Suppose now that the efficiency experts actually collect the necessary data and obtain $\bar{x} = 69.5$. Can they still assert with probability 0.99 that the error of their estimate is at most 1.30? After all, $\bar{x} = 69.5$ differs from $\mu$, the population mean, by at most 1.30 or it does not, and they have no way of knowing whether it is one or the other. Actually, they can make this assertion, but it must be understood that the 0.99 probability applies to the *method* used (getting the sample data and using the formula for $E$) and not directly to the single problem at hand.

To make this distinction, it has become the custom to use the word **confidence** here instead of "probability."

---

> **In general, we make probability statements about future values of random variables (say, the potential error of an estimate) and confidence statements once the data have been obtained.**

---

Accordingly, we would say in our example that the efficiency experts can be 99% confident that the error of their estimate, $\bar{x} = 69.5$, is at most 1.30.

Use of the formula for the maximum error entails a complication. We must know the value of the population standard deviation $\sigma$ and this is rarely the case. So, we may replace it with a plausible guess, and, being conservative, this may lead us to overstate the error. Alternatively, we can replace $\sigma$ with an estimate, usually the sample standard deviation $s$. In general, this is considered to be reasonable provided the sample size is sufficiently large, and by sufficiently large we again mean $n \geq 30$.

**EXAMPLE 11.2**    With reference to the illustration on page 263, what can we assert with 95% confidence about the maximum error if we use $\bar{x} = 19.9$ minutes as an estimate of the average time it takes an adult to assemble the given kind of toy?

**Solution**    Substituting $n = 36$, $s = 5.73$ for $\sigma$, and $z_{0.025} = 1.96$ into the formula for $E$, we find that we can assert with 95% confidence that the error is at most

$$E = 1.96 \cdot \frac{5.73}{\sqrt{36}} \approx 1.87 \text{ minutes}$$

Of course, the error is at most 1.87 minutes or it is not, and we do not know whether it is one or the other, but if we had to bet, 95 to 5 (or 19 to 1) would be fair odds that the error is at most 1.87 minutes.    ■

The formula for $E$ can also be used to determine the sample size that is needed to attain a desired degree of precision. Suppose that we want to use the mean of a large random sample to estimate the mean of a population, and we want to be able to assert with probability $1 - \alpha$ that the error of this estimate will not exceed some prescribed quantity $E$. As before, we write $E = z_{\alpha/2} \cdot \frac{\sigma}{\sqrt{n}}$, and upon solving this equation for $n$ we get

SAMPLE SIZE FOR
ESTIMATING $\mu$

$$n = \left[ \frac{z_{\alpha/2} \cdot \sigma}{E} \right]^2$$

**EXAMPLE 11.3**    The dean of a college wants to use the mean of a random sample to estimate the average amount of time students take to get from one class to the next, and she wants to be able to assert with probability 0.95 that her error will be at most 0.30 minute. If she knows from studies of a similar kind that it is reasonable to let $\sigma = 1.50$ minutes, how large a sample will she need?

**Solution**    Substituting $E = 0.30$, $\sigma = 1.50$, and $z_{0.025} = 1.96$ into the formula for $n$, we get

$$n = \left( \frac{1.96 \cdot 1.50}{0.30} \right)^2 \approx 96.04$$

which we round up to the nearest integer, 97. Thus, a random sample of size $n = 97$ is required for the estimate. (Note that the treatment would have been the same if we had said "she wants to be able to assert with 95% confidence that her error *is* at most 0.30 minute" instead of "she wants to be able to assert with probability 0.95 that her error *will be* at most 0.30 minute." It depends on when the assertion is to be made—after or before she collects the data.) ■

As can be seen from the formula for $n$ and also from the introduction to Example 11.3, it has the same shortcoming as the formula for $E$; that is, we must know (at least approximately) the value of the population standard deviation, $\sigma$. For this reason, we sometimes begin with a relatively small sample and then use its standard deviation to see whether more data are required.

Let us now introduce a different way of presenting a sample mean together with an assessment of the error we might be making if we use it to estimate the mean of the population from which the sample came. As on page 263, we shall make use of the fact that, for large random samples from infinite populations, the sampling distribution of the mean is approximately a normal distribution with the mean $\mu_{\bar{x}} = \mu$ and the standard deviation $\sigma_{\bar{x}} = \frac{\sigma}{\sqrt{n}}$, so that

$$z = \frac{\bar{x} - \mu}{\sigma / \sqrt{n}}$$

is a value of a random variable having approximately the standard normal distribution. Since the probability is $1 - \alpha$ that a random variable having this distribution will take on a value between $-z_{\alpha/2}$ and $z_{\alpha/2}$, namely, that

$$-z_{\alpha/2} < z < z_{\alpha/2}$$

we can substitute into this inequality the foregoing expression for $z$ and get

$$-z_{\alpha/2} < \frac{\overline{x} - \mu}{\sigma/\sqrt{n}} < z_{\alpha/2}$$

Then, if we multiply each term by $\sigma/\sqrt{n}$, subtract $\overline{x}$ from each term, and finally multiply each term by $-1$, we get

$$\overline{x} + z_{\alpha/2} \cdot \frac{\sigma}{\sqrt{n}} > \mu > \overline{x} - z_{\alpha/2} \cdot \frac{\sigma}{\sqrt{n}}$$

[Since we multiplied by $-1$, we had to reverse the inequality signs, as is always the case when we multiply the expressions on both sides of an inequality by a negative number. For instance, where 4 is greater than (to the right of) 3, $-4$ is less than (to the left of) $-3$.] The result obtained previously can also be written as

CONFIDENCE
INTERVAL FOR $\mu$

$$\overline{x} - z_{\alpha/2} \cdot \frac{\sigma}{\sqrt{n}} < \mu < \overline{x} + z_{\alpha/2} \cdot \frac{\sigma}{\sqrt{n}}$$

and we can assert with probability $1 - \alpha$ that it will be satisfied for any given sample. In other words, we can assert with $(1 - \alpha)100\%$ confidence that the interval from $\overline{x} - z_{\alpha/2} \cdot \frac{\sigma}{\sqrt{n}}$ to $\overline{x} + z_{\alpha/2} \cdot \frac{\sigma}{\sqrt{n}}$, determined on the basis of a large random sample, contains the population mean we are trying to estimate. When $\sigma$ is unknown and $n$ is at least 30, we replace $\sigma$ by the sample standard deviation $s$.

An interval like this is called a **confidence interval**, its endpoints are called **confidence limits**, and $1 - \alpha$ or $(1 - \alpha)100\%$ is called the **degree of confidence**. As before, the values used most often for the degree of confidence are 0.95 and 0.99 (or 95% and 99%), and the corresponding values of $z_{\alpha/2}$ are 1.96 and 2.575. In contrast to point estimates, estimates given in the form of a confidence interval are called **interval estimates**. They have the advantage of not requiring further elaboration about their reliability. This is taken care of indirectly by their width and the degree of confidence.

Since $n$ must be large to justify the normal approximation to the sampling distribution of the mean, we refer to a confidence interval calculated by means of the preceding formula as a **large-sample confidence interval** for $\mu$. It is also called a **z-interval**, being based on the $z$ statistic that has the standard normal distribution.

**EXAMPLE 11.4**    With reference to Example 11.1, where we had $n = 150$ and $\sigma = 6.2$, use the added information that the efficiency experts obtained the sample mean $\overline{x} = 69.5$ to calculate a 95% confidence interval for the average mechanical aptitude of assembly line workers in the given industry.

**Solution**    Substituting $n = 150, \sigma = 6.2, \overline{x} = 69.5$, and $z_{0.025} = 1.96$ into the confidence interval formula, we get

$$69.5 - 1.96 \cdot \frac{6.2}{\sqrt{150}} < \mu < 69.5 + 1.96 \cdot \frac{6.2}{\sqrt{150}}$$

$$68.5 < \mu < 70.5$$

where the confidence limits are rounded to one decimal. Of course, the statement that the interval from 68.5 to 70.5 contains the true average mechanical aptitude score of assembly line workers in the given industry is either true or false and we do not know whether it is true or false, but we can be 95% confident that it is true. Why? Because the method we used works 95% of the time. To put it differently, the interval may contain $\mu$ or it may not, but if we had to bet, 95 to 5 (or 19 to 1) would be fair odds that it does. ◾

Had we wanted to construct a 99% confidence interval in Example 11.4, we would have substituted 2.575 instead of 1.96 for $z_{\alpha/2}$, and we would have obtained $68.2 < \mu < 70.8$. The 99% confidence interval is wider than the 95% confidence interval—it goes from 68.2 to 70.8 instead of from 68.5 to 70.5, and this illustrates the important fact that

> **When we increase the degree of confidence, the confidence interval becomes wider and, thus, tells us less about the quantity we are trying to estimate.**

Indeed, we might say that "the surer we want to be, the less we have to be sure of."

## 11.2 THE ESTIMATION OF MEANS ($\sigma$ UNKNOWN)

In Section 11.1 we assumed that the samples were large enough, $n \geq 30$, to approximate the sampling distribution of the mean with a normal distribution and, when necessary, to replace $\sigma$ with $s$. To develop corresponding methods that apply in general when $\sigma$ is unknown, we must assume that the populations we are sampling have roughly the shape of normal distributions. We can then base our methods on the **$t$-statistic**

$$t = \frac{\overline{x} - \mu}{s/\sqrt{n}}$$

which is a value of a random variable having the **$t$-distribution**. More specifically, this distribution is called the **Student $t$-distribution** or **Student's $t$-distribution**, as it was first developed by a statistician, W. S. Gosset, who published his work under the pen name "Student." As is shown in Figure 11.2, the shape of this continuous distribution is very similar to that of the standard normal distribution—like the standard normal distribution, it is bell shaped and symmetrical with zero mean. The exact shape of the $t$ distribution depends on a parameter called the **number**

**Figure 11.2**
Standard normal distribution and $t$ distribution.

———— Standard normal distribution
- - - - $t$ distribution (4 degrees of freedom)

**of degrees of freedom**, or simply the **degrees of freedom**, which, for the methods of this section, equals $n - 1$, the sample size less one.

For the standard normal distribution, we defined $z_{\alpha/2}$ in such a way that the area under the curve to its right equals $\alpha/2$, and, hence, the area under the curve between $-z_{\alpha/2}$ and $z_{\alpha/2}$ equals $1 - \alpha$. As is shown in Figure 11.3, the corresponding values for the $t$ distribution are $-t_{\alpha/2}$ and $t_{\alpha/2}$. Since these values depend on $n - 1$, the number of degrees of freedom, they must be obtained from a special table, such as Table II at the end of this book, or perhaps a computer. Table II contains among others the values of $t_{0.025}$ and $t_{0.005}$ for 1 through 30 and selected larger degrees of freedom. As can be seen, $t_{0.025}$ and $t_{0.005}$ approach the corresponding values for the standard normal distribution when the number of degrees of freedom becomes large.

**Figure 11.3**
$t$ distribution.

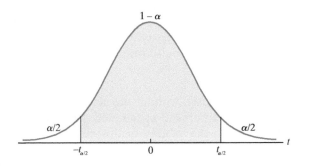

Proceeding as on page 266, we can assert with probability $1 - \alpha$ that a random variable having the $t$ distribution will take on a value between $-t_{\alpha/2}$ and $t_{\alpha/2}$, namely, that

$$-t_{\alpha/2} < t < t_{\alpha/2}$$

Then, substituting into this inequality the expression for $t$ on page 268, we get

$$-t_{\alpha/2} < \frac{\overline{x} - \mu}{s/\sqrt{n}} < t_{\alpha/2}$$

and the same steps as on pages 266 and 267 yield the following confidence interval for $\mu$:

**CONFIDENCE INTERVAL FOR $\mu$ ($\sigma$ UNKNOWN)**

$$\overline{x} - t_{\alpha/2} \cdot \frac{s}{\sqrt{n}} < \mu < \overline{x} + t_{\alpha/2} \cdot \frac{s}{\sqrt{n}}$$

The degree of confidence is $1 - \alpha$, and the only difference between this confidence interval and the $z$ interval with $s$ substituted for $\sigma$ is that $t_{\alpha/2}$ takes the place of $z_{\alpha/2}$. This confidence interval for $\mu$ is usually referred to as a *t*-**interval**, and since most tables of the $t$ distribution give the values of $t_{\alpha/2}$ only for small numbers of degrees of freedom, it has also gone by the name of a **small-sample confidence interval for $\mu$**. Nowadays, computers and other technology provide the values of $t_{\alpha/2}$ for hundreds of degrees of freedom, so this distinction no longer applies.

It must be remembered, though, that for the $t$ interval there is the added assumption that the sample comes from a normal population, or at least from a population having roughly the shape of a normal distribution. This is important, and it will be discussed further in the two examples that follow.

In Example 11.5 we will be given only the values of $n$, $\bar{x}$, and $s$, not the original data, so that there is really no way of checking the "normality" of the population sampled. All we can do in that case is hope for the best. In Example 11.6 we will be given the actual data, so that we can form a normal probability plot (see Section 9.3) to judge whether it is reasonable to look upon the data as a sample from a normal population. This requires the use of appropriate technology—computer software or a graphing calculator—but there are alternative procedures that are ordinarily taught only in more advanced courses in statistics.

**EXAMPLE 11.5** While performing a certain task under simulated weightlessness, the pulse rate of 12 astronauts increased on the average by 27.33 beats per minute with a standard deviation of 4.28 beats per minute. Construct a 99% confidence interval for the true average increase in the pulse rate of astronauts performing the given task (under the stated condition).

**Solution** As we have said previously, without the actual data there is no way of judging the "normality" of the population sampled. Nevertheless, making it clear that the result is subject to the validity of this assumption, we proceed as follows: Substituting $n = 12$, $\bar{x} = 27.33$, $s = 4.28$, and $t_{0.005} = 3.106$ (the entry in Table II for $12 - 1 = 11$ degrees of freedom) into the $t$ interval formula, we get

$$27.33 - 3.106 \cdot \frac{4.28}{\sqrt{12}} < \mu < 27.33 + 3.106 \cdot \frac{4.28}{\sqrt{12}}$$

and, hence,

$$23.49 < \mu < 31.17$$

beats per minute.    ■

In Example 11.6 we shall refer to the illustration of Section 9.3, for which we have verified already that the data can be treated as a random sample from a normal population. That was the illustration which dealt with the durability of a paint used by a highway department for center lines.

**EXAMPLE 11.6** On page 219 we showed that in the eight locations the paint deteriorated after having been crossed by 14.26, 16.78, 13.65, 11.53, 12.64, 13.37, 15.60, and 14.94 million cars, and these are the data for which the normal probability plots, Figures 9.16 and 9.17, showed that they can be treated as a random sample from a normal population. All these figures are rounded to two decimals and so are their mean $\bar{x} = 14.10$ and their standard deviation $s = 1.67$ million car crossings. Calculate a 95% confidence interval for the mean of the population sampled.

**Solution** Substituting $\bar{x} = 14.10$, $s = 1.67$, $n = 8$, and $t_{0.025} = 2.365$ for $8 - 1 = 7$ degrees of freedom into the $t$ interval formula, we get

$$14.10 - 2.365 \cdot \frac{1.67}{\sqrt{8}} < \mu < 14.10 + 2.365 \cdot \frac{1.67}{\sqrt{8}}$$

$$12.70 < \mu < 15.50$$

million car crossings.

   This is the desired 95% confidence interval for the average amount of traffic (car crossings) the paint can withstand before it deteriorates. Again, we can't tell for sure whether the interval from 12.70 million to 15.50 million contains the true average number of car crossings that the paint can withstand before it deteriorates, but 95 to 5 would be fair odds that it does. These odds are based on the fact that the method we used—taking a random sample from a normal population and using the formula given previously—works 95% of the time.
   The method we used earlier to determine the maximum error we risk with $(1 - \alpha)100\%$ confidence when we use a sample mean to estimate the mean of a population is readily adapted to problems in which $\sigma$ is unknown, provided that the population sampled has roughly the shape of a normal distribution. All we have to do is substitute $s$ for $\sigma$ and $t_{\alpha/2}$ for $z_{\alpha/2}$ in the formula for the maximum error on page 264.

**EXAMPLE  11.7**   With reference to Example 11.5, suppose that $\bar{x} = 27.33$ is being used as an estimate of the true average increase in the pulse rate of astronauts performing the given task. What can be said with 99% confidence about the maximum error?

**Solution**   Substituting $s = 4.28$, $n = 12$, and $t_{0.005} = 3.106$ (the entry in Table II for $12 - 1 = 11$ degrees of freedom) into the modified formula for $E$, we get

$$E = t_{\alpha/2} \cdot \frac{s}{\sqrt{n}} = 3.106 \cdot \frac{4.28}{\sqrt{12}} \approx 3.84$$

Thus, if we use $\bar{x} = 27.33$ beats per minute as an estimate of the true average increase in the pulse rate of astronauts performing the given task (under the stated conditions), we can assert with 99% confidence that our error is at most 3.84 beats per minute.

**EXERCISES**

**11.1**   A study made by a staff officer of an armored division showed that in a random sample of $n = 40$ days the division had on the average 1,126 vehicles in operating condition. Given that $\sigma = 135$ for such data, what can this officer assert with 95% confidence about the maximum error if he uses $\bar{x} = 1,126$ as an estimate of the actual daily average number of vehicles that this armored division has in operating condition?

**11.2**   With reference to Exercise 11.1, construct a 99% confidence interval for the actual average daily number of vehicles that this armored division has in operating condition.

**11.3**   A study of the annual growth of a certain kind of orchid showed that (under controlled conditions) a random sample of $n = 40$ of these orchids grew on the average 32.5 mm per year. Given that $\sigma = 3.2$ mm for such data, what can one conclude with 99% confidence about the maximum error if $\bar{x} = 32.5$ mm is used as an estimate of the true average annual growth of this kind of orchid (under the given controlled conditions)?

**11.4**  With reference to Exercise 11.3, construct a 98% confidence interval for the true average annual growth of this kind of orchid.

**11.5**  In a study of automobile collision insurance costs, a random sample of $n = 35$ repair costs of front-end damage caused by hitting a wall at a specified speed had a mean of $1,438. Given that $\sigma = $269$ for such data, what can be said with 98% confidence about the maximum error if $\bar{x} = $1,438$ is used as an estimate of the average cost of such repairs? Also, construct a 90% confidence interval for the average cost of such repairs.

**11.6**  What happens to the standard error of the mean if the sample size is reduced from 288 to 32?

**11.7**  A random sample of size 64 is to be taken from a population that is large enough to be treated as infinite. Given that the mean and the standard deviation of this population are $\mu = 23.5$ and $\sigma = 3.3$, find the probability that the mean of this sample will fall between 23.0 and 24.0.

**11.8**  Before bidding on a contract, a contractor wants to be 95% confident that he is in error by no more than 6 minutes when using the mean of a random sample to estimate the average time it takes for a certain kind of adobe brick to harden. How large a sample will he need if he can assume that $\sigma = 22$ minutes for the time it takes for such brick to harden?

**11.9**  It is desired to estimate the average number of hours of continuous use until a model 737 airplane will first require repairs. If it can be assumed that $\sigma = 138$ hours for such data, how large a sample is needed to be able to assert with a probability of 0.99 that the sample mean will be off by no more than 40 hours?

**11.10**  Before purchasing a large shipment of ground pork, a sausage manufacturer wants to be "pretty sure" that his error does not exceed 0.25 ounce when using the mean of a random sample to estimate the actual fat content per pound. If the standard deviation of the fat content is known to be 0.77 ounce per pound, how many one-pound samples would he need if by "pretty sure" he meant 95% confident?

**11.11**  Rework Exercise 11.10, changing "pretty sure" to mean 99% confident.

**11.12**  When a sample constitutes an appreciable portion of a finite population, we must use the second standard-error formula on page 246 and, hence, to determine the maximum error we must use the formula

$$E = z_{\alpha/2} \cdot \frac{\sigma}{\sqrt{n}} \cdot \sqrt{\frac{N - n}{N - 1}}$$

Taking a random sample of $n = 200$ from among $N = 800$ delinquent accounts, a CPA conducting an audit of a power company found that the amounts owed had a mean of $48.15 with a standard deviation of $6.19. Using $s = $6.19$ as an estimate of $\sigma$, what can she assert with 95% confidence about the maximum error if she uses $48.15 as an estimate of the average amount owed by all 800 delinquent accounts?

**11.13**  A computer is programmed to yield values of a random variable having a normal distribution whose mean and standard deviation are known only to the programmer. Each of 30 students is asked to use the computer to simulate a random sample of size $n = 5$ and use it to construct a 90% confidence interval for $\mu$. Their results are as follows:

| | | |
|---|---|---|
| $6.30 < \mu < 8.26,$ | $6.50 < \mu < 7.72,$ | $6.93 < \mu < 8.01,$ |
| $6.60 < \mu < 8.00,$ | $6.51 < \mu < 7.51,$ | $6.82 < \mu < 8.66,$ |
| $7.02 < \mu < 8.11,$ | $6.94 < \mu < 7.64,$ | $6.24 < \mu < 7.26,$ |
| $6.87 < \mu < 8.17,$ | $6.77 < \mu < 8.13,$ | $6.14 < \mu < 6.82,$ |
| $6.83 < \mu < 7.93,$ | $6.66 < \mu < 8.10,$ | $6.73 < \mu < 7.49,$ |
| $6.41 < \mu < 7.67,$ | $6.76 < \mu < 7.57,$ | $6.97 < \mu < 7.47,$ |
| $6.01 < \mu < 7.43,$ | $7.15 < \mu < 7.89,$ | $6.87 < \mu < 7.81,$ |
| $7.35 < \mu < 7.99,$ | $6.60 < \mu < 8.16,$ | $6.47 < \mu < 7.81,$ |
| $7.01 < \mu < 8.33,$ | $6.97 < \mu < 7.55,$ | $6.56 < \mu < 7.48,$ |
| $7.13 < \mu < 8.03,$ | $7.39 < \mu < 8.01,$ | $5.98 < \mu < 7.68.$ |

(a) How many of these 30 confidence intervals would we expect to contain the mean of the population sampled?

(b) Given that the computer was programmed so that $\mu = 7.30$, how many of the confidence intervals actually contain the mean of the population sampled? Discuss the result.

**11.14** With reference to Example 11.4, on page 267 where we had $n = 150$, $\bar{x} = 69.5$, and $\sigma = 6.2$, suppose that we had asked for a 97% confidence interval.

(a) Find the value of $z_{0.015}$ from Table I.

(b) Use the value of $z_{0.015}$ obtained in part (a) to calculate a 97% $z$ interval for the average mechanical aptitude of assembly-line workers in the given industry.

**11.15** In a pollution study of the air in a certain downtown area, an Environmental Protection Agency (EPA) technician obtained a mean of 2.34 micrograms of suspended benzene-soluble matter per cubic meter with a standard deviation of 0.48 microgram for a sample of size $n = 9$. Assuming that the population sampled is normal,

(a) construct a 95% confidence interval for the mean of the population sampled;

(b) what can the technician assert with 99% confidence about the maximum error if $\bar{x} = 2.34$ micrograms per cubic meter is used as an estimate of the mean of the population sampled?

**11.16** A consumer testing service wants to study the noise level of a new vacuum cleaner. Measuring the noise level of a random sample of $n = 12$ of the machines, it gets the following data (in decibels): 74.0, 78.6, 76.8, 75.5, 73.8, 75.6, 77.3, 75.8, 73.9, 70.2, 81.0, and 73.9.

(a) Use a normal probability plot to verify that it is reasonable to treat these data as a sample from a normal population.

(b) Construct a 95% confidence interval for the average noise level of such vacuum cleaners.

**11.17** A random sample of $n = 9$ pieces of Manila rope (designed for nautical use) has a mean breaking strength of 41,250 pounds and a standard deviation of 1,527 pounds. Assuming that it is reasonable to treat these data as a sample from a normal population, what can we assert with 95% confidence about the maximum error if $\bar{x} = 41,250$ pounds is used as an estimate of the mean breaking strength of such rope?

**11.18** With reference to Exercise 11.17, construct a 98% confidence interval for the mean breaking strength of the given kind of rope.

**11.19**  Use Table II to find
(a)  $t_{0.050}$ for 13 degrees of freedom;
(b)  $t_{0.025}$ for 18 degrees of freedom;
(c)  $t_{0.010}$ for 22 degrees of freedom;
(d)  $t_{0.005}$ for 15 degrees of freedom.

**11.20**  Ten bearings manufactured by a certain process have a mean diameter of 0.406 cm with a standard deviation of 0.003 cm. Construct a 99% confidence interval for the mean diameter of bearings manufactured by this process. Assume that the population sampled is normal.

**11.21**  The following are measurements of the thermal efficiency of $n = 15$ diesel engines made by a prominent manufacturer: 30.7, 35.0, 34.9, 33.6, 28.7, 32.1, 29.0, 31.4, 31.7, 31.8, 33.6, 29.7, 33.4, 28.2, and 31.6.
(a)  Use a normal probability plot to verify that it is reasonable to treat these data as a sample from a normal population.
(b)  Construct a 99% confidence interval of the average thermal efficiency of such diesel engines.

**11.22**  Five containers of a commercial solvent, randomly selected from a large production lot, weigh 19.5, 19.3, 20.0, 19.0, and 19.7 pounds. Assuming that these data can be looked upon as a sample from a normal population, construct a 99% $t$ interval for the mean weight of the containers of the solvent in the production lot.

**11.23**  In setting the type for a book, a compositor made 10, 11, 14, 8, 12, 17, 9, 12, 15, and 12 mistakes in a random sample of ten galleys. Assuming that it is reasonable to approximate the population sampled with a normal distribution, construct a 98% confidence interval for the average number of mistakes that this compositor makes per galley.

**11.24**  With reference to Exercise 11.23, change the degree of confidence to 0.93 and use a computer package or a graphing calculator to rework the exercise.

**11.25**  At seven weather observation posts in the White Mountains, the rainfall during a summer storm was measured as 0.12, 0.14, 0.18, 0.20, 0.15, 0.12, and 0.14 inch. Assuming that it is reasonable to treat these data as a random sample from a normal population, construct a 95% confidence interval for the average rainfall in the White Mountains during that storm.

**11.26**  Rework Exercise 11.25 with the degree of confidence changed to 0.98.

**11.27**  A dentist finds in a routine check that 12 prison inmates, a random sample, needed 2, 3, 6, 1, 4, 2, 4, 5, 0, 3, 5, and 1 filling. If she assumes that these data constitute a sample from a population that can be approximated closely with a normal distribution, what can she assert with 99% confidence about the maximum error if she uses the mean of this sample as an estimate of the average number of fillings needed by the inmates of this very large prison?

# 11.3 THE ESTIMATION OF STANDARD DEVIATIONS

So far we have learned in this chapter how to judge the maximum error when estimating the mean of a population and how to construct a confidence interval for a population mean. These are important techniques for we often make inferences about means, but even more important is the fact that the concepts on which they are based carry over to the estimation of other population parameters.

In this section we shall present methods of estimating population standard deviations and variances, and in Section 11.4 we shall concern ourselves with the estimation of the binomial parameter $p$; namely, with the estimation of population proportions, probabilities, and percentages.

Let us begin here with confidence intervals for $\sigma$ based on $s$, which require that the population we are sampling has roughly the shape of a normal distribution. In that case, the statistic

**CHI-SQUARE STATISTIC**

$$\chi^2 = \frac{(n-1)s^2}{\sigma^2}$$

called the **chi-square statistic** ($\chi$ is the lowercase Greek letter *chi*), is a value of a random variable having approximately the **chi-square distribution**. The parameter of this important continuous distribution is called the number of degrees of freedom, just like the parameter of the $t$ distribution, and as the chi-square distribution is used here, the number of degrees of freedom is $n - 1$. An example of a chi-square distribution is shown in Figure 11.4. Unlike the normal and $t$ distributions, its domain consists of the nonnegative real numbers.

Analogous to $z_\alpha$ and $t_\alpha$, we now define $\chi^2_\alpha$ as the value for which the area under the curve to its right (see Figure 11.4) is equal to $\alpha$; like $t_\alpha$, this value depends on the number of degrees of freedom and must be obtained from a special table, or perhaps a computer. Thus, $\chi^2_{\alpha/2}$ is such that the area under the curve to its right is $\alpha/2$, while $\chi^2_{1-\alpha/2}$ is such that the area under the curve to its left is $\alpha/2$ (see Figure 11.5). For instance, $\chi^2_{0.975}$ is the value for which the area under the curve to its left is 0.025. We made this distinction because the chi-square distribution, unlike the normal and $t$ distributions, is not symmetrical. Values of $\chi^2_{0.995}$, $\chi^2_{0.975}$, $\chi^2_{0.025}$, and $\chi^2_{0.005}$ among others are given in Table III at the end of the book for 1, 2, 3, ..., and 30 degrees of freedom.

We can now proceed as on pages 266 and 267. Since the probability is $1 - \alpha$ that a random variable having a chi-square distribution will take on a value between $\chi^2_{1-\alpha/2}$ and $\chi^2_{\alpha/2}$, namely, that

$$\chi^2_{1-\alpha/2} < \chi^2 < \chi^2_{\alpha/2}$$

we can substitute into this inequality the expression for $\chi^2$ (near the top of this page) and get

$$\chi^2_{1-\alpha/2} < \frac{(n-1)s^2}{\sigma^2} < \chi^2_{\alpha/2}$$

**Figure 11.4**
Chi-square distribution.

**Figure 11.5**
Chi-square distribution.

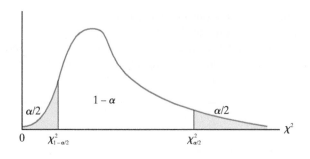

Then, applying some relatively simple algebra, we can rewrite this inequality as

CONFIDENCE
INTERVAL FOR $\sigma^2$

$$\frac{(n-1)s^2}{\chi^2_{\alpha/2}} < \sigma^2 < \frac{(n-1)s^2}{\chi^2_{1-\alpha/2}}$$

This is a $(1-\alpha)100\%$ confidence interval for $\sigma^2$, and if we take square roots, we get a corresponding $(1-\alpha)100\%$ confidence interval for $\sigma$. In the past, this kind of confidence interval has been referred to as a **small-sample confidence interval for $\sigma$** because most chi-square tables are limited to small numbers of degrees of freedom. As was the case in connection with the $t$ distribution, this no longer applies in view of the general availability of computers and other technology.

It is important to remember, however, that the population sampled must have roughly the shape of a normal distribution. In Example 11.8 we will be given only the values of $n$ and $s$, so that there is no way of checking on the "normality" of the population sampled. In Example 11.9 we will be given the original data, so that we can form a normal probability plot (see Section 9.3) to judge whether it is reasonable to look upon the data as a sample from a normal population.

**EXAMPLE  11.8**  With reference to Example 11.5, where we had $n = 12$ and $s = 4.28$ beats per minute, construct a 99% confidence interval for $\sigma$, the true standard deviation of the increase in the pulse rate of astronauts performing a given task (under stated conditions).

**Solution**  As we said on page 270, without the actual data, there is no way of judging the "normality" of the population sampled. So we shall have to state again that the result is subject to the validity of the assumption that the data came from a normal population. Then, substituting $n = 12$, $s = 4.28$, and $\chi^2_{0.995} = 2.603$ and $\chi^2_{0.005} = 26.757$ for $12 - 1 = 11$ degrees of freedom, into the confidence interval formula for $\sigma^2$, we get

$$\frac{11(4.28)^2}{26.757} < \sigma^2 < \frac{11(4.28)^2}{2.603}$$

$$7.53 < \sigma^2 < 77.41$$

Finally, taking square roots, we get $2.74 < \sigma < 8.80$ beats per minute for the desired 99% confidence interval for $\sigma$.

**EXAMPLE  11.9**   In a study of the effectiveness of a hinge lubricant, a research organization wants to investigate the variability in the number of cycles, openings and closings, before the hinge squeaks. Using $n = 15$ hinges, the number of cycles they got were

$$
\begin{array}{ccccc}
4295 & 4390 & 4338 & 4426 & 4698 \\
4405 & 4694 & 4468 & 4863 & 4230 \\
4664 & 4494 & 4535 & 4479 & 4600
\end{array}
$$

(a) Check whether it is reasonable to treat these data as a sample from a normal population.

(b) If so, construct a 95% confidence interval for $\sigma$.

**Solution**   (a) Using appropriate software, we obtained the computer-generated normal probability plot shown in Figure 11.6. As can be seen, the pattern of the fifteen dots follows that of the straight line, and this constitutes support for the assumption that the data constitute a sample from a normal population.

(b) Part of the computer printout of Figure 11.6, which we deleted together with some other nonessential information, showed that the standard deviation of the given data is $s = 172.3$. Substituting this value together with $n = 15$, and $\chi^2_{0.975} = 5.629$ and $\chi^2_{0.025} = 26.119$ for $15 - 1 = 14$ degrees of freedom, into the confidence interval formula, we get

$$
\frac{14(172.3)^2}{26.119} < \sigma^2 < \frac{14(172.3)^2}{5.629}
$$

$$
15{,}913 < \sigma^2 < 73{,}836
$$

Finally, taking square roots, we get

$$
126.1 < \sigma < 271.7 \text{ cycles}
$$

**Figure 11.6**
Normal probability
plot for Example 11.9.

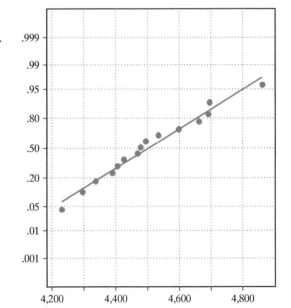

There exists another approach to the construction of confidence intervals for population standard deviations. For large samples, when $n \geq 30$, we can make use of the theory that the sampling distribution of $s$ can be approximated with a normal distribution having the mean $\sigma$ and the standard deviation

$$\sigma_s = \frac{\sigma}{\sqrt{2n}}$$

Then, converting to standard units, we can assert with probability $1 - \alpha$ that

$$-z_{\alpha/2} < \frac{s - \sigma}{\frac{\sigma}{\sqrt{2n}}} < z_{\alpha/2}$$

and fairly simple algebra leads to the following **large-sample confidence interval for** $\sigma$:

LARGE-SAMPLE
CONFIDENCE
INTERVAL FOR $\sigma$

$$\frac{s}{1 + \frac{z_{\alpha/2}}{\sqrt{2n}}} < \sigma < \frac{s}{1 - \frac{z_{\alpha/2}}{\sqrt{2n}}}$$

**EXAMPLE 11.10**    With reference to Example 4.7, where we showed that $s = 14.35$ minutes for the $n = 110$ waiting times between eruptions of Old Faithful, construct a 95% confidence interval for the standard deviation of the population (of waiting times) sampled.

**Solution**    Substituting $n = 110$, $s = 14.35$, and $z_{0.025} = 1.96$ into the large-sample confidence interval formula for $\sigma$, we get

$$\frac{14.35}{1 + \frac{1.96}{\sqrt{220}}} < \sigma < \frac{14.35}{1 - \frac{1.96}{\sqrt{220}}}$$

and, hence, $12.68 < \sigma < 16.53$ minutes. This means that we are 95% confident (and would consider it fair to give odds of 19 to 1) that the interval from 12.68 minutes to 16.53 minutes contains the true standard deviation of waiting times between eruptions of Old Faithful. ■

EXERCISES

**11.28**  The refractive indices of $n = 15$ pieces of glass, randomly selected from a large lot purchased by an optical firm, have a standard deviation of 0.012. Assuming that these measurements can be looked upon as a sample from a normal population, construct a 95% confidence interval for $\sigma$, the standard deviation of the population sampled.

**11.29**  Exercise 11.16 dealt with the noise level of some new vacuum cleaners, and in the solution the reader was asked to verify that it is reasonable to treat the given data as a sample from a normal population. Calculate $s$ for the $n = 12$ measurements and construct a 99% confidence interval for $\sigma$, which here measures the variability of the noise level of the vacuum cleaners.

**11.30**  Exercise 11.21 dealt with the thermal efficiency of certain diesel engines, and in the solution the reader was asked to verify that it is reasonable to treat the given data as a sample from a normal population. Calculate $s$ for the given $n = 18$ measurements and construct a 98% confidence interval for $\sigma$, which here measures the variability of the thermal efficiency of the given kind of engine.

**11.31** With reference to Exercise 11.22 and subject to the same assumptions, construct a 95% confidence interval for $\sigma^2$, the variance of the weights of the containers in the production lot.

**11.32** With reference to the exercises noted below and subject to the assumption that their data constitute random samples from normal populations, construct 98% confidence intervals for $\sigma$, the respective population standard deviations.

(a) Exercise 11.15, where we had $n = 9$ and $s = 0.48$ microgram;

(b) Exercise 11.20, where we had $n = 10$ and $s = 0.003$ cm.

**11.33** With reference to Exercise 11.27 and subject to the same assumption, construct a 99% confidence interval for $\sigma$, the standard deviation of the population sampled.

**11.34** With reference to parts (a) and (b), construct 95% confidence intervals for $\sigma$, the respective population standard deviations.

(a) Exercise 11.1, where $n = 40$ and $s = 135$ vehicles;

(b) Exercise 11.5, where we had $n = 35$ and $s = \$269$.

**11.35** When we deal with very small samples, good estimates of the population standard deviation can often be obtained on the basis of the sample range (the largest sample value minus the smallest). Such quick estimates of $\sigma$ are given by the sample range divided by the divisor $d$, which depends on the size of the sample. For samples from populations having roughly the shape of a normal distribution, its values are shown in the following table for $n = 2, 3, \ldots,$ and 12:

| $n$ | 2 | 3 | 4 | 5 | 6 | 7 | 8 | 9 | 10 | 11 | 12 |
|---|---|---|---|---|---|---|---|---|---|---|---|
| $d$ | 1.13 | 1.69 | 2.06 | 2.33 | 2.53 | 2.70 | 2.85 | 2.97 | 3.08 | 3.17 | 3.26 |

For instance, during the monsoon season, there were 8, 11, 9, 5, 6, 12, 7, and 9 thunderstorms in Northern Arizona in eight successive weeks. The range of this sample is $12 - 5 = 7$, and since $d = 2.85$ for $n = 8$, we can estimate the population standard deviation as $\frac{7}{2.85} = 2.46$. This is quite close to the sample standard deviation, which is $s = 2.39$ as can easily be verified.

(a) With reference to Exercise 11.16, use the range to estimate $\sigma$ for the noise level of the new kind of vacuum cleaner, and compare the result with the sample standard deviation $s$.

(b) The following are four measurements of the weight of a Phoenician tetradrachm: 14.28, 14.34, 14.26, and 14.32 grams. As can easily be verified, $s = 0.0365$ gram for these data. Use the sample range to obtain another estimate of the standard deviation of the population sampled, and compare this estimate with the value of $s$.

**11.36** Fairly reasonable estimates of the population standard deviation can often be obtained by dividing the **interquartile range**, $Q_3 - Q_1$, by 1.35. For the waiting times between eruptions of Old Faithful we obtained $Q_1 = 69.71$ and $Q_3 = 87.58$ in Example 3.24, and $s = 14.35$ in Example 4.7. Estimate the actual standard deviation of waiting times between eruptions of Old Faithful in terms of these quartiles, and compare the result with the value obtained for $s$.

# 11.4 THE ESTIMATION OF PROPORTIONS

In this section we shall deal with **count data**, namely, with data obtained by counting rather than measuring. For instance, we might concern ourselves with the number of persons who experience a side effect from a flu vaccine, the number of defectives in a shipment of manufactured product, the number of television viewers who like a certain situation comedy, the number of tires that last more than 40,000 miles, and so forth.

In particular, we shall concern ourselves with the estimation of the binomial parameter $p$, the probability of a success on an individual trial or the proportion of the time an event will occur, or with the estimation of $100p$, the corresponding percentage. Consequently, we will be able to use what we learned about the binomial distribution in Chapter 8; especially, its approximation by the normal distribution.

So far in this chapter, we have followed the convention of using lowercase Greek letters to denote the parameters of populations—$\mu$ for population means and $\sigma$ for population standard deviations. In connection with binomial populations, more rigorous texts use $\theta$ (lowercase Greek *theta*) for the probability of a success on an individual trial, but having used $p$ throughout Chapter 8, we shall continue doing so in this chapter and in Chapter 14.

The information that is usually available for the estimation of a population proportion (percentage, or probability) is a **sample proportion**, $\frac{x}{n}$, where $x$ is the number of times that an event has occurred in $n$ trials. For example, if a study shows that 54 of 120 cheerleaders, presumably a random sample, suffered what auditory experts call "moderate to severe" damage to their voices, then $\frac{x}{n} = \frac{54}{120} = 0.45$, and we can use this figure as a point estimate of the true proportion of cheerleaders who are afflicted in this way, or the probability that any one cheerleader will be afflicted in this way. Similarly, a supermarket chain might estimate the proportion of its shoppers who regularly use discount coupons as 0.68 if a random sample of 300 shoppers included 204 who regularly use discount coupons.

To be able to use methods based on the binomial distribution, it will be assumed throughout this section that there is a fixed number of independent trials and that for each trial the probability of a success—the parameter we want to estimate—has the constant value $p$. Under these conditions, we know from Chapter 8 that the sampling distribution of the number of successes has the mean $\mu = np$ and the standard deviation $\sigma = \sqrt{np(1-p)}$, and that it can be approximated by a normal distribution so long as $np$ and $n(1-p)$ are both greater than 5. Usually, this requires that $n$ must be large. For $n = 50$, for example, the normal-curve approximations used in this section may be used so long as $50p > 5$ and $50(1-p) > 5$, namely, so long as $p$ lies between 0.10 and 0.90. Similarly, for $n = 100$ they may be used so long as $p$ lies between 0.05 and 0.95, and for $n = 200$ they may be used so long as $p$ lies between 0.025 and 0.975. This should give the reader some idea of what we mean here by "$n$ being large."

If we convert to standard units, then for large values of $n$, the statistic

$$z = \frac{x - np}{\sqrt{np(1-p)}}$$

is a value of a random variable having approximately the standard normal distribution. If we substitute this expression for $z$ into the inequality

$$-z_{\alpha/2} < z < z_{\alpha/2}$$

(as on page 267), some relatively simple algebraic manipulation will yield

$$\frac{x}{n} - z_{\alpha/2} \cdot \sqrt{\frac{p(1-p)}{n}} < p < \frac{x}{n} + z_{\alpha/2} \cdot \sqrt{\frac{p(1-p)}{n}}$$

which looks like a confidence-interval formula for $p$. Indeed, if we used it repeatedly, the inequality should be satisfied $(1-\alpha)100\%$ of the time, but observe that the unknown parameter $p$ appears not only in the middle, but also in

$$\sqrt{\frac{p(1-p)}{n}}$$

to the left of the first inequality sign and to the right of the other. The quantity $\sqrt{\frac{p(1-p)}{n}}$ is called the **standard error of a proportion**, as it is, in fact, the standard deviation of the sampling distribution of a sample proportion (see Exercise 11.53).

To get around this difficulty and, at the same time, simplify the resulting formula, we substitute

$$\hat{p} = \frac{x}{n} \text{ for } p \text{ in } \sqrt{\frac{p(1-p)}{n}}$$

where $\hat{p}$ reads "$p$-hat." (This kind of notation is widely used in statistics. For instance, when we use the mean of a sample to estimate the mean of a population, we might denote it by $\hat{\mu}$, and when we use the standard deviation of a sample to estimate the standard deviation of a population, we might denote it by $\hat{\sigma}$.) Thus, we get the following **$(1-\alpha)100\%$ large-sample confidence interval for $p$**:

LARGE-SAMPLE
CONFIDENCE
INTERVAL FOR $p$

$$\hat{p} - z_{\alpha/2} \cdot \sqrt{\frac{\hat{p}(1-\hat{p})}{n}} < p < \hat{p} + z_{\alpha/2} \cdot \sqrt{\frac{\hat{p}(1-\hat{p})}{n}}$$

**EXAMPLE 11.11**    In a random sample, 136 of 400 persons given a flu vaccine experienced some discomfort. Construct a 95% confidence interval for the true proportion of persons who will experience some discomfort from the vaccine.

**Solution**    Substituting $n = 400$, $\hat{p} = \frac{136}{400} = 0.34$, and $z_{0.025} = 1.96$ into the confidence-interval formula, we get

$$0.34 - 1.96\sqrt{\frac{(0.34)(0.66)}{400}} < p < 0.34 + 1.96\sqrt{\frac{(0.34)(0.66)}{400}}$$

$$0.294 < p < 0.386$$

or, rounding to two decimals, $0.29 < p < 0.39$.

As we have pointed out before, an interval like this contains the parameter it is intended to estimate or it does not. In any particular instance we do not know which is the case, but the 95% confidence implies that the interval was obtained by a method which works 95% of the time. Note also that for $n = 400$ and $p$ on the interval from 0.29 to 0.39, $np$ and $n(1 - p)$ are both much greater than 5, so there can be no doubt that we are justified in using the normal approximation to the binomial distribution.

When it comes to small samples, we can construct confidence intervals for $p$ by using a special table, but the resulting intervals are usually so wide that they are not of much value. For example, for $x = 4$ and $n = 10$, the 95% confidence interval is $0.12 < p < 0.75$. Clearly, this interval is so wide that it does not tell us very much about the value of $p$.

The large-sample theory presented here can also be used to assess the error we may be making when we use a sample proportion to estimate a population proportion, namely, the binomial parameter $p$. Proceeding as on page 264, we can assert with probability $1 - \alpha$ that the difference between a sample proportion and $p$ will be at most

$$E = z_{\alpha/2} \cdot \sqrt{\frac{p(1 - p)}{n}}$$

However, since $p$ is unknown, we substitute for it again the sample proportion $\hat{p}$, and we arrive at the result that

> **If $\hat{p}$ is used as an estimate of $p$, we can assert with $(1 - \alpha)$ 100% confidence that the error is at most**

MAXIMUM ERROR
OF ESTIMATE

$$E = z_{\alpha/2} \cdot \sqrt{\frac{\hat{p}(1 - \hat{p})}{n}}$$

Like the confidence-interval formula on page 281, this formula requires that $n$ must be large enough to use the normal approximation to the binomial distribution.

**EXAMPLE  11.12**   In a random sample of 250 persons interviewed while exiting from polling places throughout a state, 145 said that they voted for the reelection of the incumbent governor. With 99% confidence, what can we say about the maximum error if we use $\hat{p} = \frac{145}{250} = 0.58$ as an estimate of the actual proportion of the vote that the incumbent governor will get?

**Solution**   Substituting $n = 250$, $\hat{p} = 0.58$, and $z_{0.005} = 2.575$ into the formula for $E$, we get

$$E = 2.575\sqrt{\frac{(0.58)(0.42)}{250}} \approx 0.080$$

Thus, if we use $\hat{p} = 0.58$ as an estimate of the actual proportion of the vote the incumbent governor will get, we can assert with 99% confidence that our error is at most 0.080.

With reference to this example, note also that for $n = 250$ the normal approximation to the binomial distribution is justified for any value of $p$ between 0.02 and 0.98.

As in Section 11.1, we can use the formula for the maximum error to determine how large a sample is needed to attain a desired degree of precision. If we want to use a sample proportion to estimate a population proportion $p$, and we want to be able to assert with probability $1 - \alpha$ that our error will not exceed some prescribed quantity $E$, we write as before

$$E = z_{\alpha/2} \cdot \sqrt{\frac{p(1 - p)}{n}}$$

Upon solving this equation for $n$, we get

SAMPLE SIZE

$$n = p(1 - p) \left[\frac{z_{\alpha/2}}{E}\right]^2$$

This formula cannot be used as is, because it involves the quantity $p$ we are trying to estimate. However, since $p(1 - p)$ *increases* from 0 to $\frac{1}{4}$ when $p$ increases from 0 to $\frac{1}{2}$ or decreases from 1 to $\frac{1}{2}$, we can proceed as follows:

> **If we have some information about the values that $p$ might assume, we substitute for it in the formula for $n$ whichever of these values is closest to $\frac{1}{2}$; if we have no information about the values that $p$ might assume, we substitute $\frac{1}{4}$ for $p(1 - p)$ in the formula for $n$.**

In either case, since the value we obtain for $n$ may well be larger than necessary, we can say that the probability is *at least* $1 - \alpha$ that our error will not exceed $E$.

EXAMPLE 11.13   Suppose that a state highway department wants to estimate what proportion of all trucks operating between two cities carry too heavy a load, and it wants to be able to assert with a probability of at least 0.95 that its error will not exceed 0.04. How large a sample will it need if

(a) it knows that the true proportion lies somewhere on the interval from 0.10 to 0.25;

(b) it has no idea what the true value might be?

Solution   (a) Substituting $z_{0.025} = 1.96$, $E = 0.04$, and $p = 0.25$ into the formula for $n$, we get

$$n = (0.25)(0.75)\left(\frac{1.96}{0.04}\right)^2 \approx 450.19$$

and we round this up to the next integer, 451.

**(b)** Substituting $z_{0.025} = 1.96$, $E = 0.04$, and $p(1-p) = \frac{1}{4}$ into the formula for $n$, we get

$$n = \frac{1}{4}\left(\frac{1.96}{0.04}\right)^2 = 600.25$$

and we round this up to the next integer, 601.    ◼

This shows how some knowledge about $p$ can substantially reduce the sample size needed to attain a desired degree of precision. Note also that in a problem like this we round up, if necessary, to the nearest integer.

EXERCISES

**11.37** In accordance with the rule that $np$ and $n(1-p)$ must both be greater than 5, for what values of $p$ can we use the normal approximation to the binomial distribution when

(a) $n = 400$;

(b) $n = 500$?

**11.38** In a random sample, 64 of 200 motorists stopped at a roadblock had not fastened their seat belts. Construct a 95% confidence interval for the corresponding true proportion in the population sampled

**11.39** In a random sample of 400 eligible voters interviewed in a large county, 228 objected to the use of public funds for the construction of a new professional football stadium. Construct a 95% confidence interval for the corresponding proportion of eligible voters in the whole county.

**11.40** With reference to Exercise 11.39, what can we say with 99% confidence about the maximum error if $\frac{x}{n} = \frac{228}{400} = 0.57$ is used as an estimate of the proportion of all eligible voters in the county who are against the use of public funds for the construction of a new professional football stadium?

**11.41** Among 400 fish caught in Woods Canyon Lake, 56 were inedible as a result of the chemical pollution of the environment. Construct a 99% confidence interval for the corresponding true proportion.

**11.42** With reference to Exercise 11.41, what can we say with 95% confidence about the maximum error if $\frac{x}{n} = \frac{56}{400} = 0.14$ is used as an estimate of the corresponding true proportion?

**11.43** In a random sample of 120 cheerleaders, 54 had suffered moderate to severe damage to their voices. With 90% confidence, what can we assert about the maximum error if the sample proportion $\frac{54}{120} = 0.45$ is used as an estimate of the true proportion of cheerleaders who are afflicted in this way?

**11.44** A random sample of 300 shoppers at a large supermarket includes 234 who regularly use discount coupons. Construct a 98% confidence interval for the probability that any one shopper at that supermarket, randomly chosen for an interview, will confirm that he or she regularly uses discount coupons.

**11.45** In a random sample of 1,600 adults interviewed nationwide, only 412 felt that the salaries of certain government officials should be raised. Construct a 95% confidence interval for the actual percentage of adults who share that opinion.

**11.46** In a random sample of 400 television viewers nationwide, 152 had seen a certain controversial program. With 98% confidence, what can we assert about the maximum error if we use $\frac{152}{400} \cdot 100 = 38\%$ as an estimate of the corresponding true percentage?

**11.47** In a random sample of 140 supposed UFO sightings, 119 could easily be explained in terms of natural phenomena. Construct a 99% confidence interval for the probability that a supposed UFO sighting will easily be explained in terms of natural phenomena.

**11.48** In a random sample of 80 persons convicted in U.S. District Courts on narcotics charges, 36 received probation. With 98% confidence, what can we say about the maximum error if we use $\frac{36}{80} = 0.45$ as an estimate of the probability that a person convicted in a U.S. District Court on narcotics charges will receive probation?

**11.49** When a sample constitutes more than 5% of a finite population, and the sample itself is large, we use the same finite population correction factor as in Section 10.7, and hence the following confidence limits for $p$:

$$\hat{p} \pm z_{\alpha/2} \cdot \sqrt{\frac{\hat{p}(1 - \hat{p})(N - n)}{n(N - 1)}}$$

Here $N$ is, as before, the size of the population sampled.

(a) Among the $N = 360$ families in an apartment complex, a random sample of $n = 100$ families is interviewed, and it is found that 34 of them have children of college age. Use the preceding formula to construct a 95% confidence interval for the proportion of all the families living in the apartment complex who have children of college age.

(b) With reference to Exercise 11.47, suppose that altogether there had been 350 supposed UFO sightings. Use this added information to recalculate the confidence interval asked for in that exercise.

**11.50** A political pollster is engaged by a politician to estimate the proportion of registered voters in her district who plan to vote for her in the next election. Find the sample size needed if she wants, with at least 95% confidence, the poll to be accurate to within

(a) 8 percentage points;

(b) 2 percentage points.

**11.51** Suppose that we want to estimate what proportion of all drivers exceed the maximum speed limit on a stretch of I-17 near the Rock Springs exit. How large a sample will we need so that the error of our estimate is to be at most 0.05 with at least

(a) 90% confidence;

(b) 95% confidence;

(c) 99% confidence?

**11.52** A national manufacturer wants to determine what percentage of purchases of razor blades for use by men is actually made by women. How large a sample of men shaving with razor blades will the manufacturer need to be at least 98% confident that the sample percentage will not be off by more than 2.5 percentage points and

(a) nothing is known about the true percentage;

(b) there is good reason to believe that the true percentage is at most 30%?

**11.53** Since the proportion of successes is simply the number of successes divided by $n$, the mean and the standard deviation of the sampling distribution of the proportion of successes may be obtained by dividing by $n$ the mean and the standard deviation of the sampling distribution of the number of successes. Use this argument to verify the standard error formula given on page 281.

# CHECKLIST OF KEY TERMS (with page references to their definitions)

Chi-square distribution, 275
Chi-square statistic, 275
Confidence, 265
Confidence interval, 267
Confidence limits, 267
Count data, 279
Degree of confidence, 267
Degrees of freedom, 269
Estimation, 262, 263
*Interquartile range*, 279
Interval estimate, 267
Large-sample confidence interval, 267, 278
$(1 - \alpha)100\%$ large-sample
   confidence interval for $p$, 281

Number of degrees of freedom, 269
Prediction, 262
Point estimate, 263
Sample proportion, 280
Small-sample confidence interval, 269, 276
Standard error of a proportion, 281
Student $t$-distribution, 268
$t$-distribution, 268
Tests of hypotheses, 262
$t$-interval, 269
$t$-statistic, 268
$z$-interval, 267

# REFERENCES

*An informal introduction to interval estimation is given under the heading of "How to be precise though vague" in*

MORONEY, M. J., *Facts from Figures*. London: Penguin Books, Ltd., 1956

*and also in*

GONICK, L., and SMITH, WOOLCOTT, *A Cartoon Guide to Statistics*. New York: Harper Collins Publishers, 1993.

*Detailed discussions of the chi-square and t distributions may be found in most textbooks on mathematical statistics, and more detailed tables of these distributions are given in*

PEARSON, E. S., and HARTLEY, H. O., *Biometrika Tables for Statisticians*, Vol. I. New York: John Wiley & Sons, Inc., 1968.

*Tables of confidence intervals for proportions, including those for small samples, were first published in Vol. 26 (1934) of* Biometrika. *Nowadays, they may be found, for instance, in*

MAXWELL, E. A., *Introduction to Statistical Thinking*. Englewood Cliffs, N.J.: Prentice Hall, 1983.

# TESTS OF HYPOTHESES: MEANS

**12.1** Tests of Hypotheses 287

**12.2** Significance Tests 293

**12.3** Tests Concerning Means 300

**12.4** Tests Concerning Means ($\sigma$ Unknown) 304

**12.5** Differences between Means 307

**12.6** Differences between Means ($\sigma$'s Unknown) 311

**12.7** Differences between Means (Paired Data) 313

Checklist of Key Terms 315

References 315

n the previous chapter our attention was directed to problems of *estimation*. Separate sections were devoted to the estimation of means; the estimation of means ($\sigma$ unknown); the estimation of standard deviations; and the estimation of proportions. Much of Chapter 12 and of subsequent chapters will gravitate toward *testing*. After a general introduction to tests of hypotheses in Sections 12.1 and 12.2, the remainder of this chapter will be devoted to tests concerning the mean of one population, or the means of two populations. Tests concerning population standard deviations will be treated in Chapter 13, and tests concerning percentages (proportions or probabilities) will be dealt with in Chapter 14. Subsequent chapters will deal with other, specialized, tests of hypotheses.

## 12.1 TESTS OF HYPOTHESES

In the introduction to Chapter 11 we referred to certain decision problems as tests of hypotheses without actually giving a formal definition of what we mean here by a hypothesis. In general,

> **A statistical hypothesis is an assertion or conjecture about a parameter or parameters, of a population (or populations); it may also concern the type, or nature, of a population (or populations).**

With regard to the second part of this definition, we shall see in Section 14.5 how we can test whether it is reasonable to treat a population sampled as being a binomial population, a Poisson population, or perhaps a normal population. In this chapter we shall be concerned only with hypotheses about population parameters; in particular, the mean of one population or the means of two populations.

To develop procedures for testing statistical hypotheses, we must always know exactly what to expect when a hypothesis is true, and it is for this reason that we often hypothesize the opposite of what we hope to prove. Suppose, for instance, that we suspect that a dice game is not honest. If we formulate the hypothesis that the dice are crooked, everything would depend on how crooked they are, but if we assume that they are perfectly balanced, we could calculate all the necessary probabilities and take it from there. Also, if we want to show that one method of teaching computer programming is more effective than another, we would hypothesize that the two methods are equally effective; if we want to show that one diet is healthier than another, we hypothesize that they are equally healthy; and if we want to show that a new copper-bearing steel has a higher yield strength than ordinary steel, we hypothesize that the two yield strengths are the same. Since we hypothesize that there is no difference in the effectiveness of the two teaching methods, medically no difference between the two diets, and no difference in the yield strength of the two kinds of steel, we call hypotheses like these **null hypothesis** and denote them by $H_0$. In effect, the term "null hypothesis" is used for any hypothesis set up primarily to see whether it can be rejected.

The idea of setting up a null hypothesis is common even in nonstatistical thinking. It is precisely what we do in criminal proceedings, where an accused is presumed to be innocent until his guilt has been established beyond a reasonable doubt. The presumption of innocence is a null hypothesis.

The hypothesis that we use as an alternative to the null hypothesis, namely, the hypothesis that we accept when the null hypothesis is rejected, is appropriately called an **alternative hypothesis** and is denoted by $H_A$. It must always be formulated together with the null hypothesis, for otherwise we would not know when to reject $H_0$. For instance, if a psychologist wants to test the hypothesis that it takes an adult 0.38 second to react to a visual stimulus, he tests the hypothesis against the alternative hypothesis $\mu = 0.38$ second. He would reject the hypothesis only if he gets a sample mean appreciably greater than 0.38 second. On the other hand, if he uses the alternative hypothesis $\mu \neq 0.38$ second, he would reject the null hypothesis if he gets a sample mean that is appreciably greater than, or appreciably less than, 0.38 second.

As in the preceding illustration, alternative hypotheses usually specify that the population mean (or whatever other parameter may be of concern) is less than, greater than, or not equal to the value assumed under the null hypothesis. For any given problem, the choice of one of these alternatives depends on what we hope to be able to show, or perhaps on where we want to put the burden of proof.

**EXAMPLE 12.1**   The average drying time of a manufacturer's paint is 20 minutes. Investigating the effectiveness of a modification in the chemical composition of his paint, the manufacturer wants to test the null hypothesis $\mu = 20$ minutes against a suitable alternative, where $\mu$ is the average drying time of the modified paint.

(a) What alternative hypothesis should the manufacturer use if he wants to make the modification only if it actually decreases the drying time of the paint?

(b) What alternative hypothesis should the manufacturer use if the new process is actually cheaper and he wants to make the modification unless it actually increases the drying time of the paint?

**Solution**
(a) He should use the alternative hypothesis $\mu < 20$ and make the modification only if the null hypothesis can be rejected.

(b) He should use the alternative hypothesis $\mu > 20$ and make the modification unless the null hypothesis is rejected.　■

In general, if the test of a hypothesis concerns the parameter $\mu$, its value assumed under the null hypothesis is denoted by $\mu_0$, and the null hypothesis, itself, is $\mu = \mu_0$.

To illustrate in detail the problems we face when testing a statistical hypothesis, let us refer again to the reaction-time example on page 288, and let us suppose that the psychologist wants to test the null hypothesis

$$H_0: \mu = 0.38 \text{ second}$$

against the alternative hypothesis

$$H_A: \mu \neq 0.38 \text{ second}$$

where $\mu$ is the mean reaction time of an adult to the visual stimulus. To perform this test, the psychologist decides to take a random sample of $n = 40$ adults with the intention of accepting the null hypothesis if the mean of the sample falls anywhere between 0.36 second and 0.40 second; otherwise, he will reject it.

This provides a clear-cut criterion for accepting or rejecting the null hypothesis, but unfortunately it is not infallible. Since the decision is based on a sample, there is the possibility that the sample mean may be less than 0.36 second or greater than 0.40 second even though the true mean is 0.38 second. There is also the possibility that the sample mean may fall between 0.36 second and 0.40 second, even though the true mean is, say, 0.39 second. Thus, before adopting the criterion (and, for that matter, any decision criterion) it would seem wise to investigate the chances that it may lead to a wrong decision.

Assuming that it is known from similar studies that $\sigma = 0.08$ second for this kind of data, let us first investigate the possibility of falsely rejecting the null hypothesis. Thus, assume for the sake of argument that the true average reaction time is 0.38 second; then find the probability that the sample mean will be less than or equal to 0.36 or greater than or equal to 0.40. The probability that this will happen purely due to chance is given by the sum of the areas of the two tinted regions of Figure 12.1, and it can readily be determined by approximating the sampling distribution of the mean with a normal distribution. Assuming that the population sampled may be looked upon as being infinite, which seems reasonable in this case, we have

$$\sigma_{\bar{x}} = \frac{\sigma}{\sqrt{n}} = \frac{0.08}{\sqrt{40}} \approx 0.0126$$

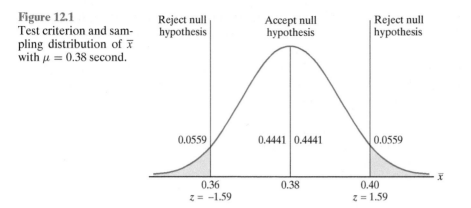

**Figure 12.1**
Test criterion and sampling distribution of $\bar{x}$ with $\mu = 0.38$ second.

and it follows that the dividing lines of the criterion, in standard units, are

$$z = \frac{0.36 - 0.38}{0.0126} \approx -1.59 \quad \text{and} \quad z = \frac{0.40 - 0.38}{0.0126} \approx 1.59$$

It follows from Table I that the area in each tail of the sampling distribution of Figure 12.1 is $0.5000 - 0.4441 = 0.0559$, and hence the probability of getting a value in either tail of the sampling distribution is $2(0.0559) = 0.1118$, or 0.11 rounded to two decimals.

Let us now consider the other possibility, where the test fails to detect that the null hypothesis is false; namely, that $\mu \neq 0.38$ second. Thus, assume for the sake of argument that the true average reaction time is 0.41 second. Now, getting a sample mean on the interval from 0.36 second to 0.40 second would lead to the erroneous acceptance of the null hypothesis that $\mu = 0.38$ second. The probability that this will happen purely due to chance is given by the area of the tinted region of Figure 12.2. The mean of the sampling distribution is now 0.41 second, but its standard deviation is, as before,

$$\sigma_{\bar{x}} = \frac{0.08}{\sqrt{40}} \approx 0.0126$$

and the dividing lines of the criterion, in standard units, are

$$z = \frac{0.36 - 0.41}{0.0126} \approx -3.97 \quad \text{and} \quad z = \frac{0.40 - 0.41}{0.0126} \approx -0.79$$

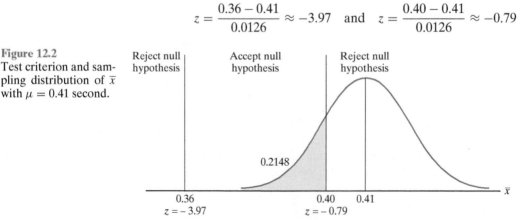

**Figure 12.2**
Test criterion and sampling distribution of $\bar{x}$ with $\mu = 0.41$ second.

Since the area under the curve to the left of $-3.97$ is negligible, it follows from Table I that the area of the tinted region of Figure 12.2 is $0.5000 - 0.2852 = 0.2148$ or 0.21 rounded to two decimals. This is the probability of erroneously accepting the null hypothesis when actually $\mu = 0.41$. It will be up to the psychologist to decide whether the 0.11 probability of erroneously rejecting the null hypothesis $\mu = 0.38$ and the 0.21 probability of erroneously accepting it when actually $\mu = 0.41$ are acceptable risks.

The situation described here is typical of tests of hypotheses, and it may be summarized as in the following table:

|  | Accept $H_0$ | Reject $H_0$ |
|---|---|---|
| $H_0$ is true | Correct decision | Type I error |
| $H_0$ is false | Type II error | Correct decision |

If the null hypothesis $H_0$ is true and accepted or false and rejected, the decision is in either case correct; if it is true and rejected or false and accepted, the decision is in either case in error. The first of these errors is called a **Type I error** and the probability of committing it is designated by the Greek letter $\alpha$ (*alpha*); the second is called a **Type II error** and the probability of committing it is designated by the Greek letter $\beta$ (*beta*). Thus, in our example we showed that for the given test criterion $\alpha = 0.11$ and $\beta = 0.21$ when $\mu = 0.41$.

The scheme just outlined is similar to what was done in Section 7.2. Analogous to the decision that the director of the research division of the pharmaceutical company had to make in Example 7.9, now the psychologist must decide whether to accept or reject the null hypothesis $\mu = 0.38$. It is difficult to carry this analogy much further, though, because in actual practice we can seldom associate cash values with the various possibilities, as we did in Example 7.9.

**EXAMPLE 12.2** Suppose that the psychologist has actually taken the sample and obtained $\bar{x} = 0.408$. What will he decide and will it be in error if

(a) $\mu = 0.38$ second;

(b) $\mu = 0.42$ second?

**Solution** Since $\bar{x} = 0.408$ exceeds 0.40, the psychologist will reject the null hypothesis $\mu = 0.38$ second.

(a) Since the null hypothesis is true and rejected, the psychologist will be making a Type I error.

(b) Since the null hypothesis is false and rejected, the psychologist will not be making an error.

In calculating the probability of a Type II error in our illustration, we arbitrarily chose the alternative value $\mu = 0.41$ second. However, in this problem, as in most others, there are infinitely many other alternatives, and for each of them there is a positive probability $\beta$ of erroneously accepting $H_0$. So, in practice, we choose some key alternative values and calculate the corresponding probabilities $\beta$ of committing a Type II error, or we sidestep the issue by proceeding in a way that will be explained in Section 12.2.

If we do calculate $\beta$ for various alternative values of $\mu$ and plot these probabilities as in Figure 12.3, we obtain a curve that is called the **operating characteristic curve**, or simply the **OC curve**, of the test criterion. Since the probability of a Type II error is the probability of accepting $H_0$ when it is false, we "completed the picture" in Figure 12.3 by labeling the vertical scale "Probability of accepting $H_0$" and plotting at $\mu = 0.38$ second the probability of accepting $H_0$ when it is true, namely, $1 - \alpha = 1 - 0.11 = 0.89$.

Examination of the curve of Figure 12.3 shows that the probability of accepting $H_0$ is greatest when $H_0$ is true, and that it is still fairly high for small departures from $\mu = 0.38$. However, for larger and larger departures from $\mu = 0.38$ in either direction, the probabilities of failing to detect them and accepting $H_0$ become smaller and smaller. In Exercise 12.10 the reader will be asked to verify some of the probabilities plotted in Figure 12.3.

If we had plotted the probabilities of rejecting $H_0$ instead of those of accepting $H_0$, we would have obtained the graph of the **power function** of the test criterion instead of its operating characteristic curve. The concept of an OC curve is used more widely in applications, especially in industrial applications, while the concept of a power function is used more widely in matters that are of theoretical interest. A detailed study of operating characteristic curves and power functions is beyond the scope of this text, and the purpose of our example is mainly to show how statistical methods can be used to measure and control the risks one is exposed to when testing hypotheses. Of course, the methods discussed here are not limited to the particular problem concerning the average reaction time to a visual stimulus—$H_0$ could have been the hypothesis that the average age at which women get a divorce is 28.5, the hypothesis that an antibiotic is 87% effective, the hypothesis that a computer-assisted method of instruction will, on

**Figure 12.3**
Operating characteristic curve.

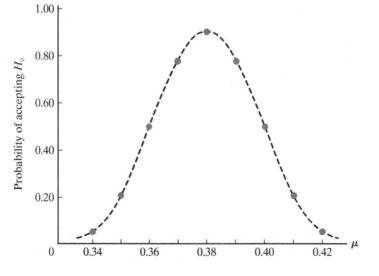

the average, raise a student's score on a standard achievement test by 7.4 points, and so forth.

# 12.2 SIGNIFICANCE TESTS

In the problem dealing with the reaction time of adults to a visual stimulus, we had less trouble with Type I errors than with Type II errors, because we formulated the null hypothesis as a **simple hypothesis** about the parameter $\mu$; that is, we formulated it so that $\mu$ took on a single value, the value $\mu = 0.38$ second, and the corresponding value of a Type I error could be calculated.[†] Had we formulated instead a **composite hypothesis** about the parameter $\mu$, say, $\mu \neq 0.38$ second, $\mu < 0.38$ second, or $\mu > 0.38$ second, where in each case $\mu$ can take on more than one possible value, we could not have calculated the probability of a Type I error without specifying how much $\mu$ differs from, is less than, or is greater than 0.38 second.

In the same illustration, the alternative hypothesis was the composite hypothesis $\mu \neq 0.38$ second, and it took quite some work to calculate the probabilities of Type II errors (for various alternative values of $\mu$) shown in the OC curve of Figure 12.3. Since this is typical of most practical situations (that is, alternative hypotheses are usually composite), let us demonstrate how Type II errors can often be sidestepped altogether.

Studies have shown that in a certain city licensed drivers average 0.9 traffic ticket per year, but a social scientist suspects that drivers over 65 years of age average more than 0.9 traffic ticket per year. So she checks the records of a random sample of licensed drivers over 65 in the given city, and bases her decision on the following criterion:

> **Reject the null hypothesis $\mu = 0.9$ (and accept the alternative hypothesis $\mu > 0.9$) if the drivers over 65 in the sample average, say, at least 1.2 traffic tickets per year; otherwise, reserve judgment (perhaps, pending further investigation).**

If one reserves judgment as in this criterion, there is no possibility of committing a Type II error—no matter what happens, the null hypothesis is never accepted. This would seem all right in the preceding example, where the social scientist wants to see primarily whether her suspicion is justified; namely, whether the null hypothesis can be rejected. If it cannot be rejected, this does not mean that she must necessarily accept it. Indeed, her suspicion may not be completely resolved.

The procedure we have outlined here is called a **significance test**, or a **test of significance**. If the difference between what we expect under the null hypothesis and what we observe in a sample is too large to be reasonably attributed to

---

[†]Note that we are applying the term "simple hypothesis" to hypotheses about specific parameters. Some statisticians use the term "simple hypothesis" only when the hypothesis completely specifies the population.

chance, we reject the null hypothesis. If the difference between what we expect and what we observe is so small that it may well be attributed to chance, we say that the result is **not statistically significant**, or simply that it is **not significant**. We then accept the null hypothesis or reserve judgment, depending on whether a definite decision one way or the other must be reached.

Since "significant" is often used interchangeably with "meaningful" or "important" in everyday language, it must be understood that we are using it here as a technical term. Specifically, the word "significant" is used in situations in which a null hypothesis is rejected. If a result is statistically significant, this does not mean that it is necessarily of any great importance, or that it is of any practical value. Suppose, for instance, that the psychologist of our reaction-time example has actually taken the sample, as in Example 12.2, and obtained $\bar{x} = 0.408$. According to the criterion on page 289 this result is statistically significant, meaning that the difference between $\bar{x} = 0.408$ and $\mu = 0.38$ is too large to be attributed to chance. It is possible, though, that no one may care about this result; not even a lawyer involved in a litigation where reaction times may be of critical relevance in determining a client's liability. His reaction might be that the whole thing is simply not worth bothering about.

Returning to the original criterion used in the reaction-time example, the one on page 288, we could convert it into that of a significance test by writing

> **Reject the null hypothesis $\mu = 0.38$ second (and accept the alternative $\mu \neq 0.38$ second) if the mean of the 40 sample values is less than or equal to 0.36 second or greater than or equal to 0.40 second; reserve judgment if the sample mean falls anywhere between 0.36 second and 0.40 second.**

So far as the rejection of the null hypothesis is concerned, the criterion has remained unchanged and the probability of a Type I error is still 0.11. However, so far as its acceptance is concerned, the psychologist is now playing it safe by reserving judgment.

Reserving judgment in a significance test is similar to what happens in court proceedings where the prosecution does not have sufficient evidence to get a conviction, but where it would be going too far to say that the defendant definitely did not commit the crime. In general, whether one can afford the luxury of reserving judgment in any given situation depends entirely on the nature of the situation. If a decision must be reached one way or the other, there is no way of avoiding the risk of committing a Type II error.

Since most of the remainder of this book is devoted to significance tests— indeed, most statistical problems that are not problems of estimation or prediction deal with tests of this kind—it will help to perform such tests by proceeding systematically as outlined in the following five steps. The first of these may look simple and straightforward, yet it presents the greatest difficulties to most beginners.

1. **We formulate a simple null hypothesis and an appropriate alternative hypothesis that is to be accepted when the null hypothesis is rejected.**

In the reaction-time example the null hypothesis was $\mu = 0.38$ second and the alternative hypothesis was $\mu \neq 0.38$ second. We choose this alternative as an illustration; in actual practice, it would reflect the psychologist's intent to reject the null hypothesis if 0.38 second is either too high or too low. We refer to this kind of alternative as a **two-sided alternative**. In the traffic-ticket example the null hypothesis was $\mu = 0.9$ ticket and the alternative hypothesis was $\mu > 0.9$ (to confirm the social scientist's suspicion that licensed drivers over 65 average more than 0.9 traffic ticket per year). This is called a **one-sided alternative**. We can also write a one-sided alternative with the inequality going the other way. For instance, if we hope to show that the average time required to do a certain job is less than 15 minutes, we would test the null hypothesis $\mu = 15$ minutes against the alternative hypothesis $\mu < 15$ minutes.

This is not the first time that we concerned ourselves with the formulation of hypotheses. Prior to Example 12.1 we mentioned some of the things that must be taken into account when choosing $H_A$, but throughout this chapter, so far, the null hypothesis has always been specified.

Basically, there are two things we must watch in connection with $H_0$. *First, whenever possible we formulate null hypotheses as simple hypotheses about the parameters with which we are concerned; second, we formulate null hypotheses in such a way that their rejection proves whatever point we hope to make.* As we have pointed out before, we choose null hypotheses as simple hypotheses so that we can calculate, or specify, the probabilities of Type I errors. We saw how this works in the reaction-time example. The reason for choosing null hypotheses so that their rejection proves whatever point we hope to make is that, in general, it is much easier to prove that something is false than to prove that it is true. Suppose, for instance, that somebody claims that all 6,000 male students attending a certain university weigh at least 145 pounds. To show that this claim is true, we literally have to weigh each of the 6,000 students; however, to show that it is false, we have only to find one student who weighs less than 145 pounds, and that should not be too difficult.

**EXAMPLE 12.3**    A bakery machine fills boxes with crackers, averaging 454 grams (roughly one pound) of crackers per box.

(a) If the management of the bakery is concerned about the possibility that the actual average is different from 454 grams, what null hypothesis and what alternative hypothesis should it use to put this to a test?

(b) If the management of the bakery is concerned about the possibility that the actual average is less than 454 grams, what null hypothesis and what alternative hypothesis should it use to put this to a test?

**Solution**    (a) The words "different from" suggest that the hypothesis $\mu \neq 454$ grams is needed together with the only other possibility, namely, the hypothesis $\mu = 454$ grams. Since the second of these hypotheses is a simple hypothesis, and its rejection (and the acceptance of the other hypothesis) confirms the management's concern, we follow the above two rules by writing

$$H_0: \quad \mu = 454 \text{ grams}$$

$$H_A: \quad \mu \neq 454 \text{ grams}$$

**(b)** The words "less than" suggest that we need the hypothesis $\mu < 454$ grams, but for the other hypothesis there are many possibilities, including $\mu \geq 454$ grams, $\mu = 454$ grams, and, say, $\mu = 456$ grams. Two of these (and many others) are simple hypotheses, but since it would be to the bakery's disadvantage to put too many crackers into the boxes, a sensible choice would be

$$H_0: \quad \mu = 454 \text{ grams}$$

$$H_A: \quad \mu < 454 \text{ grams}$$

Note that the null hypothesis is a simple hypothesis and that its rejection (and the acceptance of the alternative) confirms the management's suspicion.

It is important to add that $H_0$ and $H_A$ must be formulated before any data are actually collected, or at least without looking at the data. In particular, the choice of a one-sided alternative or a two-sided alternative should not be suggested by the data. However, it often happens that we are presented with data before we had the opportunity to contemplate the hypotheses, and in such situations we must try to assess the motives (or objectives) without using the data. If there is any doubt whether a situation calls for a one-sided or a two-sided alternative, the scrupulous action calls for a two-sided alternative.

Like the first step given on page 294, the second step looks simple and straightforward, but it is not without complications.

## 2. We specify the probability of a Type I error.

When $H_0$ is a simple hypothesis this can always be done, and we usually set the probability of a Type I error, also called the **level of significance**, at $\alpha = 0.05$ or $\alpha = 0.01$. Testing a simple hypothesis at the 0.05 (or 0.01) level of significance simply means that we are fixing the probability of rejecting $H_0$ when it is true at 0.05 (or 0.01).

The decision to use 0.05, 0.01, or some other value depends mostly on the consequences of committing a Type I error. Although it may seem desirable to make the probability of a Type I error small, we cannot make it too small, since this would tend to make the probabilities of serious Type II errors too large, and make it difficult, perhaps too difficult, to get significant results. To some extent, the choice of 0.05 or 0.01, and not, say, 0.08 and 0.03, is dictated by the availability of statistical tables. However, with the general availability of computers and various kinds of statistical calculators, this restriction no longer applies.

There are situations where we cannot, or do not want to, specify the probability of a Type I error. This could happen when we do not have enough information about the consequences of Type I errors, or when one person processes the data while another person makes the decisions. What can be done in that case is discussed on page 302.

After the null hypothesis, the alternative hypothesis, and the probability of a Type I error have been specified, the next step is

3. **Based on the sampling distribution of an appropriate statistic, we construct a criterion for testing the null hypothesis against the chosen alternative hypothesis at the specified level of significance.**

Note that in the response-time example we interchanged steps 2 and 3. First we specified the criterion and then we calculated the probability of a Type I error, but that is not what we do in actual practice. Finally,

4. **We calculate the value of the statistic on which the decision is to be based.**

and

5. **We decide whether to reject the null hypothesis, whether to accept it, or whether to reserve judgment.**

In the response-time example we rejected the null hypothesis $\mu = 0.38$ second for values of $\bar{x}$ less than or equal to 0.36 and also for values of $\bar{x}$ greater than or equal to 0.40. Such a criterion is referred to as a **two-sided criterion**, which goes here with the two-sided alternative hypothesis $\mu \neq 0.38$ second. In the traffic-ticket example we rejected the null hypothesis $\mu = 0.9$ ticket for values of $\bar{x}$ greater than or equal to 1.2, and we refer to this criterion as a **one-sided criterion**. It went with the one-sided alternative hypothesis $\mu > 0.9$ ticket.

In general, a test is called a **two-sided test** or a **two-tailed test** if the criterion on which it is based is two sided; namely, if the null hypothesis is rejected for values of the **test statistic** falling into either tail of its sampling distribution. Correspondingly, a test is called a **one-sided test** or a **one-tailed test** if the criterion on which it is based is one sided; namely, if the null hypothesis is rejected for values of the test statistic falling into one specified tail of its sampling distribution. By "test statistic" we mean the statistic (for instance, the sample mean) on which the test is based. Although there are exceptions, two-tailed tests are usually used in connection with two-sided alternative hypotheses, and one-tailed tests are usually used in connection with one-sided alternative hypotheses.

As part of the third step we must also specify whether the alternative to rejecting the null hypothesis is to accept it or to reserve judgment. This, as we have said, depends on whether we must make a decision one way or the other, or whether the circumstances permit that we delay a decision pending further study. In exercises and examples, the phrase "whether or not" will sometimes be used to indicate that a decision must be reached one way or the other.

In connection with the fifth step, let us point out that we often accept null hypotheses with the tacit hope that we are not exposed to overly high risks of committing serious Type II errors. Of course, if necessary we can calculate enough probabilities of Type II errors to get an overall picture from the operating characteristic curve of the test criterion.

Before we consider various special tests for means in the remainder of this chapter, let us point out that the concepts we have introduced here are not limited to tests concerning population means; they apply equally to tests concerning other parameters, or tests concerning the nature, or form, of populations.

**12.1** A delivery service considers replacing its vans with new equipment. If $\mu_0$ is the average weekly maintenance cost of one of the old vans and $\mu$ is the weekly maintenance cost it can expect for one of the new vans, it wants to test the null hypothesis $\mu = \mu_0$ against an appropriate alternative.

(a) What alternative hypothesis should it use if it wants to buy the new vans only if it can be shown that this will reduce the average weekly maintenance cost?

(b) What alternative hypothesis should it use if it is anxious to buy the new vans (that have some other nice features) unless it can be shown that this can be expected to increase the average weekly maintenance cost?

**12.2** A large restaurant has a waiter whom the manager suspects of making on the average more mistakes than all of its other waiters. If $\mu_0$ is the average daily number of mistakes made by all the other waiters and $\mu$ is the daily average number of mistakes made by the waiter who is under suspicion, the manager of the restaurant wants to test the null hypothesis $\mu = \mu_0$.

(a) If the manager of the restaurant has decided to let the waiter go only if the suspicion is confirmed, what alternative hypothesis should she use?

(b) If the manager of the restaurant has decided to let the waiter go unless he actually averages fewer mistakes than all the other waiters, what alternative hypothesis should she use?

**12.3** Rework Example 12.2, supposing that the mean of the psychologist's sample is $\bar{x} = 0.365$ second.

**12.4** A botanist wants to test the null hypothesis that the mean diameter of the flowers of a certain plant is 8.5 cm. He decides to take a random sample and to accept the null hypothesis if the mean of the sample falls between 8.2 cm and 8.8 cm. If the mean of the sample is less than or equal to 8.2 cm or greater than or equal to 8.8 cm, he will reject the null hypothesis and otherwise he will accept it. What decision will he make and will it be in error if

(a) $\mu = 8.5$ cm and he gets $\bar{x} = 9.1$ cm;

(b) $\mu = 8.5$ cm and he gets $\bar{x} = 8.3$ cm;

(c) $\mu = 8.7$ cm and he gets $\bar{x} = 9.1$ cm;

(d) $\mu = 8.7$ cm and he gets $\bar{x} = 8.3$ cm?

**12.5** Suppose that a psychological testing service is asked to check whether an executive is emotionally fit to assume the presidency of a large corporation. What type of error would it commit if it erroneously rejects the null hypothesis that the executive is fit for the job? What type of error would it commit if it erroneously accepts the null hypothesis that the executive is fit for the job?

**12.6** Suppose we want to test the null hypothesis that an antipollution device for cars is effective. Explain under what conditions we would commit a Type I error and under what conditions we would commit a Type II error.

**12.7** Whether an error is a Type I error or a Type II error depends on how we formulate the null hypothesis. To illustrate this, rephrase the null hypothesis of Exercise 12.6 so that the Type I error becomes a Type II error, and vice versa.

**12.8** For a given population with $\sigma = \$12$, we want to test the null hypothesis $\mu = \$75$ on the basis of a random sample of size $n = 100$. If the null hypothesis is rejected when $\bar{x}$ is greater than or equal to \$76.50 and otherwise it is accepted, find

(a) the probability of a Type I error;

(b) the probability of a Type II error when $\mu = \$75.3$;

(c) the probability of a Type II error when $\mu = \$77.22$.

**12.9** Suppose that in the response-time example the criterion is changed so that the null hypothesis $\mu = 0.38$ second is rejected if the sample mean is less than or equal to 0.355 or greater than or equal to 0.405; otherwise, the null hypothesis is accepted. The sample size is still $n = 40$ and the population standard deviation is still $\sigma = 0.08$.

(a) How does this affect the probability of a Type I error?

(b) How does this affect the probability of a Type II error when $\mu = 0.41$?

**12.10** With reference to the operating characteristic curve of Figure 12.3, verify that the probabilities of Type II errors are

(a) 0.78 when $\mu = 0.37$ second or $\mu = 0.39$ second;

(b) 0.50 when $\mu = 0.36$ second or $\mu = 0.40$ second;

(c) 0.06 when $\mu = 0.34$ second or $\mu = 0.42$ second.

**12.11** The mean age of Mr. and Mrs. Miller's three children is 15.9 years while the mean age of Mr. and Mrs. Brown's children is 12.8 years. Does it make any sense to ask whether the difference between these two means is significant?

**12.12** In a certain experiment, a null hypothesis is rejected at the 0.05 level of significance. Does this mean that the probability is at most 0.05 that the null hypothesis is true?

**12.13** In a study of extrasensory perception, 280 persons were asked to predict patterns on cards drawn at random from a deck. If two of them scored better than could be expected at the 0.01 level of significance, comment on the conclusion that these two persons must have extraordinary powers.

**12.14** During the production of spring-loaded postal scales, samples are obtained at regular intervals of time to check at the 0.05 level of significance whether the production process is under control. Is there cause for alarm if in 80 such samples the null hypothesis that the production process is under control is rejected

(a) three times;

(b) seven times?

**12.15** It has been claimed that on the average 2.6 workers are absent from an assembly line. If an efficiency expert is asked to put this to a test, what null hypothesis and what alternative hypothesis should he use?

**12.16** With reference to Exercise 12.15, would the efficiency expert use a one-tailed test or a two-tailed test if he is going to base his decision on the mean of a random sample?

**12.17** The manufacturer of a blood pressure medication claims that on the average the medication will lower a person's blood pressure by more than 20 mm. If a medical team suspects this claim, what null hypothesis and what alternative hypothesis should it use to put this to a test?

**12.18** With reference to Exercise 12.17, would the medical team use a one-tailed test or a two-tailed test if it intends to base its decision on the mean of a random sample?

**12.19** Suppose that an unscrupulous manufacturer wants "scientific proof" that a totally useless chemical additive will improve the mileage yield of gasoline.

(a) If a research group performs one experiment to investigate the additive, using the 0.05 level of significance, what is the probability that they will come up with "significant results" (which the manufacturer can use to promote the additive even though it is totally ineffective)?

(b) If two independent research groups investigate the additive, both using the 0.05 level of significance, what is the probability that at least one of them will come up with "significant results," even though the additive is totally ineffective?

(c) If 32 independent research groups investigate the additive, with all of them using the 0.05 level of significance, what is the probability that at least one of them will come up with "significant results," even though the additive is totally ineffective?

**12.20** Suppose that a manufacturer of pharmaceuticals would like to find a new ointment to reduce swellings. It tries 20 different medications and tests for each one whether it reduces swellings at the 0.10 level of significance.

(a) What is the probability that at least one of them will "prove effective," even though all of them are totally useless?

(b) What is the probability that more than one will "prove effective," even though all of them are totally useless?

# 12.3 TESTS CONCERNING MEANS

Having used tests concerning means to illustrate the basic principles of hypothesis testing, let us now demonstrate how we proceed in practice. Actually, we shall depart somewhat from the procedure used in Sections 12.1 and 12.2. In the response-time example as well as in the traffic-ticket example we stated the test criterion in terms of $\bar{x}$—in the first case we rejected the null hypothesis for $\bar{x} \leq 0.36$ or $\bar{x} \geq 0.40$, and in the second case we rejected it for $\bar{x} \geq 1.2$. Now we shall base it on the statistic

STATISTIC FOR TEST CONCERNING MEAN

$$z = \frac{\bar{x} - \mu_0}{\sigma/\sqrt{n}}$$

where $\mu_0$ is the value of the mean assumed under the null hypothesis. The reason for working with standard units, or $z$-values, is that it enables us to formulate criteria that are applicable to a great variety of problems, not just one.

The test of this section is essentially a **large-sample test**; that is, we require that the samples are large enough, $n \geq 30$, so that the sampling distribution of the mean can be approximated closely with a normal distribution and $z$ is a value of a random variable having approximately the standard normal distribution. (In the special case where we sample a normal population, $z$ is a value of a random variable having the standard normal distribution regardless of the size of $n$.) Alternatively, we refer to this test as a **z test** or as a **one-sample z test** to distinguish it from the test we shall discuss in Section 12.5. Sometimes the $z$ test is referred to as a test concerning a mean with $\sigma$ known to emphasize this essential feature.

Thus using $z$-values (standard units), we can picture tests of the null hypothesis $\mu = \mu_0$ on the test criteria shown in Figure 12.4. Depending on the alternative hypothesis, the dividing lines of a test criterion, also called its **critical values**, are $-z_\alpha$ or $z_\alpha$ for the one-sided alternatives and $-z_{\alpha/2}$ or $z_{\alpha/2}$ for the two-sided alternative: As before, $z_\alpha$ and $z_{\alpha/2}$ are such that the area to their right under the standard normal distribution are $\alpha$ and $\alpha/2$. Symbolically, these test criteria can be formulated as in the following table:

**Figure 12.4**
Test criteria for $z$ test concerning a population mean.

Alternative hypothesis $\mu < \mu_0$

Alternative hypothesis $\mu > \mu_0$

Alternative hypothesis $\mu \neq \mu_0$

| *Alternative hypothesis* | *Reject the null hypothesis if* | *Accept the null hypothesis or reserve judgment if* |
|---|---|---|
| $\mu < \mu_0$ | $z \leq -z_\alpha$ | $z > -z_\alpha$ |
| $\mu > \mu_0$ | $z \geq z_\alpha$ | $z < z_\alpha$ |
| $\mu \neq \mu_0$ | $z \leq -z_{\alpha/2}$ or $z \geq z_{\alpha/2}$ | $-z_{\alpha/2} < z < z_{\alpha/2}$ |

If the level of significance is 0.05, the dividing lines are $-1.645$ or $1.645$ for the one-sided alternatives, and $-1.96$ and $1.96$ for the two-sided alternative; if the level of significance is 0.01, the dividing lines are $-2.33$ or $2.33$ for the one-sided alternatives, and $-2.575$ and $2.575$ for the two-sided alternative. All these values come directly from Table I.

**EXAMPLE 12.4**    An oceanographer wants to test, on the basis of the mean of a random sample of size $n = 35$ and at the 0.05 level of significance, whether the average depth of the ocean in a certain area is 72.4 fathoms, as has been recorded. What will she decide if she gets $\bar{x} = 73.2$ fathoms, and she can assume from information gathered in similar studies that $\sigma = 2.1$ fathoms?

**Solution**
1.  $H_0$:   $\mu = 72.4$ fathoms

    $H_A$:   $\mu \neq 72.4$ fathoms
2.  $\alpha = 0.05$
3.  Reject the null hypothesis if $z \leq -1.96$ or $z \geq 1.96$, where

$$z = \frac{\bar{x} - \mu_0}{\sigma/\sqrt{n}}$$

and otherwise accept it (or reserve judgment).
4.  Substituting $\mu_0 = 72.4$, $\sigma = 2.1$, $n = 35$, and $\bar{x} = 73.2$ into the formula for $z$, she gets

$$z = \frac{73.2 - 72.4}{2.1/\sqrt{35}} \approx 2.25$$

5.  Since $z = 2.25$ exceeds 1.96, the null hypothesis must be rejected; to put it another way, the difference between $\bar{x} = 73.2$ and $\mu = 72.4$ is significant.

If the oceanographer had used the 0.01 level of significance in this example, she would not have been able to reject the null hypothesis because $z = 2.25$ falls between $-2.575$ and 2.575. This illustrates the importance of specifying the level of significance before any calculations have actually been made. This will spare us the temptation of later choosing a level of significance that happens to suit our purpose.

In problems like this, we often accompany the calculated value of the test statistic with the corresponding **p-value**; namely, with the probability of getting a difference between $\bar{x}$ and $\mu_0$ that is numerically greater than or equal to the one which is actually observed. For instance, in Example 12.4 the p-value is given by the total area under the standard normal curve to the left of $z = -2.25$, and to the right of $z = 2.25$, and Table I tells us that it equals $2(0.5000 - 0.4878) = 0.0244$. This practice is not new by any means, but it has been advocated more widely in recent years in view of the general availability of computers. For many distributions, computers can provide p-values that are not directly available from tables.

Quoting p-values is the method referred to on page 296 for problems where we cannot, or do not want to, specify the level of significance. This applies, for example, to problems in which we study a set of data without having to reach a decision, or when we process a set of data to enable someone else to make a decision. p-values are provided by just about all statistical software and also by graphing calculators, making it unnecessary to compare results with tabular values and making it possible to use levels of significance for which critical values are not tabulated. Of course, if it is necessary to make a decision, we still have the responsibility to specify the level of significance before we collect (or look at) the data.

In general, *p*-values may be defined as follows:

> **Corresponding to an observed value of a test statistic, the *p*-value is the lowest level of significance for which the null hypothesis could have been rejected.**

In Example 12.4 the *p*-value was 0.0244 and we could have rejected the null hypothesis at the 0.0244 level of significance. Of course, we could have rejected it for any level of significance greater than that, as we did for $\alpha = 0.05$.

If we want to base tests of significance on *p*-values instead of critical values obtained from tables, steps 1 and 2 remain the same, but steps 3, 4, and 5 must be modified as follows:

**3'.** **We specify the test statistic.**

**4'.** **We calculate the value of the specified test statistic and the corresponding *p*-value from the sample data.**

**5'.** **We compare the *p*-value obtained in step 4' with the level of significance specified in step 2. If the *p*-value is less than or equal to the level of significance, the null hypothesis must be rejected; otherwise, we accept the null hypothesis or reserve judgment.**

**EXAMPLE 12.5** Rework Example 12.4, basing the result on the *p*-value rather than using the critical value approach.

**Solution** Steps 1 and 2 remain the same as in Example 12.4, but steps 3, 4, and 5 are replaced by the following:

**3'.** The test statistic is

$$z = \frac{\bar{x} - \mu_0}{\sigma/\sqrt{n}}$$

**4'.** Substituting $\mu_0 = 72.4$, $\sigma = 2.1$, $n = 35$, and $\bar{x} = 73.2$ into the formula for $z$, we get

$$z = \frac{73.2 - 72.4}{2.1/\sqrt{35}} \approx 2.25$$

and from Table I we find that the *p*-value, the area under the curve to the left of $-2.25$ and to the right of $2.25$, is $2(0.5000 - 0.4878) = 0.0244$.

**5'.** Since 0.0244 is less than $\alpha = 0.05$, the null hypothesis must be rejected.

As we have indicated previously, the *p*-value approach can be used to advantage when we study data without having to reach a decision. To illustrate, consider the plight of a social scientist exploring the relationship between family economics and school performance. He could be testing hundreds of hypotheses involving dozens of variables. The work is very complicated, and there are no immediate policy consequences. In this situation, the social scientist can tabulate the tests of hypotheses according to their *p*-values. Those tests leading to the lowest *p*-values are the most provocative, and they will certainly be the subject of future discussion. The social scientist need not actually accept or reject the hypotheses, and the use of *p*-values furnishes a convenient alternative.

# 12.4 TESTS CONCERNING MEANS ($\sigma$ UNKNOWN)

When the population standard deviation is unknown, we proceed as in Section 11.2 and base tests concerning means on an appropriate *t* statistic. For this test we must be able to justify the assumption that the population we are sampling has roughly the shape of a normal distribution. We can then base the test of the null hypothesis $\mu = \mu_0$ on the statistic

STATISTIC FOR TEST
CONCERNING
MEAN
($\sigma$ UNKNOWN)

$$t = \frac{\bar{x} - \mu_0}{s/\sqrt{n}}$$

which is a value of a random variable having the *t* distribution (see page 268) with $n - 1$ degrees of freedom. Otherwise, we may have to use one of the alternative tests described in Chapter 18.

Tests based on the *t* statistic are referred to as *t* **tests**, and to distinguish between the one for tests concerning one mean and the one given in Section 12.6, we refer to the former as a **one-sample *t* test**. (Since most tables of critical values for the one-sample *t* test are limited to small numbers of degrees of freedom, small values of $n - 1$, the one-sample *t* test has also been referred to as a **small-sample test concerning means**. Of course, with easy access to computers and other technology, this distinction no longer applies.)

The criteria for the one-sample *t* test are very much like those shown in Figure 12.4 and in the table on page 301. Now, however, the curves represent *t* distributions instead of normal distributions, and $z$, $z_\alpha$, and $z_{\alpha/2}$ are replaced by $t$, $t_\alpha$, and $t_{\alpha/2}$. As defined on page 268, $t_\alpha$ and $t_{\alpha/2}$ are values for which the area to their right under the *t* distribution curve are $\alpha$ and $\alpha/2$. For relatively small numbers of degrees of freedom and $\alpha$ equal to 0.10, 0.05, and 0.01, the critical values may be obtained from Table II; for larger numbers of degrees of freedom and other values of $\alpha$ this requires appropriate computer software, a graphing calculator, or a special statistical calculator.

EXAMPLE 12.6 The yield of alfalfa from a random sample of six test plots is 1.4, 1.6, 0.9, 1.9, 2.2, and 1.2 tons per acre.

(a) Check whether these data can be looked upon as a sample from a normal population.

(b) If so, test at the 0.05 level of significance whether this supports the contention that the average yield for this kind of alfalfa is 1.5 tons per acre.

**Figure 12.5**
Normal probability
plot for Example 12.6
reproduced from dis-
play screen of TI-83
graphing calculator.

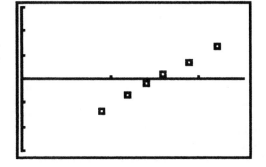

**Solution**   (a) The normal probability plot in Figure 12.5 shows no appreciable departure
from linearity, so that the data can be looked upon as a sample from a
normal population.

(b)

1. $H_0$:   $\mu = 1.5$

   $H_A$:   $\mu \neq 1.5$
2. $\alpha = 0.05$
3. Reject the null hypothesis if $t \leq -2.571$ or $t \geq 2.571$, where

$$t = \frac{\bar{x} - \mu_0}{s/\sqrt{n}}$$

and 2.571 is the value of $t_{0.025}$ for $6 - 1 = 5$ degrees of freedom;
otherwise, state that the data support the contention.
4. First calculating the mean and the standard deviation of the given
data, we get $\bar{x} = 1.533$ and $s = 0.472$. Then substituting these values
together with $n = 6$ and $\mu_0 = 1.5$ into the formula for $t$, we obtain

$$t = \frac{1.533 - 1.5}{0.472/\sqrt{6}} \approx 0.171$$

5. Since $t = 0.171$ falls between $-2.571$ and $2.571$, the null hypothesis
cannot be rejected; in other words, the data tend to support the
contention that the average yield of the given kind of alfalfa is 1.5 tons
per acre.   ■

Figure 12.6 shows the solution of Example 12.6 with the use of a graphing
calculator. Except for rounding, it confirms the values we obtained for $\bar{x}$, $s$, and

**Figure 12.6**
Solution of Example
12.6 reproduced from
the display screen of a
TI-83 graphing
calculator.

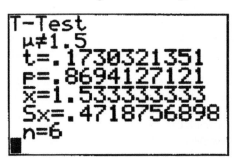

$t$, and it shows that the $p$-value is 0.869 rounded to three decimals. Since 0.869 exceeds 0.05, we conclude, as before, that the null hypothesis cannot be rejected.

**12.21** A law student wants to check on her professor's claim that convicted embezzlers spend on the average 12.3 months in jail. So she decides to test the null hypothesis $\mu = 12.3$ against the alternative hypothesis $\mu \neq 12.3$ at the 0.05 level of significance, using a random sample of $n = 35$ such cases from court files. What will she conclude if she gets $\bar{x} = 11.5$ months and uses the five-step significance test described on pages 294 through 297, knowing that $\sigma = 3.8$ months?

**12.22** Rework Exercise 12.21, basing the decision on the $p$-value instead of the sample mean.

**12.23** In a study of new sources of food, it is reported that a pound of a certain kind of fish yields on the average 3.52 ounces of FPC (fish-protein concentrate) used to enrich various food products, with the standard deviation $\sigma = 0.07$ ounce. To check whether $\mu = 3.52$ ounces is correct, a dietician decides to use the alternative hypothesis $\mu \neq 3.52$ ounces, a random sample of size $n = 32$, and the 0.05 level of significance. What will he conclude if he gets a sample mean of 3.55 ounces of FPC per pound of the fish?

**12.24** Rework Exercise 12.23, basing the decision on the $p$-value instead of the $z$ statistic.

**12.25** According to the norms established for a reading comprehension test, eighth graders should average 83.2 with the standard deviation $\sigma = 8.6$. A district superintendent feels that the eighth graders in his district are above average in reading comprehension, but he lacks proof. So he decides to test the null hypothesis $\mu = 83.2$ against a suitable alternative at the 0.01 level of significance, using the five-step format described on pages 294 through 297 and a random sample of 45 eighth graders from his district. What can he conclude if $\bar{x} = 86.7$?

**12.26** If we wish to test the null hypothesis $\mu = \mu_0$ in such a way that the probability of a Type I error is $\alpha$ and the probability of a Type II error is $\beta$ for the specified alternative value $\mu = \mu_A$, we must take a random sample of size $n$, where

$$n = \frac{\sigma^2(z_\alpha + z_\beta)^2}{(\mu_A - \mu_0)^2}$$

if the alternative hypothesis is one sided, and

$$n = \frac{\sigma^2(z_{\alpha/2} + z_\beta)^2}{(\mu_A - \mu_0)^2}$$

if the alternative hypothesis is two sided.

Suppose that we want to test the null hypothesis $\mu = 540$ mm against the alternative hypothesis $\mu < 540$ mm for a population whose standard deviation is $\sigma = 88$ mm. How large a sample will we need if the probability of a Type I error is to be 0.05 and the probability of a Type II error is to be 0.01 when $\mu = 520$ mm? Determine also for what values of $\bar{x}$ the null hypothesis will be rejected.

**12.27** A random sample of $n = 12$ graduates of a secretarial school typed on the average $\bar{x} = 78.2$ words per minutes with a standard deviation of $s = 7.9$ words per minute. Assuming that such data can be looked upon as a random sample from a normal population, use the one-sample $t$ test to test the null hypothesis $\mu = 80$ words per minute against the alternative hypothesis $\mu < 80$ words per minute for graduates of this secretarial school. Use the 0.05 level of significance.

**12.28** A coffee vending machine, tested $n = 9$ times, yielded a mean cup fill of 6.2 ounces with a standard deviation of 0.15 ounce. Assuming that these data can be looked upon as a random sample from a normal population, test the null hypothesis $\mu = 6.0$ ounces against the alternative hypothesis $\mu > 6.0$ ounces at the 0.01 level of significance.

**12.29** A random sample of five 1-quart cartons of ice cream is taken from a large production lot. If their mean fat content is 13.1% with a standard deviation of 0.51%, can we reject the null hypothesis that the mean fat content for the entire production lot is 12.5% against the alternative that it is greater than 12.5% at the 0.01 level of significance?

**12.30** A large group of senior citizens enrolled for adult evening classes at a university. To get a quick check on whether there has been an increase from last year's average age of 65.4 years, the director of the program takes a random sample of 15 of the enrollees, getting 68, 62, 70, 64, 61, 58, 65, 86, 88, 62, 60, 71, 60, 84, and 61 years. Using these data, he wants to perform a one-sample $t$ test to test the null hypothesis $\mu = 65.4$ against the alternative hypothesis $\mu > 65.4$.

   (a) Use appropriate computer software or a graphing calculator to see whether he can look upon the data as a sample from a normal population.

   (b) If so, perform the one-sample $t$ test at the 0.05 level of significance.

**12.31** A random sample from a travel agency's extensive files showed that its allotment of cabins for Panama Canal cruises from Fort Lauderdale to Acapulco were booked in 16, 16, 14, 17, 16, 19, 18, 16, 17, 14, 15, 12, 16, 18, 11, and 9 days. Assuming that the population sampled is a normal population, test the null hypothesis $\mu = 14$ against the alternative hypothesis $\mu > 14$ at the 0.05 level of significance.

**12.32** A new tranquilizer given to $n = 16$ patients reduced their pulse rate on the average by 4.36 beats per minute with a standard deviation of 0.36 beat per minute. Assuming that such data can be looked upon as a random sample from a normal population, use the 0.10 level of significance to test the pharmaceutical company's claim that on the average its new tranquilizer reduces a patient's pulse rate by 4.50 beats per minute.

**12.33** A teacher wants to determine whether the mean reading speed of certain students is at least 600 words per minute. He took a random sample of six of the students who read, respectively, 604, 615, 620, 603, 600, and 560 words in one minute.

   (a) Prior to performing a one-sample $t$ test, construct a normal probability plot.

   (b) If the normal probability plot gives no indication that the population sampled is not normal, use the one sample $t$ test to test the null hypothesis $\mu = 600$ against the alternative hypothesis $\mu < 600$ at the 0.05 level of significance.

**12.34** Five Golden Retrievers weigh 64, 66, 65, 63, and 62 pounds. Show that the mean of this sample differs significantly from $\mu = 60$ pounds, the mean of the population sampled, at the 0.05 level of significance.

**12.35** Suppose that in Exercise 12.34 the third figure is recorded incorrectly as 80 pounds instead of 65. Show that now the difference between the mean of the sample and $\mu = 60$ pounds is no longer significant. Explain the apparent paradox that even though the difference between $\bar{x}$ and $\mu$ has increased, it is no longer significant.

# 12.5 DIFFERENCES BETWEEN MEANS

There are many problems in which we must decide whether an observed difference between two sample means can be attributed to chance, or whether it is indicative of the fact that the two samples came from populations with unequal means. For instance, we may want to know whether there really is a difference in the mean gasoline consumption of two kinds of cars, when sample data show that one kind averaged 24.6 miles per gallon while, under the same conditions, the other kind averaged 25.7 miles per gallon. Similarly, we may want to decide on the basis of sample data whether men can perform a certain task faster than women, whether one kind of ceramic insulator is more brittle than another, whether the average diet in one country is more nutritious than that in another country, and so on.

The method we shall use to test whether an observed difference between two sample means can be attributed to chance, or whether it is statistically significant, is based on the following theory: If $\bar{x}_1$ and $\bar{x}_2$ are the means of two independent random samples, then the mean and the standard deviation of the sampling distribution of the statistic $\bar{x}_1 - \bar{x}_2$ are

$$\mu_{\bar{x}_1 - \bar{x}_2} = \mu_1 - \mu_2 \quad \text{and} \quad \sigma_{\bar{x}_1 - \bar{x}_2} = \sqrt{\frac{\sigma_1^2}{n_1} + \frac{\sigma_2^2}{n_2}}$$

where $\mu_1, \mu_2, \sigma_1,$ and $\sigma_2$ are the means and the standard deviations of the two populations sampled. It is customary to refer to the standard deviation of this sampling distribution as the **standard error of the difference between two means**.

By "independent" samples we mean that the selection of one sample is in no way affected by the selection of the other. Thus, the theory does not apply to "before and after" kinds of comparisons, nor does it apply, say, if we want to compare the daily caloric consumption of husbands and wives. A special method for comparing the means of dependent samples is explained in Section 12.7.

Then, if we limit ourselves to large samples, $n_1 \geq 30$ and $n_2 \geq 30$, we can base tests of the null hypothesis $\mu_1 - \mu_2 = \delta$, (*delta*, the Greek letter for lowercase *d*) on the statistic

STATISTIC FOR TEST CONCERNING DIFFERENCE BETWEEN TWO MEANS

$$z = \frac{\bar{x}_1 - \bar{x}_2 - \delta}{\sqrt{\frac{\sigma_1^2}{n_1} + \frac{\sigma_2^2}{n_2}}}$$

which is a value of a random variable having approximately the standard normal distribution. Note that we obtained this formula for $z$ by converting to standard units, namely, by subtracting from $\bar{x}_1 - \bar{x}_2$ the mean of its sampling distribution, which under the null hypothesis is $\mu_1 - \mu_2 = \delta$, and then dividing by the standard deviation of its sampling distribution.

Depending on whether the alternative hypothesis is $\mu_1 - \mu_2 < \delta$, $\mu_1 - \mu_2 > \delta$, or $\mu_1 - \mu_2 \neq \delta$, the criteria we use for the corresponding tests are shown in Figure 12.7.

Note that these criteria are like the criteria of Figure 12.4 with $\mu_1 - \mu_2$ substituted for $\mu$ and $\delta$ substituted for $\mu_0$. Analogous to the table on page 301, the criteria for tests of the null hypothesis $\mu_1 - \mu_2 = \delta$ are as follows:

**Figure 12.7**
Test criteria for two-sample $z$ test.

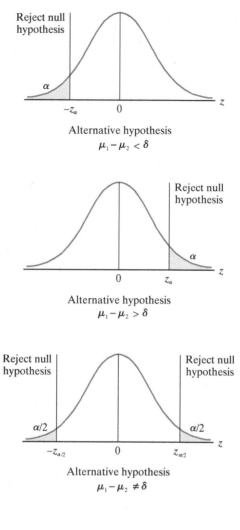

Reject null hypothesis

$\alpha$

$-z_\alpha$    $0$    $z$

Alternative hypothesis
$\mu_1 - \mu_2 < \delta$

Reject null hypothesis

$\alpha$

$0$    $z_\alpha$    $z$

Alternative hypothesis
$\mu_1 - \mu_2 > \delta$

Reject null hypothesis        Reject null hypothesis

$\alpha/2$        $\alpha/2$

$-z_{\alpha/2}$    $0$    $z_{\alpha/2}$    $z$

Alternative hypothesis
$\mu_1 - \mu_2 \neq \delta$

| Alternative hypothesis | Reject the null hypothesis if | Accept the null hypothesis or reserve judgment if |
|---|---|---|
| $\mu_1 - \mu_2 < \delta$ | $z \leq -z_\alpha$ | $z > -z_\alpha$ |
| $\mu_1 - \mu_2 > \delta$ | $z \geq z_\alpha$ | $z < z_\alpha$ |
| $\mu_1 - \mu_2 \neq \delta$ | $z \leq -z_{\alpha/2}$   or   $z \geq z_{\alpha/2}$ | $-z_{\alpha/2} < z < z_{\alpha/2}$ |

Although $\delta$ can be any constant, it is worth noting that in the great majority of problems its value is zero, and we test the null hypothesis of "no difference," namely, the null hypothesis $\mu_1 - \mu_2 = 0$ (or simply $\mu_1 = \mu_2$).

The test we have described here, the **two-sample $z$ test**, is essentially a large-sample test. It is exact only when both of the populations we are sampling are normal populations. Sometimes it is referred to as a **test concerning difference between means with $\sigma_1$ and $\sigma_2$ known** to emphasize this essential feature.

**EXAMPLE 12.7**   In a study to test whether or not there is a difference between the average heights of adult females in two different countries, random samples of size $n_1 = 120$ and $n_2 = 150$ yielded $\bar{x}_1 = 62.7$ inches and $\bar{x}_2 = 61.8$ inches. Extensive studies of a similar kind have shown that it is reasonable to let $\sigma_1 = 2.50$ inches and $\sigma_2 = 2.62$ inches. Test at the 0.05 level of significance whether the difference between these two sample means is significant.

**Solution**   1. In view of the "whether or not" in the formulation of the problem, we use

$$H_0: \mu_1 = \mu_2 \text{(namely, } \delta = 0)$$

$$H_A: \mu_1 \neq \mu_2$$

2. $\alpha = 0.05$
3. Reject the null hypothesis if $z \leq -1.96$ or $z \geq 1.96$, where

$$z = \frac{\bar{x}_1 - \bar{x}_2 - \delta}{\sqrt{\frac{\sigma_1^2}{n_1} + \frac{\sigma_2^2}{n_2}}}$$

with $\delta = 0$; otherwise, accept the null hypothesis or reserve judgment.
4. Substituting $n_1 = 120, n_2 = 150, \bar{x}_1 = 62.7, \bar{x}_2 = 61.8, \sigma_1 = 2.50, \sigma_2 = 2.62$, and $\delta = 0$ into the formula for $z$, we get

$$z = \frac{62.7 - 61.8}{\sqrt{\frac{(2.50)^2}{120} + \frac{(2.62)^2}{150}}} \approx 2.88$$

5. Since $z = 2.88$ exceeds 1.96, the null hypothesis must be rejected; in other words, the difference between $\bar{x}_1 = 62.7$ and $\bar{x}_2 = 61.8$ is statistically significant. (Whether it is also of any practical significance, say, to a manufacturer of women's clothes, is another matter.)   ■

If the person who did this analysis had not been asked to make a decision, he or she would simply have reported that the $p$-value corresponding to the value of the test statistic is $2(0.5000 - 0.4980) = 0.0040$.

Let us add that there is a certain awkwardness about comparing means when the population standard deviations are unequal. Consider, for example, two normal populations with the means $\mu_1 = 50$ and $\mu_2 = 52$ and the standard deviations $\sigma_1 = 5$ and $\sigma_2 = 15$. Although the second population has a larger mean, it is much more likely to produce a value below 40, as can easily be verified. An investigator faced with a situation like this ought to decide whether the comparison of $\mu_1$ and $\mu_2$ really addresses whatever is of any relevance.

# 12.6 DIFFERENCES BETWEEN MEANS (σ's UNKNOWN)

When the population standard deviations are unknown, we proceed as in Sections 11.2 and 12.4 and base tests concerning differences between means on an appropriate $t$ statistic. For this test we must be able to justify the assumption that the populations we are sampling have roughly the shape of normal distributions. Moreover, we must be able to justify the assumption that they have equal standard deviations. Then, we can base the test of the null hypothesis $\mu_1 - \mu_2 = \delta$, and $\mu_1 = \mu_2$ in particular, on the statistic

**STATISTIC FOR TEST CONCERNING DIFFERENCE BETWEEN MEANS (σ's UNKNOWN)**

$$t = \frac{\bar{x}_1 - \bar{x}_2 - \delta}{s_p \sqrt{\frac{1}{n_1} + \frac{1}{n_2}}} \quad \text{where} \quad s_p = \sqrt{\frac{(n_1 - 1)s_1^2 + (n_2 - 1)s_2^2}{n_1 + n_2 - 2}}$$

which is a value of a random variable having the $t$ distribution with $(n_1 - 1) + (n_2 - 1) = n_1 + n_2 - 2$ degrees of freedom. Otherwise, we may have to use one of the alternative tests described in Chapter 18.

Tests based on this new $t$ statistic are referred to as **two-sample $t$ tests**. (Since most tables of critical values for the two-sample $t$ test are limited to small numbers of degrees of freedom, small values of $n_1 + n_2 - 2$, the two-sample $t$ test has also been referred to as a **small-sample test concerning the difference between means**. As before in connection with the one-sample $t$ test, with easy access to computers and other technology, this distinction no longer applies.)

The criteria for the two-sample $t$ test are very much like those shown in Figure 12.7 and in the table on page 309; of course, the curves now represent $t$ distributions instead of normal distributions, and $z$, $z_\alpha$, and $z_{\alpha/2}$ are replaced by $t$, $t_\alpha$, and $t_{\alpha/2}$.

As can be seen from its definition, the calculation of the statistic for the two-sample $t$ test consists of two steps. First we calculate the value of $s_p$, called the **pooled standard deviation**, which is an estimate of $\sigma_1 = \sigma_2$, the by-assumption-equal population standard deviations. Then we substitute it together with the $\bar{x}$'s and $n$'s into the formula for $t$. The example that follows illustrates this procedure.

**EXAMPLE 12.8**    The following random samples are measurements of the heat-producing capacity (in millions of calories per ton) of coal from two mines:

> **Mine 1:**    8,380    8,180    8,500    7,840    7,990
> **Mine 2:**    7,660    7,510    7,910    8,070    7,790

Use the 0.05 level of significance to test whether the difference between the means of these two samples is significant.

**Solution**    Normal probability plots show that there is no reason to suspect the assumption that the data constitute samples from normal populations. Also, a test that will be described later shows in Example 13.3 that there is no reason to suspect the assumption that $\sigma_1 = \sigma_2$.

1. $H_0$:    $\mu_1 = \mu_2$

    $H_A$:    $\mu_1 \neq \mu_2$
2. $\alpha = 0.05$
3. Reject the null hypothesis if $t \leq -2.306$ or $t \geq 2.306$, where $t$ is given by the formula on page 311 with $\delta = 0$, and 2.306 is the value of $t_{0.025}$ for $5 + 5 - 2 = 8$ degrees of freedom; otherwise, state that the difference between the means of the two samples is not significant.
4. The means and the standard deviations of the two samples are $\bar{x}_1 = 8{,}178$, $\bar{x}_2 = 7{,}788$, $s_1 = 271.1$, and $s_2 = 216.8$. Substituting the values of $s_1$ and $s_2$ together with $n_1 = n_2 = 5$ into the formula for $s_p$, we get

$$s_p = \sqrt{\frac{4(271.1)^2 + 4(216.8)^2}{8}} \approx 245.5$$

and, hence,

$$t = \frac{8{,}178 - 7{,}788}{245.5\sqrt{\frac{1}{5} + \frac{1}{5}}} \approx 2.51$$

5. Since $t = 2.51$ exceeds 2.306, the null hypothesis must be rejected; in other words, we conclude that the difference between the two sample means is significant. ■

Figure 12.8 is a MINITAB printout for Example 12.8. It confirms our calculations, including the value we obtained for $t$, and it shows that the $p$-value corresponding to $t = 2.51$ (and the two-sided alternative hypothesis $\mu_1 \neq \mu_2$) is 0.036. Since this $p$-value is less than $\alpha = 0.05$, it reconfirms that the null hypothesis must be rejected.

**Figure 12.8**
Computer printout for
Example 12.8.

```
Two-Sample T-Test and CI: C1, C2

Two-sample T for C1 vs C2

        N      Mean        StDev    SE Mean
C1      5      8178          271        121
C2      5      7788          217         97

Difference = mu C1 - mu C2
Estimate for difference: 390
95% CI for difference: (32, 748)
T-Test of difference = 0 (vs not =): T-Value = 2.51
  P-Value = 0.036 DF = 8
Both use Pooled StDev = 245
```

# 12.7 DIFFERENCES BETWEEN MEANS (PAIRED DATA)

The methods of Sections 12.5 and 12.6 can be used only when the two samples are independent. Therefore, they cannot be used when we deal with "before and after" kinds of comparisons, the ages of husbands and wives, bank robber arrests and convictions in various jurisdictions, interest rates charged and paid by financial institutions, first-half and second-half pass completions by quarterbacks, cars stocked and cars sold by used-car dealers, and numerous other kinds of situations in which data are naturally paired. To handle this kind of data, we work with the (signed) differences between the pairs and test whether they can be looked upon as a random sample from a population with the mean $\mu = \delta$, usually $\mu = 0$. The tests we use for this purpose are the one-sample $z$ test of Section 12.3 or the one-sample $t$ test of Section 12.4, whichever is appropriate.

**EXAMPLE 12.9** Following are the average weekly losses of worker hours due to accidents in ten industrial plants before and after the installation of an elaborate safety program:

| | | | | |
|---|---|---|---|---|
| 45 and 36 | 73 and 60 | 46 and 44 | 124 and 119 | 33 and 35 |
| 57 and 51 | 83 and 77 | 34 and 29 | 26 and 24 | 17 and 11 |

Use the 0.05 level of significance to test whether the safety program is effective.

**Solution** The differences between the respective pairs are 9, 13, 2, 5, −2, 6, 6, 5, 2, and 6, and a normal probability plot (not displayed here) shows a distinct linear pattern. Thus, we can use the one-sample $t$ test and proceed as follows:

1. $H_0$:   $\mu = 0$

   $H_A$:   $\mu > 0$  (The alternative is that on the average there were more accidents "before" than "after.")

2. $\alpha = 0.05$
3. Reject the null hypothesis if $t \geq 1.833$, where

$$t = \frac{\bar{x} - \mu_0}{s/\sqrt{n}}$$

and 1.833 is the value of $t_{0.05}$ for $10 - 1 = 9$ degrees of freedom; otherwise, accept the null hypothesis or reserve judgment (as the situation may demand).

4. First, calculating the mean and the standard deviation of the ten differences, we get $\bar{x} = 5.2$ and $s = 4.08$. Then, substituting these values together with $n = 10$ and $\mu_0 = 0$ into the formula for $t$, we have

$$t = \frac{5.2 - 0}{4.08/\sqrt{10}} \approx 4.03$$

5. Since $t = 4.03$ exceeds 1.833, the null hypothesis must be rejected; in other words, we have shown that the industrial safety program is effective.

When the one-sample $t$ test is used in a problem like this, it is referred to as the **paired-sample $t$ test**.

**12.36** Random samples showed that 40 executives in the insurance industry claimed on the average 9.4 business lunches as deductible biweekly expenses, while 50 bank executives claimed on the average 7.9 business lunches as deductible biweekly expenses. If, on the basis of collateral information, it can be assumed that $\sigma_1 = \sigma_2 = 3.0$ for such data, test at the 0.05 level of significance whether the difference between these two sample means is significant.

**12.37** Rework Exercise 12.36, using the sample standard deviations, $s_1 = 3.3$ and $s_2 = 2.9$, instead of the assumed values of $\sigma_1$ and $\sigma_2$.

**12.38** An investigation of two kinds of photocopying equipment showed that a random sample of 60 failures of one kind of equipment took on the average 84.2 minutes to repair, while a random sample of 60 failures of another kind of equipment took on the average 91.6 minutes to repair. If, on the basis of collateral information, it can be assumed that $\sigma_1 = \sigma_2 = 19.0$ minutes for such data, test at the 0.02 level of significance whether the difference between these two sample means is significant.

**12.39** Rework Exercise 12.38, using the sample standard deviations $s_1 = 19.4$ minutes and $s_2 = 18.8$ minutes, instead of the assumed values of the population standard deviations.

**12.40** Random samples of 12 measurements each of the hydrogen content (in percent number of atoms) of gases collected from the eruptions of two volcanos yielded $\bar{x}_1 = 41.2$, $\bar{x}_2 = 45.8$, $s_1 = 5.2$, and $s_2 = 6.7$. Assuming that the conditions underlying the two-sample $t$ test can be met, decide at the 0.05 level of significance whether to accept or reject the null hypothesis that there is no difference in the mean hydrogen content of gases from the two eruptions.

**12.41** With reference to Exercise 12.40, determine the $p$-value corresponding to the value obtained for the $t$ statistic. Use it to determine whether the null hypothesis could have been rejected at the 0.10 level of significance.

**12.42** In the comparison of two kinds of paint, a consumer testing service found that four 1-gallon cans of Brand A covered on the average 514 square feet with a standard deviation of 32 square feet, while four 1-gallon cans of Brand B covered on the average 487 square feet with a standard deviation of 27 square feet. Assuming that the conditions required for the two-sample $t$ test can be met, test at the 0.02 level of significance whether the difference between these two sample means is significant.

**12.43** With reference to Exercise 12.42, what is the smallest level of significance at which the null hypothesis could have been rejected?

**12.44** Six guinea pigs injected with 0.5 mg of a medication took on the average 15.4 seconds to fall asleep with a standard deviation of 2.2 seconds, while six other guinea pigs injected with 1.5 mg of the same medication took on the average 10.6 seconds to fall asleep with a standard deviation of 2.6 seconds. Assuming that the two samples are independent random samples and that the requirements for the two-sample $t$ test can be met, test at the 0.05 level of significance whether or not this increase in dosage will, in general, reduce the average time it takes a guinea pig to fall asleep by 2.0 seconds.

**12.45** In a department store's study designed to test whether or not the mean balance outstanding on 30-day charge accounts is the same in its two suburban branch stores, random samples yielded the following results:

$$n_1 = 80 \qquad \bar{x}_1 = \$64.20 \qquad s_1 = \$16.00$$
$$n_2 = 100 \qquad \bar{x}_2 = \$71.41 \qquad s_2 = \$22.13$$

where the subscripts denote branch store 1 and branch store 2. Use the 0.05 level of significance to test the null hypothesis $\mu_1 - \mu_2 \neq 0$, where $\mu_1$ and $\mu_2$ are the actual mean balances outstanding on all 30-day charge accounts in branch stores 1 and 2.

**12.46**  With reference to Exercise 12.45, find the $p$-value corresponding to the value of the test statistic obtained in that exercise. Also, use this $p$-value to confirm the decision reached in that exercise.

**12.47**  To test the claim that the resistance of electric wire can be reduced by more than 0.050 ohm by alloying, 25 values obtained for alloyed wire yielded $\bar{x}_1 = 0.083$ ohm and $s_1 = 0.003$ ohm, and 25 values obtained for standard wire yielded $\bar{x}_2 = 0.136$ ohm and $s_2 = 0.002$ ohm. Use the level of significance $\alpha = 0.05$ to determine whether the claim has been substantiated.

**12.48**  Following are measurements of the wing span of two varieties of sparrows in millimeters:

| Variety 1: | 162 | 159 | 154 | 176 | 165 | 164 | 145 | 157 | 128 |
| Variety 2: | 147 | 180 | 153 | 135 | 157 | 153 | 141 | 138 | 161 |

Assuming that the conditions underlying the two-sample $t$ test can be met, test at the 0.05 level of significance whether the difference between the means of these two random samples is significant.

# CHECKLIST OF KEY TERMS (with page references to their definitions)

Alternative hypothesis, 288
Composite hypothesis, 293
Critical value, 300
Large-sample test, 300
Level of significance, 296
Not statistically significant, 294
Null hypothesis, 288
OC curve, 292
One-sample $t$ test, 304
One-sample $z$ test, 300
One-sided alternative, 295
One-sided criterion, 297
One-sided test, 297
One-tailed test, 297
Operating characteristic curve, 292
Paired-sample $t$ test, 313
Pooled standard deviation, 311
Power function, 292
$p$-value, 302, 303
Significance test, 293
Simple hypothesis, 293

Small-sample test concerning means, 304
Small-sample test concerning the difference
   between means, 311
Standard error of difference between means, 308
Statistical hypothesis, 287
Statistically significant, 294
$t$ test, 304
Test concerning difference between means with
   $\sigma_1$ and $\sigma_2$ known, 310
Test of significance, 293
Test statistic, 297
Two-sample $t$ test, 311
Two-sample $z$ test, 310
Two-sided alternative, 295
Two-sided criterion, 297
Two-sided test, 297
Two-tailed test, 297
Type I error, 291
Type II error, 291
$z$ test, 300

# REFERENCES

*Some easy reading on tests of hypotheses may be found in*

BROOK, R. J., ARNOLD, G. C., HASSARD, T. H., and PRINGLE, R. M., eds., *The Fascination of Statistics*. New York: Marcel Dekker, Inc., 1986.

GONICK, L., and SMITH, W., *The Cartoon Guide to Statistics*. New York: HarperCollins Publishers, Inc., 1993.

*A detailed treatment of significance tests, the choice of the level of significance, p-values, and so forth may be found in Chapters 26 and 29 of*

FREEDMAN, D., PISANI, R., and PURVES, R., *Statistics*. New York: Norton & Company, Inc., 1978.

# TESTS OF HYPOTHESES: STANDARD DEVIATIONS

**13.1** Tests Concerning Standard Deviations 317

**13.2** Tests Concerning Two Standard Deviations 321

Checklist of Key Terms 325

References 325

I n the preceding chapter we learned to judge the size of the error when estimating the mean of a population, how to construct confidence intervals for population means, and how to test hypotheses about the means of populations. These are useful statistical techniques, but even more important are the concepts on which they are based—interval estimation, degree of confidence, null and alternative hypotheses, Type I and Type II errors, tests of significance, level of significance, $p$-values, and, above all, the concept of statistical significance.

As we shall see in this chapter and in subsequent chapters, all these ideas carry over to inferences about population parameters other than the mean. In this chapter we shall concentrate on population standard deviations, which are not only important in their own right, but which must sometime be estimated or compared before inferences about other parameters can be made.

In this chapter, Section 13.1 will be devoted to tests about the standard deviation of one population, and Section 13.2 deals with tests about the standard deviations of two populations.

## 13.1 TESTS CONCERNING STANDARD DEVIATIONS

The tests we shall consider in this section concern the problem of whether a population standard deviation equals a specified constant $\sigma_0$. This kind of test may be required whenever we study the uniformity of a product, process, or operation: for instance, if we must judge whether a certain kind of glass is sufficiently homogeneous for making delicate optical equipment, whether the knowledge of a group of students is sufficiently uniform so that they can be taught in one class, whether a lack of uniformity in some workers' performance may call for stricter supervision, and so forth.

The test of the null hypothesis $\sigma = \sigma_0$, that a population standard deviation equals a specified constant, is based on the same assumption, the same statistic, and the same sampling theory as the confidence interval for $\sigma^2$ on page 276. Again assuming that we are dealing with a random sample from a normal population (or at least a population having roughly the shape of a normal distribution), we use the **chi-square statistic**

**STATISTIC FOR TEST CONCERNING STANDARD DEVIATION**

$$\chi^2 = \frac{(n-1)s^2}{\sigma_0^2}$$

which is like the one on page 273, with $\sigma$ replaced by $\sigma_0$. As before, the sampling distribution of this statistic is the chi-square distribution with $n-1$ degrees of freedom.

The test criteria are shown in Figure 13.1; depending on the alternative hypothesis, the critical values are $\chi_{1-\alpha}^2$ and $\chi_\alpha^2$ for the one-sided alternatives, and they are $\chi_{1-\alpha/2}^2$ and $\chi_{\alpha/2}^2$ for the two-sided alternative. Symbolically, we can formulate these criteria for testing the null hypothesis $\sigma = \sigma_0$ as follows:

| Alternative hypothesis | Reject the null hypothesis if | Accept the null hypothesis or reserve judgment if |
|---|---|---|
| $\sigma < \sigma_0$ | $\chi^2 \le \chi_{1-\alpha}^2$ | $\chi^2 > \chi_{1-\alpha}^2$ |
| $\sigma > \sigma_0$ | $\chi^2 \ge \chi_\alpha^2$ | $\chi^2 < \chi_\alpha^2$ |
| $\sigma \ne \sigma_0$ | $\chi^2 \le \chi_{1-\alpha/2}^2$ or $\chi^2 \ge \chi_{\alpha/2}^2$ | $\chi_{1-\alpha/2}^2 < \chi^2 < \chi_{\alpha/2}^2$ |

The values of $\chi_{0.995}^2$, $\chi_{0.99}^2$, $\chi_{0.975}^2$, $\chi_{0.95}^2$, $\chi_{0.05}^2$, $\chi_{0.025}^2$, $\chi_{0.01}^2$, and $\chi_{0.005}^2$ are given in Table III at the end of the book for 1, 2, 3, ..., and 30 degrees of freedom.

**EXAMPLE 13.1**  To judge certain safety features of a car, an engineer must know whether the reaction time of drivers to a given emergency situation has a standard deviation of 0.010 second, or whether it is greater than 0.010 second. What can she conclude at the 0.05 level of significance if she gets the following random sample of $n = 15$ reaction times?

$$\begin{array}{ccccc} 0.32 & 0.30 & 0.31 & 0.28 & 0.30 \\ 0.31 & 0.28 & 0.31 & 0.29 & 0.28 \\ 0.30 & 0.29 & 0.27 & 0.29 & 0.29 \end{array}$$

**Figure 13.1**
Criteria for tests concerning standard deviations.

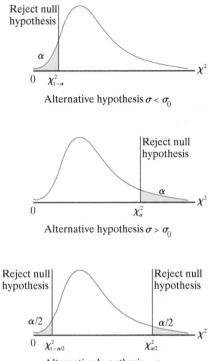

Alternative hypothesis $\sigma < \sigma_0$

Alternative hypothesis $\sigma > \sigma_0$

Alternative hypothesis $\sigma \neq \sigma_0$

**Solution**  The normal probability plot of Figure 13.2 shows a distinct linear pattern, and hence justifies the assumption that the population sampled has the shape of a normal distribution.

1. $H_0$:   $\sigma = 0.010$
   $H_A$:   $\sigma > 0.010$
2. $\alpha = 0.05$
3. Reject the null hypothesis if $\chi^2 \geq 23.685$, where

$$\chi^2 = \frac{(n-1)s^2}{\sigma_0^2}$$

and 23.685 is the value of $\chi_{0.05}^2$ for $15 - 1 = 14$ degrees of freedom; otherwise, accept it.

**Figure 13.2**
Normal probability plot for Example 13.1 reproduced from display screen of TI-83 graphing calculator.

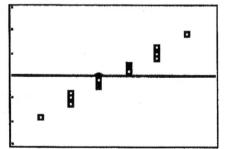

4. Calculating the standard deviation of the sample data, we get $s = 0.014$, and substituting this value together with $n = 15$ and $\sigma_0 = 0.010$ into the formula for $\chi^2$, we get

$$\chi^2 = \frac{14(0.014)^2}{(0.010)^2} \approx 27.44$$

5. Since $\chi^2 = 27.44$ exceeds 23.685, the null hypothesis must be rejected; in other words, the engineer can conclude that the standard deviation of the reaction time of drivers to the given emergency situation is greater than 0.010 second. ■

Since most tables of critical values for chi-square tests are limited to small numbers of degrees of freedom, the test we have described here has often been referred to as a **small-sample test concerning standard deviations**. As we have said before, with easy access to computers and other technology, this distinction no longer applies.

When $n$ is large, $n \geq 30$, tests of the null hypothesis $\sigma = \sigma_0$ can be based on the same theory as the large-sample confidence interval for $\sigma$ given in Section 11.3. That is, we use the statistic

### STATISTIC FOR LARGE-SAMPLE TEST CONCERNING STANDARD DEVIATION

$$z = \frac{s - \sigma_0}{\sigma_0/\sqrt{2n}}$$

which is a value of a random variable having the standard normal distribution. Thus, the criteria for this large-sample test of the null hypothesis $\sigma = \sigma_0$ are like those shown in Figure 12.4 and in the table on page 301; the only difference is that $\mu$ and $\mu_0$ are replaced by $\sigma$ and $\sigma_0$.

**EXAMPLE 13.2**   The specifications for the mass production of certain springs require, among other things, that the standard deviation of their compressed lengths should not exceed 0.040 cm. If a random sample of size $n = 35$ from a certain production lot has $s = 0.053$, does this constitute evidence at the 0.01 level of significance for the null hypothesis $\sigma = 0.040$ or for the alternative hypothesis $\sigma > 0.040$?

**Solution**   1. $H_0:$   $\sigma = 0.040$
$H_A:$   $\sigma > 0.040$
2. $\alpha = 0.01$
3. The null hypothesis must be rejected if $z \geq 2.33$, where

$$z = \frac{s - \sigma_0}{\sigma_0/\sqrt{2n}}$$

and otherwise it must be accepted.
4. Substituting $n = 35$, $s = 0.053$, and $\sigma_0 = 0.040$ into the formula for $z$, we get

$$z = \frac{0.053 - 0.040}{0.040/\sqrt{70}} \approx 2.72$$

5. Since $z = 2.72$ exceeds 2.33, the null hypothesis must be rejected; in other words, the data show that the production lot does not meet specifications.

■

The $p$-value corresponding to $z = 2.72$ is $0.5000 - 0.4967 = 0.0033$, and since this is less than 0.01, it would also have led to the rejection of the null hypothesis.

# 13.2 TESTS CONCERNING TWO STANDARD DEVIATIONS

In this section we shall discuss tests concerning the equality of two standard deviations. Among other applications, it is often used in connection with the two-sample $t$ test, where it has to be assumed that the two populations sampled have equal standard deviations. For instance, in Example 12.8, which dealt with the heat-producing capacity of coal from two mines, we had $s_1 = 271.1$ (millions of calories per ton) and $s_2 = 216.8$. Despite what may seem to be a large difference, we assumed that the corresponding population standard deviations were equal. Now we shall put this to a rigorous test.

Given independent random samples of size $n_1$ and $n_2$ from populations having roughly the shape of normal distributions and the standard deviations $\sigma_1$ and $\sigma_2$, we base tests of the null hypothesis $\sigma_1 = \sigma_2$ on the **F statistic**:

**S**TATISTICS FOR TEST CONCERNING THE EQUALITY OF TWO STANDARD DEVIATIONS

$$F = \frac{s_1^2}{s_2^2} \quad \text{or} \quad F = \frac{s_2^2}{s_1^2} \text{ depending on } H_A$$

where $s_1$ and $s_2$ are the corresponding sample standard deviations. Based on the assumption that the populations sampled have roughly the shape of normal distributions and that the null hypothesis is $\sigma_1 = \sigma_2$, it can be shown that such ratios, appropriately called **variance ratios**, are values of random variables having the **F distribution**. This continuous distribution depends on two parameters, called the **numerator and denominator degrees of freedom**. They are $n_1 - 1$ and $n_2 - 1$, or $n_2 - 1$ and $n_1 - 1$, depending on which of the two sample variances go into the numerator and the denominator of the $F$ statistic.

If we based all the tests on the statistic

$$F = \frac{s_1^2}{s_2^2}$$

we could reject the null hypothesis $\sigma_1 = \sigma_2$ for $F \le F_{1-\alpha}$ when the alternative hypothesis is $\sigma_1 < \sigma_2$ and for $F \ge F_\alpha$ when the alternative hypothesis is $\sigma_1 > \sigma_2$. In this notation, $F_{1-\alpha}$ and $F_\alpha$ are defined in the same way in which we defined the critical values $\chi_{1-\alpha}^2$ and $\chi_\alpha^2$ for the chi-square distribution. Unfortunately, things are not as simple as that. Since there exists a fairly straightforward mathematical relationship between $F_{1-\alpha}$ and $F_\alpha$, most $F$ tables give only values corresponding to right-hand tails with $\alpha$ less than 0.50; for instance, Table IV at the end of the book contains only values of $F_{0.05}$ and $F_{0.01}$.

**Figure 13.3**
Criteria for tests con-
cerning the equality of
two standard deviations.

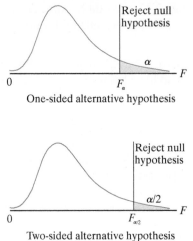

For this reason, we use

$$F = \frac{s_2^2}{s_1^2} \quad \text{or} \quad F = \frac{s_1^2}{s_2^2}$$

depending on whether the alternative hypothesis is $\sigma_1 < \sigma_2$ or $\sigma_1 > \sigma_2$, and in either case we reject the null hypothesis for $F \geq F_\alpha$ (see Figure 13.3). When the alternative hypothesis is $\sigma_1 \neq \sigma_2$, we use the greater of the two variance ratios,

$$F = \frac{s_1^2}{s_2^2} \quad \text{or} \quad F = \frac{s_2^2}{s_1^2}$$

and reject the null hypothesis for $F \geq F_{\alpha/2}$ (see Figure 13.3). In all these tests the degrees of freedom are $n_1 - 1$ and $n_2 - 1$, or $n_2 - 1$ and $n_1 - 1$, depending on which sample variance goes into the numerator and which one goes into the denominator. Symbolically, these criteria for testing the null hypothesis $\sigma_1 = \sigma_2$ are summarized in the following table:

| Alternative hypothesis | Test statistic | Reject the null hypothesis if | Accept the null hypothesis or reserve judgment if |
|---|---|---|---|
| $\sigma_1 < \sigma_2$ | $F = \frac{s_2^2}{s_1^2}$ | $F \geq F_\alpha$ | $F < F_\alpha$ |
| $\sigma_1 > \sigma_2$ | $F = \frac{s_1^2}{s_2^2}$ | $F \geq F_\alpha$ | $F < F_\alpha$ |
| $\sigma_1 \neq \sigma_2$ | The larger of the two ratios | $F \geq F_{\alpha/2}$ | $F < F_{\alpha/2}$ |

The degrees of freedom are as indicated previously.

**EXAMPLE 13.3**   In Example 12.8 we had $s_1 = 271.1$ and $s_2 = 216.8$ for two independent random samples of size $n_1 = 5$ and $n_2 = 5$, and in the solution we justified the assumption that the data constitute samples from normal populations. Use the 0.02 level of significance to test whether there is any evidence that the standard deviations of the two populations sampled are not equal.

**Solution**   Having already justified the assumption about normality, we proceed as follows:

1. $H_0:$   $\sigma_1 = \sigma_2$
   $H_A:$   $\sigma_1 \neq \sigma_2$
2. $\alpha = 0.02$
3. Reject the null hypothesis if $F \geq 16.0$, where

$$F = \frac{s_1^2}{s_2^2} \quad \text{or} \quad F = \frac{s_2^2}{s_1^2}$$

whichever is larger, and 16.0 is the value of $F_{0.01}$ for $5 - 1 = 4$ and $5 - 1 = 4$ degrees of freedom; otherwise, accept the null hypothesis.
4. Since $s_1 = 271.1$ and $s_2 = 216.8$, we substitute these values into the first of the two variance ratios and get

$$F = \frac{(271.1)^2}{(216.8)^2} \approx 1.56$$

5. Since $F = 1.56$ does not exceed 16.0, the null hypothesis cannot be rejected; there is no reason for not using the two-sample $t$ test in Example 12.8.   ◼

In a problem like this where the alternative hypothesis is $\sigma_1 \neq \sigma_2$, Table IV limits us to the 0.02 and 0.10 levels of significance. Had we wanted to use another level of significance, we would have had to refer to a more extensive table, a computer, or some other technology. A HEWLETT PACKARD STAT/MATH calculator yields the $p$-value 0.3386, so the result would have been the same for just about any reasonable level of significance.

**EXAMPLE 13.4**   It is desired to determine whether there is less variability in the gold plating done by company 1 than in the gold plating done by company 2. If independent random samples yielded $s_1 = 0.033$ mil (based on $n_1 = 12$) and $s_2 = 0.061$ mil (based on $n_2 = 10$), test the null hypothesis $\sigma_1 = \sigma_2$ against the alternative hypothesis $\sigma_1 < \sigma_2$ at the 0.05 level of significance.

**Solution**   Assuming that the populations sampled have roughly the shape of normal distributions, we proceed as follows.

1. $H_0:$   $\sigma_1 = \sigma_2$
   $H_A:$   $\sigma_1 < \sigma_2$
2. $\alpha = 0.05$
3. Reject the null hypothesis if $F \geq 2.90$, where

$$F = \frac{s_2^2}{s_1^2}$$

and 2.90 is the value of $F_{0.05}$ for $10 - 1 = 9$ and $12 - 1 = 11$ degrees of freedom; otherwise, accept the null hypothesis or reserve judgment.

4. Substituting $s_1 = 0.033$ and $s_2 = 0.061$ in to the formula for $F$, we get

$$F = \frac{(0.061)^2}{(0.033)^2} \approx 3.42$$

5. Since $F = 3.42$ exceeds 2.90, the null hypothesis must be rejected; in other words, we conclude that there is less variability in the gold plating done by company 1.

Since the procedure described in this section is very sensitive to departures from the underlying assumptions, it must be used with considerable caution. To put it differently, we say that the test is not **robust**.

**13.1**  In a laboratory experiment, $s = 0.0086$ for $n = 10$ determinations of the specific heat of iron. Assuming that the population sampled has roughly the shape of a normal distribution, use the 0.05 level of significance to test the null hypothesis $\sigma = 0.0100$ for such determinations against the alternative hypothesis $\sigma < 0.0100$.

**13.2**  In a random sample of the amounts of time that $n = 18$ women took to complete the written test for their driver's licenses, the standard deviation was $s = 3.8$ minutes. Assuming that these data can be looked upon as a sample from a population having roughly the shape of a normal distribution, test the null hypothesis $\sigma = 2.7$ minutes against the alternative hypothesis $\sigma \neq 2.7$ minutes at the 0.01 level of significance.

**13.3**  Use appropriate technology to find the $p$-value corresponding to the test statistic obtained in Exercise 13.2, and use it to rework that exercise at the 0.03 level of significance.

**13.4**  A dietician took a random sample of size $n = 35$ (one pound pieces of a certain kind of fish) to study the variability of its yield of fish-protein concentrate. Given that the standard deviation of her sample was reported as $s = 0.082$, test the null hypothesis $\sigma = 0.065$ against the alternative hypothesis $\sigma > 0.065$ at the 0.05 level of significance.

**13.5**  It has been reported that the constant annual growth of certain miniature fruit trees in their fifth to tenth years has the standard deviation $\sigma = 0.80$ inch. Given that a nursery owner got $s = 0.74$ inch for $n = 40$ such trees, test the null hypothesis $\sigma = 0.80$ against the alternative $\sigma < 0.80$ at the 0.01 level of significance.

**13.6**  Given an experiment consisting of a random sample of $n = 10$ observations, for which $s^2 = 1.44$ oz$^2$, what else has to be assumed if one wants to test the null hypothesis $\sigma = 1.4$ oz against the alternative $\sigma < 1.4$ oz?

**13.7**  Two different lighting techniques are compared by measuring the intensity of light at selected locations by both methods. If $n_1 = 12$ measurements of the first technique have the standard deviation $s_1 = 2.6$ foot-candles, $n_2 = 16$ measurements of the second technique have the standard deviation $s_2 = 4.4$ foot-candles, and it can be assumed that both samples may be regarded as independent random samples from normal populations, test at the 0.05 level of significance whether the two lighting techniques are equally variable or whether the first technique is less variable than the second.

**13.8**  The amounts of time required by Dr. L. to do routine insurance checkups on $n_1 = 25$ patients have the standard deviation $s_1 = 4.2$ minutes, while the amounts of time required by Dr. M. to do the same procedure on $n_2 = 21$ patients have the standard deviation $s_2 = 3.0$ minutes. Assuming that these data constitute independent random samples from normal populations, test at the 0.05 level of significance whether the amounts of time required by these two doctors for this procedure are equally variable or whether they are more variable for Dr. L.

# CHECKLIST OF KEY TERMS (with page references to their definitions)

Chi-square statistic, 318
Denominator degrees of freedom, 321
F distribution, 321
F statistic, 321
Numerator degrees of freedom, 321

Robust, 324
Small-sample test concerning
  standard deviations, 320
Variance ratio, 321

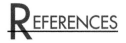

# REFERENCES

*Theoretical discussions of the chi-square and F distributions may be found in most textbooks on mathematical statistics; for instance, in*

MILLER, I., and MILLER, M., *John E. Freund's Mathematical Statistics*, 6th ed. Upper Saddle River, N.J.: Prentice Hall, 1999.

*For more detailed tables of the chi-square and F distributions, see, for example,*

PEARSON, E. S., and HARTLEY, H. O., *Biometrika Tables for Statisticians*, Vol. I. New York: John Wiley & Sons, Inc., 1968.

# TESTS OF HYPOTHESES BASED ON COUNT DATA

**14.1**  Tests Concerning Proportions  327

**14.2**  Tests Concerning Proportions (Large Samples)  328

**14.3**  Differences between Proportions  329

**14.4**  The Analysis of an $r \times c$ Table  333

**14.5**  Goodness of Fit  345

Checklist of Key Terms  350

References  350

Many problems in business, science, and daily life deal with **count data** (that is, data obtained by enumeration as compared with measurement), which are used to estimate or test hypotheses about proportions, percentages, or probabilities. In principle, the work of this chapter will be very similar to that of Chapters 11 through 13. In these chapters we used measurements to estimate the means of populations and their standard deviations, and we also used them to test hypotheses about these parameters. The only exception was Section 11.4, where we used sample proportions to estimate population proportions, percentages and probabilities. Such data were referred to as count data, since they were obtained by performing counts rather than measurements.

In this chapter we shall use count data in tests of hypotheses. Sections 14.1 and 14.2 deal with tests concerning proportions, which serve also as tests concerning percentages (proportions multiplied by 100) and as tests concerning probabilities (proportions in the long run). These tests are based on the observed number of successes in $n$ trials, or the observed proportion of successes in $n$ trials, and it will be assumed throughout that these trials are independent and that the probability of a success is the same for each trial. In other words, it will be assumed that we are testing hypotheses about the parameter $p$ of binomial populations.

In Section 14.3 and the beginning of Section 14.4 we shall study tests about two or more population proportions. Then, in the remainder of Section 14.4 and in Section 14.5 we shall generalize the discussion to the multinomial case, where there are more than two possible outcomes for each trial. This kind of problem would arise, for example, if one is interested in the relationship between a person's score on a qualifying test for a job (say, below average, average, or above average) and his or her performance on the job (say, poor, fair, good, or excellent). Finally, Section 14.6 deals with the comparison of observed frequency distributions and distributions that might be expected according to theory or assumptions.

# 14.1 TESTS CONCERNING PROPORTIONS

The tests we shall discuss here and in Section 14.2 make it possible, for example, to decide on the basis of sample data whether it is true that the proportion of tenth graders who can name the two senators of their state is only 0.28, whether it is true that 12% of the information supplied by the IRS to taxpayers is in error, or whether the probability is really 0.25 that a flight from Seattle to San Francisco will be late.

Whenever possible, such tests are based directly on tables of binomial probabilities, or on information about binomial probabilities obtained with the use of computers or other technology. Furthermore, these tests are simplest when using the $p$-value approach.

**EXAMPLE 14.1** It has been claimed that more than 70% of the students attending a large state university are opposed to a plan to increase student fees in order to build new parking facilities. If 15 of 18 students selected at random at that university are opposed to the plan, test the claim at the 0.05 level of significance.

**Solution**
1. $H_0$: $p = 0.70$
   $H_A$: $p > 0.70$
2. $\alpha = 0.05$
3'. The test statistic is the observed number of students in the sample who oppose the plan.
4'. The test statistic is $x = 15$, and Table V shows that the $p$-value, the probability of 15 or more "successes" for $n = 18$ and $p = 0.70$, is $0.105 + 0.046 + 0.013 + 0.002 = 0.166$.
5'. Since 0.166 is greater than 0.05, the null hypothesis cannot be rejected; in other words, the data do not support the claim that more than 70% of the students at the given university are opposed to the plan. ■

**EXAMPLE 14.2** It has been claimed that 38% of all shoppers can identify a highly advertised trade mark. If, in a random sample, 25 of 45 shoppers were able to identify the trade mark, test at the 0.05 level of significance whether to accept or reject the null hypothesis $p = 0.38$.

**Solution** Since Table V does not give the binomial probabilities for $p = 0.38$ or $n > 20$, we could use National Bureau of Standards table referred to on page 205 or appropriate technology.

1. $H_0$: $p = 0.38$
   $H_A$: $p \neq 0.38$
2. $\alpha = 0.05$

**Figure 14.1**
Computer printout for
Example 14.2.

| **Cumulative Distribution Function** |
|---|

Binomial with n = 45 and p = 0.380000

| x | P( X <= x) |
|---|---|
| 24.00 | 0.9875 |
| 25.00 | 0.9944 |

**3′.** The test statistic is $x = 25$, the number of shoppers in the sample who can identity the trade mark.

**4′.** For a two-tailed test like this, the $p$-value is twice the smaller of the probabilities for $x \le 25$ and for $x \ge 25$. Since Table V does not give binomial probabilities for $p = 0.38$ or for $n > 20$, we use the computer printout of Figure 14.1. Accordingly, the probability of $x \le 25$ is 0.9944 and the probability of $x \ge 25$ is 1 minus the probability of $x \le 24$, namely, $1 - 0.9875 = 0.0125$. Thus, the $p$-value is $2(0.0125) = 0.0250$.

**5′.** Since 0.025 is less than $\alpha = 0.05$, the null hypothesis must be rejected. The correct percentage of shoppers who can identify the trade mark is not 38%. In fact, since $\frac{25}{45} \cdot 100 = 55.6\%$, it is greater than 38%. ■

# 14.2 TESTS CONCERNING PROPORTIONS (LARGE SAMPLES)

When $n$ is large enough to justify the normal-curve approximation to the binomial distribution, $np > 5$ and $n(1 - p) > 5$, tests of the null hypothesis $p = p_0$ can be based on the statistic

**STATISTIC FOR LARGE-SAMPLE TEST CONCERNING PROPORTION**

$$z = \frac{x - np_0}{\sqrt{np_0(1 - p_0)}}$$

which is a value of a random variable having approximately the standard normal distribution. Since $x$ is a discrete random variable, many statisticians prefer to make the continuity correction that represents $x$ by the interval from $x - \frac{1}{2}$ to $x + \frac{1}{2}$, and hence use the alternative statistic

**STATISTIC FOR LARGE-SAMPLE TEST CONCERNING PROPORTION (WITH CONTINUITY CORRECTION)**

$$z = \frac{x \pm \frac{1}{2} - np_0}{\sqrt{np_0(1 - p_0)}}$$

where the $+$ sign is used when $x < np_0$ and the $-$ sign is used when $x > np_0$. Note that the continuity correction does not even have to be considered when, without it, the null hypothesis cannot be rejected. Otherwise, it will have to be considered mainly when, without it, the value we obtain for $z$ is very close to the critical value (or one of the critical values) of the test criterion.

The criteria for this large-sample test are again like those of Figure 12.4 with $p$ and $p_0$ substituted for $\mu$ and $\mu_0$. Analogous to the table on page 301, the criteria for tests of the null hypothesis $p = p_0$ are as follows:

| Alternative hypothesis | Reject the null hypothesis if | Accept the null hypothesis or reserve judgment if |
|:---:|:---:|:---:|
| $p < p_0$ | $z \leq -z_\alpha$ | $z > -z_\alpha$ |
| $p > p_0$ | $z \geq z_\alpha$ | $z < z_\alpha$ |
| $p \neq p_0$ | $z \leq -z_{\alpha/2}$ or $z \geq z_{\alpha/2}$ | $-z_{\alpha/2} < z < z_{\alpha/2}$ |

**EXAMPLE 14.3**  To test a nutritionist's claim that at least 75% of the preschool children in a certain large county have protein-deficient diets, a sample survey revealed that 206 of 300 preschool children in that county had protein-deficient diets. Test the null hypothesis $p = 0.75$ against the alternative hypothesis $p < 0.75$ at the 0.01 level of significance.

**Solution**  Since $np = 300(0.75) = 225$ and $n(1 - p) = 300(0.25) = 75$ are both greater than 5, we can use the large-sample test based on the normal approximation to the binomial distribution.

1. $H_0$: $p = 0.75$
   $H_A$: $p < 0.75$
2. $\alpha = 0.01$
3. Reject the null hypothesis if $z \leq -2.33$, where

$$z = \frac{x - np_0}{\sqrt{np_0(1 - p_0)}}$$

otherwise, accept the null hypothesis or reserve judgment.
4. Substituting $x = 206$, $n = 300$, and $p_0 = 0.75$ into the formula for $z$, we get

$$z = \frac{206 - 300(0.75)}{\sqrt{300(0.75)(0.25)}} \approx -2.53$$

5. Since $-2.53$ is less than $-2.33$, the null hypothesis must be rejected. In other words, we conclude that less than 75% of the preschool children in that county have protein deficient diets. (Had we used the continuity correction, we would have obtained $z = -2.47$, and the conclusion would have been the same.)

## 14.3 DIFFERENCES BETWEEN PROPORTIONS

After we presented tests concerning means in Chapter 12 and tests concerning standard deviations in Chapter 13, we learned how to perform tests concerning the means of two populations and tests concerning the standard deviations of two populations. Continuing this pattern, we shall now present a test concerning two population proportions.

There are many problems where we must decide whether an observed difference between two sample proportions can be attributed to chance, or whether it is indicative of the fact that the corresponding population proportions are not equal. For instance, we may want to decide on the basis of sample data whether there is a difference between the actual proportions of persons with and without flu shots who catch the disease, or we may want to test on the basis of samples whether two manufacturers of electronic equipment ship equal proportions of defectives.

The method we shall use here to test whether an observed difference between two sample proportions is significant, or whether it can be attributed to chance, is based on the following theory: If $x_1$ and $x_2$ are the numbers of successes obtained in $n_1$ trials of one kind and $n_2$ of another, the trials are all independent, and the corresponding probabilities of a success are, respectively, $p_1$ and $p_2$, then the sampling distribution of the difference

$$\frac{x_1}{n_1} - \frac{x_2}{n_2}$$

has the mean $p_1 - p_2$ and the standard deviation

$$\sqrt{\frac{p_1(1-p_1)}{n_1} + \frac{p_2(1-p_2)}{n_2}}$$

It is customary to refer to this standard deviation as the **standard error of the difference between two proportions**.

When we test the null hypothesis $p_1 = p_2(= p)$ against an appropriate alternative hypothesis, the sampling distribution of the difference between two sample proportions has the mean $p_1 - p_2 = 0$, and its standard deviation can be written as

$$\sqrt{p(1-p)\left(\frac{1}{n_1} + \frac{1}{n_2}\right)}$$

where $p$ is usually estimated by **pooling** the data and substituting for $p$ the combined sample proportion

$$\hat{p} = \frac{x_1 + x_2}{n_1 + n_2}$$

which, as before, reads "$p$-hat." Then, converting to standard units, we obtain the statistic

**STATISTIC FOR TEST CONCERNING DIFFERENCE BETWEEN TWO PROPORTIONS**

$$z = \frac{\dfrac{x_1}{n_1} - \dfrac{x_2}{n_2}}{\sqrt{\hat{p}(1-\hat{p})\left(\dfrac{1}{n_1} + \dfrac{1}{n_2}\right)}} \qquad \text{with} \quad \hat{p} = \frac{x_1 + x_2}{n_1 + n_2}$$

which, for large samples, is a value of a random variable having approximately the standard normal distribution. To make this formula appear more compact, we can substitute in the numerator $\hat{p}_1$ for $x_1/n_1$ and $\hat{p}_2$ for $x_2/n_2$.

The test criteria are again like those shown in Figure 12.4 with $p_1$ and $p_2$ substituted for $\mu$ and $\mu_0$. Analogous to the table on page 301, the criteria for tests of the null hypothesis $p_1 = p_2$ are as follows:

| Alternative hypothesis | Reject the null hypothesis if | Accept the null hypothesis or reserve judgment if |
|:---:|:---:|:---:|
| $p_1 < p_2$ | $z \leq -z_\alpha$ | $z > -z_\alpha$ |
| $p_1 > p_2$ | $z \geq z_\alpha$ | $z < z_\alpha$ |
| $p_1 \neq p_2$ | $z \leq -z_{\alpha/2}$ or $z \geq z_{\alpha/2}$ | $-z_{\alpha/2} < z < z_{\alpha/2}$ |

**EXAMPLE 14.4** A study showed that 56 of 80 persons who saw a spaghetti sauce advertised on television during a situation comedy and 38 of 80 other persons who saw it advertised during a football game remembered the brand name two hours later. At the 0.01 level of significance, what can we conclude about the claim that it is more cost effective to advertise this product during a situation comedy rather than during a football game? Assume that the cost of running the ad on the two programs is the same.

**Solution**

1. $H_0$: $p_1 = p_2$
   $H_A$: $p_1 > p_2$
2. $\alpha = 0.01$
3. Reject the null hypothesis if $z \geq 2.33$, where $z$ is given by the formula on page 330; otherwise, accept it or reserve judgment.
4. Substituting $x_1 = 56$, $x_2 = 38$, $n_1 = 80$, $n_2 = 80$, and

$$\hat{p} = \frac{56 + 38}{80 + 80} = 0.5875$$

into the formula for $z$, we get

$$z = \frac{\dfrac{56}{80} - \dfrac{38}{80}}{\sqrt{(0.5875)(0.4125)\left(\dfrac{1}{80} + \dfrac{1}{80}\right)}} \approx 2.89$$

5. Since $z = 2.89$ exceeds 2.33, the null hypothesis must be rejected; in other words, we conclude that advertising the spaghetti sauce during a situation comedy is more cost effective than advertising it during a football game.

**14.1** A travel agent claims that at most 5% of all persons requesting brochures for transatlantic cruises will actually take such a cruise within a year's time. If, in a random sample of 16 persons who requested a brochure for such a cruise, 3 actually took it, is this sufficient evidence to reject the travel agent's claim $p = 0.05$ against the alternative $p > 0.05$ at the 0.01 level of significance?

**14.2** With reference to Exercise 14.1, could the travel agent's claim have been rejected at the 0.05 level of significance?

**14.3** A social scientist claims that among persons living in urban areas 50% are opposed to capital punishment (while the others are in favor of it or undecided). Test the null hypothesis $p = 0.50$ against the alternative hypothesis $p \neq 0.50$ at the 0.10 level of significance if in a random sample of $n = 20$ persons living in urban areas, 14 are against capital punishment.

**14.4** A physicist claims that at most 19% of all persons exposed to a certain amount of radiation will feel any effects. If, in a random sample, 4 of 13 persons feel an effect, test the claim at the 0.05 level of significance.

**14.5** A committee investigating accidents in elementary schools claims that at least 36% of all accidents in elementary schools are due at least in part to improper supervision. If a random sample of 300 such accidents included 94 that were due at least in part to improper supervision, does this support the committee's claim? To answer this question, test the null hypothesis $p = 0.36$ against the alternative hypothesis $p < 0.36$ at the 0.05 level of significance, using
(a) the formula for $z$ without the continuity correction;
(b) the formula for $z$ with the continuity correction.

**14.6** In the construction of tables of random numbers there are various ways of testing for possible departures from randomness. One of these consists of checking whether there are as many even digits (0, 2, 4, 6, or 8) as there are odd digits (1, 3, 5, 7, or 9). Thus, count the number of even digits among the 350 digits in the first 10 rows of the sample page of random numbers reproduced in Figure 10.2, and test at the 0.05 level of significance whether there is any significant sign of a lack of randomness.

**14.7** For each of 500 simulated random samples, a statistics class determined a 95% confidence interval for the mean, and it found that only 464 of them contained the mean of the population sampled; the other 36 did not. At the 0.01 level of significance, is there any real evidence to doubt that the method employed yields 95% confidence intervals?

**14.8** In a random sample of 600 persons interviewed at a baseball game, 157 complained that their seats were uncomfortable. Test the claim that 30% of the persons at the game would feel that way, using the
(a) 0.05 level of significance;
(b) 0.01 level of significance.

**14.9** One method of seeding clouds was successful in 54 of 150 attempts, while another method was successful in 33 of 100 attempts. At the 0.05 level of significance, can we conclude that the first method is more effective than the second?

**14.10** One mail solicitation for a charity brought 412 responses to 5,000 letters and another, more expensive, mail solicitation brought 312 responses to 3,000 letters. Use the 0.01 level of significance to test the null hypothesis that the two solicitations are equally effective against the alternative that the more expensive one is more effective.

**14.11** In a random sample of visitors to the Heard Museum in Phoenix, Arizona, 22 of 100 families from New England and 33 of 120 families from California purchased some Indian jewelry in the gift shop. Use the 0.05 level of significance to test the null hypothesis $p_1 = p_2$, that there is no difference between the corresponding population proportions, against the alternative hypothesis $p_1 \neq p_2$.

**14.12** The service department of a Chrysler dealership offers fresh donuts to customers in its waiting room. When it supplied 12 dozen donuts from Bakery $A$, it found that 96 were eaten completely while the others were partially eaten and discarded. When it supplied 12 dozen donuts from Bakery $B$ it found that 105 were eaten completely while the others were partially eaten and discarded. At the 0.05 level of significance test whether the difference between the corresponding sample proportions is significant.

**14.13** A random sample of 100 high school students were asked whether they would turn to their parents for help with a homework assignment in mathematics, and another random sample of 100 high school students were asked the same question with regard to a homework assignment in English. If 62 students in the first sample and 44 students in the second sample would turn to their parents for help, test at the 0.05 level of significance whether the difference between the two sample proportions, $\frac{62}{100}$ and $\frac{44}{100}$, may be attributed to chance.

**14.14** In a random sample of 200 marriage license applications recorded in 2000, 62 of the women were at least one year older than the men, and in a random sample of 300 marriage license applications recorded in 2005, 99 of the women were at least one year older than the men. At the 0.01 level of significance, is the upward trend statistically significant?

# 14.4 THE ANALYSIS OF AN $r \times c$ TABLE

The method we shall describe in this section applies to several kinds of problems that differ conceptually but are analyzed in the same way. First let us consider a problem that is an immediate generalization of the kind of problem we studied in Section 14.3. Suppose that independent random samples of single, married, and widowed or divorced persons were asked whether "friends and social life" or "job or primary activity" contributes most to their general well-being, and that the results were as follows:

|  | Single | Married | Widowed or divorced |
|---|---|---|---|
| Friends and social life | 47 | 59 | 56 |
| Job or primary activity | 33 | 61 | 44 |
| Total | 80 | 120 | 100 |

Here we have samples of size $n_1 = 80$, $n_2 = 120$, and $n_3 = 100$ from three binomial populations, and we want to determine whether the differences among the proportions of persons choosing "friends and social life" are statistically significant. The hypothesis we shall have to test is the null hypothesis $p_1 = p_2 = p_3$, where $p_1$, $p_2$, and $p_3$ are the corresponding true proportions for the three binomial populations. The alternative hypothesis will have to be that $p_1$, $p_2$, and $p_3$ are not all equal.

The binomial distribution applies only when each trial has two possible outcomes. When there are more than two possible outcomes, we use the multinomial distribution (see Section 8.6) instead of the binomial distribution, provided the trials are all independent, the number of trials is fixed, and for each possible outcome the probability does not change from trial to trial. To illustrate such a multinomial situation, suppose that in the preceding example there had been the third alternative "health and physical condition," and the result had been as shown in the following table:

|  | Single | Married | Widowed or divorced |  |
|---|---|---|---|---|
| Friends and social life | 41 | 49 | 42 | 132 |
| Job or primary activity | 27 | 50 | 33 | 110 |
| Health and physical condition | 12 | 21 | 25 | 58 |
|  | 80 | 120 | 100 | 300 |

As before, there are three separate samples, the column totals are the fixed sample sizes, but each trial (each person interviewed) allows for three different outcomes. Note that the row totals, $41 + 49 + 42 = 132$, $27 + 50 + 33 = 110$, and $12 + 21 + 25 = 58$ depend on the responses of the persons interviewed, and hence on chance. In general, a table like this, with $r$ rows and $c$ columns, is called an $r \times c$ ("$r$ by $c$") **table**. In particular, the preceding one is referred to as a $3 \times 3$ table.

The null hypothesis we shall want to test in this multinomial situation is that for each of the three choices ("friends and social life," "job or primary activity," and "health and physical condition") the probabilities are the same for each of the three groups of persons interviewed. Symbolically, if $p_{ij}$ is the probability of obtaining a response belonging to the $i$th row and the $j$th column of the table, the null hypothesis is

$$H_0: p_{11} = p_{12} = p_{13}, \quad p_{21} = p_{22} = p_{23}, \text{ and}$$

$$p_{31} = p_{32} = p_{33}$$

where the $p$'s must add up to 1 for each column. More compactly, we can write this null hypothesis as

$$H_0: p_{i1} = p_{i2} = p_{i3} \quad \text{for} \quad i = 1, 2, \text{ and } 3$$

The alternative hypothesis is that the $p$'s are not all equal for at least one row, namely,

$$H_A: p_{i1}, p_{i2}, \text{ and } p_{i3} \text{ are not all equal for at least one value of } i$$

Before we show how all these problems are analyzed, let us mention one more situation where the method of this section applies. What distinguishes it from the preceding examples is that the column totals as well as the row totals are left to chance. To illustrate, suppose that we want to investigate whether there is a relationship between the test scores of persons who have gone through a certain job-training program and their subsequent performance on the job. Suppose, furthermore, that a random sample of 400 cases taken from very extensive files yielded the following result:

**Performance**

|  |  | Poor | Fair | Good |  |
|---|---|---|---|---|---|
|  | Below average | 67 | 64 | 25 | 156 |
| **Test score** | Average | 42 | 76 | 56 | 174 |
|  | Above average | 10 | 23 | 37 | 70 |
|  |  | 119 | 163 | 118 | 400 |

Here there is only one sample, the **grand total** of 400 is its fixed size, and each trial (each case chosen from the files) permits nine different outcomes. It is mainly in connection with problems like this that $r \times c$ tables are referred to as **contingency tables**.

The purpose of the investigation that led to the table immediately preceding was to see whether there is a relationship between the test scores of persons who have gone through the job-training program and their subsequent performance on the job. In general, the hypotheses we test in the analysis of a contingency table are

$H_0$:  The two variables under consideration are independent.

$H_A$:  The two variables are not independent.

Despite the differences we have described, the analysis of an $r \times c$ table is the same for all three of our examples, and we shall illustrate it here in detail by analyzing the second example, the one that dealt with the different factors contributing to one's well-being. If the null hypothesis is true, we can combine the three samples and estimate the probability that any one person will choose "friends and social life" as the factor that contributes most to his or her well-being as

$$\frac{41 + 49 + 42}{300} = \frac{132}{300}$$

Hence, among the 80 single persons and the 120 married persons we can expect, respectively, $\frac{132}{300} \cdot 80 = \frac{132 \cdot 80}{300} = 35.2$ and $\frac{132}{300} \cdot 120 = \frac{132 \cdot 120}{300} = 52.8$ to choose "friends and social life" as the factor contributing most to their well-being. Note that in both cases we obtained the expected frequency by multiplying the row total by the column total and then dividing by the grand total for the entire table. Indeed, the argument that led to this result can be used to show that in general

> **The expected frequency for any cell of an $r \times c$ table can be obtained by multiplying the total of the row to which it belongs by the total of the column to which it belongs and then dividing by the grand total for the entire table.**

With this rule we get expected frequencies of $\frac{110 \cdot 80}{300} \approx 29.3$ and $\frac{110 \cdot 120}{300} = 44.0$ for the first and second cells of the second row.

It is not necessary to calculate all the expected frequencies in this way, as it can be shown (see Exercises 14.34 and 14.35) that the sum of the expected frequencies for any row or column equals the sum of the corresponding observed frequencies. Thus, we can get some of the expected frequencies by subtraction from row or column totals. For instance, for our example we get

$$132 - 35.2 - 52.8 = 44.0$$

for the third cell of the first row

$$110 - 29.3 - 44.0 = 36.7$$

for the third cell of the second row, and

$$80 - 35.2 - 29.3 = 15.5$$
$$120 - 52.8 - 44.0 = 23.2$$
$$100 - 44.0 - 36.7 = 19.3$$

for the three cells of the third row. These results are summarized in the following table, where the expected frequencies are shown in parentheses below the corresponding observed frequencies:

|  | Single | Married | Widowed or divorced |
|---|---|---|---|
| *Friends and social life* | 41 (35.2) | 49 (52.8) | 42 (44.0) |
| *Job or primary activity* | 27 (29.3) | 50 (44.0) | 33 (36.7) |
| *Health and physical condition* | 12 (15.5) | 21 (23.2) | 25 (19.3) |

To test the null hypothesis under which the **expected cell frequencies** were calculated, we compare them with the **observed cell frequencies**. It stands to reason that the null hypothesis should be rejected if the discrepancies between the observed and expected frequencies are large, and that it should be accepted (or at least that we reserve judgment) if the discrepancies between the observed and expected frequencies are small.

Denoting the observed frequencies by the letter $o$ and the expected frequencies by the letter $e$, we base this comparison on the following **chi-square statistic**:

**STATISTIC FOR ANALYSIS OF $r \times c$ TABLE**

$$\chi^2 = \sum \frac{(o - e)^2}{e}$$

If the null hypothesis is true, this statistic is a value of a random variable having approximately the chi-square distribution (see page 275) with $(r - 1)(c - 1)$

degrees of freedom. When $r = 3$ and $c = 3$ as in our example, the number of degrees of freedom is $(3 - 1)(3 - 1) = 4$. Note that after we had calculated four of the expected frequencies with the rule on page 336, all the remaining ones were automatically determined by subtraction from row or column totals.

Since we shall want to reject the null hypothesis when the discrepancies between the $o$'s and $e$'s are large, we use the one-tailed test criterion of Figure 14.2; symbolically, we reject the null hypothesis at the level of significance $\alpha$ if $\chi^2 \geq \chi^2_\alpha$ for $(r - 1)(c - 1)$ degrees of freedom. Remember, though, that this test is only an approximate large-sample test, and it is recommended that it not be used when one (or more) of the expected frequencies is less than 5. (When this is the case, we can sometimes salvage the situation by combining some of the cells, rows, or columns so that none of the expected cell frequencies will be less than 5. In that case there is a corresponding loss in the number of degrees of freedom.)

**Figure 14.2**
Criterion for $\chi^2$ test.

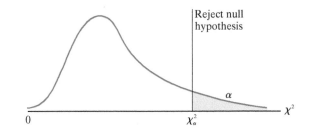

Reject null hypothesis

$\alpha$

$\chi^2$

$\chi^2_\alpha$

0

**EXAMPLE 14.5**   With reference to the problem dealing with the factors contributing most to one's well-being, test at the 0.01 level of significance whether for each of the three alternatives the probabilities are the same for persons who are single, married, or widowed or divorced.

**Solution**
1.  $H_0$: $p_{i1} = p_{i2} = p_{i3}$ for $i = 1, 2$, and 3.
    $H_A$: $p_{i1}$, $p_{i2}$, and $p_{i3}$ are not all equal for at least one value of $i$.
2.  $\alpha = 0.01$
3.  Reject the null hypothesis if $\chi^2 \geq 13.277$, where

$$\chi^2 = \sum \frac{(o - e)^2}{e}$$

and 13.277 is the value of $\chi^2_{0.01}$ for $(3 - 1)(3 - 1) = 4$ degrees of freedom; otherwise, accept the null hypothesis or reserve judgment.
4.  Substituting the observed and expected frequencies summarized in the table on page 336 into the formula for $\chi^2$, we get

$$\chi^2 = \frac{(41 - 35.2)^2}{35.2} + \frac{(49 - 52.8)^2}{52.8} + \frac{(42 - 44.0)^2}{44.0}$$
$$+ \frac{(27 - 29.3)^2}{29.3} + \frac{(50 - 44.0)^2}{44.0} + \frac{(33 - 36.7)^2}{36.7}$$
$$+ \frac{(12 - 15.5)^2}{15.5} + \frac{(21 - 23.2)^2}{23.2} + \frac{(25 - 19.3)^2}{19.3}$$
$$\approx 5.37$$

**5.** Since $\chi^2 = 5.37$ is less than 13.277, the null hypothesis cannot be rejected; that is, we conclude that for each of the three alternatives the probabilities are the same for persons who are single, married, or widowed or divorced, or we reserve judgment. ◼

A MINITAB printout of the preceding chi-square analysis is shown in Figure 14.3. The difference between the values of $\chi^2$ obtained previously and in Figure 14.3, 5.37 and 5.337, is due to rounding. The printout also shows that the *p*-value corresponding to the value of the chi-square statistic is 0.254. Since this exceeds 0.05, the specified level of significance, we conclude as before that the null hypothesis cannot be rejected.

**Figure 14.3**
Computer printout for the analysis of Example 14.5.

```
Chi-Square Test: Single, Married, Widowed or Divorced

Expected counts are printed below observed counts
Chi-Square contributions are printed below expected counts

                               Widowed
                                  or
           Single   Married   Divorced   Total
      1       41        49         42      132
            35.20     52.80      44.00
            0.956     0.273      0.091

      2       27        50         33      110
            29.33     44.00      36.67
            0.186     0.818      0.367

      3       12        21         25       58
            15.47     23.20      19.33
            0.777     0.209      1.661

Total         80       120        100      300

Chi-Sq = 5.337, DF = 4, P-Value = 0.254
```

Some statisticians prefer the alternative formula

**ALTERNATIVE FORMULA FOR CHI-SQUARE STATISTIC**

$$\chi^2 = \sum \frac{o^2}{e} - n$$

where *n* is the grand total of the frequencies for the entire table. This alternative formula does simplify the calculations, but the original formula shows more clearly how $\chi^2$ is actually affected by the discrepancies between the *o*'s and the *e*'s.

EXAMPLE **14.6**   Use this alternative formula to recalculate $\chi^2$ for Example 14.5.

**Solution**

$$\chi^2 = \frac{41^2}{35.2} + \frac{49^2}{52.8} + \frac{42^2}{44.0} + \frac{27^2}{29.3} + \frac{50^2}{44.0}$$
$$+ \frac{33^2}{36.7} + \frac{12^2}{15.5} + \frac{21^2}{23.2} + \frac{25^2}{19.3} - 300$$
$$\approx 5.37$$

This agrees with the result obtained before.

Before we test for independence in our third example, the one dealing with test scores and on-the-job performance, let us demonstrate first that the rule on page 335 for calculating expected cell frequencies applies also to this kind of situation. Under the null hypothesis of independence, the probability of randomly choosing the file of a person whose test score is below average and whose on-the-job performance is poor is given by the product of the probability of choosing the file of a person whose test score is below average and the probability of choosing the file of a person whose on-the-job performance is poor. Using the total of the first row, the total of the first column, and the grand total for the entire table to estimate these probabilities, we get

$$\frac{67 + 64 + 25}{400} = \frac{156}{400}$$

for the probability of choosing the file of a person whose test score is below average. Similarly, we get

$$\frac{67 + 42 + 10}{400} = \frac{119}{400}$$

for the probability of choosing the file of a person whose on-the-job performance is poor. Hence we estimate the probability of choosing the file of a person whose test score is below average and whose on-the-job performance is poor as $\frac{156}{400} \cdot \frac{119}{400}$, and in a sample of size 400 we would expect to find

$$400 \cdot \frac{156}{400} \cdot \frac{119}{400} = \frac{156 \cdot 119}{400} \approx 46.4$$

persons who fit this distribution. Observe that in the final step of these calculations, $\frac{156 \cdot 119}{400}$ is precisely the product of the total of the first row and the total of the first column divided by the grand total for the entire table. This illustrates that the rule on page 335 for calculating expected cell frequencies applies also to the example where the row totals as well as the column totals depend on chance.

EXAMPLE **14.7**   With reference to the problem dealing with test scores in a job-training program and subsequent on-the-job performance, test at the 0.01 level of significance whether these two kinds of assessments are independent.

**Solution**   1. $H_0$: Test scores and on-the-job performance are independent.
   $H_A$: Test scores and on-the-job performance are not independent.

**2.** $\alpha = 0.01$

**3.** Reject the null hypothesis if $\chi^2 \geq 13.277$, where

$$\chi^2 = \sum \frac{(o-e)^2}{e}$$

and 13.277 is the value of $\chi^2_{0.01}$ for $(3-1)(3-1) = 4$ degrees of freedom; otherwise, accept the null hypothesis or reserve judgment.

**4.** Multiplying row totals by column totals and then dividing by the grand total for the entire table, we obtain the expected cell frequencies shown in the following table in parentheses underneath the corresponding observed cell frequencies:

|  |  | Performance | | |
| --- | --- | --- | --- | --- |
|  |  | *Poor* | *Fair* | *Good* |
|  | *Below average* | 67 (46.4) | 64 (63.6) | 25 (46.0) |
| **Qualification test score** | *Average* | 42 (51.8) | 76 (70.9) | 56 (51.3) |
|  | *Above average* | 10 (20.8) | 23 (28.5) | 37 (20.7) |

Then, substituting the observed frequencies and the expected frequencies into the original formula for $\chi^2$, we get

$$\chi^2 = \frac{(67-46.4)^2}{46.4} + \frac{(64-63.6)^2}{63.6} + \frac{(25-46.0)^2}{46.0}$$
$$+ \frac{(42-51.8)^2}{51.8} + \frac{(76-70.9)^2}{70.9} + \frac{(56-51.3)^2}{51.3}$$
$$+ \frac{(10-20.8)^2}{20.8} + \frac{(23-28.5)^2}{28.5} + \frac{(37-20.7)^2}{20.7}$$
$$\approx 40.89$$

**5.** Since $\chi^2 = 40.89$ exceeds 13.277, the null hypothesis must be rejected; that is, we conclude that there is a relationship between the test scores and on-the-job performance. ∎

Right after Example 14.5 we repeated the work with a computer printout of the chi-square analysis. This time we shall repeat the work with the use of a graphing calculator. At the top of Figure 14.4 we find the observed data entered into **MATRIX [A]**, and at the bottom we find the results of the chi-square analysis, $\chi^2 = 41.01$ rounded to two decimals and the corresponding $p$-value of 0.000000027. (This value of chi-square differs from the one obtained previously due to rounding, and the $p$-value cannot be taken too literally since our statistic has only approximately the chi-square distribution with 4 degrees of freedom.)

**Figure 14.4**
Analysis of the $r \times c$ table of Example 14.7 reproduced from the display screen of a TI-83 graphing calculator.

Although the middle part of Figure 14.4 is not really needed, it does show the expected cell frequencies in **MATRIX [B]**.

In the analysis of $r \times c$ tables, the special case where $r = 2$ and the column totals are fixed sample sizes has many important applications. Here, we are testing, in fact, for significant differences among $c$ sample proportions, and we can simplify the notation by letting $p_1, p_2, \ldots,$ and $p_c$ denote the corresponding population proportions. Also, for $c = 2$ we have here an alternative method for testing the significance of the difference between two proportions (as in Section 14.3), but it applies only when the alternative hypothesis is $p_1 \neq p_2$. In that case, the relationship between the $z$ statistic of Section 14.3 and the $\chi^2$ statistic of this section is $z^2 = \chi^2$, as the reader will be asked to verify in Exercise 14.32 for the data of Example 14.4.

A noteworthy, though undesirable, feature of the chi-square analysis of an $r \times c$ table is that the $\chi^2$ statistic is not affected by interchanges of rows and/or columns. This makes it wasteful of information whenever the row categories and/or column categories reflect a definite order; that is, when we deal with ordered categorical data. This was the case, for instance, in Example 14.7, where the test scores ranged from below average to average to above average, and the on-the-job performance ratings ranged from poor to fair to good. To avoid this shortcoming of the chi-square analysis of $r \times c$ tables, statisticians have developed alternative procedures. In these procedures, numbers replace the ordered categories. Usually, but not necessarily, these numbers are consecutive integers, preferably integers that will make the arithmetic as simple as possible. (For instance, for three ordered categories we might use the integers $-1, 0,$ and $1$.) We shall not go into any details about this, but an illustration may be found in Example 17.2. Also, two books dealing with the analysis of ordinal categorical data are listed among the references at the end of this chapter.

**EXERCISES**

**14.15** Use the alternative formula for chi-square on page 338 to recalculate the value of the chi-square statistic obtained in Example 14.7.

**14.16** Suppose that we interview 50 Chevrolet mechanics, 50 Ford mechanics, and 50 Chrysler mechanics and ask them whether their latest models are very easy, easy, fairly difficult, and very difficult to work on. What hypotheses shall we want to test if we intend to perform a chi-square analysis of the resulting $3 \times 4$ table?

**14.17** Suppose that we take a random sample of 400 persons living in federal housing projects and classify each one according to whether he or she has part-time employment, full-time employment, or no employment, and also according to whether he or she has 0, 1, 2, 3, or 4 or more children. What hypotheses shall we want to test if we are going to perform a chi-square analysis of the resulting $3 \times 5$ table?

**14.18** Of a group of 200 persons suffering anxiety disorders, 100 received psychotherapy and 100 received psychological counseling. A panel of psychiatrists determined after six months whether their condition had deteriorated, remained unchanged, or improved. The results are shown in the following table. Use the 0.05 level of significance to test whether the two kinds of treatments are equally effective.

|  | Psychotherapy | Psychological counseling |
|---|---|---|
| Deteriorated | 8 | 11 |
| Unchanged | 58 | 62 |
| Improved | 34 | 27 |

**14.19** Use a computer to rework the chi-square analysis of Exercise 14.18.

**14.20** A research group, interested in whether the proportions of sons taking up the occupations of their fathers are equal for a selected group of occupations, took random samples of size 200, 150, 180, and 100, respectively, in which the fathers are doctors, bankers, teachers, and lawyers, and obtained the following results:

|  | Doctors | Bankers | Teachers | Lawyers |
|---|---|---|---|---|
| Same Occupation | 37 | 22 | 26 | 23 |
| Different occupation | 163 | 128 | 154 | 77 |

Use the 0.05 level of significance to test whether the differences among the four sample proportions, $\frac{37}{200} = 0.185$, $\frac{22}{150} \approx 0.147$, $\frac{26}{180} \approx 0.144$, and $\frac{23}{100} = 0.23$, are significant.

**14.21** The dean of a large university wants to determine whether there is a connection between academic rank and a faculty member's opinion concerning a proposed curriculum change. Interviewing a sample of 80 instructors, 140 assistant professors, 100 associate professors, and 80 professors, he obtains the results shown in the following table:

|  | Instructor | Assistant Professor | Associate Professor | Professor |
|---|---|---|---|---|
| *Against* | 8 | 19 | 15 | 12 |
| *Indifferent* | 40 | 41 | 24 | 16 |
| *For* | 32 | 80 | 61 | 52 |

Use the 0.01 level of significance to test the null hypothesis that there are really no differences in opinion concerning the curriculum change among the four groups.

**14.22** Use a computer or a graphing calculator to repeat Exercise 14.21.

**14.23** Decide on the basis of the information given in the following table, the result of a sample survey conducted at a large state university, whether there is a relationship between students' interest and ability in studying a foreign language. Use the 0.05 level of significance.

|  |  | Ability | | |
|---|---|---|---|---|
|  |  | Low | Average | High |
| **Interest** | *Low* | 28 | 17 | 15 |
|  | *Average* | 20 | 40 | 20 |
|  | *High* | 12 | 28 | 40 |

**14.24** In a study to determine whether there is a relationship between bank employees' standard of dress and their professional advancement, a random sample of size $n = 500$ yielded the results shown in the following table:

|  | Speed of advancement | | |
|---|---|---|---|
|  | Slow | Average | Fast |
| *Very well dressed* | 38 | 135 | 129 |
| *Well dressed* | 32 | 68 | 43 |
| *Poorly dressed* | 13 | 25 | 17 |

Use the 0.025 level of significance to test whether there really is a relationship between bank employees' standard of dress and their professional advancement.

**14.25** Use a computer or a graphing calculator to rework Exercise 14.24 with the level of significance changed to 0.01.

**14.26** A large manufacturer hires many handicapped workers. To see whether their handicaps affect their performance, the personnel manager obtained the following sample data, where the column totals are fixed sample sizes:

|  | | Deaf | Blind | Other handicap | No handicap |
|---|---|---|---|---|---|
| | Above average | 11 | 3 | 14 | 36 |
| **Performance** | Average | 24 | 11 | 39 | 134 |
| | Below average | 5 | 6 | 7 | 30 |

Explain why a "standard" chi-square analysis with $(3-1)(4-1) = 6$ degrees of freedom cannot be performed.

**14.27**  With reference to Exercise 14.26, combine the first three columns and then test at the 0.05 level of significance whether handicaps affect the workers' performance.

**14.28**  If the analysis of a contingency table shows that there is a relationship between the two variables under consideration, the strength of the relationship may be measured by the **contingency coefficient**

$$C = \sqrt{\frac{\chi^2}{\chi^2 + n}}$$

where $\chi^2$ is the value of the chi-square statistic obtained for the table and $n$ is the grand total of all the frequencies. This coefficient takes on values between 0 (corresponding to independence) and a maximum value less than 1 depending on the size of the table; for instance, it can be shown that for a $3 \times 3$ table the maximum value of $C$ is $\sqrt{2/3} \approx 0.82$.

(a)  Calculate $C$ for Example 14.7, which dealt with the test scores and the on-the-job performance of persons who have gone through the job-training program, where we had $n = 400$ and $\chi^2 = 40.89$.

(b)  Find the contingency coefficient for the contingency table of Exercise 14.24.

**14.29**  Find the contingency coefficient for the following contingency table pertaining to the performance of 190 radios:

|  | | | Fidelity | |
|---|---|---|---|---|
| | | Low | Average | High |
| | Low | 7 | 12 | 31 |
| **Selectivity** | Average | 35 | 59 | 18 |
| | High | 15 | 13 | 0 |

**14.30**  Rework Exercise 14.11 by analyzing the data like a $2 \times 2$ table, and verify that the value obtained here for $\chi^2$ equals the square of the value obtained originally for $z$.

**14.31**  Rework Exercise 14.12 by analyzing the data like a $2 \times 2$ table, and verify that the value obtained here for $\chi^2$ equals the square of the value obtained originally for $z$.

**14.32** Rework Example 14.4 on page 331 by analyzing the data like a 2 × 2 table, and verify that the value obtained here for $\chi^2$ equals the square of the value obtained originally for $z$.

**14.33** The following table shows the results of a study in which random samples of the members of five large unions were asked whether they are for or against a certain political candidate:

|  | Union 1 | Union 2 | Union 3 | Union 4 | Union 5 |
|---|---|---|---|---|---|
| *For the candidate* | 74 | 81 | 69 | 75 | 91 |
| *Against the candidate* | 26 | 19 | 31 | 25 | 9 |

At the 0.01 level of significance, can we conclude that the differences among the five sample proportions are significant?

**14.34** Verify that if the expected frequencies for an $r \times c$ table are calculated with the rule on page 335, the sum of the expected frequencies for any row equals the sum of the corresponding observed frequencies.

**14.35** Verify that if the expected frequencies for an $r \times c$ table are calculated with the rule on page 335, the sum of the expected frequencies for any column equals the sum of the corresponding observed frequencies.

# 14.5 GOODNESS OF FIT

In this section we shall consider another application of the chi-square criterion, in which we compare an observed frequency distribution with a distribution we might expect according to theory or assumptions. We refer to such a comparison as a test of **goodness of fit**.

To illustrate, suppose that the management of an airport wants to check an air-traffic controller's claim that the number of radio messages received per minute is a random variable having the Poisson distribution with the mean $\lambda = 1.5$. If correct, this might require hiring additional personnel. Appropriate recording devices yielded the following data on the number of radio messages received in a random sample of 200 one-minute intervals:

```
0 0 1 1 5   3 1 1 2 0   1 0 3 1 0   2 2 2 2 0   1 2 2 0 1
0 2 1 0 2   3 1 1 3 1   0 0 0 1 0   1 2 0 1 3   1 0 0 0 3
0 0 0 1 2   2 0 0 3 0   3 1 1 1 5   2 2 0 2 4   1 1 1 4 2
3 0 0 0 1   0 1 1 1 0   1 0 2 0 2   2 0 1 1 0   2 2 2 1 4
0 0 2 2 0   1 2 0 2 0   0 2 1 1 1   2 1 2 4 0   0 2 0 0 2
0 2 0 0 2   1 3 0 2 0   1 1 3 0 1   0 2 1 0 1   2 2 3 1 1
4 0 1 2 0   0 1 0 2 2   1 0 3 1 1   0 1 0 0 3   2 1 3 1 0
0 2 3 0 0   3 3 0 2 1   0 3 0 0 2   1 1 1 0 1   0 0 2 2 3
```

Summarized, these data yield

| Number of radio messages | Observed frequency |
|:---:|:---:|
| 0 | 70 |
| 1 | 57 |
| 2 | 46 |
| 3 | 20 |
| 4 | 5 |
| 5 | 2 |

**Figure 14.5**
Computer printout for Poisson probabilities with $\lambda = 1.5$.

**Probability Density Function**

```
Poisson with mu = 1.50000

       x     P( X = x)
    0.00       0.2231
    1.00       0.3347
    2.00       0.2510
    3.00       0.1255
    4.00       0.0471
    5.00       0.0141
    6.00       0.0035
    7.00       0.0008
    8.00       0.0001
```

If the air-traffic controller's claim is true, the corresponding expected frequencies are obtained by multiplying by 200 each of the probabilities shown in Figure 14.5. This yields

| Number of radio messages | Expected frequency |
|:---:|:---:|
| 0 | 44.6 |
| 1 | 66.9 |
| 2 | 50.2 |
| 3 | 25.1 |
| 4 or more | 13.1 |

where we combined "4 or more" into one class, since the expected frequency for "5 or more" is 3.7, which is less than 5 and, hence, too small.

To test whether the discrepancies between the observed frequencies and the expected frequencies can be attributed to chance, we use the same chi-square statistic as in Section 14.4:

$$\chi^2 = \sum \frac{(o - e)^2}{e}$$

calculating $\frac{(o-e)^2}{e}$ separately for each class of the distribution. Then, if the value we get for $\chi^2$ is greater than or equal to $\chi^2_\alpha$, we reject the null hypothesis on which the expected frequencies are based at the level of significance $\alpha$. The number of degrees of freedom is $k - m - 1$, where $k$ is the number of terms

$$\frac{(o - e)^2}{e}$$

added in the formula for $\chi^2$, and $m$ is the number of parameters of the probability distribution (in this case the Poisson distribution) that have to be estimated from the sample data.

**EXAMPLE 14.8** Based on the two sets of frequencies given on page 346, the ones that were observed and the ones that were expected for a Poisson population with $\lambda = 1.5$, test the air–traffic controller's claim at the 0.01 level of significance.

**Solution**
1. $H_0$: The population sampled has the Poisson distribution with $\lambda = 1.5$.
   $H_A$: The population sampled does not have the Poisson distribution with $\lambda = 1.5$.
2. $\alpha = 0.01$
3. With "4 or more" combined into one class, $k = 5$ in the formula for the degrees of freedom, and since none of the parameters of the Poisson distribution had to be estimated from the data, $m = 0$. Therefore, reject the null hypothesis if

   $$\chi^2 \geq 13.277, \quad \text{where} \quad \chi^2 = \sum \frac{(o - e)^2}{e}$$

   and 13.277 is the value of $\chi^2_{0.01}$ for $k - m - 1 = 5 - 0 - 1 = 4$ degrees of freedom; otherwise, accept the null hypothesis or reserve judgment.
4. Substituting the observed frequencies and the expected frequencies into the formula for $\chi^2$, we get

   $$\chi^2 = \frac{(70 - 44.6)^2}{44.6} + \frac{(57 - 66.9)^2}{66.9} + \frac{(46 - 50.2)^2}{50.2}$$
   $$+ \frac{(20 - 25.1)^2}{25.1} + \frac{(7 - 13.1)^2}{13.1}$$
   $$\approx 20.2$$

5. Since 20.2 exceeds 13.277, the null hypothesis must be rejected; we conclude that either the population does not have a Poisson distribution or it has a Poisson distribution with $\lambda$ different from 1.5. ∎

To check whether a Poisson distribution with $\lambda$ different from 1.5 might provide a better fit, let us calculate the mean of the observed distribution, getting

$$\frac{0 \cdot 70 + 1 \cdot 57 + 2 \cdot 46 + 3 \cdot 20 + 4 \cdot 5 + 5 \cdot 2}{200} = \frac{239}{200} \approx 1.2$$

Thus, let us see what happens when we use $\lambda = 1.2$ instead of $\lambda = 1.5$.

**Figure 14.6**
Computer printout of
Poisson probabilities
with $\lambda = 1.2$.

**Probability Density Function**

Poisson with mu = 1.20000

| x | P( X = x) |
|------|-----------|
| 0.00 | 0.3012 |
| 1.00 | 0.3614 |
| 2.00 | 0.2169 |
| 3.00 | 0.0867 |
| 4.00 | 0.0260 |
| 5.00 | 0.0062 |
| 6.00 | 0.0012 |
| 7.00 | 0.0002 |
| 8.00 | 0.0000 |

For $\lambda = 1.2$, we get the expected frequencies by multiplying the probabilities in Figure 14.6 by 200 (after combining "4 or more" into one category). This yields

| Number of radio messages | Expected frequency |
|--------------------------|--------------------|
| 0 | 60.2 |
| 1 | 72.3 |
| 2 | 43.4 |
| 3 | 17.3 |
| 4 or more | 6.7 |

**EXAMPLE 14.9** Based on the observed frequencies on page 346 and the expected frequencies given immediately above test at the 0.01 level of significance whether the data constitute a sample from a Poisson population.

**Solution**
1. $H_0$: The population sampled has a Poisson distribution.
   $H_A$: The population sampled does not have a Poisson distribution.
2. $\alpha = 0.01$
3. Since the expected frequency corresponding to "5 or more" is again less than 5, the classes corresponding to 4 and "5 or more" must be combined and $k = 5$; also, since the parameter $\lambda$ was estimated from the data, $m = 1$. Therefore, reject the null hypothesis if $\chi^2 \geq 11.345$, where

$$\chi^2 = \sum \frac{(o - e)^2}{e}$$

and 11.345 is the value of $\chi^2_{0.01}$ for $k - m - 1 = 5 - 1 - 1 = 3$ degrees of freedom; otherwise, accept the null hypothesis or reserve judgment.
4. Substituting the observed frequencies and the expected frequencies into the formula for $\chi^2$, we get

$$\chi^2 = \frac{(70 - 60.2)^2}{60.2} + \frac{(57 - 72.3)^2}{72.3} + \frac{(46 - 43.4)^2}{43.4}$$
$$+ \frac{(20 - 17.3)^2}{17.3} + \frac{(7 - 6.7)^2}{6.7}$$
$$\approx 5.4$$

5. Since 5.4 is less than 11.345, the null hypothesis cannot be rejected. Using a HEWLETT PACKARD statistical calculator, we found that the $p$-value corresponding to $\chi^2 = 5.4$ and 3 degrees of freedom is 0.1272, so the fit is not too bad, but we would be inclined to reserve judgment about the nature of the population. ■

The method illustrated in this section is used quite generally to test how well distributions, expected on the basis of theory or assumptions, fit, or describe, observed data. In the exercises that follow we shall test also whether observed distributions have (at least approximately) the shape of binomial and normal distributions.

EXERCISES

**14.36** Ten years' data show that in a given city there were no bank robberies in 57 months, one bank robbery in 36 months, two bank robberies in 15 months, and three or more bank robberies in 12 months. At the 0.05 level of significance, does this substantiate the claim that the probabilities of 0, 1, 2, or 3 or more bank robberies in any one month are 0.40, 0.30, 0.20, and 0.10?

**14.37** Following is the distribution of the number of females in 160 litters, each consisting of four mice:

| Number of females | Number of litters |
|:---:|:---:|
| 0 | 12 |
| 1 | 37 |
| 2 | 55 |
| 3 | 47 |
| 4 | 9 |

Test at the 0.01 level of significance whether these data may be looked upon as random samples from a binomial population with $n = 4$ and $p = 0.50$.

**14.38** Each day a surgeon schedules at most four operations, and his activities on 300 days are summarized in the following table:

| Number of surgeries | Number of days |
|:---:|:---:|
| 0 | 2 |
| 1 | 10 |
| 2 | 33 |
| 3 | 136 |
| 4 | 119 |

Test at the 0.05 level of significance whether these data can be looked upon as random samples from a binomial population with $n = 4$ and $p = 0.70$.

**14.39** With reference to Exercise 14.38, test at the 0.05 level of significance whether the data can be looked upon as random samples from a binomial population. (*Hint*: Calculate the mean of the given distribution and use the formula $\mu = np$ to estimate $p$.)

**14.40** Following is the distribution of the numbers of bears spotted on 100 sightseeing tours in Denali National Park:

| Number of bears | Number of tours |
|---|---|
| 0 | 70 |
| 1 | 23 |
| 2 | 7 |
| 3 | 0 |

Given that for the Poisson distribution with $\lambda = 0.5$ the probabilities of 0, 1, 2, and 3 "successes" are, respectively, 0.61, 0.30, 0.08, and 0.01, test at the 0.05 level of significance whether the data on bear sightings may be looked upon as a random sample from a Poisson population with $\lambda = 0.5$.

# 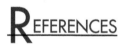HECKLIST OF KEY TERMS (with page references to their definitions)

Cell, 336
Chi-square statistic, 336
Contingency coefficient, 344
Contingency table, 335
Count data, 326
Expected cell frequencies, 336
Goodness of fit, 345

Grand total, 335
Observed cell frequencies, 336
Pooling, 330
$r \times c$ table, 334
Standard error of difference between two proportions, 330

# REFERENCES

*The theory that underlies the various tests of this chapter is discussed in most textbooks on mathematical statistics; for instance, in*

MILLER, I., and MILLER, M., *John E. Freund's Mathematical Statistics*, 6th ed. Upper Saddle River, N.J.: Prentice Hall, 1999.

*Details about contingency tables may be found in*

EVERITT, B. S., *The Analysis of Contingency Tables*. New York: John Wiley & Sons, Inc., 1977.

*Research on the analysis of r × c tables with ordered categories can be found in*

AGRESTI, A., *Analysis of Ordinal Categorical Data*. New York: John Wiley & Sons, Inc., 1984.

GOODMAN, L. A., *The Analysis of Cross-Classified Data Having Ordered Categories*. Cambridge, Mass.: Harvard University Press, 1984.

**R.131** The objective of a research project is to determine what percentage of the farm workers in southern Arizona, where cotton is grown, are illegal aliens. How large a sample will be required if it is felt that the actual percentage is at most 22% and the director of the project wants to be able to assert with probability 0.95 that the estimate will be off by at most 2.5%?

**R.132** What null hypothesis and what alternative hypothesis should we use if we want to test the claim that on the average children attending elementary schools in an urban school district live more than 1.5 miles from the school they attend?

**R.133** Six packages of sunflower seeds randomly selected from a large shipment weighed, respectively, 15.9, 15.5, 16.2, 15.8, 16.0, and 15.7 ounces. Use a normal probability plot generated by means of a computer or a graphing calculator to justify the assumption that these data constitute a sample from a normal population.

**R.134** With reference to Exercise R.133, what can we assert with 95% confidence about the maximum error if we use the mean of the sample, $\bar{x} = 15.85$ ounces, to estimate the mean of the population sampled?

**R.135** In five track meets, a high jumper cleared 84, $81\frac{1}{2}$, 82, $80\frac{1}{2}$, and 83 inches. Show that the null hypothesis $\mu = 78$ can be rejected against the alternative hypothesis $\mu > 78$ at the 0.01 level of significance.

**R.136** With reference to Exercise R.135, show that the null hypothesis $\mu = 78$ cannot be rejected against the alternative hypothesis $\mu > 78$ at the 0.01 level of significance if the fifth figure is recorded incorrectly as 93 instead of 83. Explain the apparent paradox that even though the difference between $\bar{x}$ and $\mu$ has increased, it is no longer significant.

**R.137** In a study of parents' feelings about a required course in sex education taught to their children, a random sample of 360 parents are classified according to whether they have one, two, or three or more children in the school system and also whether they feel that the course is poor, adequate, or good. The results are shown in the following table:

|  | *1* | *2* | *3 or more* |
|---|---|---|---|
| *Poor* | 48 | 40 | 12 |
| *Adequate* | 55 | 53 | 29 |
| *Good* | 57 | 46 | 20 |

Test at the 0.05 level of significance whether there is a relationship between parents' reaction to the course and the number of children they have in the school system.

**R.138** In a study conducted at a large airport, 81 persons in a random sample of 300 persons who had just gotten off a plane and 32 persons in a random sample of 200 persons who were about to board a plane admitted that they were afraid of flying. Use the $z$ statistic to test at the 0.01 level of significance whether the difference between the corresponding sample proportions is significant.

**R.139** Use the $\chi^2$ statistic to rework Exercise R.138, and verify that the value obtained here for $\chi^2$ equals the square of the value obtained for $z$ in Exercise R.138.

**R.140** The following table shows how many times the departure of a daily flight from Vancouver to San Francisco was delayed in 50 weeks:

| Delays per week | Number of weeks |
|:---:|:---:|
| 0 | 12 |
| 1 | 16 |
| 2 | 13 |
| 3 | 8 |
| 4 | 1 |

Use the 0.05 level of significance to test the null hypothesis that the departure of the flight is delayed 10% of the time; namely, the null hypothesis that the data constitute a random sample from a binomial population with $n = 7$ and $p = 0.10$.

**R.141** With reference to Exercise R.140, use the 0.05 level of significance to test the null hypothesis that the data constitute a random sample from a binomial population with $n = 7$. (*Hint*: Estimate $p$ by calculating the mean of the given distribution and then using the formula $\mu = np$.)

**R.142** In order to evaluate the clinical effects of a certain steroid in treating chronically underweight persons, a random sample of 60 such persons was given 25-mg dosages over a period of 12 weeks, while another random sample of 60 such persons was given 50-mg dosages over the same period of time. The results showed that those in the first group gained on the average 8.5 pounds while those in the second group gained on the average 11.3 pounds. If previous tests have shown that $\sigma_1 = \sigma_2 = 1.4$ pounds, test at the 0.05 level of significance whether the difference between the two sample means is significant.

**R.143** A study was designed to compare the effectiveness of the weight-reducing programs of three health spas. The table that follows shows the weight losses of patrons who took the respective programs of diet and exercise for six weeks:

|  |  | Health Spa 1 | Health Spa 2 | Health Spa 3 |
|---|---|:---:|:---:|:---:|
| **Weight loss** | Less than ten pounds | 86 | 91 | 125 |
|  | Ten or more pounds | 18 | 21 | 38 |

Use the 0.05 level of significance to test the null hypothesis that the three programs are equally effective.

**R.144** A random sample taken as part of a study in nutrition showed that 400 young adults in an Australian city averaged a protein intake of 1.274 g per kilogram of body weight. Given that $\sigma = 0.22$ g per kilogram of body weight, what can one assert with 95% confidence about the maximum error if $\bar{x} = 1.274$ g is used as an estimate of the mean protein intake per kilogram of body weight for the population sampled?

**R.145** A microbiologist found 13, 17, 7, 11, 15, and 9 microorganisms in six cultures. Use appropriate computer software or a graphing calculator to obtain a normal probability plot and, thus, verify that these data may be looked upon as a sample from a normal population.

**R.146** In a random sample of 150 persons shopping at Scottsdale's Fashion Square, 118 made at least one purchase. Construct a 95% confidence interval for the probability that a person randomly chosen from among persons shopping at Scottsdale's Fashion Square will make at least one purchase.

**R.147** In an election for County Treasurer, the Independent candidate received 10,361 votes (about 48%) and the Republican candidate received 11,225 votes (about 52%). Is it reasonable to ask whether the difference between these two percentages is significant?

**R.148** In accordance with the rule that $np$ and $n(1 - p)$ must both be greater than 5, for what values of $p$ can we use the normal approximation to the binomial distribution when $n = 400$?

**R.149** In a study of the length of young-of-the-year fresh drumfish in Lake Erie, it was found that the lengths of 60 of them had the standard deviation $s = 10.4$ mm. Construct a 99% confidence interval for the variance of the population sampled.

**R.150** A team of physicians is asked to check whether a highly paid athlete is physically fit to play in the NFL. What type of error would be committed if the hypothesis that the athlete is physically fit to play in the NFL is erroneously accepted? What type of error would be committed if the hypothesis that the athlete is physically fit to play in the NFL is erroneously rejected?

**R.151** In a multiple-choice test administered to high school sophomores after a visit to a natural history museum, 23 of 80 boys and 19 of 80 girls, both random samples, identified a geode as an Italian pastry. Test at the 0.05 level of significance whether the difference between the corresponding sample proportions is significant.

**R.152** In a random sample of $n = 25$ servings of a breakfast cereal, the sugar content averaged 10.42 grams with a standard deviation of 1.76 grams. Assuming that these data constitute a sample from a normal population, construct a 95% confidence interval for $\sigma$, the standard deviation of the population sampled.

**R.153** The following table shows how samples of the residents of three federally financed housing projects replied to the question whether they would continue to live there if they had the choice:

|     | Project 1 | Project 2 | Project 3 |
|-----|-----------|-----------|-----------|
| Yes | 63        | 84        | 69        |
| No  | 37        | 16        | 31        |

Test at the 0.01 level of significance whether the differences among the three sample proportions (of "yes" answers) may be attributed to chance.

**R.154** A testing laboratory wants to estimate the average lifetime of a multitoothed cutting tool. If a random sample of size $n = 6$ showed tool lives of 2,470, 2,520, 2,425, 2,505, 2,440, and 2,400 cuts, and it can be assumed that these data constitute a random sample from a normal population, what can they assert with 99% confidence about the maximum error if the mean of this sample is used as an estimate of the mean of the population sampled?

**R.155** In a study of the reading habits of financial advisors, it is desired to estimate the average number of financial reports they read per week. Assuming that it is reasonable to use $\sigma = 3.4$, how large a sample would be required if one wants to be able to assert with probability 0.99 that the sample mean will not be off by more than 1.2?

**R.156** A geneticist found that in independent random samples of 100 men and 100 women there were 31 men and 24 women with an inherited blood disorder. Can she conclude at the 0.01 level of significance that the corresponding true proportion for men is significantly greater than that for women?

(a) Comment on the formulation of this question.

(b) Restate the question as it should have been asked, and answer it by performing the appropriate test.

**R.157** Following is the distribution of the daily emission of sulfur oxides by an industrial plant:

| Tons of sulfur oxides | Frequency |
|---|---|
| 5.0–8.9 | 3 |
| 9.0–12.9 | 10 |
| 13.0–16.9 | 14 |
| 17.0–20.9 | 25 |
| 21.0–24.9 | 17 |
| 25.0–28.9 | 9 |
| 29.0–32.9 | 2 |

As can easily be verified, the mean of this distribution is $\bar{x} = 18.85$ and its standard deviation is $s = 5.55$. To test the null hypothesis that these data constitute a random sample from a normal population, proceed with the following steps:

(a) Find the probabilities that a random variable having the normal distribution with $\mu = 18.85$ and $\sigma = 5.5$ will take on a value less than 8.95, between 8.95 and 12.95, between 12.95 and 16.95, between 16.95 and 20.95, between 20.95 and 24.95, between 24.95 and 28.95, and greater than 28.95.

(b) Changing the first and last classes of the distribution to "8.95 or less" and "28.95 or more," find the expected normal curve frequencies corresponding to the seven classes of the distribution by multiplying the probabilities obtained in part (a) by the total frequency of 80.

(c) Test at the 0.05 level of significance whether the given data may be looked upon as a random sample from a normal population.

**R.158** Among 210 persons with alcohol problems admitted to the psychiatric emergency room of a hospital, 36 were admitted on a Monday, 19 on a Tuesday, 18 on a Wednesday, 24 on a Thursday, 33 on a Friday, 40 on a Saturday, and 40 on a Sunday. Use the 0.05 level of significance to test the null hypothesis that this psychiatric emergency room can expect equally many persons with alcohol problems on each day of the week.

**R.159** The following are the pull strengths (in pounds) required to break the bond of two kinds of glue:

**Glue 1:**　25.3　20.2　21.1　27.0　16.9　30.1
　　　　　　17.8　22.9　27.2　20.0

**Glue 2:**　24.9　22.5　21.8　23.6　19.8　21.6
　　　　　　20.4　22.1

As a preliminary to the two-sample $t$ test, use the 0.02 level of significance to test whether it is reasonable to assume that the corresponding population standard deviations are equal.

**R.160** Based on the results of $n = 14$ trials, we want to test the null hypothesis $p = 0.30$ against the alternative hypothesis $p > 0.30$. If we reject the null hypothesis when the number of successes is eight or more and otherwise we accept it, find the probability of a

(a) Type I error;

(b) Type II error when $p = 0.40$;

(c) Type II error when $p = 0.50$;

(d) Type II error when $p = 0.60$.

**R.161** Test runs with eight models of an experimental engine showed that they operated for 25, 18, 31, 19, 32, 27, 24, and 28 minutes with a gallon of a certain kind of fuel. Estimate the standard deviation of the population sampled using

(a) the sample standard deviation;

(b) the sample range and the method described in Exercise 11.35.

**R.162** An undercover government agent wants to determine what percentage of vendors at flea markets keep records for income tax purposes. How large a random sample will he need if he wants to be able to assert with 95% confidence that the error of his estimate, the sample percentage, is at most 6% and

(a) he has no idea about the true value,

(b) he is certain that the true percentage is at least 60%?

**R.163** To find out whether the inhabitants of two South Pacific islands may be regarded as having the same racial ancestry, an anthropologist determined the cephalic indices of six adult males from each island, getting $\bar{x}_1 = 77.4$ and $\bar{x}_2 = 72.2$ and the corresponding standard deviations $s_1 = 3.3$ and $s_2 = 2.1$. Assuming that the data constitute independent random samples from normal populations, test the null hypothesis $\sigma_1 = \sigma_2$ against the alternative hypothesis $\sigma_1 \neq \sigma_2$ at the 0.10 level of significance (as a preliminary to the two-sample $t$ test).

**R.164** In an experiment, an interviewer of job applicants is asked to write down her initial impression (favorable or unfavorable) after two minutes and her final impression at the end of the interview. Use the following data and the 0.01 level of significance to test the interviewer's claim that her initial and final impressions are the same 85% of the time:

|  |  | Initial impression | |
|---|---|---|---|
|  |  | *Favorable* | *Unfavorable* |
| **Final impression** | *Favorable* | 184 | 32 |
|  | *Unfavorable* | 56 | 128 |

**R.165** In a random sample of 10 rounds of golf played on her home course, a golf professional averaged 70.8 with a standard deviation of 1.28. Assuming that her scores can be looked upon as a random sample from a normal population, use the 0.01 level of significance to test the null hypothesis $\sigma = 1.0$ against the alternative hypothesis that her game is actually less consistent.

**R.166** A political pollster wants to determine the proportion of the population that favors a regulatory change in the use of marijuana by cancer patients. How large a sample will she need if she wants to be able to assert with a probability of at least 0.90 that the sample proportion will differ from the population proportion by at most 0.04?

**R.167** A bank is considering replacing its ATMs with a new model. If $\mu_0$ is the average length of time that its old machines function between repairs, against what alternative hypothesis should it test the null hypothesis $\mu = \mu_0$, where $\mu$ is the corresponding average length of time for the new machines, if

(a) it does not want to replace the old machines unless the new machines prove to be superior;

(b) it wants to replace the old machines unless the new machines actually turn out to be inferior?

**R.168** In a study of the relationship between family size and intelligence, 40 "only children" had an average IQ of 101.5 and 50 "first borns" in families with two children had an average IQ of 105.9. If it can be assumed that $\sigma_1 = \sigma_2 = 5.9$ for such data, test at the 0.01 level of significance whether the difference between the two sample means is significant.

# 15 ANALYSIS OF VARIANCE

**15.1**  Differences Among $k$ Means: An Illustration  358

**15.2**  The Design of Experiments: Randomization  362

**15.3**  One-Way Analysis of Variance  363

**15.4**  Multiple Comparisons  370

**15.5**  The Design of Experiments: Blocking  376

**15.6**  Two-Way Analysis of Variance  377

**15.7**  Two-Way Analysis of Variance without Interaction  378

**15.8**  The Design of Experiments: Replication  382

**15.9**  Two-Way Analysis of Variance with Interaction  382

**15.10**  The Design of Experiments: Further Considerations  387

Checklist of Key Terms  394

References  395

n this chapter we shall generalize the material in Sections 12.5 and 12.6 by considering problems in which we must decide whether observed differences among more than two sample means can be attributed to chance, or whether they are indicative of real differences among the means of the populations sampled. For instance, we may want to decide on the basis of sample data whether there really is a difference in the effectiveness of three methods of teaching a foreign language. We may also want to compare the average yield per acre of eight varieties of wheat, we may want to judge whether there really is a difference in the average mileage obtained with five kinds of gasoline, or we may want to determine whether there really is a difference in the durability of four exterior house paints. The method we use for the analysis of problems like this is called an **analysis of variance**, or an **ANOVA** for short.

Beyond this, an analysis of variance can be used to sort out several questions at the same time. For instance, with regard to the first of our four examples in the preceding paragraph, we might also ask whether the observed differences among the sample means are really due to the differences in teaching the foreign language and not due to the quality of the teaching or the merits of the textbooks being used, or perhaps the intelligence of the students being taught. Similarly, with regard to the different varieties of wheat, we might ask whether the differences we observe in their yield are really due to their quality and not due to the use of different fertilizers, differences in the composition of the soil, or perhaps differences in the amount of irrigation that is applied to the soil. Considerations like these lead to the important subject of **experimental design**; namely, to the problem of planning experiments in such a way that meaningful questions can be asked and put to a test.

Following an introductory example in Section 15.1 and the discussion of **randomization** in Section 15.2, we shall present the **one-way analysis of variance** in Section 15.3, followed by a discussion of **multiple comparisons** in Section 15.4. The latter are designed to sort out interpretations when an analysis of variance leads to significant results. Subsequently, the notion of **blocking** in Section 15.5 leads to the analysis of **two-way experiments** in Section 15.6. Various related topics are introduced in the remainder of the chapter.

# 15.1 DIFFERENCES AMONG $k$ MEANS: AN ILLUSTRATION

To illustrate the kind of situation in which we might perform an analysis of variance, consider the following part of a study of the calcium contamination of river water. The data are the amounts of calcium (average parts per million) measured at three different locations along the Mississippi River:

| **Location M:** | 42 | 37 | 41 | 39 | 43 | 41 |
|---|---|---|---|---|---|---|
| **Location N:** | 37 | 40 | 39 | 38 | 41 | 39 |
| **Location O:** | 32 | 28 | 34 | 32 | 30 | 33 |

As can easily be verified, the means of these three samples are 40.5, 39.0, and 31.5, and what we would like to know is whether the differences among them are significant or whether they can be attributed to chance.

In a problem like this, we denote the means of the $k$ populations sampled by $\mu_1, \mu_2, \ldots,$ and $\mu_k$, and test the null hypothesis $\mu_1 = \mu_2 = \cdots = \mu_k$ against the alternative hypothesis that these $\mu$'s are not all equal.[†] This null hypothesis would be supported when the differences among the sample means are small, and the alternative hypothesis would be supported when at least some of the differences among the sample means are large. Thus, we need a measure of the discrepancies among the $\bar{x}$'s, and with it a rule that tells us when the discrepancies are so large that the null hypothesis can be rejected.

To begin with, let us make two assumptions that are critical to the method by which we shall analyze our problem:

---

**The data constitute random samples from normal populations. These normal populations all have the same standard deviation.**

---

[†] For work later in this chapter, it will be desirable to write the $k$ means as $\mu_1 = \mu + \alpha_1$, $\mu_2 = \mu + \alpha_2, \ldots,$ and $\mu_k = \mu + \alpha_k$, where

$$\mu = \frac{\mu_1 + \mu_2 + \cdots + \mu_k}{k}$$

is called the **grand mean** and the $\alpha$'s, whose sum is zero (see Exercise 15.27), are called the **treatment effects**. In this notation, we test the null hypothesis $\alpha_1 = \alpha_2 = \cdots = \alpha_k = 0$ against the alternative hypothesis that the $\alpha$'s are not all equal to zero.

In our illustration we have samples from $k = 3$ populations, one for each location, and we shall assume that these populations are normal populations with the same standard deviation $\sigma$. The three populations may or may not have equal means; indeed, this is precisely what we hope to discover with an analysis of variance.

Of course, the value of $\sigma$ is unknown, but in an analysis of variance we shall estimate $\sigma^2$, the population variance, in two different ways, and then base the decision whether or not to reject the null hypothesis on the ratio of these two estimates. The first of these two estimates will be based on the variation *among* the $\bar{x}$'s, and it will tend to be greater than what we might expect when the null hypothesis is *false*. The second estimate will be based on the variation *within* the samples, and hence it will not be affected by the null hypothesis being true or false.

Let us begin with the first of the two estimates of $\sigma^2$ by calculating the variance of the $\bar{x}$'s. Since the mean of the three $\bar{x}$'s is

$$\frac{40.5 + 39.0 + 31.5}{3} = 37.0$$

substitution into the formula for the sample standard deviation and squaring yields

$$s_{\bar{x}}^2 = \frac{(40.5 - 37.0)^2 + (39.0 - 37.0)^2 + (31.5 - 37.0)^2}{3 - 1}$$

$$= 23.25$$

where the subscript $\bar{x}$ serves to indicate that $s_{\bar{x}}^2$ is the variance of the sample means.

If the null hypothesis is true, we can look upon the three samples as samples from one and the same population and, hence, upon $s_{\bar{x}}^2$ as an estimate of $\sigma_{\bar{x}}^2$, the square of the standard error of the mean. Now, since

$$\sigma_{\bar{x}} = \frac{\sigma}{\sqrt{n}}$$

for random samples of size $n$ from an infinite population, we can look upon $s_{\bar{x}}^2$ as an estimate of

$$\sigma_{\bar{x}}^2 = \left(\frac{\sigma}{\sqrt{n}}\right)^2 = \frac{\sigma^2}{n}$$

and, therefore, upon $n \cdot s_{\bar{x}}^2$ as an estimate of $\sigma^2$. For our illustration, we thus have $n \cdot s_{\bar{x}}^2 = 6(23.25) = 139.5$ as an estimate of $\sigma^2$, the common variance of the three populations sampled.

This is the estimate of $\sigma^2$ based on the variation among the sample means, and if $\sigma^2$ were known, we could compare $n \cdot s_{\bar{x}}^2$ with $\sigma^2$ and reject the null hypothesis if $n \cdot s_{\bar{x}}^2$ is much greater than $\sigma^2$. However, in actual practice $\sigma^2$ is unknown, and we have no choice but to obtain another estimate of $\sigma^2$ that is not affected by the null hypothesis being true or false. As we said before, such a second estimate would be based on the variation within the samples, as measured by $s_1^2$, $s_2^2$, and $s_3^2$. The values of these three sample variances are $s_1^2 = 4.7$, $s_2^2 = 2.0$, and $s_3^2 = 4.7$ for our example, but rather than choose one of them, we *pool* (average) them, getting

$$\frac{s_1^2 + s_2^2 + s_3^2}{3} = \frac{4.7 + 2 + 4.7}{3} = 3.8$$

as our second estimate of $\sigma^2$.

We now have two estimates of $\sigma^2$, $n \cdot s_{\bar{x}}^2 = 139.5$ and $\frac{s_1^2 + s_2^2 + s_3^2}{3} = 3.8$, and it should be observed that the first estimate, based on the variation among the sample means, is much greater than the second, based on the variation within the samples. This suggests that the three population means are probably not all equal; namely, that the null hypothesis ought to be rejected. To put this comparison on a rigorous basis, we use the **F statistic**

**S**TATISTIC FOR TEST CONCERNING DIFFERENCES AMONG MEANS

$$F = \frac{estimate\ of\ \sigma^2\ based\ on\ the\ variation\ among\ the\ \bar{x}'s}{estimate\ of\ \sigma^2\ based\ on\ the\ variation\ within\ the\ samples}$$

If the null hypothesis is true and the assumptions we made are valid, the sampling distribution of this statistic is the **F distribution**, which we met earlier in Chapter 13, where we also used it to compare two variances and referred to the $F$ statistic as a **variance ratio**. Since the null hypothesis will be rejected only when $F$ is large (that is, when the variation among the $\bar{x}$'s is too great to be attributed to chance), we base our decision on the criterion of Figure 15.1. For $\alpha = 0.05$ or 0.01, the values of $F_\alpha$ may be looked up in Table IV at the end of the book, and if we compare the means of $k$ random samples of size $n$, the **numerator and denominator degrees of freedom** are, respectively, $k - 1$ and $k(n - 1)$.

Returning to our numerical illustration, we find that $F = \frac{139.5}{3.8} \approx 36.7$ rounded to one decimal, and since this exceeds 6.36, the value of $F_{0.01}$ for $k - 1 = 3 - 1 = 2$ and $k(n - 1) = 3(6 - 1) = 15$ degrees of freedom, the null hypothesis must be rejected at the 0.01 level of significance. In other words, the differences among the three sample means are too large to be attributed to chance. (As we shall see later, the $p$-value for $F = 36.7$ and 2 and 15 degrees of freedom is actually less than 0.000002.)

The technique we have described in this section is the simplest form of an analysis of variance. Although we could go ahead and perform $F$ tests for differences among $k$ means without further discussion, it will be instructive to

**Figure 15.1**
Test criterion based on
$F$ distribution.

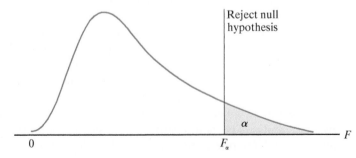

look at the problem from a somewhat different analysis-of-variance point of view, and we shall do so in Section 15.3.

**15.1**  Samples of peanut butter produced by three different companies were tested for aflatoxin content (parts per billion) with the following results:

| **Company 1:** | 4.4 | 0.6 | 6.4 | 1.2 | 2.8 | 4.4 |
| **Company 2:** | 0.8 | 2.6 | 1.9 | 3.7 | 5.3 | 1.3 |
| **Company 3:** | 1.1 | 3.4 | 1.6 | 0.5 | 4.3 | 2.3 |

(a)  Calculate $n \cdot s_{\bar{x}}^2$ for these data, the mean of the variances of the three samples, and the value of $F$.

(b)  Assuming that the data constitute random samples from three normal populations with the same standard deviation, test at the 0.05 level of significance whether the differences among the three sample means can be attributed to chance.

**15.2**  What are the numerator and denominator degrees of freedom of the $F$ distribution when we compare the means of

(a)  $k = 4$ random samples of size $n = 20$;

(b)  $k = 8$ random samples of size $n = 15$?

**15.3**  An agronomist planted three test plots each with four varieties of wheat and obtained the following yields (in pounds per plot):

| **Variety A:** | 65 | 64 | 60 |
| **Variety B:** | 55 | 56 | 63 |
| **Variety C:** | 56 | 59 | 59 |
| **Variety D:** | 62 | 59 | 62 |

(a)  Calculate $n \cdot s_{\bar{x}}^2$ for these data, the mean of the variances of the four samples, and the value of $F$.

(b)  Assuming that these data constitute random samples from four normal populations with the same standard deviation, test at the 0.01 level of significance whether the differences among the four sample means can be attributed to chance.

**15.4**  Following are the caloric values of the fat content of meals served at three elementary schools:

| **School 1:** | 127 | 143 | 142 | 117 | 140 | 146 | 141 | 148 |
| **School 2:** | 127 | 146 | 138 | 143 | 142 | 124 | 130 | 130 |
| **School 3:** | 147 | 132 | 132 | 134 | 157 | 137 | 144 | 145 |

(a)  Given that $\bar{x}_1 = 138, \bar{x}_2 = 135, \bar{x}_3 = 141, s_1^2 = 111.43, s_2^2 = 68.29$, and $s_3^2 = 77.71$, calculate $n \cdot s_{\bar{x}}^2$ for these data, the mean of the variances of the three samples, and the value of $F$.

(b)  Assuming that these data constitute random samples from three normal populations with the same standard deviation, test at the 0.05 level of significance whether the differences among the three sample means can be attributed to chance.

**15.5**  Following are the fourth-grade reading comprehension scores on a standardized test obtained by random samples of students from three large schools:

| School 1: | 81 | 83 | 77 | 72 | 86 | 92 | 83 | 78 | 80 | 75 |
| School 2: | 73 | 112 | 66 | 104 | 95 | 81 | 62 | 76 | 129 | 90 |
| School 3: | 84 | 89 | 81 | 76 | 79 | 83 | 85 | 74 | 80 | 78 |

Explain why the method described in Section 15.1 should probably not be used to test for significant differences among the three sample means.

## 15.2 THE DESIGN OF EXPERIMENTS: RANDOMIZATION

Suppose that we are asked to compare the cleansing action of three detergents on the basis of the following whiteness readings made on 15 swatches of white cloth, which were first soiled with India ink and then washed in an agitator-type machine with the respective detergents:

| Detergent X: | 77 | 81 | 71 | 76 | 80 |
| Detergent Y: | 72 | 58 | 74 | 66 | 70 |
| Detergent Z: | 76 | 85 | 82 | 80 | 77 |

The means of the three samples are 77, 68, and 80, and an analysis of variance showed that the means of the three populations sampled are not all equal.

It would seem natural to conclude that the three detergents are not equally effective, but a moment's reflection will show that this conclusion is not so "natural" at all. For all we know, the swatches cleaned with detergent Y may have been more soiled than the others, the washing times may have been longer for detergent Z, there may have been differences in water hardness or water temperature, and even the instruments used to make the whiteness readings may have gone out of adjustment after the readings for detergents X and Z were made.

It is entirely possible, of course, that the differences among the three sample means, 77, 68, and 80, are due largely to differences in the quality of the three detergents, but we have just listed several other factors that could be responsible. It is important to remember that *a significance test may show that differences among sample means are too large to be attributed to chance, but it cannot say why the differences occurred.*

In general, if we want to show that one factor (among various others) can be considered the cause of an observed phenomenon, we must somehow make sure that none of the other factors can reasonably be held responsible. There are various ways in which this can be done; for instance, we can conduct a rigorously **controlled experiment** in which all variables except the one of concern are held fixed. To do this in the example dealing with the three detergents, we might soil the swatches with exactly equal amounts of India ink, always use exactly the same washing time, use water of exactly the same temperature and hardness, and inspect the measuring instruments after each use. Under these rigid conditions, significant differences among the sample means cannot be due to differently soiled swatches, or differences in washing time, water temperature, water hardness, or measuring instruments. On the positive side, the differences show that the detergents are not all equally effective if they are used in this narrowly restricted way. Of course, we cannot say whether the same differences would exist if the washing time is longer or shorter, if the water has a different temperature or hardness, and so on.

In most cases, "overcontrolled" experiments like the one just described do not really provide us with the kind of information we want. Also, such experiments are rarely possible in actual practice; for example, it would have been difficult in our illustration to be sure that the instruments really were measuring identically on repeated washings or that some other factor, not thought of or properly controlled, was not responsible for the observed differences in whiteness. So, we look for alternatives. At the other extreme we can conduct an experiment in which none of the extraneous factors is controlled, but in which we protect ourselves against their effects by **randomization**. That is, we design, or plan, the experiment in such a way that the variations caused by these extraneous factors can all be combined under the general heading of "chance."

In our example dealing with the three detergents, we could accomplish this by numbering the swatches (which need not be equally soiled) from 1 to 15, specifying the random order in which they are to be washed and measured, and randomly selecting the five swatches that are to be washed with each of the three detergents. When all the variations due to uncontrolled extraneous factors can thus be included under the heading of chance variation, we refer to the design of the experiment as a **completely randomized design**.

As should be apparent, randomization protects against the effects of the extraneous factors only in a probabilistic sort of way. For instance, in our example it is possible, though very unlikely, that detergent X will be randomly assigned to the five swatches that happen to be the least soiled, or that the water happens to be coldest when we wash the five swatches with detergent Y. It is partly for this reason that we often try to control some of the factors and randomize the others, and thus use designs that are somewhere between the two extremes that we have described.

Randomization protects against the effects of factors that cannot be completely controlled, but it does not relieve a person designing an experiment from the responsibility of designing it carefully simply because randomization will be used. In our example, a serious effort should be made to prepare the swatches as equally soiled as possible.

Finally, we must point out that randomization should be used even when all extraneous factors are carefully controlled. In our example, even if careful steps have been taken to control the amount of India ink with which the swatches are soiled, the wash temperature, the water hardness, and so on, assigning the swatches to the detergents should still be randomized.

# 15.3 ONE-WAY ANALYSIS OF VARIANCE

An **analysis of variance** expresses a measure of the total variation in a set of data as a sum of terms, each attributed to a specific source, or cause, of variation. Here, we shall describe this with reference to the calcium contamination example and, as we shall see, *the approach is different, but otherwise we accomplish exactly what we accomplished in Section 15.1.* In that example there were two such sources of variation: (1) the differences in location along the Mississippi River, and (2) chance, which in problems of this kind is called the **experimental error**. When there is only one source of variation other than chance, we refer to the analysis as a **one-way analysis of variance**. Other versions of the analysis of variance will be treated later in this chapter.

As a measure of the total variation of $kn$ observations consisting of $k$ samples of size $n$, we shall use the **total sum of squares**[†]

$$SST = \sum_{i=1}^{k} \sum_{j=1}^{n} (x_{ij} - \bar{x}..)^2$$

where $x_{ij}$ is the $j$th observation of the $i$th sample ($i = 1, 2, \ldots, k$ and $j = 1, 2, \ldots, n$), and $\bar{x}..$ is the **grand mean**, the mean of all the $kn$ measurements or observations. Note that if we divide the total sum of squares $SST$ by $kn - 1$, we get the variance of the combined data.

If we let $\bar{x}_i.$ denote the mean of the $i$th sample (for $i = 1, 2, \ldots, k$), we can now write the following identity that forms the basis for a one-way analysis of variance:[‡]

**IDENTITY FOR ONE-WAY ANALYSIS OF VARIANCE**

$$SST = n \cdot \sum_{i=1}^{k} (\bar{x}_i. - \bar{x}..)^2 + \sum_{i=1}^{k} \sum_{j=1}^{n} (x_{ij} - \bar{x}_i.)^2$$

It is customary to refer to the first term on the right, the quantity that measures the variation among the sample means, as the **treatment sum of squares** $SS(Tr)$, and to the second term, which measures the variation within the individual samples, as the **error sum of squares** $SSE$. The choice of the word "treatment" is explained by the origin of many analysis-of-variance techniques in agricultural experiments where different fertilizers, for example, are regarded as different **treatments** applied to the soil. So, we shall refer to the three locations along the Mississippi River as three treatments, and in other problems we may refer to four different nationalities as four different treatments, five different advertising campaigns as five different treatments, and so on. The word "error" in "error sum of squares" pertains to the experimental error, or chance.

In this notation, the identity for a one-way analysis of variance reads

$$SST = SS(Tr) + SSE$$

and since its proof requires quite a bit of algebraic manipulation, let us merely verify it numerically for the example of Section 15.1. Substituting the original

---

[†] The use of double subscripts and double summations is treated briefly in Section 3.8.
[‡] This identity may be derived by writing the total sum of squares as

$$SST = \sum_{i=1}^{k} \sum_{j=1}^{n} (x_{ij} - \bar{x}..)^2 = \sum_{i=1}^{k} \sum_{j=1}^{n} [(\bar{x}_i. - \bar{x}..) + (x_{ij} - \bar{x}_i.)]^2$$

and then expanding the squares $[(\bar{x}_i. - \bar{x}..) + (x_{ij} - \bar{x}_i.)]^2$ by means of the binomial theorem and simplifying algebraically.

data, the three sample means, and the grand mean (see pages 358 and 359) into the formulas for the three sums of squares, we get

$$
\begin{aligned}
SST = {} & (42 - 37)^2 + (37 - 37)^2 + (41 - 37)^2 + (39 - 37)^2 \\
& + (43 - 37)^2 + (41 - 37)^2 + (37 - 37)^2 + (40 - 37)^2 \\
& + (39 - 37)^2 + (38 - 37)^2 + (41 - 37)^2 + (39 - 37)^2 \\
& + (32 - 37)^2 + (28 - 37)^2 + (34 - 37)^2 + (32 - 37)^2 \\
& + (30 - 37)^2 + (33 - 37)^2 \\
= {} & 336
\end{aligned}
$$

$$
\begin{aligned}
SS(Tr) & = 6[(40.5 - 37)^2 + (39.0 - 37)^2 + (31.5 - 37)^2] \\
& = 279
\end{aligned}
$$

$$
\begin{aligned}
SSE = {} & (42 - 40.5)^2 + (37 - 40.5)^2 + (41 - 40.5)^2 + (39 - 40.5)^2 \\
& + (43 - 40.5)^2 + (41 - 40.5)^2 + (37 - 39.0)^2 + (40 - 39.0)^2 \\
& + (39 - 39.0)^2 + (38 - 39.0)^2 + (41 - 39.0)^2 + (39 - 39.0)^2 \\
& + (32 - 31.5)^2 + (28 - 31.5)^2 + (34 - 31.5)^2 + (32 - 31.5)^2 \\
& + (30 - 31.5)^2 + (33 - 31.5)^2 \\
= {} & 57
\end{aligned}
$$

and it can be seen that

$$
SS(Tr) + SSE = 279 + 57 = 336 = SST
$$

Although this may not be apparent immediately, what we have done here is very similar to what we did in Section 15.1. Indeed, $SS(Tr)$ divided by $k - 1$ equals the quantity that we denoted by $n \cdot s_{\bar{x}}^2$ and put into the numerator of the $F$ statistic on page 360. Called the **treatment mean square**, it measures the variation among the sample means and it is denoted by $MS(Tr)$. Thus,

$$
MS(Tr) = \frac{SS(Tr)}{k - 1}
$$

and for the calcium contamination example we get $MS(Tr) = \frac{279}{2} = 139.5$. This equals the value we got for $n \cdot s_{\bar{x}}^2$ on page 359.

Similarly, $SSE$ divided by $k(n - 1)$ equals the mean of the $k$ sample variances, $\frac{1}{3}(s_1^2 + s_2^2 + s_3^2)$ in our example, which we put into the denominator of the $F$ statistic on page 360. Called the **error mean square**, it measures the variation within the samples and it is denoted by $MSE$. Thus,

$$
MSE = \frac{SSE}{k(n - 1)}
$$

and for the calcium contamination example we get $MSE = \frac{57}{3(6-1)} = 3.8$. This equals the value we got for $\frac{1}{3}(s_1^2 + s_2^2 + s_3^2)$ on page 360.

Since $F$ was defined on page 360 as the ratio of these two measures of the variation among the sample means and within the samples, we can now write

**STATISTIC FOR TEST CONCERNING DIFFERENCES AMONG MEANS**

$$F = \frac{MS(Tr)}{MSE}$$

In practice, we display the work required for the determination of $F$ in the following kind of table, called an **analysis of variance table**:

| Source of variation | Degrees of freedom | Sum of squares | Mean square | $F$ |
|---|---|---|---|---|
| Treatments | $k - 1$ | $SS(Tr)$ | $MS(Tr)$ | $\dfrac{MS(Tr)}{MSE}$ |
| Error | $k(n - 1)$ | $SSE$ | $MSE$ | |
| Total | $kn - 1$ | $SST$ | | |

The degrees of freedom for treatments and error are the numerator and denominator degrees of freedom referred to on page 360. Note that they equal the quantities we divide into the sums of squares to obtain the corresponding mean squares.

After we have calculated $F$, we proceed as in Section 15.1. Assuming again that the data constitute samples from normal populations with the same standard deviation, we reject the null hypothesis

$$\mu_1 = \mu_2 = \cdots = \mu_k$$

against the alternative hypothesis that these $\mu$'s are not all equal, or the null hypothesis

$$\alpha_1 = \alpha_2 = \cdots = \alpha_k = 0$$

against the alternative hypothesis that these treatment effects are not all equal to zero, if the value of $F$ is greater than or equal to $F_\alpha$ for $k - 1$ and $k(n - 1)$ degrees of freedom.

**EXAMPLE 15.1**  Use the sums of squares calculated on page 365 to construct an analysis of variance table for the calcium contamination example, and test at the 0.01 level of significance whether the differences among the means obtained for the three locations along the Mississippi River are significant.

**Solution**  Since $k = 3, n = 6, SST = 336, SS(Tr) = 279$, and $SSE = 57$, we get $k - 1 = 2$, $k(n - 1) = 15$, $MS(Tr) = \frac{279}{2} = 139.5$, $MSE = \frac{57}{15} = 3.8$, and $F = \frac{139.5}{3.8} = 36.71$ rounded to two decimals. All these results are summarized in the following table:

| Source of variation | Degrees of freedom | Sum of squares | Mean square | F |
|---|---|---|---|---|
| Treatments | 2 | 279 | 139.5 | 36.71 |
| Error | 15 | 57 | 3.8 | |
| Total | 17 | 336 | | |

Note that the total degrees of freedom, $kn - 1$, is the sum of the degrees of freedom for treatments and error.

Finally, since $F = 36.71$ exceeds 6.36, the value of $F_{0.01}$ for 2 and 15 degrees of freedom, we conclude as in Section 15.1 that the null hypothesis must be rejected.

The numbers that we used in the example dealing with the calcium contamination of Mississippi River water were intentionally rounded so that the calculations would be easy. In actual practice, the calculation of the sums of squares can be quite tedious unless we use the following computing formulas in which $T_i.$ denotes the sum of the values for the $i$th treatment (that is, the sum of the values in the $i$th sample), and $T..$ denotes the **grand total** of all the data:

COMPUTING FORMULAS FOR SUMS OF SQUARES (EQUAL SAMPLE SIZES)

$$SST = \sum_{i=1}^{k} \sum_{j=1}^{n} x_{ij}^2 - \frac{1}{kn} \cdot T_{..}^2$$

$$SS(Tr) = \frac{1}{n} \cdot \sum_{i=1}^{k} T_{i.}^2 - \frac{1}{kn} \cdot T_{..}^2$$

and by subtraction

$$SSE = SST - SS(Tr)$$

**EXAMPLE 15.2** Use these computing formulas to verify the sums of squares obtained on page 365.

**Solution** First calculating the various totals, we get

$$T_1. = 42 + 37 + 41 + 39 + 43 + 41 = 243$$
$$T_2. = 37 + 40 + 39 + 38 + 41 + 39 = 234$$
$$T_3. = 32 + 28 + 34 + 32 + 30 + 33 = 189$$
$$T.. = 243 + 234 + 189 = 666$$

and

$$\sum\sum x^2 = 42^2 + 37^2 + 41^2 + \cdots + 32^2 + 30^2 + 33^2$$
$$= 24{,}978$$

Then, substituting these totals together with $k = 3$ and $n = 6$ into the formulas for the sums of squares, we get

$$SST = 24{,}978 - \frac{1}{18}(666)^2$$

$$= 24{,}978 - 24{,}642$$

$$= 336$$

$$SS(Tr) = \frac{1}{6}(243^2 + 234^2 + 189^2) - 24{,}642$$

$$= 24{,}921 - 24{,}642$$

$$= 279$$

and

$$SSE = 336 - 279 = 57$$

As can be seen, these results are identical with the ones obtained before.

A MINITAB printout of the analysis of variance of Example 15.2 is shown in Figure 15.2. Besides the degrees of freedom, the sums of squares, the mean squares, the value of $F$ and the corresponding $p$-value, it provides some further information of no relevance here that has been deleted. The $p$-value is given as 0.000 rounded to three decimals; our HEWLETT PACKARD STAT/MATH calculator gave it as 0.0000017 rounded to seven decimals.

**Figure 15.2**
Analysis of variance of the calcium contamination data.

**One-way ANOVA: M, N, O**

```
Source   DF      SS       MS       F       P
Factor    2   279.00   139.50   36.71   0.000
Error    15    57.00     3.80
Total    17   336.00

S = 1.949   R-Sq = 83.04%   R-Sq(adj) = 80.77%


                             Individual 95% CIs For Mean Based
on
                             Pooled StDev
Level  N    Mean   StDev   -----+---------+---------+--------
-+----
M      6   40.500  2.168                                 (----*--
--)
N      6   39.000  1.414                              (---*----)
O      6   31.500  2.168   (----*----)
                           -----+---------+---------+--------

-+----
                             31.5       35.0       38.5
42.0

Pooled StDev = 1.949
```

The method discussed so far in this section applies only when the sample sizes are all equal, but minor modifications make it applicable also when the sample sizes are not all equal. If the $i$th sample is of size $n_i$, the computing formulas become

<table>
<tr><td>COMPUTING FORMULAS FOR SUMS OF SQUARES (UNEQUAL SAMPLE SIZES)</td><td>

$$SST = \sum_{i=1}^{k} \sum_{j=1}^{n_i} x_{ij}^2 - \frac{1}{N} \cdot T_{..}^2$$

$$SS(Tr) = \sum_{i=1}^{k} \frac{T_{i.}^2}{n_i} - \frac{1}{N} \cdot T_{..}^2$$

$$SSE = SST - SS(Tr)$$

</td></tr>
</table>

where $N = n_1 + n_2 + \cdots + n_k$. The only other change is that the total number of degrees of freedom is $N - 1$, and the degrees of freedom for treatments and error are $k - 1$ and $N - k$.

**EXAMPLE 15.3**   A laboratory technician wants to compare the breaking strength of three kinds of thread and originally he had planned to repeat each determination six times. Not having enough time, however, he has to base his analysis on the following results (in ounces):

**Thread 1:**  18.0   16.4   15.7   19.6   16.5   18.2
**Thread 2:**  21.1   17.8   18.6   20.8   17.9   19.0
**Thread 3:**  16.5   17.8   16.1

Assuming that these data constitute random samples from three normal populations with the same standard deviation, perform an analysis of variance to test at the 0.05 level of significance whether the differences among the sample means are significant.

**Solution**   1.  $H_0:$    $\mu_1 = \mu_2 = \mu_3$
$H_A:$    The $\mu$'s are not all equal.
2.  $\alpha = 0.05$
3.  Reject the null hypothesis if $F \geq 3.89$, where $F$ is to be determined by an analysis of variance and 3.89 is the value of $F_{0.05}$ for $k - 1 = 3 - 1 = 2$ and $N - k = 15 - 3 = 12$ degrees of freedom; otherwise, accept the null hypothesis or reserve judgment.
4.  Substituting  $n_1 = 6, n_2 = 6, n_3 = 3, N = 15, T_{1.} = 104.4, T_{2.} = 115.2, T_{3.} = 50.4, T_{..} = 270.0,$ and $\sum \sum x^2 = 4{,}897.46$ into the computing formulas for the sums of squares, we get

$$SST = 4{,}897.46 - \frac{1}{15}(270.0)^2 = 37.46$$

$$SS(Tr) = \frac{104.4^2}{6} + \frac{115.2^2}{6} + \frac{50.4^2}{3} - \frac{1}{15}(270.0)^2$$

$$= 15.12$$

and

$$SSE = 37.46 - 15.12 = 22.34$$

Since the degrees of freedom are $k - 1 = 3 - 1 = 2$, $N - k = 15 - 3 = 12$, and $N - 1 = 14$, we then get

$$MS(Tr) = \frac{15.12}{2} = 7.56; \quad MSE = \frac{22.34}{12} = 1.86; \quad \text{and} \quad F = \frac{7.56}{1.86} = 4.06;$$

and all these results are summarized in the following analysis-of-variance table:

| Source of variation | Degrees of freedom | Sum of squares | Mean square | F |
|---|---|---|---|---|
| Treatments | 2 | 15.12 | 7.56 | 4.06 |
| Error | 12 | 22.34 | 1.86 | |
| Total | 14 | 37.46 | | |

5. Since $F = 4.06$ exceeds 3.89, the null hypothesis must be rejected; in other words, we conclude that there is a difference in the strength of the three kinds of thread. ■

If the level of significance had not been specified in this example, we could have noted that $F = 4.06$ falls between 3.89 and 6.93, the values of $F_{0.05}$ and $F_{0.01}$ for 2 and 12 degrees of freedom, and we could simply have stated for the $p$-value that $0.01 < p < 0.05$. Or, using the same calculator as in Example 15.2, we would have found that the $p$-value is 0.0450 rounded to four decimals.

# 15.4 MULTIPLE COMPARISONS

An analysis of variance provides a method for determining whether the differences among $k$ sample means are significant. It does not, however, tell us which means are significantly different from which others. Consider, for example, the following data on the amounts of time (in minutes) it took a certain person to drive to work on five days, selected at random, along each of four different routes:

**Route 1:**   25   26   25   25   28
**Route 2:**   27   27   28   26   26
**Route 3:**   28   29   33   30   30
**Route 4:**   28   29   27   30   27

**Figure 15.3**
Analysis of variance of the travel-time data reproduced from the display screen of a TI-83 graphing calculator.

First we shall have to see whether the differences among the four sample means 25.8, 26.8, 30.0, and 28.2, are significant. To this end we performed an analysis of variance with the use of a graphing calculator, with the results shown in Figure 15.3. Since the display screen of the TI-83 graphing calculator is fairly small, only part of the results are shown immediately, as in the upper part of Figure 15.3. The rest of the results are obtained by scrolling down, and they are shown in the lower part of Figure 15.3. Thus, we find that $F = 8.74$ rounded to two decimals and that the corresponding $p$-value for the $F$ distribution with 3 and 16 degrees of freedom is 0.001 rounded to three decimals. This means that the differences among the four sample means are significant, most certainly at the 0.01 level of significance.

While the route with the lowest average driving time, Route 1, certainly appears to be faster than Route 3, we are not so sure whether we can declare Route 1 to be faster also than Route 2. It is true that 25.8 is less than 26.8, but at this time we have no idea whether the difference between these two means is significant. Of course, we could perform an ordinary two-sample $t$ test to compare these two routes, but there are altogether $\binom{4}{2} = 6$ possible pairs, and if we perform that many tests there is a good chance that we may commit at least one Type I error. [If the $t$ tests are performed at the 0.05 level of significance, the probability is $1 - (0.95)^6$ or about 0.26 of committing at least one Type I error.]

To control the Type I error probabilities when conducting comparisons like these, an area of study called **multiple comparisons** has been developed in fairly recent years. This is a complicated topic that is widely misunderstood, and there are a few issues that have not been clarified even by the experts. Here we present one such method, based on what is called the **studentized range**.[†]

---

[†] **Studentizing** is the process of dividing a statistic by a statistically independent estimate of scale. The expression is derived from the *nom de plume* of W. S. Gosset, who first introduced the process in 1907 by discussing the distribution of the mean divided by the sample standard deviation.

As we shall explain it here, this test applies only when the samples are all of the same size. The books referred to on page 395 also explain how to handle the case where the sample sizes are not all equal, and they discuss various alternative multiple comparisons tests, named after the statisticians who contributed most to their development.

The studentized range procedure is designed to control the overall probability of making at least one Type I error when comparing the various pairs of means. It is based on the argument that the difference between the means of two treatments (say, treatments $i$ and $j$) is significant if

$$|\bar{x}_i - \bar{x}_j| \geq \frac{q_\alpha}{\sqrt{n}} \cdot s$$

where $s$ is the square root of $MSE$ in the analysis of variance, $\alpha$ is the overall level of significance, and $q_\alpha$ is obtained from Table IX for the given values of $k$ (the number of treatments in the analysis of variance) and df (the number of degrees of freedom for error in the analysis of variance table).

When using the studentized range technique (or, for that matter, any one of the many multiple comparison tests), we begin by arranging the treatments according to the size of their means, ranked from low to high. For our driving-time example, we thus get

| Route 1 | Route 2 | Route 4 | Route 3 |
|---------|---------|---------|---------|
| 25.8 | 26.8 | 28.2 | 30.0 |

Then we calculate the least significant range, $\frac{q_\alpha}{\sqrt{n}} \cdot s$, for the studentized range technique. Since $n = 5$ for our example, $s = \sqrt{1.9}$ according to Figure 15.3, or 1.38 rounded to two decimals, and $q_{0.05} = 4.05$ for $k = 4$ and df $= 16$ in Table IX, we get

$$\frac{q_\alpha}{\sqrt{n}} \cdot s = \frac{4.05}{\sqrt{5}} \cdot 1.38 = 2.50$$

rounded to two decimals.

Calculating the absolute values of the differences between the means of all possible pairs of routes, we get 1.0 for Routes 1 and 2, 2.4 for Routes 1 and 4, 4.2 for Routes 1 and 3, 1.4 for Routes 2 and 4, 3.2 for Routes 2 and 3, and 1.8 for Routes 4 and 3. As can be seen, only those for Routes 1 and 3 and those for Routes 2 and 3 exceed 2.50 and, hence, are significant. To summarize all this information we draw a line under all sets of treatments for which the differences between the means are not significant, and for our example we thus get

| Route 1 | Route 2 | Route 4 | Route 3 |
|---------|---------|---------|---------|
| 25.8 | 26.8 | 28.2 | 30.0 |

Being interested in minimizing the driving time, this tells us that Routes 1, 2, and 4 *as a group* are preferable to Route 3, and that Routes 3 and 4 *as a group* are less desirable than the other two. To go further than that, we may have to consider some other factors; perhaps, the beauty of the scenery along the way.

**15.6**  An experiment is performed to determine which of three golf ball brands, *A*, *B*, or *C*, will attain the greatest distance when driven from a tee. Criticize the experiment if
(a) one golf pro hits all the brand *A* balls, another hits all the brand *B* balls, and a third hits all the brand *C* balls;
(b) all brand *A* balls are hit first, brand *B* balls next, and brand *C* balls last.

**15.7**  A botanist wants to compare three kinds of tulip bulbs having, respectively, red, white, and yellow flowers. She has four bulbs of each kind and plants them in a flower bed in the following pattern, where R, W, and Y denote the three colors:

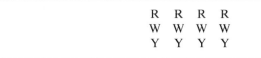

```
R   R   R   R
W   W   W   W
Y   Y   Y   Y
```

When the plants reach maturity, she measures their height and performs an analysis of variance. Criticize this experiment and indicate how it might be improved.

**15.8**  To compare three weight-reducing diets, five of 15 persons are assigned randomly to each of the diets. After they have been on these diets for two weeks, a one-way analysis of variance is performed on their weight losses to test the null hypothesis that the three diets are equally effective. It has been claimed that this procedure cannot yield a valid conclusion because the five persons who originally weighed the most might be assigned to the same diet. Verify that the probability of this happening by chance is about 0.001.

**15.9**  With reference to the preceding exercise, suppose that five of the 15 persons are assigned randomly to each of the three diets, but it is discovered subsequently that the five persons who originally weighed the most are all assigned to the same diet. Should the one-way analysis of variance still be performed?

**15.10**  Rework part (b) of Exercise 15.1 by performing an analysis of variance, using the computing formulas to obtain the necessary sums of squares. Compare the values of *F* obtained here and in part (a) of Exercise 15.1.

**15.11**  Use appropriate computer software or a graphing calculator to rework Exercise 15.1.

**15.12**  Rework part (b) of Exercise 15.4 by performing an analysis of variance, using the computing formulas to obtain the necessary sums of squares. Compare the values of *F* obtained here and in part (a) of Exercise 15.4.

**15.13**  Use appropriate computer software or a graphing calculator to rework Exercise 15.4.

**15.14**  The following are the numbers of mistakes made on five occasions by four word-processor operators, setting a technical report:

| | | | | | |
|---|---|---|---|---|---|
| **Operator 1:** | 10 | 13 | 9 | 11 | 12 |
| **Operator 2:** | 11 | 13 | 8 | 16 | 12 |
| **Operator 3:** | 10 | 15 | 13 | 11 | 15 |
| **Operator 4:** | 15 | 7 | 11 | 12 | 9 |

Assuming that the necessary assumptions can be met, perform an analysis of variance and decide at the 0.05 level of significance whether the differences among the four sample means can be attributed to chance.

**15.15** The following data show the yields of soybeans (in bushels per acre) planted two inches apart on essentially similar plots with the rows 20, 24, 28, and 32 inches apart:

| 20 in. | 24 in. | 28 in. | 32 in. |
|--------|--------|--------|--------|
| 23.1 | 21.7 | 21.9 | 19.8 |
| 22.8 | 23.0 | 21.3 | 20.4 |
| 23.2 | 22.4 | 21.6 | 19.3 |
| 23.4 | 21.1 | 20.2 | 18.5 |
| 23.6 | 21.9 | 21.6 | 19.1 |
| 21.7 | 23.4 | 23.8 | 21.9 |

Assuming that these data constitute random samples from four normal populations with the same standard deviation, perform an analysis of variance to test at the 0.01 level of significance whether the differences among the four sample means can be attributed to chance.

**15.16** Use appropriate computer software or a graphing calculator to rework Exercise 15.15.

**15.17** A large marketing firm uses many photocopy machines, several of each of four different models. During the last six months, the office manager has tabulated for each machine the average number of minutes per week that it is out of service due to repairs, resulting in the following data:

| | | | | | | |
|---|---|---|---|---|---|---|
| **Model G:** | 56 | 61 | 68 | 42 | 82 | 70 |
| **Model H:** | 74 | 77 | 92 | 63 | 54 | |
| **Model K:** | 25 | 36 | 29 | 56 | 44 | 48 | 38 |
| **Model M:** | 78 | 105 | 89 | 112 | 61 | |

Assuming that the necessary assumptions can be met, perform an analysis of variance to decide whether the differences among the means of the four samples can be attributed to chance. Use $\alpha = 0.01$. (*Hint:* Make use of the results that the totals for the four samples are 379, 360, 276, and 445, that the grand total is 1,460, and that $\sum \sum x^2 = 104,500$.)

**15.18** When used with three different lubricants, a specific group of machine parts show the following weight losses (in milligrams) due to friction:

| | | | | | | | | | | |
|---|---|---|---|---|---|---|---|---|---|---|
| **Lubricant X:** | 10 | 13 | 12 | 10 | 14 | 8 | 12 | 13 | | |
| **Lubricant Y:** | 9 | 8 | 12 | 9 | 8 | 11 | 7 | 6 | 8 | 11 | 9 |
| **Lubricant Z:** | 6 | 7 | 7 | 5 | 9 | 8 | 4 | 10 | | |

Assuming that these data constitute random samples from three normal populations with the same standard deviation, perform an analysis of variance to decide whether the differences among the three sample means can be attributed to chance. Use the 0.01 level of significance.

**15.19** To study its performance, a newly designed motorboat was timed over a marked course under various wind and water conditions. Assuming that the necessary conditions can be met, use the following data (in minutes) to test at the 0.05 level of significance whether the difference among the three sample means is significant:

| Calm conditions: | 26 | 19 | 16 | 22 | | |
|---|---|---|---|---|---|---|
| Moderate conditions: | 25 | 27 | 25 | 20 | 18 | 23 |
| Choppy conditions: | 23 | 25 | 28 | 31 | 26 | |

**15.20** Use appropriate computer software or a graphing calculator to rework
 (a) Exercise 15.18;
 (b) Exercise 15.19.

**15.21** The following values are the percentages of the previous year's fruit yield for apple trees managed under eight different spraying schedules.

| Schedule | | | | | | |
|---|---|---|---|---|---|---|
| A | 130 | 98 | 128 | 106 | 139 | 121 |
| B | 142 | 133 | 122 | 131 | 132 | 141 |
| C | 114 | 141 | 95 | 123 | 118 | 140 |
| D | 77 | 99 | 84 | 76 | 70 | 75 |
| E | 109 | 86 | 113 | 101 | 103 | 112 |
| F | 148 | 143 | 111 | 142 | 131 | 100 |
| G | 149 | 129 | 134 | 108 | 119 | 126 |
| H | 92 | 129 | 111 | 103 | 107 | 125 |

Assuming that the necessary assumptions can be met, use appropriate computer software to conduct an analysis of variance with $\alpha = 0.05$.

**15.22** In Example 15.1 we did an analysis of variance for the data given originally on page 358, where the means for the three locations along the Mississippi River were 40.5, 39.0, and 31.5. Use the studentized range method to perform a multiple comparisons test at the 0.01 level of significance, and discuss the results assuming that low calcium contamination is desirable.

**15.23** As a continuation of Exercise 15.15, use the studentized range method to perform a multiple comparisons test at the 0.01 level of significance and interpret the results.

**15.24** As a continuation of Exercise 15.21, use the studentized range method to perform a multiple comparisons test at the 0.05 level of significance and interpret the results.

**15.25** An analysis of variance and a subsequent multiple comparisons test of the performance of four real estate persons yielded the following results:

Mr. Brown   Ms. Jones   Mr. Black   Mrs. Smith

where Mrs. Smith had the highest average sales. Interpret the results.

**15.26** An analysis of variance and a subsequent multiple comparisons test of the fat content of five frozen dinners yielded the following results:

A   C   B   F   D   E

where A has the most fat and E has the least. Interpret these results, given that these foods are on a list of recommendations for a low-fat diet.

**15.27** With reference to the footnote on page 358, verify that the sum of the treatment effects, the $\alpha$'s, is equal to zero.

**15.28** Verify symbolically that for a one-way analysis of variance
 (a) $\dfrac{SS(Tr)}{k-1} = n \cdot s_{\bar{x}}^2$;
 (b) $\dfrac{SSE}{k(n-1)} = \dfrac{1}{k} \cdot \sum\limits_{i=1}^{k} s_i^2$, where $s_i^2$ is the variance of the $i$th sample.

# 15.5 THE DESIGN OF EXPERIMENTS: BLOCKING

To introduce another concept that is of importance in the design of experiments, suppose that a reading comprehension test is given to random samples of eighth graders from each of four schools, and that the results are

**School A:** 87   70   92
**School B:** 43   75   56
**School C:** 70   66   50
**School D:** 67   85   79

The means of these four samples are 83, 58, 62, and 77, and since the differences among them are very large, it would seem reasonable to conclude that there are some real differences in the average reading comprehension of eighth graders in the four schools. This does not follow, however, from a one-way analysis of variance. We get

| Source of variation | Degrees of freedom | Sum of squares | Mean square | F |
|---|---|---|---|---|
| Treatments | 3 | 1,278 | 426 | 2.90 |
| Error | 8 | 1,176 | 147 | |
| Total | 11 | 2,454 | | |

and since $F = 2.90$ is less than 4.07, the value of $F_{0.05}$ for 3 and 8 degrees of freedom, the null hypothesis (that the population means are all equal) cannot be rejected at the 0.05 level of significance.

The reason for this is that there are not only considerable differences among the four means, but also very large differences among the values within the samples. In the first sample they range from 70 to 92, in the second sample from 43 to 75, in the third sample from 50 to 70, and in the fourth sample from 67 to 85. Giving this some thought, it would seem reasonable to conclude that these differences within the samples may well be due to differences in ability, an extraneous factor (we might call it a "nuisance" factor) that was randomized by taking a random sample of eighth graders from each school. Thus, variations due to differences in ability were included in the experimental error; this "inflated" the error sum of squares that went into the denominator of the $F$ statistic, and the results were not significant.

To avoid this kind of situation, we could hold the extraneous factor fixed, but this will seldom give us the information we want. In our example, we could limit the study to eighth graders with a scholastic grade-point average (GPA) of 90 or above, but then the results would apply only to eighth graders with a GPA of 90 or above. Another possibility is to vary the known source of variability (the extraneous factor) deliberately over as wide a range as necessary, and to do it in such a way that the variability it causes can be measured and, hence, eliminated from the experimental error. This means that we should plan the experiment in such a way that we can perform a **two-way analysis of variance**, in which the total variability of the data is partitioned into three components attributed, respectively, to treatments (in our example, the four schools), the extraneous factor, and experimental error.

As we shall see later, this can be accomplished in our example by randomly selecting from each school one eighth grader with a low GPA, one eighth grader with a typical GPA, and one eighth grader with a high GPA, where "low," "typical," and "high" are presumably defined in a rigorous way. Suppose, then, that we proceed in this way and get the results shown in the following table:

|          | Low GPA | Typical GPA | High GPA |
|----------|---------|-------------|----------|
| School A | 71      | 92          | 89       |
| School B | 44      | 51          | 85       |
| School C | 50      | 64          | 72       |
| School D | 67      | 81          | 86       |

What we have done here is called **blocking**, and the three levels of GPA are called **blocks**. In general, blocks are the levels at which we hold an extraneous factor fixed, so that we can measure its contribution to the total variability of the data by means of a two-way analysis of variance. In the scheme we chose for our example, we are dealing with **complete blocks**—they are complete in the sense that each treatment appears the same number of times in each block. There is one eighth grader from each school in each block.

Suppose, furthermore, that the order in which the students are tested may have some effect on the results. If the order is randomized within each block (that is, for each level of GPA), we refer to the design of the experiment as a **randomized block design**.

# 15.6 TWO-WAY ANALYSIS OF VARIANCE

The analysis of experiments where blocking is used to reduce the error sum of squares requires a **two-way analysis of variance**. In this kind of analysis the variables under consideration are referred to as "treatments" and "blocks" even though it applies also to **two-factor experiments**, where both variables are of material concern.

Before we go into any details, let us point out that there are essentially two ways of analyzing such two-variable experiments, and they depend on whether the two variables are independent, or whether there is an **interaction**. Suppose, for instance, that a tire manufacturer is experimenting with different kinds of treads, and he finds that one kind is especially good for use on dirt roads while another kind is especially good for use on hard pavement. If this is the case, we say that there is an interaction between road conditions and tread design. On the other hand, if each of the treads is affected equally by the different road conditions, we would say that there is no interaction and that the two variables (road conditions and tread design) are independent. The latter case will be taken up first in Section 15.7, while a method that is suitable also for testing for interactions will be described in Section 15.9.

# $15.7$ TWO-WAY ANALYSIS OF VARIANCE WITHOUT INTERACTION

To formulate the hypotheses to be tested in the two-variable case, let us write $\mu_{ij}$ for the population mean that corresponds to the $i$th treatment and the $j$th block. In our earlier example, $\mu_{ij}$ is the average reading comprehension score in the $i$th school for eighth graders with grade point average level $j$. We express this as

$$\mu_{ij} = \mu + \alpha_i + \beta_j$$

As in the footnote to page 358, $\mu$ is the grand mean (the average of all the population means $\mu_{ij}$), and the $\alpha_i$ are the treatment effects (whose sum is zero). Correspondingly, we refer to the $\beta_j$ as the **block effects** (whose sum is also zero), and write the two null hypotheses we want to test as

$$\alpha_1 = \alpha_2 = \cdots = \alpha_k = 0 \quad \text{and} \quad \beta_1 = \beta_2 = \cdots = \beta_n = 0$$

The alternative to the first null hypothesis (which in our illustration amounts to the hypothesis that the average reading comprehension of eighth graders is the same in all four schools) is that the treatment effects $\alpha_i$ are not all zero; the alternative to the second null hypothesis (which in our illustration amounts to the hypothesis that the average reading comprehension of eighth graders is the same for all three levels of GPA) is that the block effects $\beta_j$ are not all zero.

To test the second of the null hypotheses, we need a quantity, similar to the treatment sum of squares, that measures the variation among the block means (58, 72, and 83 for the data on page 377). So, if we let $T_{\cdot j}$ denote the total of all the values in the $j$th block, substitute it for $T_{i\cdot}$ in the computing formula for $SS(Tr)$ on page 367, sum on $j$ instead of $i$, and interchange $n$ and $k$, we obtain, analogous to $SS(Tr)$ the **block sum of squares**.

COMPUTING
FORMULA FOR
BLOCK SUM OF
SQUARES

$$SSB = \frac{1}{k} \cdot \sum_{j=1}^{n} T_{\cdot j}^2 - \frac{1}{kn} \cdot T_{\cdot \cdot}^2$$

In a two-way analysis of variance (with no interaction) we compute $SST$ and $SS(Tr)$ according to the formulas on page 369, $SSB$ according to the formula immediately above, and then we get $SSE$ by subtraction. Since

$$SST = SS(Tr) + SSB + SSE$$

we have

**ERROR SUM OF SQUARES (TWO-WAY ANALYSIS OF VARIANCE)**

$$SSE = SST - [SS(Tr) + SSB]$$

Observe that the error sum of squares for a two-way analysis of variance does not equal the error sum of squares for a one-way analysis of variance performed on the same data, even though we denote both with the symbol $SSE$. In fact, we are now partitioning the error sum of squares for the one-way analysis of variance into two terms: the block sum of squares, $SSB$, and the remainder that is the new error sum of squares, $SSE$.

We can now construct the following analysis-of-variance table for a two-way analysis of variance (with no interaction):

| Source of variation | Degrees of freedom | Sum of squares | Mean square | F |
|---|---|---|---|---|
| Treatments | $k-1$ | $SS(Tr)$ | $MS(Tr) = \dfrac{SS(Tr)}{k-1}$ | $\dfrac{MS(Tr)}{MSE}$ |
| Blocks | $n-1$ | $SSB$ | $MSB = \dfrac{SSB}{n-1}$ | $\dfrac{MSB}{MSE}$ |
| Error | $(k-1)(n-1)$ | $SSE$ | $MSE = \dfrac{SSE}{(k-1)(n-1)}$ | |
| Total | $kn-1$ | $SST$ | | |

The mean squares are again the sums of squares divided by their respective degrees of freedom, and the two $F$ values are the mean squares for treatments and blocks divided by the mean square for error. Also, the degrees of freedom for blocks are $n-1$ (like those for treatments with $n$ substituted for $k$), and the degrees of freedom for error are found by subtracting the degrees of freedom for treatments and blocks from $kn-1$, the total number of degrees of freedom:

$$(kn-1) - (k-1) - (n-1) = kn - k - n + 1$$
$$= (k-1)(n-1)$$

Thus, in the significance test for treatments the numerator and denominator degrees of freedom for $F$ are $k-1$ and $(k-1)(n-1)$, and in the significance test for blocks the numerator and denominator degrees of freedom for $F$ are $n-1$ and $(k-1)(n-1)$.

**EXAMPLE 15.4**   In the example that we used to illustrate the need for blocking, we gave the following data to compare the reading comprehension scores of eighth graders in four different schools using low, typical, and high grade-point averages as blocks:

|          | Low GPA | Typical GPA | High GPA |
|----------|---------|-------------|----------|
| School A | 71      | 92          | 89       |
| School B | 44      | 51          | 85       |
| School C | 50      | 64          | 72       |
| School D | 67      | 81          | 86       |

Assuming that the data consist of independent random samples from normal populations all having the same standard deviation, test at the 0.05 level of significance whether the differences among the means obtained for the four schools (treatments) are significant, and also whether the differences among the means obtained for the three levels of GPA (blocks) are significant.

**Solution**

1. $H_0$'s:   $\alpha_1 = \alpha_2 = \alpha_3 = \alpha_4 = 0$
   $\beta_1 = \beta_2 = \beta_3 = 0$
   $H_A$'s:   The treatment effects are not all equal to zero; the block effects are not all equal to zero.

2. $\alpha = 0.05$ for both tests.

3. For treatments, reject the null hypothesis if $F \geq 4.76$, where $F$ is to be determined by a two-way analysis of variance and 4.76 is the value of $F_{0.05}$ for $k - 1 = 4 - 1 = 3$ and $(k - 1)(n - 1) = (4 - 1)(3 - 1) = 6$ degrees of freedom. For blocks, reject the null hypothesis if $F \geq 5.14$, where $F$ is to be determined by a two-way analysis of variance and 5.14 is the value of $F_{0.05}$ for $n - 1 = 3 - 1 = 2$ and $(k - 1)(n - 1) = (4 - 1)(3 - 1) = 6$ degrees of freedom. If either null hypothesis cannot be rejected, accept it or reserve judgment.

4. Substituting   $k = 4, n = 3, T_{1.} = 252, T_{2.} = 180, T_{3.} = 186, T_{4.} = 234, \quad T_{.1} = 232, T_{.2} = 288, T_{.3} = 332, T_{..} = 852$, and $\sum \sum x^2 = 63,414$ into the computing formulas for the sums of squares, we get

$$SST = 63,414 - \frac{1}{12}(852)^2$$
$$= 63,414 - 60,492$$
$$= 2,922$$

$$SS(Tr) = \frac{1}{3}(252^2 + 180^2 + 186^2 + 234^2) - 60,492$$
$$= 1,260$$

$$SSB = \frac{1}{4}(232^2 + 288^2 + 332^2) - 60,492$$
$$= 1,256$$

and

$$SSE = 2,922 - (1,260 + 1,256)$$
$$= 406$$

Since the degrees of freedom are $k - 1 = 4 - 1 = 3, n - 1 = 3 - 1 = 2,$ $(k - 1)(n - 1) = (4 - 1)(3 - 1) = 6,$ and $kn - 1 = 4 \cdot 3 - 1 = 11,$ we then get $MS(Tr) = \frac{1,260}{3} = 420, MSB = \frac{1,256}{2} = 628, MSE = \frac{406}{6} \approx 67.67, F \approx \frac{420}{67.67} \approx$ 6.21 for treatments, and $F \approx \frac{628}{67.67} \approx 9.28$ for blocks. All these results are summarized in the following analysis-of-variance table:

| Source of variation | Degrees of freedom | Sum of squares | Mean square | F |
|---|---|---|---|---|
| Treatments | 3 | 1,260 | 420 | 6.21 |
| Blocks | 2 | 1,256 | 628 | 9.28 |
| Error | 6 | 406 | 67.67 | |
| Total | 11 | 2,922 | | |

5. Since $F = 6.21$ exceeds 4.76, the null hypothesis for treatments must be rejected; and since $F = 9.28$ exceeds 5.14, the null hypothesis for blocks must be rejected. In other words, we conclude that the average reading comprehension of eighth graders at the four schools is not the same, and that the average reading comprehension of eighth graders at the three levels of GPA is not the same. Observe that by blocking we got significant differences in the average reading comprehension of eighth graders at the four schools, whereas without blocking we did not. ■

When we double checked the calculations in our solution of Example 15.4, we found that the most recent version of MINITAB uses the *p*-value approach. As can be seen from Figure 15.4, the columns headed DF (degrees of freedom), SS (sums of squares), MS (mean squares), and *F* are the same as before; but Figure 15.4 also shows a column of *p*-values, which directly lead to decisions about the hypotheses. For treatments, the *p*-value is 0.029, and since 0.029 is less than 0.05, the null hypothesis for treatments must be rejected. For blocks, the *p*-value is 0.015, and since 0.015 is less than 0.05, the null hypothesis for blocks must be rejected.

As we pointed out earlier, a two-way analysis of variance can also be used in the analysis of a two-factor experiment, where both variables (factors) are of material concern. It could be used, for example, in the analysis of the following data collected in an experiment designed to test whether the range of a missile

**Figure 15.4**
Computer printout for Example 15.4.

```
Two-Way ANOVA: C3 versus C1, C2

Analysis of Variance for C3
Source    DF      SS       MS      F        P
C1         3   1260.0    420.0   6.21    0.029
C2         2   1256.0    628.0   9.28    0.015
Error      6    406.0     67.7
Total     11   2922.0
```

flight (in miles) is affected by differences among launchers and also by differences among fuels.

|  | Fuel 1 | Fuel 2 | Fuel 3 | Fuel 4 |
|---|---|---|---|---|
| Launcher X | 45.9 | 57.6 | 52.2 | 41.7 |
| Launcher Y | 46.0 | 51.0 | 50.1 | 38.8 |
| Launcher Z | 45.7 | 56.9 | 55.3 | 48.1 |

Note that we used a different format for this table to distinguish between two-factor experiments with extraneous factors randomized over the entire experiment, and experiments with extraneous factors randomized separately over each block.

Also, when a two-way analysis of variance is used in this way, we usually call the two variables **factor A** and **factor B** (instead of treatments and blocks and write $SSA$ and $MSA$ instead of $SS(Tr)$ and $MS(Tr)$; we still use $SSB$ and $MSB$ but now the B stands for factor B instead of blocks.

# 15.8 THE DESIGN OF EXPERIMENTS: REPLICATION

In Section 15.5 we showed how we can increase the amount of information to be gained from an experiment by blocking, that is, by eliminating the effect of an extraneous factor. Another way to increase the amount of information to be gained from an experiment is to increase the volume of the data. For instance, in the example on page 376, we might increase the size of the samples and give the reading comprehension test to 20 eighth graders from each school instead of three. For more complicated designs, the same thing can be accomplished by executing the entire experiment more than once, and this is called **replication**. With reference to the example on page 376, we might conduct the experiment (select and test 12 eighth graders) in one week, and then replicate (repeat) the entire experiment in the next week.

Conceptually, replication does not present any difficulties, but computationally it does, and we mentioned it here only because it is required for the work in Section 15.9. Furthermore, if an experiment requiring a two-way analysis of variance is replicated, it may then require a three-way analysis of variance, since replication, itself, could be a source of variation in the data. This would be the case, for instance, in our example dealing with the reading comprehension scores, if it got very hot and humid during the second week, thus making it difficult for the students to concentrate.

# 15.9 TWO-WAY ANALYSIS OF VARIANCE WITH INTERACTION

When we first mentioned the concept of interaction, we cited an experiment where a tire manufacturer discovers that one kind of tread is especially good for use on dirt roads while another kind of tread is especially good for use on hard pavement. A similar situation arises when a farmer finds that one variety of corn does best with one kind of fertilizer while another variety does best with a different kind of fertilizer; or when it is observed that one person makes the fewest mistakes with one word processor while another person makes the fewest mistakes with a different word processor.

To consider a numerical example, let us refer to the two-factor experiment on page 382, the one that dealt with the effect of three different launchers and four different fuels on the range of certain missiles. If we analyzed these data by the method of Section 15.7, we would partition SST, a measure of the total variation among the data, into three components that are attributed, respectively, to the different launchers, the different fuels, and error (or chance). If there are interactions, which is quite possible, the variations they cause would be concealed because they are included as part of SSE, the error sum of squares. To isolate a sum of squares that can be attributed to interaction, we need some other way of measuring chance variation, and we shall do this by repeating the entire experiment. Suppose, then, that this yields the data shown in the following table:

|  | Fuel 1 | Fuel 2 | Fuel 3 | Fuel 4 |
|---|---|---|---|---|
| Launcher X | 46.1 | 55.9 | 52.6 | 44.3 |
| Launcher Y | 46.3 | 52.1 | 51.4 | 39.6 |
| Launcher Z | 45.8 | 57.9 | 56.2 | 47.6 |

which we call Replicate 2 to distinguish it from the data on page 382 which we now call Replicate 1. Combining the two replications in one table, we get

|  | Fuel 1 | Fuel 2 | Fuel 3 | Fuel 4 |
|---|---|---|---|---|
| Launcher X | 45.9, 46.1 | 57.6, 55.9 | 52.2, 52.6 | 41.7, 44.3 |
| Launcher Y | 46.0, 46.3 | 51.0, 52.1 | 50.1, 51.4 | 38.8, 39.6 |
| Launcher Z | 45.7, 45.8 | 56.9, 57.9 | 55.3, 56.2 | 48.1, 47.6 |

where the first value in each cell comes from Replicate 1 and the second value comes from Replicate 2.

Now we can represent chance variation by the variation *within* the 12 cells of the table, and in general our new error sum of squares becomes

$$SSE = \sum_{i=1}^{k} \sum_{j=1}^{n} \sum_{h=1}^{r} (x_{ijh} - \bar{x}_{ij\cdot})^2$$

where $x_{ijh}$ is the value corresponding to the $i$th treatment, the $j$th block, and the $h$th replicate, and $\bar{x}_{ij\cdot}$ is the mean of the values in the cell corresponding to the $i$th treatment and the $j$th block.

Replacing the two values in each cell of the table immediately preceding by their mean, we get

|  | Fuel 1 | Fuel 2 | Fuel 3 | Fuel 4 |
|---|---|---|---|---|
| Launcher X | 46 | 56.75 | 52.4 | 43 |
| Launcher Y | 46.15 | 51.55 | 50.75 | 39.2 |
| Launcher Z | 45.75 | 57.40 | 55.75 | 47.85 |

and this is what our data would look like after chance variation has been removed. In other words, the only variation that is left is due to treatments, blocks, and interaction, and if we performed a two-way analysis of variance as in Section 15.7, we would get corresponding treatment, block, and **interaction sums of squares**, with the latter being what used to be the error sum of squares. Actually, all this is done only conceptually. If we actually performed a two-way analysis of variance with the means replacing the two values in each cell, we would find that each of the sums of squares is divided by a factor of 2. Correspondingly, if there had been $r$ replications, each sum of squares would have been divided by a factor of $r$.

As in Sections 15.3 and 15.7, there exist computing formulas for the various sums of squares in a two-way analysis of variance with interaction. However, since the necessary calculations are unwieldy to say the least, this work is just about always done with the use of a computer. This is precisely what we shall do here, getting the various degrees of freedom, sums of squares, mean squares, values of $F$, and $p$-values from the MINITAB printout shown in Figure 15.5.

**Figure 15.5**
Computer printout for a two-way analysis of variance with interaction.

```
Two-Way ANOVA: C3 versus C1, C2

Analysis of Variance for C3
Source          DF        SS        MS        F        P
C1               2    91.503    45.752    70.61    0.000
C2               3   570.825   190.275   293.67    0.000
Interaction      6    50.937     8.489    13.10    0.000
Error           12     7.775     0.648
Total           23   721.040
```

Actually, all we need are the $p$-values, given in the printout as 0.000, which means 0.000 rounded to three decimals. Since they are all less than 0.05, the null hypotheses for launchers, fuels, and launcher/fuel interactions must all be rejected. (The actual $p$-values, obtained by means of a HEWLETT PACKARD STAT/MATH calculator, are 0.00000023, 0.000000000017, and 0.0001.)

EXERCISES

**15.29** To compare the amounts of time that three television stations allot to commercials, a research worker measured the time devoted to commercials in random samples of 15 shows on each station. To her dismay, she discovered that there is so much variation within the samples—for one station the figures vary from 8 to 35 minutes—that it is virtually impossible to get significant results. Is there a way in which she might be able to overcome this obstacle?

**15.30** To compare five word processors, $A$, $B$, $C$, $D$, and $E$, four persons, 1, 2, 3, and 4, were timed in preparing a certain report on each of the machines. The results (in minutes) are shown in the following table:

|   | 1 | 2 | 3 | 4 |
|---|---|---|---|---|
| A | 49.1 | 48.2 | 52.3 | 57.0 |
| B | 47.5 | 40.9 | 44.6 | 49.5 |
| C | 76.2 | 46.8 | 50.1 | 55.3 |
| D | 50.7 | 43.4 | 47.0 | 52.6 |
| E | 55.8 | 48.3 | 82.6 | 57.8 |

Explain why these data should not be analyzed by the method of Section 15.7.

**15.31** The following are the cholesterol contents (in milligrams per package) that four laboratories obtained for 6-ounce packages of three very similar diet foods:

|   | Laboratory 1 | Laboratory 2 | Laboratory 3 | Laboratory 4 |
|---|---|---|---|---|
| Diet food A | 3.7 | 2.8 | 3.1 | 3.4 |
| Diet food B | 3.1 | 2.6 | 2.7 | 3.0 |
| Diet food C | 3.5 | 3.4 | 3.0 | 3.3 |

Perform a two-way analysis of variance, using the 0.01 level of significance for both tests.

**15.32** Four different, although supposedly equivalent, forms of a standardized achievement test in science were given to each of five students, and the following are the scores that they obtained:

|   | Student C | Student D | Student E | Student F | Student G |
|---|---|---|---|---|---|
| Form 1 | 77 | 62 | 52 | 66 | 68 |
| Form 2 | 85 | 63 | 49 | 65 | 76 |
| Form 3 | 81 | 65 | 46 | 64 | 79 |
| Form 4 | 88 | 72 | 55 | 60 | 66 |

Perform a two-way analysis of variance at the 0.01 level of significance for both tests.

**15.33** A laboratory technician measured the breaking strength of each of five kinds of linen threads by using four different measuring instruments, $I_1$, $I_2$, $I_3$, and $I_4$, and obtained the following results (in ounces):

|          | $I_1$ | $I_2$ | $I_3$ | $I_4$ |
|----------|-------|-------|-------|-------|
| Thread 1 | 20.9  | 20.4  | 19.9  | 21.9  |
| Thread 2 | 25.0  | 26.2  | 27.0  | 24.8  |
| Thread 3 | 25.5  | 23.1  | 21.5  | 24.4  |
| Thread 4 | 24.8  | 21.2  | 23.5  | 25.7  |
| Thread 5 | 19.6  | 21.2  | 22.1  | 21.1  |

Perform a two-way analysis of variance, using the 0.05 level of significance for both tests.

**15.34** The following are the numbers of defectives produced by four workmen operating, in turn, three different machines:

|         |       |  $B_1$ | $B_2$ | $B_3$ | $B_4$ |
|---------|-------|--------|-------|-------|-------|
|         | $A_1$ | 35     | 38    | 41    | 32    |
| Machine | $A_2$ | 31     | 40    | 38    | 31    |
|         | $A_3$ | 36     | 35    | 43    | 25    |

Workman ($B_1$–$B_4$ above)

Perform a two-way analysis of variance, using the 0.05 level of significance for both tests.

**15.35** In an experiment designed to evaluate three detergents, a laboratory ran three loads of washing at each combination of detergents and water temperatures and obtained the following whiteness readings:

|            | Detergent A  | Detergent B  | Detergent C  |
|------------|--------------|--------------|--------------|
| Cold Water | 45, 39, 46   | 43, 46, 41   | 55, 48, 53   |
| Warm Water | 37, 32, 43   | 40, 37, 46   | 56, 51, 53   |
| Hot Water  | 42, 42, 46   | 44, 45, 38   | 46, 49, 42   |

Use the 0.01 level of significance to test for differences among the detergents, differences due to water temperature, and differences due to interactions.

**15.36** A consumer-products-testing service wants to compare the quality of 24 cakes baked in its kitchen with each of four different mixes prepared according to three different recipes (varying the amounts of fresh ingredients added), once by Chef X and once by Chef Y. They ask a taste-tester to rate the cakes on a scale from 1 to 100, yielding the following results, where in each case the first figure pertains to the cake baked by Chef X and the second figure pertains to the cake baked by Chef Y:

|          | Mix A   | Mix B   | Mix C   | Mix D   |
|----------|---------|---------|---------|---------|
| Recipe 1 | 66, 62  | 70, 68  | 74, 68  | 73, 67  |
| Recipe 2 | 68, 61  | 71, 73  | 74, 70  | 66, 61  |
| Recipe 3 | 75, 68  | 69, 71  | 67, 63  | 70, 66  |

Use the 0.05 level of significance to test for differences due to the different recipes, differences due to the different mixes, and differences due to a recipe–mix interaction.

# 15.10 THE DESIGN OF EXPERIMENTS: FURTHER CONSIDERATIONS

In Section 15.5 we saw how blocking can be used to eliminate the variability due to one extraneous factor from the experimental error, and, in principle, several extraneous sources of variation can be handled in the same way. The only real problem is that this may inflate the size of an experiment beyond practical bounds. Suppose, for instance, that in the example dealing with the reading comprehension of eighth graders we would also like to eliminate whatever variability there may be due to differences in age (12, 13, or 14) and in sex. Allowing for all possible combinations of GPA, age, and sex, we will have to use $3 \cdot 3 \cdot 2 = 18$ different blocks, and if there is to be one eighth grader from each school in each block, we will have to select and test $18 \cdot 4 = 72$ eighth graders in all. If we also wanted to eliminate whatever variability there may be due to ethnic background, for which we might consider five categories, this would raise the required number of eighth graders to $72 \cdot 5 = 360$.

In this section we will show how problems like this can sometimes be resolved, at least in part, by planning experiments as **Latin squares**. At the same time, we also hope to impress upon the reader that it is through proper design that experiments can be made to yield a wealth of information. To give an example, suppose that a market research organization wants to compare four ways of packaging a breakfast food, but it is concerned about possible regional differences in the popularity of the breakfast food, and also about the effects of promoting the breakfast food in different ways. So, it decides to test market the different kinds of packaging in the northeastern, southeastern, northwestern, and southwestern parts of the United States and to promote them with discounts, lotteries, coupons, and two-for-one sales. Thus, there are $4 \cdot 4 = 16$ blocks (combinations of regions and methods of promotion) and it would take $16 \cdot 4 = 64$ market areas (cities) to promote each kind of packaging once within each block. Moreover, the test markets must be separated from each other so that the promotion methods do not interfere with each other, and the United States simply does not have 64 sufficiently widely separated test markets. It is of interest to note, however, that with proper planning 16 market areas (cities) will suffice. To illustrate, let us consider the following arrangement, called a Latin square, in which the letters $A$, $B$, $C$, and $D$ represent the four kinds of packaging:

|  | Discounts | Lotteries | Coupons | 2-for-1 sales |
|---|---|---|---|---|
| *Northeast* | A | B | C | D |
| *Southeast* | B | C | D | A |
| *Northwest* | C | D | A | B |
| *Southwest* | D | A | B | C |

In general, a Latin square is a square array of the letters $A, B, C, D, \ldots$, of the English (Latin) alphabet, which is such that each letter occurs once and only once in each row and in each column.

The preceding Latin square, looked upon as an experimental design, requires that discounts be used with packaging $A$ in a city in the Northeast, with packaging $B$ in a city in the Southeast, with packaging $C$ in a city in the Northwest, and with packaging $D$ in a city in the Southwest; that lotteries be used with packaging $B$ in a city in the Northeast, with packaging $C$ in a city in the Southeast, with packaging $D$ in a city in the Northwest, and with packaging $A$ in a city in the Southwest; and so on. Note that each kind of promotion is used once in each region and once with each kind of packaging; each kind of packaging is used once in each region and once with each kind of promotion; and each region is used once with each kind of packaging and once with each kind of promotion. As we shall see, this will enable us to perform an analysis of variance leading to significance tests for all three variables.

The analysis of an $r \times r$ Latin square is very similar to a two-way analysis of variance. The total sum of squares and the sums of squares for rows and columns are calculated in the same way in which we previously calculated $SST$, $SS(Tr)$, and $SSB$, but we must find an extra sum of squares that measures the variability due to the variable represented by the letters $A, B, C, D, \ldots$, namely, a new treatment sum of squares. The formula for this sum of squares is

**TREATMENT SUM OF SQUARES FOR LATIN SQUARE**

$$SS(Tr) = \frac{1}{r} \cdot (T_A^2 + T_B^2 + T_C^2 + \cdots) - \frac{1}{r^2} \cdot T^2$$

where $T_A$ is the total of the observations corresponding to treatment $A$, $T_B$ is the total of the observations corresponding to treatment $B$, and so forth. Finally, the error sum of squares is again obtained by subtraction:

**ERROR SUM OF SQUARES FOR LATIN SQUARE**

$$SSE = SST - [SSR + SSC + SS(Tr)]$$

where $SSR$ and $SSC$ are the sums of squares for rows and columns.

We can now construct an analysis-of-variance table for the analysis of an $r \times r$ Latin square. The mean squares are again the sums of squares divided by their respective degrees of freedom, and the three $F$-values are the mean squares for rows, columns, and treatments divided by the mean square for error. The degrees of freedom for rows, columns, and treatments are all $r - 1$, and, by subtraction, the degree of freedom for error is

$$(r^2 - 1) - (r - 1) - (r - 1) - (r - 1) = r^2 - 3r + 2 = (r - 1)(r - 2)$$

Thus, for each of the three significance tests the numerator and denominator degrees of freedom for $F$ are $r - 1$ and $(r - 1)(r - 2)$.

| Source of variation | Degrees of freedom | Sum of squares | Mean square | F |
|---|---|---|---|---|
| Rows | $r - 1$ | SSR | $MSR = \dfrac{SSR}{r - 1}$ | $\dfrac{MSR}{MSE}$ |
| Columns | $r - 1$ | SSC | $MSC = \dfrac{SSC}{r - 1}$ | $\dfrac{MSC}{MSE}$ |
| Treatments | $r - 1$ | SS(Tr) | $MS(Tr) = \dfrac{SS(Tr)}{r - 1}$ | $\dfrac{MS(Tr)}{MSE}$ |
| Error | $(r - 1)(r - 2)$ | SSE | $MSE = \dfrac{SSE}{(r - 1)(r - 2)}$ | |
| Total | $r^2 - 1$ | SST | | |

**EXAMPLE 15.5**  Suppose that in the breakfast-food study referred to in this section, the market research organization gets the data shown in the following table, where the figures are a week's sales in 10 thousands:

|  | Discounts | Lotteries | Coupons | 2-for-1 sales |
|---|---|---|---|---|
| Northeast | A 48 | B 38 | C 42 | D 53 |
| Southeast | B 39 | C 43 | D 50 | A 54 |
| Northwest | C 42 | D 50 | A 47 | B 44 |
| Southwest | D 46 | A 48 | B 46 | C 52 |

Assuming that the necessary assumptions can be met, analyze this Latin square with each of the tests performed at the 0.05 level of significance.

**Solution**

1. $H_0$'s: The row, column, and treatment effects (defined as footnote to page 358 and to page 388) are all equal to zero.
   $H_A$'s: The respective effects are not all equal to zero.
2. $\alpha = 0.05$ for each test.
3. For rows, columns, or treatments, reject the null hypothesis if $F \geq 4.76$, where the $F$'s are obtained by means of an analysis of variance, and 4.76 is the value of $F_{0.05}$ for $r - 1 = 4 - 1 = 3$ and $(r - 1)(r - 2) = (4 - 1)(4 - 2) = 6$ degrees of freedom.

4. Substituting $r = 4$, $T_1. = 181$, $T_2. = 186$, $T_3. = 183$, $T_4. = 192$, $T_{\cdot 1} = 175$, $T_{\cdot 2} = 179$, $T_{\cdot 3} = 185$, $T_{\cdot 4} = 203$, $T_A = 197$, $T_B = 167$, $T_C = 179$, $T_D = 199$, $T.. = 742$, and $\sum\sum x^2 = 34{,}756$ into the computing formulas for the sums of squares, we get

$$SST = 34{,}756 - \frac{1}{16}(742)^2 = 34{,}756 - 34{,}410.25 = 345.75$$

$$SSR = \frac{1}{4}(181^2 + 186^2 + 183^2 + 192^2) - 34{,}410.25 = 17.25$$

$$SSC = \frac{1}{4}(175^2 + 179^2 + 185^2 + 203^2) - 34{,}410.25 = 114.75$$

$$SS(Tr) = \frac{1}{4}(197^2 + 167^2 + 179^2 + 199^2) - 34{,}410.25 = 174.75$$

$$SSE = 345.75 - (17.25 + 114.75 + 174.75) = 39.00$$

The remainder of the work is shown in the following analysis-of-variance table:

| Source of variation | Degrees of freedom | Sum of squares | Mean square | F |
|---|---|---|---|---|
| Rows (regions) | 3 | 17.25 | $\frac{17.25}{3} = 5.75$ | $\frac{5.75}{6.5} \approx 0.88$ |
| Columns (promotion methods) | 3 | 114.75 | $\frac{114.75}{3} = 38.25$ | $\frac{38.25}{6.5} \approx 5.88$ |
| Treatments (packaging) | 3 | 174.75 | $\frac{174.75}{3} = 58.25$ | $\frac{58.25}{6.5} \approx 8.96$ |
| Error | 6 | 39.00 | $\frac{39.00}{6} = 6.5$ | |
| Total | 15 | 345.75 | | |

5. For rows, since $F = 0.88$ is less than 4.76, the null hypothesis cannot be rejected; for columns, since $F = 5.88$ exceeds 4.76, the null hypothesis must be rejected; for treatments, since $F = 8.96$ exceeds 4.76, the null hypothesis must be rejected. In other words, we conclude that differences in promotion and packaging, but not the different regions, affect the breakfast food's sales. ■

There are many other experimental designs besides the ones we have discussed, and they serve a great variety of special purposes. Widely used, for example, are the **incomplete block designs**, which apply when it is impossible to have each treatment in each block.

The need for such a design arises, for example, when we want to compare 13 kinds of tires but, of course, cannot put them all on a test car at the same time. Numbering the tires from 1 to 13, we might use the following experimental design:

| Test run | Tires | | | | Test run | Tires | | | |
|----------|----|----|----|----|----------|----|----|----|----|
| *1* | 1 | 2 | 4 | 10 | *8* | 8 | 9 | 11 | 4 |
| *2* | 2 | 3 | 5 | 11 | *9* | 9 | 10 | 12 | 5 |
| *3* | 3 | 4 | 6 | 12 | *10* | 10 | 11 | 13 | 6 |
| *4* | 4 | 5 | 7 | 13 | *11* | 11 | 12 | 1 | 7 |
| *5* | 5 | 6 | 8 | 1 | *12* | 12 | 13 | 2 | 8 |
| *6* | 6 | 7 | 9 | 2 | *13* | 13 | 1 | 3 | 9 |
| *7* | 7 | 8 | 10 | 3 | | | | | |

Here there are 13 test runs, or blocks, and since each kind of tire appears together with each other kind of tire once within the same block, the design is referred to as a **balanced incomplete block design**. The fact that each kind of tire appears together with each other kind of tire once within the same block is important; it facilitates the statistical analysis because it assures that we have the same amount of information for comparing each two kinds of tires. In general, the analysis of incomplete block designs is fairly complicated, and we shall not go into it here, as it has been our purpose only to demonstrate what can be accomplished by the careful design of an experiment.

EXERCISES

**15.37** An agronomist wants to compare the yield of 15 varieties of corn, and at the same time study the effects of four different fertilizers and three methods of irrigation. How many test plots must he plant if each variety of corn is to be grown in one test plot with each possible combination of fertilizers and methods of irrigation?

**15.38** A midwestern potato farmer wants to compare the yield of four varieties of potatoes, and at the same time study the effect of six different methods of fertilization and two methods of irrigation. How many test plots must he plant for a complete factorial experiment with only one observation of each kind?

**15.39** A pharmaceutical manufacturer wants to market a new cold remedy that is actually a combination of four medications, and wants to experiment first with two dosages for each medication. If $A_L$ and $A_H$ denote the low and high dosage of medication $A$, $B_L$ and $B_H$ the low and high dosage of medication $B$, $C_L$ and $C_H$ the low and high dosage of medication $C$, and $D_L$ and $D_H$ the low and high dosage of medication $D$, list the 16 preparations that must be tested if each dosage of each medication is to be used once in combination with each dosage of each of the other medications.

**15.40** Making use of the fact that each of the letters must occur once and only once in each row and each column, complete the following Latin squares:

**15.41** To compare four different brands of golf balls, $A$, $B$, $C$, and $D$, each kind was driven by each of four golf pros, $P_1$, $P_2$, $P_3$, and $P_4$, using once each four different drivers, $D_1$, $D_2$, $D_3$, and $D_4$. The distances from the tee to the points where the balls came to rest (in yards) were as shown in the following table:

|       | $D_1$ | $D_2$ | $D_3$ | $D_4$ |
|-------|-------|-------|-------|-------|
| $P_1$ | D 231 | B 215 | A 261 | C 199 |
| $P_2$ | C 234 | A 300 | B 280 | D 266 |
| $P_3$ | A 301 | C 208 | D 247 | B 255 |
| $P_4$ | B 253 | D 258 | C 210 | A 290 |

Assuming that the necessary assumptions can be met, use a computer to analyze this Latin square, using the 0.05 level of significance for each test.

**15.42** The sample data in the following $3 \times 3$ Latin square are the scores in an American history test obtained by nine college students of various ethnic backgrounds and of various professional interests, who were taught by instructors $A$, $B$, and $C$:

**Ethnic background**

|  | Hispanic | German | Polish |
|---|---|---|---|
| Law | A 75 | B 86 | C 69 |
| Medicine | B 95 | C 79 | A 86 |
| Engineering | C 70 | A 83 | B 93 |

Assuming that the necessary assumptions can be met, use a computer to analyze this Latin square, using the 0.05 level of significance for each test.

**15.43** Among nine persons interviewed in a poll, three are Easterners, three are Southerners, and three are Westerners. By profession, three of them are teachers, three are lawyers, and three are doctors, and no two of the same profession come from the same part of the United States. Also, three are Democrats, three are Republicans, and three are Independents, and no two of the same political affiliation are of the same profession or come from the same part of the United States. If one of the teachers is an Easterner and an Independent, another teacher is a Southerner and a Republican, and one of the lawyers is a Southerner and a Democrat, what is the political affiliation of the doctor who is a Westerner? (*Hint*: Construct a 3 × 3 Latin square. This exercise is a simplified version of a famous problem posed by R. A. Fisher in his classical work *The Design of Experiments*.)

**15.44** To test their ability to make decisions under pressure, the nine senior executives of a company are to be interviewed by each of four psychologists. As it takes a psychologist a full day to interview three of the executives, the schedule for the interviews is arranged as follows, where the nine executives are denoted by *A, B, C, D, E, F, G, H,* and *I*:

| Day | Psychologist | Executives | | |
|---|---|---|---|---|
| March 2 | I | B | C | ? |
| March 3 | I | E | F | G |
| March 4 | I | H | I | A |
| March 5 | II | C | ? | H |
| March 6 | II | B | F | A |
| March 9 | II | D | E | ? |
| March 10 | III | D | G | A |
| March 11 | III | C | F | ? |
| March 12 | III | B | E | H |
| March 13 | IV | B | ? | I |
| March 16 | IV | C | ? | A |
| March 17 | IV | D | F | H |

Replace the six question marks with the appropriate letters, given that each of the nine executives is to be interviewed together with each of the other executives once and only once on the same day. Note that this will make the arrangement a balanced incomplete block design, which may be important because each executive is tested together with each other executive once under identical conditions.

**15.45** A newspaper regularly prints the columns of seven writers but has room for only three in each edition. Complete the following schedule, in which the writers are numbered 1–7, so that each writer's column appears three times per week, and a column of each writer appears together with a column of each other writer once per week.

| Day | Writers | | |
|-----------|---|---|---|
| Monday | 1 | 2 | 3 |
| Tuesday | 4 | | |
| Wednesday | 1 | 4 | 5 |
| Thursday | 2 | | |
| Friday | 1 | 6 | 7 |
| Saturday | 5 | | |
| Sunday | 2 | 4 | 6 |

# CHECKLIST OF KEY TERMS (with page references to their definitions)

Analysis of variance, 357, 363
Analysis of variance table, 366
ANOVA, 357
Balanced incomplete block design, 391
Block effects, 378
Block sum of squares, 378
Blocking, 358, 377
Blocks, 377
Complete blocks, 377
Completely randomized design, 363
Controlled experiment, 362
Denominator degrees of freedom, 360
Error mean square, 365
Error sum of squares, 364
Experimental design, 357
Experimental error, 363
Factors, 382
$F$ statistic, 360
$F$ distribution, 360
Grand mean, 358, 364
Grand total, 367

Incomplete block design, 390
Interaction, 377
Interaction sums of squares, 384
Latin square, 387
Multiple comparisons, 358, 371
Numerator degrees of freedom, 360
One-way analysis of variance, 358, 364
Randomization, 358, 363
Randomized block design, 377
Replication, 382
Studentizing, 371
Studentized range, 371
Total sum of squares, 364
Treatment effects, 358
Treatment mean square, 365
Treatment sum of squares, 364
Treatments, 364
Two-factor experiment, 377
Two-way analysis of variance, 377
Two-way experiments, 358
Variance ratio, 360

# REFERENCES

*The following are some of the many books that have been written on the subject of analysis of variance:*

GUENTHER, W. C., *Analysis of Variance*. Upper Saddle River, N.J.: Prentice Hall, Inc., 1964.

SNEDECOR, G. W., and COCHRAN, W. G., *Statistical Methods*, 6th ed. Ames: Iowa State University Press, 1973.

*Problems relating to the design of experiments are also treated in the preceding books and in*

ANDERSON, V. L., and MCLEAN, R. A., *Design of Experiments: A Realistic Approach*. New York: Marcel Dekker. Inc., 1974.

BOX, G. E. P., HUNTER, W. G., and HUNTER, J. S., *Statistics for Experimenters*. New York: John Wiley & Sons, Inc., 1978.

COCHRAN, W. G., and COX, G. M., *Experimental Design*, 2nd ed. New York: John Wiley & Sons, Inc., 1957.

FINNEY, D. J., *An Introduction to the Theory of Experimental Design*. Chicago: University of Chicago Press, 1960.

FLEISS, J., *The Design and Analysis of Clinical Experiments*, New York: John Wiley & Sons, Inc., 1986.

HICKS, C. R., *Fundamental Concepts in the Design of Experiments*, 2nd ed. New York: Holt, Rinehart and Winston, 1973.

ROMANO, A., *Applied Statistics for Science and Industry*. Boston: Allyn and Bacon, Inc., 1977.

*A table of Latin squares for $r = 3, 4, 5, \ldots,$ and 12 may be found in the aforementioned book by W. G. Cochran and G. M. Cox.*

*Informally, some questions of experimental design are discussed in Chapters 18 and 19 of*

BROOK, R. J., ARNOLD, G. C., HASSARD, T. H., and PRINGLE, R. M., eds., *The Fascination of Statistics*. New York: Marcel Dekker, Inc., 1986.

*The topic of multiple comparisons is treated in detail in*

FEDERER, W. T., *Experimental Design, Theory and Application*. New York: Macmillan Publishing Co., Inc., 1955.

HOCHBERG, Y., and TAMHANE, A., *Multiple Comparison Procedures*. New York: John Wiley & Sons, Inc., 1987.

# 16 REGRESSION

**16.1**  Curve Fitting 397

**16.2**  The Method of Least Squares 399

**16.3**  Regression Analysis 410

\***16.4**  Multiple Regression 418

\***16.5**  Nonlinear Regression 422

Checklist of Key Terms 430

References 430

n many statistical investigations, the main goal is to establish relationships that make it possible to predict one or more variables in terms of other variables which are known. For instance, studies are made to predict the future sales of a product in terms of its price, a person's weight loss in terms of the number of weeks he or she has been on an 800-calories-per-day diet, family expenditures on medical care in terms of family income, the per capita consumption of certain food items in terms of their nutritional value and the amount of money spent advertising them on television, and so forth.

Of course, it would be ideal if we could predict one quantity exactly in terms of another, but this is seldom possible. In most instances we must be satisfied with predicting averages or expected values. For instance, we cannot predict exactly how much money a specific college graduate will earn ten years after graduation, but given suitable data we can predict the average earnings of all college graduates ten years after graduation. Similarly, we can predict the average yield of a variety of wheat in terms of the total rainfall in July, and we can predict the expected grade-point average of a student starting law school in terms of his or her IQ. This problem of predicting the average value of one variable in terms of the known value of another variable (or the known values of other variables) is called the problem of **regression**. This term dates back to Francis Galton (1822–1911), who used it first in 1877 in a study that showed that the heights of persons born to short and tall parents will tend to move back or "regress" toward the mean height of the population.

In Sections 16.1 and 16.2 we present a general introduction to curve fitting and the method that is most widely used, the **method of least squares**. Then, in Section 16.3, we discuss questions concerning inferences based on straight lines fit to paired data. Problems in which predictions are based on several variables and problems in which the relationship between two variables is not linear are treated in the two optional sections, Sections 16.4 and 16.5.

# **16.1** CURVE FITTING

Whenever possible, we try to express, or approximate, relationships between known quantities and quantities that are to be predicted in terms of mathematical equations. This has been very successful in the natural sciences, where it is known, for instance, that at a constant temperature the relationship between the volume, $y$, and the pressure, $x$, of a gas is given by the formula

$$y = \frac{k}{x}$$

where $k$ is a numerical constant. Also, it has been shown that the relationship between the size of a culture of bacteria, $y$, and the length of time, $x$, it has been exposed to certain environmental conditions is given by the formula

$$y = a \cdot b^x$$

where $a$ and $b$ are numerical constants. More recently, equations like these have also been used to describe relationships in the behavioral sciences, the social sciences, and other fields. For instance, the first of the preceding equations is often used in economics to describe the relationship between price and demand, and the second has been used to describe the growth of one's vocabulary or the accumulation of wealth.

Whenever we use observed data to arrive at a mathematical equation that describes the relationship between two variables—a procedure known as **curve fitting**—we must face three kinds of problems:

---

**We must decide what kind of curve, and hence what kind of "predicting" equation to use.**

**We must find the particular equation that is "best" in some sense.**

**We must investigate certain questions regarding the merits of the particular equation, and of predictions made from it.**

---

The second of these problems is discussed in some detail in Section 16.2, and the third in Section 16.3.

The first kind of problem is usually decided by direct inspection of the data. We plot the data on ordinary (arithmetic) graph paper, sometimes on special graph paper with special scales, and we decide by visual inspection upon the kind of curve (a straight line, a parabola, ...) that best describes the overall pattern of the data. There are methods by which this can be done more objectively, but they are fairly advanced and they will not be discussed in this book.

So far as our work here is concerned, we shall concentrate mainly on **linear equations** in two unknowns. They are of the form

$$y = a + bx$$

where $a$ is the $y$-intercept (the value of $y$ for $x = 0$) and $b$ is the slope of the line (namely, the change in $y$ which accompanies an increase of one unit in $x$).[†] Linear equations are useful and important not only because many relationships are actually of this form, but also because they often provide close approximations to relationships that would otherwise be difficult to describe in mathematical terms.

The term "linear equation" arises from the fact that the graph of $y = a + bx$ is a straight line. That is, all pairs of values of $x$ and $y$ that satisfy an equation of the form $y = a + bx$ constitute points that fall on a straight line. In practice, the values of $a$ and $b$ are estimated from observed data, and once they have been determined, we can substitute values of $x$ into the equation and calculate the corresponding predicted values of $y$.

To illustrate, suppose that we are given data on a midwestern county's production of wheat, $y$ (in bushels per acre), and its annual rainfall, $x$ (in inches measured from September through August), and that by the method of Section 16.2 we obtain the predicting equation

$$y = 0.23 + 4.42x$$

(see Exercise 16.8). The corresponding graph is shown in Figure 16.1, and it should be observed that for any pair of values of $x$ and $y$ that are such that $y = 0.23 + 4.42x$, we get a point $(x, y)$ that falls on the line. Substituting $x = 6$, for instance, we find that when there is an annual rainfall of 6 inches, we can expect a yield of

$$y = 0.23 + 4.42 \cdot 6 = 26.75$$

**Figure 16.1**
Graph of linear equation.

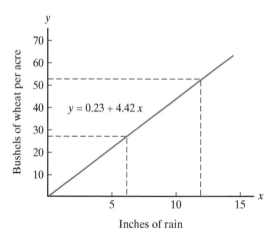

$y = 0.23 + 4.42\,x$

Bushels of wheat per acre

Inches of rain

---

[†]In other branches of mathematics, linear equations in two unknowns are often written as $y = mx + b$, but $y = a + bx$ has the advantage that it lends itself more easily to generalization—for instance, as in $y = a + bx + cx^2$ or as in $y = a + b_1x_1 + b_2x_2$.

bushels per acre; similarly, substituting $x = 12$, we find that when there is an annual rainfall of 12 inches, we can expect a yield of

$$y = 0.23 + 4.42 \cdot 12 = 53.27$$

bushels per acre. The points (6, 26.75) and (12, 53.27) lie on the straight line of Figure 16.1, and this is true for any other points obtained in the same way.

# 16.2 THE METHOD OF LEAST SQUARES

Once we have decided to fit a straight line to a given set of data, we face the second kind of problem, namely, that of finding the equation of the particular line which in some sense provides the best possible fit. To illustrate what is involved, let us consider the following sample data obtained in a study of the relationship between the length of time that a person has been exposed to a high level of noise and the sound frequency range to which his or her ears will respond. Here $x$ is the length of time (rounded to the nearest week) that a person has been living near a major airport directly in the flight path of departing jets, and $y$ is his or her hearing range (in thousands of cycles per second):

| Number of weeks $x$ | Hearing range $y$ |
|---|---|
| 47 | 15.1 |
| 56 | 14.1 |
| 116 | 13.2 |
| 178 | 12.7 |
| 19 | 14.6 |
| 75 | 13.8 |
| 160 | 11.9 |
| 31 | 14.8 |
| 12 | 15.3 |
| 164 | 12.6 |
| 43 | 14.7 |
| 74 | 14.0 |

In Figure 16.2, these 12 **data points**, $(x, y)$, are plotted in what is called a **scattergram**. We did this with the use of a computer, even though it would have been easy enough to do by hand. As can be seen, the points do not all fall on a straight line, but the overall pattern of the relationship is reasonably well described as being linear. At least, there is no noticeable departure from linearity, so we feel justified in deciding that a straight line is a suitable description of the underlying relationship.

We now face the problem of finding the equation of the line that in some sense provides the best fit to the data and that, it is hoped, will later yield the best possible predictions of $y$ from $x$. Logically speaking, there is no limit to the number of straight lines that can be drawn on a piece of graph paper. Some of these lines would fit the data so poorly that we could not consider them seriously, but many others would seem to provide more or less good fits, and the problem

**Figure 16.2**
Computer printout of
the hearing data.

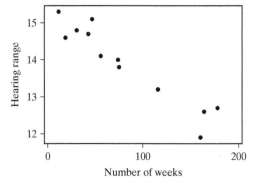

is to find the one line that fits the data best in some well-defined sense. If all the points actually fall on a straight line there is no problem, but this is an extreme case rarely encountered in practice. In general, we have to be satisfied with a line having certain desirable properties, short of perfection.

The criterion that, today, is used almost exclusively for defining a "best" fit dates back to the early part of the nineteenth century and the work of the French mathematician Adrien Legendre; it is known as the **method of least squares**. As it will be used here, this method requires that the line that we fit to our data be such that the sum of the squares of the vertical distances from the points to the line is a minimum.

To explain why this is done, let us refer to the following data, which might represent the numbers of correct answers, $x$ and $y$, that four students got on two parts of a multiple-choice test:

| $x$ | $y$ |
|-----|-----|
| 4   | 6   |
| 9   | 10  |
| 1   | 2   |
| 6   | 2   |

In Figure 16.3 we plotted the corresponding data points, and through them we drew two lines to describe the overall pattern.

If we use the horizontal line in the diagram below to "predict" $y$ for the given values of $x$, we would get $y = 5$ in each case, and the errors of these "predictions"

**Figure 16.3**
Two lines fit to the four
data points.

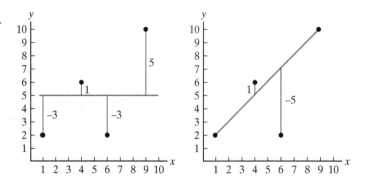

are $6 - 5 = 1, 10 - 5 = 5, 2 - 5 = -3$, and $2 - 5 = -3$. In Figure 16.3, they are the vertical deviations from the points to the line.

The sum of these errors is $1 + 5 + (-3) + (-3) = 0$, but this is no indication of their size, and we find ourselves in a position similar to that on page 76 that led to the definition of the standard deviation. Squaring the errors as we squared the deviations from the mean on page 76, we find that the sum of the squares of the errors is $1^2 + 5^2 + (-3)^2 + (-3)^2 = 44$.

Now let us consider the line in diagram 16.3 on the right, which was drawn so that it passes through the points $(1, 2)$ and $(9, 10)$; as can easily be verified, its equation is $y = 1 + x$. Judging by eye, this line seems to provide a much better fit than the horizontal line in the diagram on the left, and if we use it to "predict" $y$ for the given values of $x$, we would get $1 + 4 = 5, 1 + 9 = 10, 1 + 1 = 2$, and $1 + 6 = 7$. The errors of these "predictions," which in the figure on the right are also the vertical distances from the points to the line, are $6 - 5 = 1, 10 - 10 = 0, 2 - 2 = 0$, and $2 - 7 = -5$.

The sum of these errors is $1 + 0 + 0 + (-5) = -4$, which is numerically greater than the sum we obtained for the errors made with the other line of Figure 16.3, but this is of no consequence. The sum of the squares of the errors is now $1^2 + 0^2 + 0^2 + (-5)^2 = 26$, and this is much less than the 44 that we obtained before. In this sense, the line on the right provides a much better fit to the data than the horizontal line on the left.

We can even go one step further and ask for the equation of the line for which the sum of the squares of the errors (the sum of the squares of the vertical deviations from the points to the line) is a minimum. In Exercise 16.11 the reader will be asked to verify that the equation of such a line is $y = \frac{15}{17} + \frac{14}{17}x$ for our example. We refer to it as a **least-squares line**.

To show how the equation of such a line is actually obtained for a given set of **data points**, let us consider $n$ pairs of numbers $(x_1, y_1), (x_2, y_2), \ldots$, and $(x_n, y_n)$ that might represent the thrust and the speed of $n$ rockets, the height and weight of $n$ persons, the reading rate and the reading comprehension of $n$ students, the cost of salvaging $n$ ship wrecks and the values of the recovered treasure, or the age and the repair costs of $n$ automobiles. If we write the equation of the line as $\hat{y} = a + bx$, where the symbol $\hat{y}$ ($y$-hat) is used to distinguish between observed values of $y$ and the corresponding values $\hat{y}$ on the line, the least-squares criterion requires that we minimize the sum of the squares of the differences between the $y$'s and the $\hat{y}$'s (see Figure 16.4). This means that we must find the numerical values of the constants $a$ and $b$ appearing in the equation $\hat{y} = a + bx$ for which

$$\sum (y - \hat{y})^2 = \sum [y - (a + bx)]^2$$

is as small as possible. As it takes calculus or fairly tedious algebra to find the expressions for $a$ and $b$ that minimize $\sum (y - \hat{y})^2$, let us merely state the result that they are given by the solutions for $a$ and $b$ of the following system of two linear equations:

$$\sum y = na + b \left( \sum x \right)$$
$$\sum xy = a \left( \sum x \right) + b \left( \sum x^2 \right)$$

**Figure 16.4**
The difference between
$y$ and $\hat{y}$.

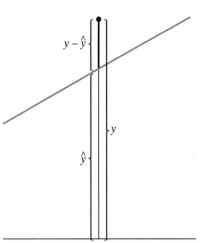

In these equations, called the **normal equations**, $n$ is the number of pairs of observations, $\sum x$ and $\sum y$ are the sums of the observed $x$'s and $y$'s, $\sum x^2$ is the sum of the squares of the $x$'s, and $\sum xy$ is the sum of the products obtained by multiplying each $x$ by the corresponding $y$.

**EXAMPLE 16.1**  Given that $n = 12$, $\sum x = 975$, $\sum x^2 = 117,397$, $\sum y = 166.8$, $\sum y^2 = 2,331.54$, and $\sum xy = 12,884.4$ for the hearing-range data on page 399, set up the normal equations for getting a least-squares line.

**Solution**  Substituting $n = 12$ and four of the five sums of squares into the expressions for the normal equations, we get

$$166.8 = 12a + 975b$$

$$12,884.4 = 975a + 117,397b$$

(Note that we did not need $\sum y^2$ for this example, but we gave it here together with the other summations for use in a future example.) ■

If the reader has had some experience solving systems of linear equations in elementary algebra, he or she can continue by solving these two equations for $a$ and $b$ using either the **method of elimination** or the method based on the use of **determinants**. Alternatively, one can solve the two normal equations symbolically for $a$ and $b$, and then substitute the value of $n$ and the required summations into the resulting formulas. Among the various ways in which we can write these formulas, most convenient, perhaps, is the format where we use as building blocks the quantities

$$S_{xx} = \sum x^2 - \frac{1}{n}\left(\sum x\right)^2 \quad \text{and} \quad S_{xy} = \sum xy - \frac{1}{n}\left(\sum x\right)\left(\sum y\right)$$

and then write the computing formulas for $a$ and $b$ as

SOLUTIONS
OF NORMAL
EQUATIONS

$$b = \frac{S_{xy}}{S_{xx}}$$

$$a = \frac{\sum y - b\left(\sum x\right)}{n}$$

where we first calculate $b$ and then substitute its value into the formula for $a$. (Note that this is the same $S_{xx}$ that we used on page 78 in the computing formula for the sample standard deviation.)

EXAMPLE **16.2**   Use these formulas for solving normal equations to find $a$ and $b$ for the hearing-range example.

**Solution**   First substituting $n = 12$ and the required summation given in Example 16.1 into the formulas for $S_{xx}$ and $S_{xy}$, we get

$$S_{xx} = 117{,}397 - \frac{1}{12}(975)^2 = 38{,}178.25$$

and

$$S_{xy} = 12{,}884.4 - \frac{1}{12}(975)(166.8) = -668.1$$

Then, $b = \dfrac{-668.1}{38{,}178.25} \approx -0.0175$ and $a = \dfrac{166.8 - (-0.0175)(975)}{12} \approx 15.3$, both rounded to three significant figures, and the equation of the least-squares line can be written as

$$\hat{y} = 15.3 - 0.0175x$$ ■

What we have done here may well be described as a mere exercise in arithmetic, for we seldom, if ever, go through all these details in determining a least-squares line. Nowadays, the required summations can be obtained even with the most primitive garden-variety kind of handheld calculator, and the values of $a$ and $b$ can be obtained with any kind of statistical technology. Indeed, the trickiest part of the whole operation is that of entering the data and, if necessary, making corrections, unless we are using a computer or a graphing calculator, where the data can be displayed and edited.

Observe also that when $b$ is negative, as it is in Example 16.2, the least-squares line has a *downward slope* going from left to right. In other words, the relationship between $x$ and $y$ is such that $y$ decreases when $x$ increases, as can be seen also from Figure 16.2. On the other hand, when $b$ is positive, this means that the least-squares line has an *upward slope* going from left to right, namely, that $y$ increases when $x$ increases. Finally, when $b$ equals zero, the least-squares line is horizontal and knowledge of $x$ is of no help in estimating, or predicting, a value of $y$.

EXAMPLE **16.3**   Use a graphing calculator to rework Example 16.2, without actually utilizing the summations given in Example 16.1.

**Figure 16.5**
Data for Example 16.3.

**Figure 16.6**
Solution of
Example 16.3.

**Solution**   Figure 16.5 shows the original data entered in a graphing calculator. The display screen is too small to show all the data, but the remainder of the data was obtained by scrolling. Then the command **STAT CALC 8** yields the results shown in Figure 16.6. Rounding to three significant figures, as before, we get $a = 15.3$ and $b = -0.0175$, and the equation of the least-squares line is, of course, the same

$$\hat{y} = 15.3 - 0.0175x$$

Had we used a computer for this example, MINITAB would have yielded the printout shown in Figure 16.7. The equation of the least-squares line, called here the **regression equation** (which will be explained later), is again $y = 15.3 - 0.0175x$, and the coefficients $a$ and $b$ are given in the column headed "Coef" as 15.3218 and $-0.017499$. Some of the additional details in this printout will be used later on.

Figure 16.7
MINITAB printout
Example 16.3.

```
Regression Analysis: y versus x

The regression equation is
y = 15.3 - 0.0175 x

Predictor        Coef        SE Coef         T         P
Constant       15.3218        0.1845       83.4     0.000
x             -0.017499       0.001865     -9.38     0.000

S = 0.3645          R-Sq = 89.8%    R-Sq(adj) = 88.8%

Analysis of Variance

Source           DF       SS         MS         F         P
Regression        1     11.691     11.691     88.00     0.000
Residual Error   10      1.329      0.133
Total            11     13.020
```

**EXAMPLE 16.4**   Use the least-squares equation obtained in Example 16.2 or in Example 16.3 to estimate the hearing range of a person who has been exposed to the airport noise (as described on page 399) for

  **(a)** one year;

  **(b)** two years.

**Solution**   **(a)** Substituting $x = 52$ into $\hat{y} = 15.3 - 0.0175x$, we get $\hat{y} = 15.3 - 0.0175(52)$ $= 14.4$ thousand cycles per second rounded to three significant figures.

  **(b)** Substituting $x = 104$ into this equation, we get $\hat{y} = 15.3 - 0.0175(104)$ $= 13.5$ thousand cycles per second rounded to three significant figures.

When we make an estimate like this, or a prediction, we cannot really expect that we will always hit the answer right on the nose. With reference to our example, it would be very unreasonable to expect that every person who has been exposed to airport noise for a given length of time will have exactly the same hearing range. To make meaningful estimates or predictions based on least-squares lines, we must look upon the values of $\hat{y}$ obtained by substituting given values of $x$ as averages, or expected values. Interpreted in this way, we refer to least-squares lines as **regression lines**, or better as **estimated regression lines**, since the values of $a$ and $b$ are estimates based on sample data and, hence, can be expected to vary from sample to sample. Questions relating to the goodness of such estimates will be discussed in Section 16.3.

In the discussion of this section we have considered only the problem of fitting a straight line to paired data. More generally, the method of least squares can also be used to fit other kinds of curves and to derive predicting equations in more than two unknowns. The problem of fitting curves other than straight lines by the method of least squares will be discussed briefly in Section 16.5 and some examples of predicting equations in more than two unknowns will be given in Section 16.4. Both of these sections are marked optional.

**16.1** A dog that had six hours of obedience training made five mistakes at a dog show, a dog that had twelve hours of obedience training made six mistakes, and a dog that had eighteen hours of obedience training made only one mistake. If we let $x$ denote the number of hours of obedience training and $y$ the number of mistakes, which of the two equations,

$$y = 10 - \frac{1}{2}x \quad \text{or} \quad y = 8 - \frac{1}{3}x$$

provides a better fit to the three data points, $(6, 5)$, $(12, 6)$, and $(18, 1)$, in the sense of least squares?

**16.2** With reference to Exercise 16.1, use a computer or a graphing calculator to check whether the line providing the better fit is a least-squares line.

**16.3** To see whether a widely used food preservative contributes to the hyperactivity of preschool children, a dietician chose a random sample of 10 four-year-olds known to be fairly hyperactive from various nursery schools and observed their behavior 45 minutes after they had eaten measured amounts of food containing the preservative. In the table that follows, $x$ is the amount of food consumed containing the preservative (in grams) and $y$ is a subjective rating of hyperactivity (on a scale from 1 to 20) based on a child's restlessness and interaction with other children:

| $x$ | $y$ |
| --- | --- |
| 36 | 6 |
| 82 | 14 |
| 45 | 5 |
| 49 | 13 |
| 21 | 5 |
| 24 | 8 |
| 58 | 14 |
| 73 | 11 |
| 85 | 18 |
| 52 | 6 |

(a) Draw a scattergram to judge whether a straight line might reasonably describe the overall pattern of the data.

(b) Use a ruler to draw a straight line that, judging by eye, should be fairly close to a least-squares line.

(c) Use the line drawn in part (b) to estimate the hyperactivity rating of such a child that had 65 grams of food with the preservative 45 minutes earlier.

**16.4** With reference to Exercise 16.3, use appropriate software or a graphing calculator to verify that the least-squares equation for estimating $y$ in terms of $x$ is $\hat{y} = 1.5 + 0.16x$ rounded to two significant figures. Also, use this equation to estimate the hyperactivity rating of such a child that had 65 grams of food with the preservative 45 minutes earlier and compare the result with that of part (c) of Exercise 16.3.

**16.5**  With reference to Exercise 16.3, where $\sum x = 525$, $\sum y = 100$, $\sum x^2 = 32,085$, and $\sum xy = 5,980$, set up the two normal equations and solve them by using either the method of elimination or that based on determinants.

**16.6**  The following table shows how many weeks six persons have worked at an automobile inspection station and the number of cars each one inspected between noon and 2 P.M. on a given day:

| Number of weeks employed $x$ | Number of cars inspected $y$ |
|:---:|:---:|
| 2 | 13 |
| 7 | 21 |
| 9 | 23 |
| 1 | 14 |
| 5 | 15 |
| 12 | 21 |

Given that $\sum x = 36$, $\sum y = 107$, $\sum x^2 = 304$, and $\sum xy = 721$, use the computing formulas on page 403 to find $a$ and $b$, and hence the equation of the least-squares line.

**16.7**  Use the result of Exercise 16.6 to estimate how many cars a person can be expected to inspect during the same two-hour period if he or she has been working at the inspection station for eight weeks.

**16.8**  Verify that the equation of the example on page 398 can be obtained by fitting a least-squares line to the following data:

| Rainfall (inches) | Yield of wheat (bushels per acre) |
|:---:|:---:|
| 12.9 | 62.5 |
| 7.2 | 28.7 |
| 11.3 | 52.2 |
| 18.6 | 80.6 |
| 8.8 | 41.6 |
| 10.3 | 44.5 |
| 15.9 | 71.3 |
| 13.1 | 54.4 |

**16.9**  The following data pertain to the chlorine residue in a swimming pool at various times after it has been treated with chemicals:

| Number of hours | Chlorine residue (parts per million) |
|:---:|:---:|
| 0 | 2.2 |
| 2 | 1.8 |
| 4 | 1.5 |
| 6 | 1.4 |
| 8 | 1.1 |
| 10 | 1.1 |
| 12 | 0.9 |

where the reading at 0 hour was taken immediately after the chemical treatment was completed.

(a) Use the computing formulas on page 403 to fit a least-squares line from which we can predict the chlorine residue in terms of the number of hours since the pool has been treated with chemicals.

(b) Use the equation of the least-squares line obtained in part (a) to estimate the chlorine residue in the pool five hours after it has been treated with chemicals.

(c) Suppose you discover that the data for this exercise were obtained on a very hot day. Explain why the results of parts (a) and (b) might be quite misleading.

**16.10** Use appropriate software or a graphing calculator to rework part (a) of Exercise 16.9.

**16.11** With reference to the four data points on page 400, which were $(4, 6)$, $(9, 10)$, $(1, 2)$, and $(6, 2)$, verify that the equation of the least squares line is

$$\hat{y} = \frac{15}{17} + \frac{14}{17}x$$

Also, calculate the sum of the squares of the vertical deviations from the four points to this line, and compare the result with 44 and 26, the corresponding sums of squares obtained for the two lines shown in Figure 16.3.

**16.12** Raw material used in the production of a synthetic fiber is stored in a place that has no humidity control. Measurements of the relative humidity in the storage place and the moisture content of a sample of the raw material (both in percentages) on 12 days yielded the following results:

| Humidity $x$ | Moisture content $y$ |
|---|---|
| 46 | 12 |
| 53 | 14 |
| 37 | 11 |
| 42 | 13 |
| 34 | 10 |
| 29 | 8 |
| 60 | 17 |
| 44 | 12 |
| 41 | 10 |
| 48 | 15 |
| 33 | 9 |
| 40 | 13 |

(a) Draw a scattergram to verify that a straight line pretty well describes the overall relationship between the two variables.

(b) Given that $\sum x = 507$, $\sum y = 144$, $\sum x^2 = 22{,}265$, and $\sum xy = 6{,}314$, set up the two normal equations.

(c) Solve the two normal equations, using either the method of elimination or the method based on determinants.

**16.13**  With reference to Exercise 16.12, use the summations given in part (b) and the computing formulas on page 403 to find the equation of the least-squares line.

**16.14**  Use appropriate software or a graphing calculator to find the equation of the least-squares line for the relative humidity and moisture content data of Exercise 16.12.

**16.15**  Use the equation obtained in Exercises 16.12, 16.13, or 16.14 to estimate the moisture content when the relative humidity is 38%.

**16.16**  Suppose that in Exercise 16.12 we had wanted to estimate what relative humidity will yield a moisture content of 10%. We could substitute $\hat{y} = 10$ into the equation obtained in either of Exercises 16.12, 16.13, or 16.14, and solve for $x$, but this would not provide an estimate in the least-squares sense. To obtain a least-squares estimate of the relative humidity in terms of moisture content, we would have to denote moisture content by $x$, humidity by $y$, and then fit a least-squares line to these data. Use appropriate software or a graphing calculator to obtain such a least-squares line and use it to estimate the relative humidity that will yield a moisture content of 10%.

**16.17**  When the $x$'s are equally spaced (that is, when the differences between successive values of $x$ are all equal), finding the equation of a least-squares line can be simplified a great deal by coding the $x$'s by assigning them the values $\ldots, -3, -2, -1, 0, 1, 2, 3, \ldots$ when $n$ is odd or $\ldots, -5, -3, -1, 1, 3, 5, \ldots$ when $n$ is even. With this kind of coding, the sum of the coded $x$'s, call them $u$'s, is zero, and the computing formulas for $a$ and $b$ on page 403 become

$$a = \frac{\sum y}{n} \quad \text{and} \quad b = \frac{\sum uy}{\sum u^2}$$

Of course, the equation of the resulting least-squares line expresses $y$ in terms of $u$, and we have to account for this when we use the equation to make estimates or predictions.

(a)  During its first five years of operation, a company's gross income from sales was 1.4, 2.1, 2.6, 3.5, and 3.7 million dollars. Fit a least-squares line and, assuming that the trend continues, predict the company's gross income from sales during its sixth year of operation.

(b)  At the end of eight successive years, a manufacturing company had 1.0, 1.7, 2.3, 3.1, 3.5, 3.4, 3.9, and 4.7 million dollars invested in plants and equipment. Fit a least-squares line and, assuming that the trend continues, predict the company's investment in plants and equipment at the end of the tenth year.

**\*16.18**  Verify that if we solve the normal equations *symbolically* by using determinants, we obtain the following alternative computing formulas for $a$ and $b$:

$$a = \frac{\left(\sum y\right)\left(\sum x^2\right) - \left(\sum x\right)\left(\sum xy\right)}{n\left(\sum x^2\right) - \left(\sum x\right)^2}$$

$$b = \frac{n\left(\sum xy\right) - \left(\sum x\right)\left(\sum y\right)}{n\left(\sum x^2\right) - \left(\sum x\right)^2}$$

**\*16.19**  Use the computing formulas of Exercise 16.18 to rework
(a)  Exercise 16.6;
(b)  Exercise 16.13.

# 16.3 REGRESSION ANALYSIS

In Example 16.4 we used a least-squares line to estimate, or predict, the hearing range of a person who has been exposed to the airport noise for two years as 13.5 thousand cycles per second. Even if we interpret the least-squares line correctly as a regression line (that is, treat estimates based on it as averages or expected values), there are questions that remain to be answered. For instance,

> **How good are the values we obtained for $a$ and $b$ in the least-squares equation $\hat{y} = 15.3 - 0.0175x$?**
>
> **How good an estimate is $\hat{y} = 13.5$ thousand cycles per second of the average hearing range of persons who have been exposed to the airport noise for two years?**

After all, $a = 15.3$ and $b = -0.0175$, as well as $\hat{y} = 13.5$, are only estimates based on sample data, and if we base our calculations on a different sample, the method of least squares would probably yield different values for $a$ and $b$, and a different value for $\hat{y}$ for $x = 104$. Also, for making predictions, we might ask

> **Can we give an interval for which we can assert with some degree of confidence that it will contain the hearing range of a person who will have been exposed to the airport noise for two years?**

With regard to the first of these questions, we said that $a = 15.3$ and $b = -0.0175$ are "only estimates based on sample data," and this implies the existence of corresponding true values, usually denoted by $\alpha$ and $\beta$ and referred to as the true **regression coefficients**. Accordingly, there is also a true regression line $\mu_{y|x} = \alpha + \beta x$, where $\mu_{y|x}$ is the true mean of $y$ for a given value of $x$. To distinguish between $a$ and $\alpha$ and $b$ and $\beta$, we refer to $a$ and $b$ as the **estimated regression coefficients**. They are often denoted by $\hat{\alpha}$ and $\hat{\beta}$ instead of $a$ and $b$.

To clarify the idea of a true regression line, let us consider Figure 16.8, where we have drawn the distributions of $y$ for several values of $x$. With reference to our numerical example, these curves are the distributions of the hearing range of persons who have been exposed to the airport noise for one, two, and three weeks, and to complete the picture we can visualize similar curves for all other values of $x$ within the range of values under consideration. Note that the means of all the distributions of Figure 16.8 lie on the true regression line $\mu_{y|x} = \alpha + \beta x$.

In **linear regression analysis** we assume that the $x$'s are constants, not values of random variables, and that for each value of $x$ the variable to be predicted, $y$, has a certain distribution (as pictured in Figure 16.8) whose mean is $\alpha + \beta x$. In **normal regression analysis** we assume, furthermore, that these distributions are all normal distributions with the same standard deviations $\sigma$.

Based on these assumptions, it can be shown that the estimated regression coefficients $a$ and $b$, obtained by the method of least squares, are values of

**Figure 16.8**
Distributions of $y$ for
given values of $x$.

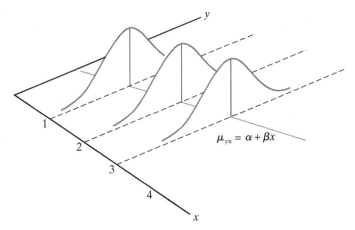

$$\mu_{y|x} = \alpha + \beta x$$

random variables having normal distributions with the means $\alpha$ and $\beta$ and the
standard deviations

$$\sigma\sqrt{\frac{1}{n} + \frac{\bar{x}^2}{S_{xx}}} \quad \text{and} \quad \frac{\sigma}{\sqrt{S_{xx}}}$$

The estimated regression coefficients $a$ and $b$ are, however, not statistically
independent. Note that both of these standard error formulas require that we
estimate $\sigma$, the common standard deviation of the normal distributions pictured
in Figure 16.8. Otherwise, since the $x$'s are assumed to be constants, there is no
problem in determining $\bar{x}$ and $S_{xx}$. The estimate of $\sigma$ we shall use here is called
the **standard error of estimate** and it is denoted by $s_e$. Its formula is

$$s_e = \sqrt{\frac{\sum(y - \hat{y})^2}{n - 2}}$$

where, again, the $y$'s are the observed values of $y$ and the $\hat{y}$'s are the corresponding
values on the least-squares line. Observe that $s_e^2$ is the sum of the squares of
the vertical deviations from the points to the line (namely, the quantity that we
minimized by the method of least squares) divided by $n - 2$.

The preceding formula defines $s_e$, but in practice we calculate its value by
means of the computing formula

**S**TANDARD ERROR
OF ESTIMATE

$$s_e = \sqrt{\frac{S_{yy} - bS_{xy}}{n - 2}}$$

where

$$S_{yy} = \sum y^2 - \frac{1}{n}\left(\sum y\right)^2$$

analogous to the formula for $S_{xx}$ on page 402.

**EXAMPLE  16.5**  Calculate $s_e$ for the least-squares line that we fit to the data on page 399.

**Solution**  Since $n = 12$ and we have already shown that $S_{xy} = -668.1$, the only other quantity needed is $S_{yy}$. Since we gave $\sum y = 166.8$ and $\sum y^2 = 2,331.54$ in Example 16.1, it follows that

$$S_{yy} = 2,331.54 - \frac{1}{12}(166.8)^2 = 13.02$$

and, hence, that

$$s_e = \sqrt{\frac{13.02 - (-0.0175)(-668.1)}{10}}$$

$$\approx 0.3645$$

Actually, all this work was not really necessary; the result is given in the computer printout of Figure 16.7, where it says $s = 0.3645$. Also, our graphing calculator could have yielded $s = 0.3644981554$, but we did not display this detail in Figure 16.6.

If we make all the assumptions of normal regression analysis, that the $x$'s are constants and the $y$'s are values of random variables having normal distributions with the means $\mu_{y|x} = \alpha + \beta x$ and the same standard deviation $\sigma$, inferences about the regression coefficients $\alpha$ and $\beta$ can be based on the statistics

**STATISTICS FOR INFERENCES ABOUT REGRESSION COEFFICIENTS**

$$t = \frac{a - \alpha}{s_e\sqrt{\dfrac{1}{n} + \dfrac{\bar{x}^2}{S_{xx}}}}$$

$$t = \frac{b - \beta}{s_e/\sqrt{S_{xx}}}$$

whose sampling distributions are $t$ distributions with $n - 2$ degrees of freedom. Note that the quantities in the denominators are estimates of the corresponding standard errors with $s_e$ substituted for $\sigma$.

The example that follows illustrates how we test hypotheses about either of the regression coefficients $\alpha$ and $\beta$.

**EXAMPLE  16.6**  Suppose that it has been claimed that a person's hearing range decreases by 0.02 thousand cycle per second for each week that a person has lived near the airport directly in the flight path of departing jets, and that the data on page 399 were obtained to test this claim at the 0.05 level of significance.

**Solution**  For what follows, it must be assumed that all the assumptions underlying a normal regression analysis are satisfied.
1. $H_0: \beta = -0.02$
   $H_A: \beta \neq -0.02$
2. $\alpha = 0.05$

3. Reject the null hypothesis if $t \leq -2.228$ or $t \geq 2.228$, where

$$t = \frac{b - \beta}{s_e / \sqrt{S_{xx}}}$$

and 2.228 is the value of $t_{0.025}$ for $12 - 2 = 10$ degrees of freedom; otherwise, accept the null hypothesis or reserve judgment.

4. Since we already know from Examples 16.1, 16.2, and 16.5 that $S_{xx} = 38,178.25$, $b = -0.0175$, and $s_e = 0.3645$, substitution of these values together with $\beta = -0.02$ yields

$$t = \frac{-0.0175 - (-0.02)}{0.3645 / \sqrt{38,178.25}} \approx 1.340$$

5. Since $t = 1.340$ falls on the interval from $-2.228$ to $2.228$, the null hypothesis cannot be rejected; there is no real evidence to refute the claim. ■

Again, we could have saved ourselves some work by referring to the computer printout of Figure 16.7. In the column headed SE Coef it shows that the estimated standard error of $b$, the quantity that goes into the denominator of the $t$ statistic, is 0.001865, so we can write directly

$$t = \frac{-0.0175 - (-0.02)}{0.001865} = 1.340$$

Tests concerning the regression coefficient $\alpha$ are performed in the same way, except that we use the first, instead of the second, of the two $t$ statistics. In most practical applications, however, the regression coefficient $\alpha$ is not of much interest—it is just the $y$-intercept, namely, the value of $y$ that corresponds to $x = 0$. In many cases it has no real meaning.

To construct confidence intervals for the regression coefficients $\alpha$ and $\beta$, we substitute for the middle term of $-t_{\alpha/2} < t < t_{\alpha/2}$ the appropriate $t$ statistic from page 412. Then, relatively simple algebra leads to the formulas

CONFIDENCE LIMITS FOR REGRESSION COEFFICIENTS

and

$$a \pm t_{\alpha/2} \cdot s_e \sqrt{\frac{1}{n} + \frac{\bar{x}^2}{S_{xx}}}$$

$$b \pm t_{\alpha/2} \cdot \frac{s_e}{\sqrt{S_{xx}}}$$

where the degree of confidence is $(1 - \alpha)100\%$ and $t_{\alpha/2}$ is the entry in Table II for $n - 2$ degrees of freedom.

EXAMPLE **16.7**   The following data show the average number of hours that six students spent on homework per week and their grade-point indexes for the courses they took in that semester:

| Hours spent on homework $x$ | Grade-point index $y$ |
|:---:|:---:|
| 15 | 2.0 |
| 28 | 2.7 |
| 13 | 1.3 |
| 20 | 1.9 |
| 4 | 0.9 |
| 10 | 1.7 |

Assuming that all the assumptions underlying a normal regression analysis are satisfied, construct a 99% confidence interval for $\beta$, the amount by which a student in the population sampled could have raised his or her grade-point index by studying an extra hour per week.

**Figure 16.9**
MINITAB printout for Example 16.7.

---
**Regression Analysis: y versus x**

```
The regression equation is
C2 = 0.721 + 0.0686 x

Predictor        Coef          SE Coef           T            P
Constant       0.7209          0.2464         2.93        0.043
x              0.06860         0.01467        4.68        0.009

S = 0.2720            R-Sq = 84.5%      R-Sq(adj) = 80.7%
```
---

**Solution**  Using the computer printout shown in Figure 16.9, we find that $b = 0.06860$ and that the estimate of the standard error of $b$, by which we have to multiply $t_{\alpha/2}$, is 0.01467. Since $t_{0.005} = 4.604$ for $6 - 2 = 4$ degrees of freedom, we get $0.0686 \pm 4.604(0.01467)$ and, hence,

$$0.0011 < \beta < 0.1361$$

This confidence interval is rather wide, and this is due to two things—the very small size of the sample and the relatively large variation measured by $s_e$, namely, the variation among the grade-point indexes of students doing the same amount of homework.  ∎

To answer the second kind of question asked on page 410 concerning the estimation, or prediction, of the average value of $y$ for a given value of $x$, we use a method that is very similar to the one just discussed. With the same assumptions as before, we base our argument on another $t$ statistic, arriving at the following $(1 - \alpha)100\%$ confidence interval for $\mu_{y|x_0}$, the mean of $y$ when $x = x_0$:

**C**ONFIDENCE LIMITS FOR MEAN OF $y$ WHEN $x = x_0$

$$(a + bx_0) \pm t_{\alpha/2} \cdot s_e \sqrt{\frac{1}{n} + \frac{(x_0 - \bar{x})^2}{S_{xx}}}$$

As before, the number of degrees of freedom is $n - 2$ and the corresponding value of $t_{\alpha/2}$ may be read from Table II.

**EXAMPLE 16.8**    Referring again to the data on page 399, suppose that we want to estimate the average hearing range of persons who have lived near the airport directly in the flight path of departing jets for two years. Construct a 95% confidence interval.

**Solution**    Assuming that all the assumptions underlying a normal regression analysis are satisfied, we substitute $n = 12$, $x_0 = 104$ weeks, $\sum x = 975$ (from Example 16.1) and hence $\bar{x} = 975/12 = 81.25$, $S_{xx} = 38{,}178.25$ (from Example 16.2), $a + bx_0 = 13.5$ (from Example 16.4), $s_e = 0.3645$ (from Example 16.5), and $t_{0.025} = 2.228$ for $12 - 2 = 10$ degrees of freedom into the preceding confidence-interval formula, getting

$$13.5 \pm 2.228(0.3645)\sqrt{\frac{1}{12} + \frac{(104 - 81.25)^2}{38{,}178.25}}$$

and hence

$$13.25 < \mu_{y|x_0} < 13.75$$

thousand cycles per second when $x = 104$ weeks. (Had we used $a = 15.32$ instead of $a = 15.3$ in Example 16.4, we would have obtained $a + bx_0 = 13.50$ instead of 13.48, which we rounded up to 13.5. So, the result would have been the same.)

The third question asked on page 410 differs from the other two. It does not concern the estimation of a population parameter, but the prediction of a single future observation. The endpoints of an interval for which we can assert with a given degree of confidence that it will contain such an observation are called **limits of prediction**, and the calculation of such limits will answer the third kind of question. Basing our argument on yet another $t$ statistic, we arrive at the following $(1 - \alpha)100\%$ limits of prediction for a value of $y$ when $x = x_0$:

**LIMITS OF PREDICTION**

$$(a + bx_0) \pm t_{\alpha/2} \cdot s_e \sqrt{1 + \frac{1}{n} + \frac{(x_0 - \bar{x})^2}{S_{xx}}}$$

Again, the number of degrees of freedom is $n - 2$ and the corresponding value of $t_{\alpha/2}$ may be read from Table II.

Note that the only difference between these limits of prediction and the confidence limits for $\mu_{y|x_0}$ given previously is the 1 that we added to the quantity under the square root sign. Thus, it will be left to the reader to verify in Exercise 16.24 that for the hearing-range example and $x_0 = 104$ the 95% limits of prediction are 12.65 and 14.35. It should not come as a surprise that this interval is much wider than the one obtained in Example 16.8. Whereas the limits of prediction apply to a prediction for one person, the confidence limits obtained in Example 16.8 apply to the mean of all persons who live or have lived near the airport directly in the flight path of departing jets for two years.

Let us remind the reader that all these methods are based on the very stringent assumptions of normal regression analysis. Furthermore, if we base more than one inference on the same data, we will run into problems with regard to the levels of significance and/or degrees of confidence. The random variables on which the various procedures are based are clearly not independent.

**16.20** Assume that the data of Exercise 16.3 satisfy the assumptions required by a normal regression analysis.

(a) If the work in Exercise 16.4 was done with a computer, use the information provided by the software to test the null hypothesis $\beta = 0.15$ against the alternative hypothesis $\beta \neq 0.15$ at the 0.05 level of significance.

(b) For tests concerning the regression coefficient $\beta$, the TI-83 graphing calculator provides the value of $t$ only for tests of the null hypothesis $\beta = 0$. Since the difference is only in the numerator, the value of $t$ for tests of the null hypothesis $\beta = \beta_0$ may be obtained by multiplying the value of $t$ provided by the calculator by

$$\frac{b - \beta_0}{b}$$

If the work in Exercise 16.4 was done with a graphing calculator, use this method for calculating $t$ to test the null hypothesis $\beta = 0.15$ against the alternative hypothesis $\beta \neq 0.15$ at the 0.05 level of significance.

**16.21** Assume that the data of Exercise 16.6 satisfy the assumptions required by a normal regression analysis.

(a) Use the sums given in that exercise, $\sum y^2 = 2,001$, and the result that $b = 0.898$, to calculate the value of $s_e$.

(b) Use the information given in part (a) as well as its result to test the null hypothesis $\beta = 1.5$ against the alternative hypothesis $\beta < 1.5$ at the 0.05 level of significance.

**16.22** Use a computer or a graphing calculator to rework both parts of Exercise 16.21. If a graphing calculator is used, follow the suggestion given in part (b) of Exercise 16.20.

**16.23** Assuming that the data of Exercise 16.8 satisfy the assumptions required by a normal regression analysis, use the result of that exercise and a computer or a graphing calculator to test the null hypothesis $\beta = 3.5$ against the alternative hypothesis $\beta > 3.5$ at the 0.01 level of significance. If a graphing calculator is used, follow the suggestion given in part (b) of Exercise 16.20.

**16.24** With reference to the data on page 399 and the calculations in Example 16.8, show that for $x_0 = 104$ the 95% limits of prediction for the hearing range are 12.65 and 14.35 thousand cycles per second.

**16.25** With reference to Exercise 16.9, use a computer or a graphing calculator to test the null hypothesis $\beta = -0.15$ against the alternative hypothesis $\beta \neq -0.15$ at the 0.01 level of significance. It must be assumed, of course, that the data of Exercise 16.9 satisfy the assumptions required by a normal regression analysis. Also, if a graphing calculator is used, follow the suggestion of part (b) of Exercise 16.20.

**16.26** With reference to the preceding exercise and the same assumptions, construct a 95% confidence interval for the hourly reduction of the chlorine residue.

**16.27** Assume that the data of Exercise 16.12 satisfy the assumptions required by a normal regression analysis.

(a) Use the sums given in that exercise, $\sum y^2 = 1,802$, and the result that $b = 0.272$, to calculate the value of $s_e$.

(b) Use the information given in part (a) as well as its result to test the null hypothesis $\beta = 0.40$ against the alternative hypothesis $\beta < 0.40$ at the 0.05 level of significance.

**16.28** Use a computer or a graphing calculator to rework both parts of Exercise 16.27. If a graphing calculator is used, follow the suggestion given in part (b) of Exercise 16.20.

**16.29** Assuming that the data of Exercise 16.12 satisfy the required assumptions for a normal regression analysis, use a computer or a graphing calculator to determine a 95% confidence interval for the mean moisture content when the humidity is 50%.

**16.30** The following table shows the assessed values and the selling prices of eight houses, constituting a random sample of all the houses sold recently in a rural area:

| Assessed value (thousands of dollars) | Selling price (thousands of dollars) |
|---|---|
| 70.3 | 114.4 |
| 102.0 | 169.3 |
| 62.5 | 106.2 |
| 74.8 | 125.0 |
| 57.9 | 99.8 |
| 81.6 | 132.1 |
| 110.4 | 174.2 |
| 88.0 | 143.5 |

Assuming that these data satisfy the required assumptions for a normal regression analysis, use a computer or a graphing calculator to find

(a) a 95% confidence interval for the mean selling price of a house in this rural area that is assessed at $90,000;

(b) 95% limits of prediction for a house in this rural area that has been assessed at $90,000.

**16.31** Assuming that the data of Exercise 16.3 satisfy the assumptions required by a normal regression analysis, use a computer or a graphing calculator to determine

(a) a 99% confidence interval for the mean hyperactivity rating of a four-year-old at one of the nursery schools 45 minutes after he or she has eaten 60 grams of food containing the preservative;

(b) 99% limits of prediction for the hyperactivity rating of one of these children who ate 60 grams of food with the preservative 45 minutes earlier.

**16.32** Assuming that the data of Exercise 16.8 satisfy the assumptions required by a normal regression analysis, use a computer or a graphing calculator to determine

(a) a 98% confidence interval for the average yield of wheat when there are only 10 inches of rain;

(b) 98% limits of prediction for the yield of wheat when there are only 10 inches of rain.

**16.33** Assuming that the data of Example 16.7 satisfy the assumptions required by a normal regression analysis, use a computer or a graphing calculator to determine

(a) a 95% confidence interval for the average grade-point index of students who average only five hours of homework per week during the semester;

(b) 95% limits of prediction for the grade-point index of a student who averages only five hours of homework per week during the semester.

# *16.4 MULTIPLE REGRESSION†

Although there are many problems where one variable can be predicted quite accurately in terms of another, it stands to reason that predictions should improve if one considers additional relevant information. For instance, we should be able to make better predictions of the performance of newly hired teachers if we consider not only their education, but also their years of experience and their personality. Also, we should be able to make better predictions of a new textbook's success if we consider not only the quality of the work, but also the potential demand and the competition.

Many mathematical formulas can serve to express relationships among more than two variables, but most commonly used in statistics (partly for reasons of convenience) are linear equations of the form

$$y = b_0 + b_1x_1 + b_2x_2 + \cdots + b_kx_k$$

Here $y$ is the variable that is to be predicted, $x_1, x_2, \ldots,$ and $x_k$ are the $k$ known variables on which predictions are to be based, and $b_0, b_1, b_2, \ldots,$ and $b_k$ are numerical constants that must be determined from observed data.

To illustrate, consider the following equation, which was obtained in a study of the demand for different meats:

$$\hat{y} = 3.489 - 0.090x_1 + 0.064x_2 + 0.019x_3$$

Here $y$ denotes the total consumption of federally inspected beef and veal in millions of pounds, $x_1$ denotes a composite retail price of beef in cents per pound, $x_2$ denotes a composite retail price of pork in cents per pound, and $x_3$ denotes income as measured by a certain payroll index. With this equation, we can predict the total consumption of federally inspected beef and veal corresponding to specified values of $x_1, x_2,$ and $x_3$.

The problem of determining a linear equation in more than two variables that best describes a given set of data is that of finding numerical values for $b_0, b_1, b_2, \ldots,$ and $b_k$. This is usually done by the method of least squares; that is, we minimize the sum of squares $\sum(y - \hat{y})^2$, where as before the $y$'s are the observed values and the $\hat{y}$'s are the values calculated by means of the linear equation. In principle, the problem of determining the values of $b_0, b_1, b_2, \ldots,$ and $b_k$ is the same as it is in the two-variable case, but manual solutions may be very tedious because the method of least squares requires that we solve as many normal equations as there are unknown constants $b_0, b_1, b_2, \ldots,$ and $b_k$. For instance, when there are two independent variables $x_1$ and $x_2$, and we want to fit the equation

$$y = b_0 + b_1x_1 + b_2x_2$$

---

†This section is marked optional because the calculations, while possible with a calculator for very simple problems, generally require special computer software.

we must solve the three normal equations

**N**ORMAL
EQUATIONS (TWO
INDEPENDENT
VARIABLES)

$$\sum y = n \cdot b_0 + b_1 \left( \sum x_1 \right) + b_2 \left( \sum x_2 \right)$$

$$\sum x_1 y = b_0 \left( \sum x_1 \right) + b_1 \left( \sum x_1^2 \right) + b_2 \left( \sum x_1 x_2 \right)$$

$$\sum x_2 y = b_0 \left( \sum x_2 \right) + b_1 \left( \sum x_1 x_2 \right) + b_2 \left( \sum x_2^2 \right)$$

Here $\sum x_1 y$ is the sum of the products obtained by multiplying each given value of $x_1$ by the corresponding value of $y$, $\sum x_1 x_2$ is the sum of the products obtained by multiplying each given value of $x_1$ by the corresponding value of $x_2$, and so on.

**EXAMPLE 16.9**  The following data show the number of bedrooms, the number of baths, and the prices at which eight one-family houses sold recently in a certain community:

| Number of bedrooms $x_1$ | Number of baths $x_2$ | Price (dollars) $y$ |
|---|---|---|
| 3 | 2 | 143,800 |
| 2 | 1 | 109,300 |
| 4 | 3 | 158,800 |
| 2 | 1 | 109,200 |
| 3 | 2 | 154,700 |
| 2 | 2 | 114,900 |
| 5 | 3 | 188,400 |
| 4 | 2 | 142,900 |

Find a linear equation that will enable one to predict the average sales price of a one-family house in the given community in terms of the number of bedrooms and the number of baths.

**Solution**  The quantities needed for substitution into the three normal equations are $n = 8$, $\sum x_1 = 25$, $\sum x_2 = 16$, $\sum y = 1{,}122{,}000$, $\sum x_1^2 = 87$, $\sum x_1 x_2 = 55$, $\sum x_2^2 = 36$, $\sum x_1 y = 3{,}711{,}100$, and $\sum x_2 y = 2{,}372{,}700$, and we get

$$1{,}122{,}000 = 8b_0 + 25b_1 + 16b_2$$

$$3{,}711{,}100 = 25b_0 + 87b_1 + 55b_2$$

$$2{,}372{,}700 = 16b_0 + 55b_1 + 36b_2$$

We could solve these equations by the method of elimination or by using determinants, but in view of the rather tedious calculations, such work is nowadays left to computers. So, let us refer to the computer printout of Figure 16.10, where we find in the column headed "Coef" that $b_0 = 65{,}430$, $b_1 = 16{,}752$, and $b_2 = 11{,}235$. In the line immediately above the coefficients we find that the least-squares equation is

$$\hat{y} = 65{,}430 + 16{,}752\, x_1 + 11{,}235\, x_2$$

**Figure 16.10**
MINITAB printout for
Example 16.9.

**Regression Analysis: y versus x1, x2**

```
The regression equation is
y = 65430 + 16752  x1 + 11235  x2

Predictor          Coef         SE Coef           T           P
Constant          65430           12134        5.39       0.003
x1                16752            6636        2.52       0.053
x2                11235            9885        1.14       0.307
```

This tells us that (in the given community at the time the study was being made) each extra bedroom added on the average $16,752, and each bath $11,235, to the sales price of a house.  ◼

**EXAMPLE 16.10**   Based on the result of Example 16.9, determine the average sales price of a house with three bedrooms and two baths (in the given community at the time the study was being made).

**Solution**   Substituting $x_1 = 3$ and $x_2 = 2$ into the least-squares equation obtained in Example 16.9, we get

$$\hat{y} = 65,430 + 16,752(3) + 11,235(2)$$

$$= 138,156$$

or approximately $138,200.  ◼

*16.34  The following are data on the ages and incomes of a random sample of executives working for a large multinational corporation, and the number of years they did postgraduate work at a university:

| Age $x_1$ | Years postgraduate work $x_2$ | Income (dollars) $y$ |
|---|---|---|
| 38 | 4 | 181,700 |
| 46 | 0 | 173,300 |
| 39 | 5 | 189,500 |
| 43 | 2 | 179,800 |
| 32 | 4 | 169,900 |
| 52 | 7 | 212,500 |

(a) Use appropriate computer software to fit an equation of the form $y = b_0 + b_1 x_1 + b_2 x_2$ to the given data.
(b) Use the equation obtained in part (a) to estimate the average income of 39-year-old executives of the corporation who did three years of postgraduate work at a university.

**\*16.35** The following data were collected to determine the relationship between two processing variables and the hardness of a certain kind of steel:

| Hardness (Rockwell 30-T) | Copper content (percent) | Annealing temperature (degrees F) |
|---|---|---|
| $y$ | $x_1$ | $x_2$ |
| 78.9 | 0.02 | 1,000 |
| 55.2 | 0.02 | 1,200 |
| 80.9 | 0.10 | 1,000 |
| 57.4 | 0.10 | 1,200 |
| 85.3 | 0.18 | 1,000 |
| 60.7 | 0.18 | 1,200 |

(a) Use appropriate computer software to fit an equation of the form $y = b_0 + b_1x_1 + b_2x_2$ to the given data.

(b) Use the equation obtained in part (a) to estimate the hardness of steel when its copper content is 0.14% and the annealing temperature is 1,100 degrees Fahrenheit.

**\*16.36** When the $x_1$'s and/or the $x_2$'s are equally spaced, the calculation of the regression coefficients can be simplified considerably by using the kind of coding described in Exercise 16.17. Rework Exercise 16.35 without a computer after coding the three $x_1$ values $-1, 0$, and 1, and the two $x_2$ values $-1$ and 1. (Note that, when coded, the 0.14% copper content becomes 0.50 and the 1,100 annealing temperature becomes 0.)

**\*16.37** The following are data on the percent effectiveness of a pain reliever and the amounts of three medications (in milligrams) present in each capsule:

| Medication A | Medication B | Medication C | Percent effective |
|---|---|---|---|
| $x_1$ | $x_2$ | $x_3$ | $y$ |
| 15 | 20 | 10 | 47 |
| 15 | 20 | 20 | 54 |
| 15 | 30 | 10 | 58 |
| 15 | 30 | 20 | 66 |
| 30 | 20 | 10 | 59 |
| 30 | 20 | 20 | 67 |
| 30 | 30 | 10 | 71 |
| 30 | 30 | 20 | 83 |
| 45 | 20 | 10 | 72 |
| 45 | 20 | 20 | 82 |
| 45 | 30 | 10 | 85 |
| 45 | 30 | 20 | 94 |

(a) Use appropriate computer software to fit an equation of the form $y = b_0 + b_1x_1 + b_2x_2 + b_3x_3$ to the given data.

(b) Use the equation obtained in part (a) to estimate the average percent effectiveness of capsules that contain 12.5 milligrams of Medication A, 25 milligrams of Medication B, and 15 milligrams of Medication C.

**\*16.38** Rework Exercise 16.37 without a computer after coding the three $x_1$ values $-1, 0$, and 1, the two $x_2$ values $-1$ and 1, and the two $x_3$ values $-1$ and 1.

# *16.5 NONLINEAR REGRESSION

When the pattern of a set of data points departs appreciably from a straight line, we must consider fitting some other kind of curve. In this section we shall first describe two situations where the relationship between $x$ and $y$ is not linear, but the method of Section 16.2 can nevertheless be employed. Then we shall give an example of **polynomial curve fitting** by fitting a parabola.

We usually plot paired data on various kinds of graph paper to see whether there are scales for which the points fall close to a straight line. Of course, when this is the case for ordinary graph paper, we proceed as in Section 16.2. If it is the case when we use **semilog paper** (with equal subdivisions for $x$ and a logarithmic scale for $y$, as shown in Figure 16.11), this indicates that an **exponential curve** will provide a good fit. The equation of such a curve is

$$y = a \cdot b^x$$

or in logarithmic form

$$\log y = \log a + x(\log b)$$

where "log" stands for logarithm to the base 10. (Actually, we could use any base including the irrational number $e$, in which case the equation is often written as $y = a \cdot e^{bx}$, or in logarithmic form as $\ln y = \ln a + bx$.)

Observe that if we write $A$ for $\log a$, $B$ for $\log b$, and $Y$ for $\log y$, the original equation in logarithmic form becomes $Y = A + Bx$, which is the usual equation of a straight line. Thus, to fit an exponential curve to a given set of paired data, we simply apply the method of Section 16.2 to the data points $(x, Y)$.

EXAMPLE   16.11   The following are data on a company's net profits during the first six years that it has been in business:

| Year | Net profit (thousands of dollars) |
|------|-----------------------------------|
| 1 | 112 |
| 2 | 149 |
| 3 | 238 |
| 4 | 354 |
| 5 | 580 |
| 6 | 867 |

In Figure 16.11 these data are plotted on ordinary graph paper on the left and on semilog paper (with a logarithmic scale for $y$) on the right. As can be seen, the overall pattern is remarkably "straightened out" in the figure on the right, and this suggests that we ought to fit an exponential curve.

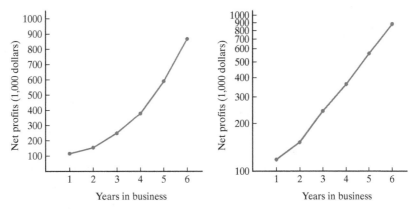

**Figure 16.11**
Data plotted on ordinary and semilog graph paper.

**Solution**   Getting the logarithms of the y's with a calculator, or perhaps from a table of logarithms, we find that

| $x$ | $y$ | $Y = \log y$ |
|---|---|---|
| 1 | 112 | 2.0492 |
| 2 | 149 | 2.1732 |
| 3 | 238 | 2.3766 |
| 4 | 354 | 2.5490 |
| 5 | 580 | 2.7634 |
| 6 | 867 | 2.9380 |

Then, for these data we get $n = 6$, $\sum x = 21$, $\sum x^2 = 91$, $\sum Y = 14.8494$, and $\sum xY = 55.1664$, and hence $S_{xx} = 91 - \frac{1}{6}(21)^2 = 17.5$ and $S_{xY} = 55.1664 - \frac{1}{6}(21)(14.8494) = 3.1935$. Finally, substitution into the two formulas on page 403. yields

$$B = \frac{3.1935}{17.5} \approx 0.1825$$

$$A = \frac{14.8494 - 0.1825(21)}{6} \approx 1.8362$$

and the equation that describes the relationship is

$$\hat{Y} = 1.8362 + 0.1825x$$

Since 1.8362 and 0.1825 are the estimates corresponding to log a and log b, we find by taking antilogarithms that $a = 68.58$ and $b = 1.52$. Thus, the equation of the exponential curve that best describes the relationship between the company's net profit and the number of years it has been in business is given by

$$\hat{y} = 68.58(1.52)^x$$

where $\hat{y}$ is in thousands of dollars.   ▪

Although the calculations in Example 16.11 were quite easy, we could, of course, have used a computer. Entering the values of x and Y in columns c1 and c2, we got the printout shown in Figure 16.12. As can be seen, the values we had calculated for A and B are shown in the column headed "Coef."

**Figure 16.12**
Computer printout for
Example 16.11.

> **Regression Analysis: log y versus x**
>
> ```
> The regression equation is
> logy = 1.84 + 0.182 x
>
> Predictor        Coef    SE Coef      T       P
> Constant      1.83619    0.02245   81.79   0.000
> x             0.182490   0.005764  31.66   0.000
>
>
> S = 0.0241142    R-Sq = 99.6%    R-Sq(adj) = 99.5%
> ```

**Figure 16.13**
Values of $a$ and $b$ reproduced from display
screen of TI-83 graph-
ing calculator.

To get the exponential equation in its final form, it would have been even easier to use a graphing calculator. After entering the original $x$'s and $y$'s, the command **STAT CALC ExpReg** yields the display shown in Figure 16.13. As can be seen, the constants $a$ and $b$ rounded to two decimals are identical with those in the exponential equation given at the end of Example 16.11.

Once we fit an exponential curve to a set of paired data, we can predict a future value of $y$ by substituting into its equation the corresponding value of $x$. However, it is usually much more convenient to substitute $x$ into the logarithmic form of the equation, namely, into

$$\log \hat{y} = \log a + x(\log b)$$

**EXAMPLE 16.12**    With reference to Example 16.11, predict the company's net profit for the eighth year that it will have been in business.

**Solution**    Substituting $x = 8$ into the logarithmic form of the equation for the exponential curve, we get

$$\log \hat{y} = 1.8362 + 8(0.1825)$$

$$= 3.2962$$

and hence $\hat{y} = 1,980$ or $\$1,980,000$.    ◼

If data points fall close to a straight line when plotted on **log-log paper** (with logarithmic scales for both $x$ and $y$), this indicates that an equation of the form

$$y = a \cdot x^b$$

will provide a good fit. In the logarithmic form, the equation of such a **power function** is

$$\log y = \log a + b(\log x)$$

which is a linear equation in $\log x$ and $\log y$. (Writing $A$, $X$, and $Y$ for $\log a$, $\log x$, and $\log y$, the equation becomes

$$Y = A + bX$$

which is the usual equation of a straight line.) For fitting a power curve, we can thus, apply the method of Section 16.2 to the problem expressed as $Y = A + bX$. The work that is required for fitting a power function is very similar to what we did in Example 16.11, and we shall not illustrate it by means of an example. However, in Exercise 16.47 the reader will find a problem to which the method can be applied.

When the values of $y$ first increase and then decrease, or first decrease and then increase, a **parabola** having the equation

$$y = a + bx + cx^2$$

will often provide a good fit. This equation can also be written as

$$y = b_0 + b_1 x + b_2 x^2$$

to conform with the notation of Section 16.4. Thus, it can be seen that parabolas can be looked upon as linear equations in the two unknowns $x_1 = x$ and $x_2 = x^2$, and fitting a parabola to a set of paired data is nothing new—we simply use the method of Section 16.4. If we actually wanted to use the normal equations on page 402 with $x_1 = x$ and $x_2 = x^2$, this would require that we determine $\sum x, \sum x^2, \sum x^3, \sum x^4, \sum y, \sum xy$, and $\sum x^2 y$, and the subsequent solution of three simultaneous linear equations. As can well be imagined, this would require a great deal of arithmetic and it is rarely done without the use of appropriate technology. In the two examples that follow, we shall first illustrate fitting a parabola with the use of a computer and then repeat the problem with the use of a graphing calculator.

EXAMPLE **16.13**    The following are data on the drying time of a varnish and the amount of a certain chemical additive:

| Amount of additive (grams) $x$ | Drying time (hours) $y$ |
|---|---|
| 1 | 7.2 |
| 2 | 6.7 |
| 3 | 4.7 |
| 4 | 3.7 |
| 5 | 4.7 |
| 6 | 4.2 |
| 7 | 5.2 |
| 8 | 5.7 |

(a) Fit a parabola that, as would appear from Figure 16.14, is the right kind of curve to fit to the given data.

**Figure 16.14**
Scattergram of the
varnish-drying-time
data.

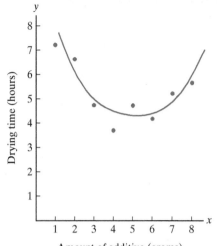

**Figure 16.15**
Computer printout for
fitting parabola.

> **Regression Analysis: y versus x, x2**
>
> The regression equation is
> y = 9.24 - 2.01 x + 0.199 x2
>
> Predictor     Coef     SE Coef        T        p
> Constant    9.2446     0.7645     12.09    0.000
> x          -2.0149     0.3898     -5.17    0.004
> x2          0.19940    0.04228     4.72    0.005
>
> S = 0.5480     R-Sq = 85.3%   R-Sq(adj) = 79.4%

**(b)** Use the result of part (a) to predict the drying time of the varnish when 6.5 grams of the chemical are added.

**Solution**

**(a)** Using the MINITAB printout shown in Figure 16.15, we find that $b_0 = 9.2446$, $b_1 = -2.0149$, and $b_2 = 0.19940$ (in the column headed "Coef"). Rounding to two decimals, we can, thus, write the equation of the parabola as

$$\hat{y} = 9.24 - 2.01x + 0.20x^2$$

(Note that we had entered the values of $x$ in column c1, the values of $x^2$ in column c2, and the values of $y$ in column c3.)

**(b)** Substituting $x = 6.5$ into the equation obtained in part (a), we get

$$\hat{y} = 9.24 - 2.01(6.5) + 0.20(6.5)^2$$

$$\approx 4.62 \text{ hours}$$

**EXAMPLE 16.14** Rework part (a) of Example 16.13 with the use of a graphing calculator.

**Figure 16.16**
Fit of parabola repro-
duced from the display
screen of a TI-83 graph-
ing calculator.

```
QuadReg
 y=ax²+bx+c
 a=.1994047619
 b=-2.014880952
 c=9.244642857
 R²=.8530653852
```

**Solution**    To avoid confusion, let us point out that the TI-83 uses the equation $y = ax^2 + bx + c$, with $a$ and $c$ interchanged from the version we gave on page 425. After we entered the $x$'s and the $y$'s, the command **STAT CALC QuadReg** yielded the result shown in Figure 16.16. Rounding to two decimals, we get

$$\hat{y} = 0.20x^2 - 2.01x + 9.24$$

which, except for the order of the terms, is the same as before.    ■

On page 425, we introduced parabolas as curves that bend once—that is, their values first increase and then decrease, or first decrease and then increase. For patterns that "bend" more than once, **polynomial equations** of higher degree, such as $y = a + bx + cx^2 + dx^3$ or $y = a + bx + cx^2 + dx^3 + ex^4$ can be fitted by the technique illustrated in Example 16.13. In practice, we often use sections of such curves, especially parts of parabolas, when there is only a slight curvature in the pattern we want to describe.

# EXERCISES

**\*16.39**    The following data pertain to the growth of a colony of bacteria in a culture medium:

| Days since inoculation $x$ | Bacteria count (thousands) $y$ |
|:---:|:---:|
| 2 | 112 |
| 4 | 148 |
| 6 | 241 |
| 8 | 363 |
| 10 | 585 |

(a) Given that $\sum x = 30, \sum x^2 = 220, \sum Y = 11.9286$ (where $Y = \log y$), and $\sum xY = 75.2228$, set up the two normal equations for fitting an exponential curve, solve them for $\log a$ and $\log b$ by the method of elimination or by the method based on determinants, and write the equation of the exponential curve in logarithmic form.

(b) Convert the equation obtained in part (a) into the form $y = ab^x$.

(c) Use the equation obtained in part (a) to estimate the bacteria count five days after inoculation.

**\*16.40**    Use a computer or a graphing calculator to rework Exercise 16.39.

*16.41  Show that the use of a computer or a graphing calculator fits the exponential equation

$$\hat{y} = (101.17)(0.9575)^x$$

to the following data on the percentage of the radial tires made by a manufacturer that are still usable after having been driven the given numbers of miles:

| Miles driven (thousands) x | Percentage usable y |
|---|---|
| 1 | 97.2 |
| 2 | 91.8 |
| 5 | 82.5 |
| 10 | 64.4 |
| 20 | 41.0 |
| 30 | 29.9 |
| 40 | 17.6 |
| 50 | 11.3 |

*16.42  Convert the equation obtained in Exercise 16.41 into logarithmic form and use it to estimate what percentage of the tires will still be usable after having been driven for 25,000 miles.

*16.43  Show that the use of a computer or a graphing calculator fits the exponential equation,

$$y = (1.178)(2.855)^x$$

to the following data on the curing time of test samples of concrete, $x$, and their tensile strength, $y$:

| x (hours) | y (psi) |
|---|---|
| 1 | 3.54 |
| 2 | 8.92 |
| 3 | 27.5 |
| 4 | 78.8 |
| 5 | 225.0 |
| 6 | 639.0 |

*16.44  Assuming that the exponential trend continues, use the equation obtained in Exercise 16.43 (in logarithmic form) to estimate the tensile strength of a test sample of the concrete that has been cured for eight hours.

*16.45  A small piece of a rare slow-growing cactus was grafted to another cactus with a strong root system and its height, measured annually, was as shown in the following table:

| Years after grafting<br>x | Height (millimeters)<br>y |
|---|---|
| 1 | 22 |
| 2 | 25 |
| 3 | 29 |
| 4 | 34 |
| 5 | 38 |
| 6 | 44 |
| 7 | 51 |
| 8 | 59 |
| 9 | 68 |

Use a computer or a graphing calculator to fit an exponential curve.

**\*16.46** The following are data on the amount of fertilizer applied to the soil, $x$ (in pounds per square yard), and the yield of a certain food crop, $y$ (in pounds per square yard):

| x | y |
|---|---|
| 0.5 | 32.0 |
| 1.1 | 34.3 |
| 2.2 | 15.7 |
| 0.2 | 20.8 |
| 1.6 | 33.5 |
| 2.0 | 21.5 |

(a) Draw a scattergram to verify that it is reasonable to describe the overall pattern with a parabola.

(b) Show that the use of a computer or a graphing calculator fits the parabola

$$y = 14.79 + 38.9x - 17.56x^2$$

to the data.

(c) Use the equation obtained in part (b) to estimate the yield when 1.5 pounds of the fertilizer are applied per square yard.

**\*16.47** The following data pertain to the demand for a product, $y$ (in thousands of units), and its price, $x$ (in dollars), in five very similar market areas:

| Price<br>x | Demand<br>y |
|---|---|
| 20 | 22 |
| 16 | 41 |
| 10 | 120 |
| 11 | 89 |
| 14 | 56 |

Show that the use of a computer or graphing calculator fits the parabola

$$y = 384.4 - 36.0x + 0.896x^2$$

to the data, and use the equation to estimate the demand when the price of the product is $12.

# CHECKLIST OF KEY TERMS (with page references to their definitions)

Curve fitting, 397
Data points, 399, 401
Determinants, 402
Estimated regression coefficients, 410
Estimated regression line, 405
*Exponential curve, 422
Least-squares line, 401
Limits of prediction, 415
Linear equation, 398
Linear regression analysis, 410
*Log-log paper, 424
Method of elimination, 402
Method of least squares, 397, 400
*Multiple regression, 418
Nonlinear regression, 422

Normal equations, 402
Normal regression analysis, 410
*Parabola, 425
*Polynomial curve fitting, 422
*Polynomial equation, 427
*Power function, 424
Regression, 396
Regression analysis, 410
Regression coefficients, 410
Regression equation, 404
Regression line, 405
Scattergram, 399
*Semilog paper, 422
Standard error of estimate, 411

# REFERENCES

*Methods of deciding which kind of curve to fit to a given set of paired data may be found in books on numerical analysis and in more advanced texts in statistics. Further information about the material of this chapter may be found in*

CHATTERJEE, S., and PRICE, B., *Regression Analysis by Example*, 2nd ed. New York: John Wiley & Sons, Inc., 1991.

DANIEL, C., and WOOD, F., *Fitting Equations to Data*, 2nd ed. New York: John Wiley & Sons, Inc., 1980.

DRAPER, N. R., and SMITH, H., *Applied Regression Analysis*, 2nd ed. New York: John Wiley & Sons, Inc., 1981.

EZEKIEL, M., and FOX, K. A., *Methods of Correlation and Regression Analysis*, 3rd ed. New York: John Wiley & Sons, Inc., 1959.

WEISBERG, S., *Applied Linear Regression*, 2nd ed. New York: John Wiley & Sons, Inc., 1985.

WONNACOTT, T. H., and WONNACOTT, R. J., *Regression: A Second Course in Statistics*. New York: John Wiley & Sons, Inc., 1981.

# CORRELATION

**17.1** The Coefficient of Correlation 432

**17.2** The Interpretation of r 438

**17.3** Correlation Analysis 443

*__17.4__ Multiple and Partial Correlation 447

Checklist of Key Terms 450

References 451

I n Chapter 16 we learned how to fit a least-squares line to paired data. Now we will determine how well this line actually fits the data. Of course we can get a fair idea by inspection, say by inspecting a scatter diagram that shows the line together with the data. But to be more objective, we shall analyze the total variation of the observed $y$'s (in our example, the total variation among the hearing ranges) and look for possible causes or explanations.[†]

| Number of weeks x | Hearing range y |
|---|---|
| 47 | 15.1 |
| 56 | 14.1 |
| 116 | 13.2 |
| 178 | 12.7 |
| 19 | 14.6 |
| 75 | 13.8 |
| 160 | 11.9 |
| 31 | 14.8 |
| 12 | 15.3 |
| 164 | 12.6 |
| 43 | 14.7 |
| 74 | 14.0 |

Table reproduced from page 399.

[†] As in Chapter 16, $x$ is the length of time (rounded to the nearest week) that a person has been living near a major airport directly in the path of departing jets. $y$ is his or her hearing range (in thousands of cycles per second.)

As can be seen from this table, there are substantial differences among the $y$'s, the smallest being 11.9 and the largest being 15.3. However, we also see that the hearing range of 11.9 thousand cycles per second was that of a person who had lived at that location for 160 weeks, while the hearing range of 15.3 thousand cycles per second was that of a person who had lived there for only 12 weeks. This suggests that the differences in hearing range may well be due, at least in part, to differences in the length of time that the persons had been exposed to the airport noise. This raises the following question, which we shall answer in this chapter: *Of the total variation among the y's, how much can be attributed to the relationship between the two variables x and y (that is, to the fact that the observed values of y correspond to different values of x), and how much can be attributed to chance?*

In Section 16.2 we learned how to fit a least-squares line to paired data. It should be observed, however, that the least-squares line is only an estimated regression line because it is based on sample data. If we repeated the study of Section 16.2 with data for major airports in other cities or for a sample of 12 persons at the same airport, we would probably get different values for values of the equation $y = a + bx$.

In Section 17.1 we shall introduce the coefficient of correlation as a measure of the strength of the linear relationship between two variables, in Section 17.2 we shall learn how to interpret it, and in Section 17.3 we shall study related problems of inference. The problems of multiple and partial correlation will be touched upon lightly in the optional Section 17.4.

# 17.1 THE COEFFICIENT OF CORRELATION

In reply to the question raised above, let us point out that we are faced here with an analysis of variance. Figure 17.1 shows what we mean. As can be seen from the diagram, the deviation of an observed value of $y$ from the mean of all the $y$'s, $y - \bar{y}$, can be written as the sum of two parts. One part is the deviation of $\hat{y}$ (the value on the line corresponding to an observed value of $x$) from the mean of all the $y$'s, $\hat{y} - \bar{y}$; the other part is the deviation of the observed value of $y$ from the corresponding value on the line, $y - \hat{y}$. Symbolically, we write

$$y - \bar{y} = (\hat{y} - \bar{y}) + (y - \hat{y})$$

**Figure 17.1**
Illustration showing that $y - \bar{y} = (y - \hat{y}) + (\hat{y} - \bar{y})$.

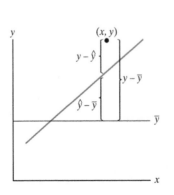

for any observed value $y$, and if we square the expressions on both sides of this identity and sum over all $n$ values of $y$, we find that algebraic simplifications lead to

$$\sum(y - \bar{y})^2 = \sum(\hat{y} - \bar{y})^2 + \sum(y - \hat{y})^2$$

The quantity on the left measures the total variation of the $y$'s and we call it the **total sum of squares**; note that $\sum(y - \bar{y})^2$ is just the variance of the $y$'s multiplied by $n - 1$. The first of the two sums on the right, $\sum(\hat{y} - \bar{y})^2$, is called the **regression sum of squares** and it measures that part of the total variation of the $y$'s that can be ascribed to the relationship between the two variables $x$ and $y$; indeed, if all the points lie on the least-squares line, then $y = \hat{y}$ and the regression sum of squares equals the total sum of squares. In practice, this is hardly, if ever, the case, and the fact that all the points do not lie on a least-squares line is an indication that there are other factors than differences among the $x$'s that affect the values of $y$. It is customary to combine all these other factors under the general heading of "chance." Chance variation is thus measured by the amounts by which the points deviate from the line; specifically, it is measured by $\sum(y - \hat{y})^2$, called the **residual sum of squares**, which is the second of the two components into which we partitioned the total sum of squares.

To determine these sums of squares for the hearing-range example, we could substitute the observed values of $y$, $\bar{y}$, and the values of $\hat{y}$ obtained by substituting the $x$'s into $\hat{y} = 15.3 - 0.0175x$ (see page 403), but there are simplifications. First, for $\sum(y - \bar{y})^2$ we have the computing formula

$$S_{yy} = \sum y^2 - \frac{1}{n}\left(\sum y\right)^2$$

and on page 412 we showed that it equals 13.02 for our example. Second, $\sum(y - \hat{y})^2$ is the quantity we minimized by the method of least squares, and divided by $n - 2$ it defined $s_e^2$ on page 412. Thus, it equals $(n - 2)s_e^2$ and $(12 - 2)(0.3645)^2 \approx 1.329$ in our example, for which we showed in Example 16.5 that $s_e = 0.3645$. Finally, by subtraction, the regression sum of squares is given by

$$\sum(\hat{y} - \bar{y})^2 = \sum(y - \bar{y})^2 - \sum(y - \hat{y})^2$$

and we get $13.02 - 1.329 = 11.69$ (rounded to two decimals) for our example.

It is of interest to note that all the sums of squares we have calculated here could have been obtained directly from the computer printout of Figure 16.7, which is reproduced in Figure 17.2. Under "Analysis of Variance" in the column headed SS, we find that the total sum of squares is 13.020, the error (residual) sum of squares is 1.329, and the regression sum of squares is 11.691. The minor differences between the values shown here and calculated previously are, of course, due to rounding.

We are now ready to examine the sums of squares. Comparing the regression sum of squares with the total sum of squares, we find that

$$\frac{\sum (\hat{y} - \bar{y})^2}{\sum (y - \bar{y})^2} = \frac{11.69}{13.02} \approx 0.898$$

is the proportion of the total variation of the hearing ranges that can be attributed to the relationship with $x$, namely, to the differences in the length of time that the 12 persons in the sample had been exposed to the airport noise. This quantity is referred to as the **coefficient of determination** and it is denoted by $r^2$. Note that the coefficient of determination is also given in the printout of Figure 17.2, where it says near the middle that $R - sq = 89.8\%$.

**Figure 17.2**
Copy of Figure 16.7.

```
Regression Analysis: y versus x

The regression equation is
y = 15.3 - 0.0175 x

Predictor        Coef      SE Coef         T          P
Constant      15.3218       0.1845     83.04      0.000
x            -0.017499     0.001865     -9.38      0.000

S = 0.3645    R-Sq = 89.8%  R-Sq(adj) = 88.8%

Analysis of Variance

Source             DF          SS         MS         F         p
Regression          1      11.691     11.691     88.00     0.000
Residual Error     10       1.329      0.133
Total              11      13.020
```

If we take the square root of the coefficient of determination, we get the **coefficient of correlation** that is denoted by the letter $r$. Its sign is chosen so that it is the same as that of the estimated regression coefficient $b$, and for our example, where $b$ is negative, we get

$$r = -\sqrt{0.898} \approx -0.95$$

rounded to two decimals.

It follows that the correlation coefficient is positive when the least-squares line has an upward slope, namely, when the relationship between $x$ and $y$ is such that small values of $y$ tend to go with small values of $x$ and large values of $y$ tend to go with large values of $x$. Also, the correlation coefficient is negative when the least-squares line has a downward slope, namely, when large values of $y$ tend to go with small values of $x$ and small values of $y$ tend to go with large values of $x$. Examples of **positive** and **negative correlations** are shown in the first two diagrams of Figure 17.3.

**Figure 17.3**
Types of correlation.

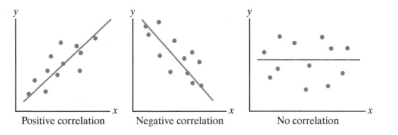

Positive correlation    Negative correlation    No correlation

Since part of the variation of the $y$'s cannot exceed their total variation, $\sum(y - \hat{y})^2$ cannot exceed $\sum(y - \bar{y})^2$, and it follows from the formula defining $r$ that correlation coefficients must lie on the interval from $-1$ to $+1$. If all the points actually fall on a straight line, the residual sum of squares, $\sum(y - \hat{y})^2$, is zero,

$$\sum(\hat{y} - \bar{y})^2 = \sum(y - \bar{y})^2$$

and the resulting value of $r$, $-1$ or $+1$, is indicative of a perfect fit. If, however, the scatter of the points is such that the least-squares line is a horizontal line coincident with $\bar{y}$ (that is, a line with slope 0 that intersects the $y$-axis at $a = \bar{y}$), then

$$\sum(y - \hat{y})^2 = \sum(y - \bar{y})^2 \quad \text{and} \quad r = 0$$

In that case none of the variation of the $y$'s can be attributed to their relationship with $x$, and the fit is so poor that knowledge of $x$ is of no help in predicting $y$. The predicted value of $y$ is $\bar{y}$ regardless of $x$. An example of this is shown in the third diagram of Figure 17.3.

The formula that defines $r$ shows clearly the nature, or essence, of the coefficient of correlation, but in actual practice it is seldom used to determine its value. Instead we use the computing formula

**C**OMPUTING
FORMULA FOR
COEFFICIENT OF
CORRELATION

$$r = \frac{S_{xy}}{\sqrt{S_{xx} \cdot S_{yy}}}$$

which has the added advantage that it automatically gives $r$ the correct sign. The quantities needed to calculate $r$ with this formula were previously defined, but for convenient reference, we remind the reader that

$$S_{xx} = \sum x^2 - \frac{1}{n}\left(\sum x\right)^2$$

$$S_{yy} = \sum y^2 - \frac{1}{n}\left(\sum y\right)^2$$

$$S_{xy} = \sum xy - \frac{1}{n}\left(\sum x\right)\left(\sum y\right)$$

**EXAMPLE 17.1**   Following are the scores that 12 students received in final examinations in economics and anthropology:

| Economics | Anthropology |
|-----------|--------------|
| 51 | 74 |
| 68 | 70 |
| 72 | 88 |
| 97 | 93 |
| 55 | 67 |
| 73 | 73 |
| 95 | 99 |
| 74 | 73 |
| 20 | 33 |
| 91 | 91 |
| 74 | 80 |
| 80 | 86 |

Use the computing formula to calculate $r$.

**Solution**   First calculating the necessary sums, we get $\sum x = 850$, $\sum x^2 = 65{,}230$, $\sum y = 927$, $\sum y^2 = 74{,}883$, and $\sum xy = 69{,}453$. Then, substituting these values together with $n = 12$ into the formulas for $S_{xx}$, $S_{yy}$, and $S_{xy}$, we find that

$$S_{xx} = 65{,}230 - \frac{1}{12}(850)^2 \approx 5{,}021.67$$

$$S_{yy} = 74{,}883 - \frac{1}{12}(927)^2 = 3{,}272.25$$

$$S_{xy} = 69{,}453 - \frac{1}{12}(850)(927) = 3{,}790.5$$

and

$$r = \frac{3{,}790.5}{\sqrt{(5{,}021.67)(3{,}272.25)}} \approx 0.935$$

The quantity $S_{xy}$ in the numerator of the formula for $r$ is actually a computing formula for $\sum(x - \bar{x})(y - \bar{y})$, which, divided by $n$, is called the first **product moment**. For this reason, $r$ is also referred to at times as the **product-moment coefficient of correlation**. Note that in $\sum(x - \bar{x})(y - \bar{y})$ we add the products obtained by multiplying the deviation of each $x$ from $\bar{x}$ by the deviation of the corresponding $y$ from $\bar{y}$. In this way we literally measure how the $x$'s and $y$'s vary together. If their relationship is such that large values of $x$ tend to go with large values of $y$, and small values of $x$ with small values of $y$, the deviations $x - \bar{x}$ and $y - \bar{y}$ tend to be both positive or both negative, and most of the products $(x - \bar{x})(y - \bar{y})$ will be positive. On the other hand, if the relationship is such that large values of $x$ tend to go with small values of $y$, and small values of $x$ with large values of $y$, the deviations $x - \bar{x}$ and $y - \bar{y}$ tend to be of opposite sign, and most of the products $(x - \bar{x})(y - \bar{y})$

will be negative. For this reason, $\sum(x - \bar{x})(y - \bar{y})$, divided by $n - 1$, is called the **sample covariance**.

Correlation coefficients are sometimes calculated in the analysis of $r \times c$ tables, provided that the row categories as well as the column categories are ordered. This is the kind of alternative to the chi-square analysis we suggested at the end of Section 14.4, where we pointed out that the ordering of the categories is not taken into account in the calculation of the $\chi^2$ statistic. To use $r$ in a problem like this, we replace the ordered categories by similarly ordered sets of numbers. As we said on page 342, such numbers are usually, though not necessarily, consecutive integers, preferably integers that will make the arithmetic as simple as possible. For three categories we might use 1, 2, and 3, or $-1$, 0, and 1; for four categories we might use 1, 2, 3, and 4, or $-1$, 0, 1, and 2, or perhaps $-3$, $-1$, 1, and 3. The calculation of $r$ as a measure of the strength of the relationship between two categorical variables is illustrated by the following example.

**EXAMPLE  17.2**    In Example 14.7 we analyzed the following $3 \times 3$ table to see whether there is a relationship between the test scores of persons who have gone through a certain job-training program and their subsequent performance on the job:

|  |  | Poor | Fair | Good |  |
|---|---|---|---|---|---|
|  |  | \multicolumn{3}{c}{Performance} |  |  |
|  | Below average | 67 | 64 | 25 | 156 |
| **Test score** | Average | 42 | 76 | 56 | 174 |
|  | Above average | 10 | 23 | 37 | 70 |
|  |  | 119 | 163 | 118 | 400 |

Label the test scores $x = -1$, $x = 0$, and $x = 1$, the performance ratings $y = -1$, $y = 0$, and $y = 1$, and calculate $r$.

**Solution**    Labeling the rows and the columns as indicated, we get

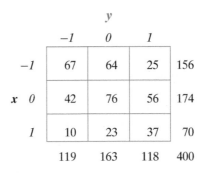

| $x$ | $y$ | $-1$ | $0$ | $1$ |  |
|---|---|---|---|---|---|
|  | $-1$ | 67 | 64 | 25 | 156 |
|  | $0$ | 42 | 76 | 56 | 174 |
|  | $1$ | 10 | 23 | 37 | 70 |
|  |  | 119 | 163 | 118 | 400 |

where the row totals tell us how many times $x$ equals $-1, 0,$ and $1,$ and the column totals tell us how many times $y$ equals $-1, 0,$ and $1.$ Thus,

$$\sum x = 156(-1) + 174 \cdot 0 + 70 \cdot 1 = -86$$

$$\sum x^2 = 156(-1)^2 + 174 \cdot 0^2 + 70 \cdot 1^2 = 226$$

$$\sum y = 119(-1) + 163 \cdot 0 + 118 \cdot 1 = -1$$

$$\sum y^2 = 119(-1)^2 + 163 \cdot 0^2 + 118 \cdot 1^2 = 237$$

and for $\sum xy$ we must add the products obtained by multiplying each cell frequency by the corresponding values of $x$ and $y.$ Omitting all cells where either $x = 0$ or $y = 0,$ we get

$$\sum xy = 67(-1)(-1) + 25(-1)1 + 10 \cdot 1(-1) + 37 \cdot 1 \cdot 1$$

$$= 69$$

Then, substitution into the formulas for $S_{xx}, S_{yy},$ and $S_{xy}$ yields

$$S_{xx} = 226 - \frac{1}{400}(-86)^2 = 207.51$$

$$S_{yy} = 237 - \frac{1}{400}(-1)^2 = 237.00$$

$$S_{xy} = 69 - \frac{1}{400}(-86)(-1) = 68.78$$

all rounded to two decimals, and finally

$$r = \frac{68.78}{\sqrt{(207.51)(237.00)}} \approx 0.31$$

# 17.2 THE INTERPRETATION OF $r$

When $r$ equals $+1, -1,$ or $0,$ there is no problem about the interpretation of the coefficient of correlation. As we have already indicated, it is $+1$ or $-1,$ when all the points actually fall on a straight line, and it is zero when the fit of the least-squares line is so poor that knowledge of $x$ does not help in the prediction of $y.$ In general, the definition of $r$ tells us that $100r^2$ is the percentage of the total variation of the $y$'s that is explained by, or is due to, their relationship with $x.$ This itself is an important measure of the relationship between two variables; beyond this, it permits valid comparisons of the strength of several relationships.

**EXAMPLE 17.3**   If $r = 0.80$ in one study and $r = 0.40$ in another, would it be correct to say that the 0.80 correlation is twice as strong as the 0.40 correlation?

**Solution**   No! When $r = 0.80,$ then $100(0.80)^2 = 64\%$ of the variation of the $y$'s is accounted for by the relationship with $x,$ and when $r = 0.40,$ then $100(0.40)^2 = 16\%$ of the variation of the $y$'s is accounted for by the relationship with $x$. Thus, in the sense of "percentage of variation accounted for" we can say that the 0.80 correlation is four times as strong as the 0.40 correlation.

In the same way, we say that a relationship for which $r = 0.60$, is nine times as strong as a relationship for which $r = 0.20$.

There are several pitfalls in the interpretation of the coefficient of correlation. First, it is often overlooked that $r$ measures only the strength of linear relationships; second, it should be remembered that a strong correlation (a value of $r$ close to $+1$ or $-1$) does not necessarily imply a cause–effect relationship.

If $r$ is calculated indiscriminately, for instance, for the three sets of data of Figure 17.4, we get $r = 0.75$ in each case, but it is a meaningful measure of the strength of the relationship only in the first case. In the second case there is a very strong curvilinear relationship between the two variables, and in the third case six of the seven points actually fall on a straight line, but the seventh point is so far off that it suggests the possibility of a gross error of measurement or an error in recording the data. Thus, before we calculate $r$, we should always plot the data to see whether there is reason to believe that the relationship is, in fact, linear.

**Figure 17.4**
Three sets of paired data for which $r = 0.75$.

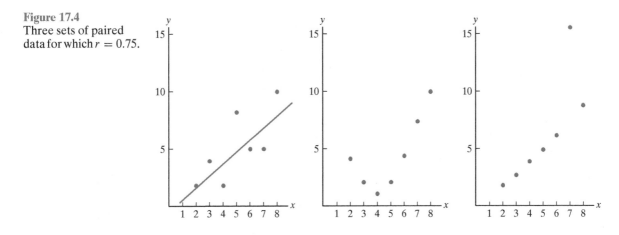

The fallacy of interpreting a high value of $r$ (that is, a value close to $+1$ or $-1$), as an indication of a cause–effect relationship, is best explained with a few examples. Frequently used as an illustration, is the high positive correlation between the annual sales of chewing gum and the incidence of crime in the United States. Obviously, one cannot conclude that crime might be reduced by prohibiting the sale of chewing gum; both variables depend upon the size of the population, and it is this mutual relationship with a third variable (population size) that produces the positive correlation. Another example is the strong positive correlation that was observed between the number of storks seen nesting in English villages and the number of children born in the same villages. We leave it to the reader's ingenuity to explain why there might be a strong correlation in this case in the absence of any cause–effect relationship.

**17.1** In Example 16.7 we gave the following data on the average number of hours that six students spent on homework per week and their grade-point indexes for the courses they took in that semester:

| Hours spent on homework $x$ | Grade-point index $y$ |
|---|---|
| 15 | 2.0 |
| 28 | 2.7 |
| 13 | 1.3 |
| 20 | 1.9 |
| 4 | 0.9 |
| 10 | 1.7 |

Calculate $r$ and compare the result with the square root of the value of $r^2$ given in the printout of Figure 16.9.

**17.2** In Exercise 16.3 we gave data on the amount of food containing a certain preservative that $n = 10$ four-year-olds had consumed and their hyperactivity rating 45 minutes later. Given that $\sum x = 525$, $\sum x^2 = 32,085$, $\sum y = 100$, $\sum y^2 = 1,192$, and $\sum xy = 5,980$, calculate $r$.

**17.3** With reference to Exercise 17.2, what percentage of the total variation of the $y$'s (hyperactivity ratings) is accounted for by the relationship with $x$ (the amount of food eaten containing the preservative)?

**17.4** Following are the numbers of minutes it took $n = 12$ mechanics to assemble a piece of machinery in the morning, $x$, and in the late afternoon, $y$:

| $x$ | $y$ |
|---|---|
| 12 | 14 |
| 11 | 11 |
| 9 | 14 |
| 13 | 11 |
| 10 | 12 |
| 11 | 15 |
| 12 | 12 |
| 14 | 13 |
| 10 | 16 |
| 9 | 10 |
| 11 | 10 |
| 12 | 14 |

Given that $S_{xx} = 25.67$, $S_{xy} = -0.33$, and $S_{yy} = 42.67$, calculate $r$.

**17.5** Use a computer or a graphing calculator to calculate $r$ from the original data given in Exercise 17.4.

**17.6** The following data were obtained in a study of the relationship between the resistance (ohms) and the failure time (minutes) of certain overloaded resistors:

| Resistance | Failure time |
|:---:|:---:|
| 48 | 45 |
| 28 | 25 |
| 33 | 39 |
| 40 | 45 |
| 36 | 36 |
| 39 | 35 |
| 46 | 36 |
| 40 | 45 |
| 30 | 34 |
| 42 | 39 |
| 44 | 51 |
| 48 | 41 |
| 39 | 38 |
| 34 | 32 |
| 47 | 45 |

Demonstrate that the use of a computer or a graphing calculator gives the value of $r$ as 0.704. Also determine what percentage of the variation in failure times is due to differences in resistance.

**17.7** Following are the elevations (feet) and average high temperatures (degrees Fahrenheit) on Labor Day of eight cities in Arizona:

| Elevation | High temperature |
|:---:|:---:|
| 1,418 | 92 |
| 6,905 | 70 |
| 735 | 98 |
| 1,092 | 94 |
| 5,280 | 79 |
| 2,372 | 88 |
| 2,093 | 90 |
| 196 | 96 |

Demonstrate that the use of a computer or a graphing calculator gives the value of $r$ as −0.99. Also determine what percentage of the variation in the high temperatures is due to differences in elevation.

**17.8** After a student calculated $r$ for a large set of paired data, she discovered to her dismay that the variable that should have been labeled $x$ was labeled $y$ and the variable that should have been labeled $y$ was labeled $x$. Is there any reason for being dismayed?

**17.9** After a student had computed $r$ for the heights and weights of a large number of persons, he realized that the weights were given in kilograms and the heights were given in centimeters. He had wanted to obtain $r$ for the weights in pounds and the heights in inches. Given that there are 0.393 inch per centimeter and 2.2 pounds per kilogram, how should he correct his calculations?

**17.10**  If we calculate $r$ for each of the following sets of data, should we be surprised if we get $r = 1$ for (a) and $r = -1$ for (b)? Explain your answers.

| (a) | $x$ | $y$ | | (b) | $x$ | $y$ |
|-----|-----|-----|---|-----|-----|-----|
| | 6 | 9 | | | 12 | 5 |
| | 14 | 11 | | | 8 | 15 |

**17.11**  State in each case whether you would expect a positive correlation, a negative correlation, or no correlation:
(a) the ages of husbands and wives;
(b) the amount of rubber on tires and the number of miles they have been driven;
(c) the number of hours that golfers practice and their scores;
(d) shoe size and IQ;
(e) the weight of the load of trucks and their gasoline consumption.

**17.12**  State in each case whether you would expect a positive correlation, a negative correlation, or no correlation:
(a) pollen count and the sale of anti-allergy drugs;
(b) income and education;
(c) the number of sunny days in August in Detroit and the attendance at the Detroit Zoo;
(d) shirt size and sense of humor;
(e) number of persons getting flu shots and number of persons catching the flu.

**17.13**  If $r = 0.41$ for one set of paired data and $r = 0.29$ for another set of paired data, compare the strengths of the two relationships.

**17.14**  In a medical study it was found that $r = 0.70$ for the weight of babies six months old and their weight at birth, and that $r = 0.60$ for the weight of babies six months old and their average daily intake of food. Give a counterexample to show that it is not valid to conclude that weight at birth and food intake together account for

$$(0.70)^2 100\% + (0.60)^2 100\% = 85\%$$

of the variation of babies' weight when they are six months old. Can you explain why the conclusion is not valid?

**17.15**  Working with various socioeconomic data for recent years, a research worker got $r = 0.9225$ for the number of foreign language degrees offered by U.S. colleges and universities and the mileage of railroad track owned by U.S. railroads. Can we conclude that

$$(0.9225)^2 100\% \approx 85.1\%$$

of the variation in the foreign language degrees is accounted for by differences in the ownership of railroad track?

**17.16**  A student computed the correlation between height and weight for a large group of third-grade school children and obtained a value of $r = 0.32$. She was unable to decide whether she should conclude that being tall causes a child to put on more weight or that having excess weight enables a child to grow taller. Help her solve this dilemma.

**17.17**  In Example 17.2 we illustrated the use of the correlation coefficient in the analysis of a contingency table with ordered categories. Use the same procedure to analyze the following table reproduced from Exercise 14.29, where the relationship between the fidelity and the selectivity of radios was analyzed by means of the chi-square criterion:

| | | Fidelity | |
|---|---|---|---|
| | Low | Average | High |
| Low | 7 | 12 | 31 |
| Selectivity   Average | 35 | 59 | 18 |
| High | 15 | 13 | 0 |

**17.18** Use the same procedure as in Example 17.2 to analyze the following table reproduced from Exercise 14.24, where the relationship between bank employees' standard of dress and their professional advancement was analyzed by means of the chi-square criterion:

| | | Speed of advancement | |
|---|---|---|---|
| | Slow | Average | Fast |
| Very well dressed | 38 | 135 | 129 |
| **Standard of dress**   Well dressed | 32 | 68 | 43 |
| Poorly dressed | 13 | 25 | 17 |

Note that the speed of advancement goes from low to high, whereas the standard of dress goes from high to low.

# 17.3 CORRELATION ANALYSIS

When $r$ is calculated on the basis of sample data, we may get a strong positive or negative correlation purely by chance, even though there is actually no relationship whatever between the two variables under consideration.

Suppose, for instance, that we take a pair of dice, one red and one green, roll them five times, and get the following results:

| Red die | Green die |
|---|---|
| $x$ | $y$ |
| 4 | 5 |
| 2 | 2 |
| 4 | 6 |
| 2 | 1 |
| 6 | 4 |

Presumably, there is no relationship between $x$ and $y$, the numbers of points showing on the two dice. It is hard to see why large values of $x$ should go with large values of $y$ and small values of $x$ with small values of $y$, but calculating $r$, we get the surprisingly high value $r = 0.66$. This raises the question of whether something is wrong with the assumption that there is no relationship between $x$ and $y$, and to answer it we shall have to see whether the high value of $r$ may be attributed to chance.

When a correlation coefficient is calculated from sample data, as in the preceding example, the value we get for $r$ is only an estimate of a corresponding parameter, the **population correlation coefficient**, which we denote by $\rho$ (the Greek letter *rho*). What $r$ measures for a sample, $\rho$ measures for a population.

To make inferences about $\rho$ on the basis of $r$, we shall have to make several assumptions about the distributions of the random variables whose values we observe. In **normal correlation analysis** we make the same assumptions as in normal regression analysis (see page 410), except that the $x$'s are not constants, but values of a random variable having a normal distribution.

Since the sampling distribution of $r$ is rather complicated under these assumptions, it is common practice to base inferences about $\rho$ on the **Fisher Z transformation**, a change of scale from $r$ to $Z$, which is given by

$$Z = \frac{1}{2} \cdot \ln \frac{1+r}{1-r}$$

Here the abbreviation ln denotes "natural logarithm," that is, logarithm to the base $e$, where $e = 2.71828\ldots$. This transformation is named after R. A. Fisher, a prominent statistician who showed that under the assumptions of normal correlation analysis and for any value of $\rho$, the distribution of $Z$ is approximately normal with

$$\mu_Z = \frac{1}{2} \cdot \ln \frac{1+\rho}{1-\rho} \quad \text{and} \quad \sigma_Z = \frac{1}{\sqrt{n-3}}$$

Converting $Z$ to standard units (that is, subtracting $\mu_Z$ and then dividing by $\sigma_Z$), we arrive at the result that

STATISTIC FOR
INFERENCES
ABOUT $\rho$

$$z = (Z - \mu_Z)\sqrt{n-3}$$

has approximately the standard normal distribution. The application of this theory is greatly facilitated by the use of Table X at the end of the book, which gives the values of $Z$ corresponding to $r = 0.00, 0.01, 0.02, 0.03, \ldots$, and 0.99. Observe that only positive values are given in this table; if $r$ is negative, we simply look up $-r$ and take the negative of the corresponding $Z$. Note also that the formula for $\mu_Z$ is like that for $Z$ with $r$ replaced by $\rho$; therefore, Table X can be used to look up values of $\mu_Z$ corresponding to given values of $\rho$.

EXAMPLE  **17.4**  Using the 0.05 level of significance, test the null hypothesis of no correlation (that is, the null hypothesis $\rho = 0$) for the illustration on page 443, where we rolled a pair of dice five times and got $r = 0.66$.

**Solution**

1. $H_0 : \rho = 0$
   $H_A : \rho \neq 0$
2. $\alpha = 0.05$
3. Since $\mu_Z = 0$ for $\rho = 0$, reject the null hypothesis if $z \leq -1.96$ or $z \geq 1.96$, where

$$z = Z \cdot \sqrt{n-3}$$

   Otherwise, state that the value of $r$ is not significant.
4. Substituting $n = 5$ and $Z = 0.793$, the value of $Z$ corresponding to $r = 0.66$ according to Table X, we get

$$z = 0.793\sqrt{5-3}$$

$$= 1.12$$

5. Since $z = 1.12$ falls between $-1.96$ and $1.96$, the null hypothesis cannot be rejected. In other words, the value obtained for $r$ is not significant, as we should, of course, have expected. An alternative way of handling this kind of problem (namely, testing the null hypothesis $\rho = 0$) is given in Exercise 17.22. ■

**EXAMPLE 17.5** With reference to the hearing-range and airport-noise example, where we showed that $r = -0.95$ for $n = 12$, test the null hypothesis $\rho = -0.80$ against the alternative hypothesis $\rho < -0.80$ at the 0.01 level of significance.

**Solution**

1. $H_0 : \rho = -0.80$
   $H_A : \rho < -0.80$
2. $\alpha = 0.01$
3. Reject the null hypothesis if $z \geq -2.33$, where

$$z = (Z - \mu_Z)\sqrt{n-3}$$

4. Substituting $n = 12$, $Z = -1.832$ corresponding to $r = -0.95$, and $\mu_Z = -1.099$ corresponding to $\rho = -0.80$, we get

$$z = [-1.832 - (-1.099)]\sqrt{12-3}$$

$$\approx -2.20$$

5. Since $-2.20$ falls between $-2.33$ and $2.33$, the null hypothesis cannot be rejected. ■

To construct confidence intervals for $\rho$, we first construct confidence intervals for $\mu_Z$, and then convert to $r$ and $\rho$ by means of Table X. A confidence interval formula for $\mu_Z$ may be obtained by substituting

$$z = (Z - \mu_Z)\sqrt{n-3}$$

for the middle term of the double inequality $-z_{\alpha/2} < z < z_{\alpha/2}$, and then manipulating the terms algebraically so that the middle term is $\mu_Z$. This leads to the following $(1 - \alpha)100\%$ confidence interval for $\mu_Z$:

CONFIDENCE
INTERVAL FOR $\mu_Z$

$$Z - \frac{z_{\alpha/2}}{\sqrt{n-3}} < \mu_Z < Z + \frac{z_{\alpha/2}}{\sqrt{n-3}}$$

**EXAMPLE 17.6** If $r = 0.62$ for the cost estimates of two mechanics for a random sample of $n = 30$ repair jobs, construct a 95% confidence interval for the population correlation coefficient $\rho$.

**Solution** Getting $Z = 0.725$, corresponding to $r = 0.62$, from Table X, and substituting it together with $n = 30$ and $z_{0.025} = 1.96$ into the preceding confidence interval formula for $\mu_Z$, we find that

$$0.725 - \frac{1.96}{\sqrt{27}} < \mu_Z < 0.725 + \frac{1.96}{\sqrt{27}}$$

or

$$0.348 < \mu_Z < 1.102$$

Finally, looking up the values of $r$ that come closest to $Z = 0.348$ and $Z = 1.102$ in Table X, we get the 95% confidence interval

$$0.33 < \rho < 0.80$$

for the true strength of the linear relationship between cost estimates made by the two mechanics. ∎

**17.19** Assuming that the assumptions for a normal correlation analysis are met, test the null hypothesis $\rho = 0$ against the alternative hypothesis $\rho \neq 0$ at the 0.05 level of significance, given that
(a) $n = 15$ and $r = 0.59$;
(b) $n = 20$ and $r = 0.41$;
(c) $n = 40$ and $r = 0.36$.

**17.20** Assuming that the assumptions for a normal correlation analysis are met, test the null hypothesis $\rho = 0$ against the alternative hypothesis $\rho \neq 0$ at the 0.01 level of significance, given that
(a) $n = 14$ and $r = 0.54$;
(b) $n = 22$ and $r = -0.61$;
(c) $n = 44$ and $r = 0.42$.

**17.21** Assuming that the assumptions for a normal correlation analysis are met, test the null hypothesis $\rho = -0.50$ against the alternative hypothesis $\rho > -0.50$ at the 0.01 level of significance, given that
(a) $n = 17$ and $r = -0.22$;
(b) $n = 34$ and $r = -0.43$.

**17.22** Under the assumptions of normal correlation analysis, the test of the null hypothesis $\rho = 0$ may also be based on the statistic

$$t = \frac{r\sqrt{n-2}}{\sqrt{1-r^2}}$$

which has the $t$ distribution with $n - 2$ degrees of freedom. Use this statistic to test in each case whether the value of $r$ is significant at the 0.05 level of significance:

(a) $n = 12$ and $r = 0.77$;

(b) $n = 16$ and $r = 0.49$.

**17.23** Rework Exercise 17.22 with the level of significance changed to 0.01.

**17.24** The $t$ statistic given in Exercise 17.22 is identical with the $t$ statistic that tests $\beta = 0$ in linear regression analysis, and its value is provided by the linear regression programs of statistical software and graphing calculators. This is illustrated by the following:

(a) Use the linear regression program of a statistical software package or a graphing calculator to obtain the values of $t$ and $r$ for the data of Exercise 16.8.

(b) Substitute the value obtained for $r$ in part (a) and $n = 8$ into the formula for $t$ given in Exercise 17.22.

(c) Compare the two values of $t$.

**17.25** In a study of the relationship between the death rate from lung cancer and the per capita consumption of cigarettes 20 years earlier, data for $n = 9$ countries yielded $r = 0.73$. At the 0.05 level of significance, test the null hypothesis $\rho = 0.50$ against the alternative hypothesis $\rho > 0.50$.

**17.26** In a study of the relationship between the available heat (per cord) of green wood and air-dried wood, data for $n = 13$ kinds of wood yielded $r = 0.94$. Use the 0.01 level of significance to test the null hypothesis $\rho = 0.75$ against the alternative hypothesis $\rho \neq 0.75$.

**17.27** Assuming that the assumptions for a normal correlation analysis are met, use the Fisher $Z$ transformation to construct 95% confidence intervals for $\rho$, given that

(a) $n = 15$ and $r = 0.80$;

(b) $n = 28$ and $r = -0.24$;

(c) $n = 63$ and $r = 0.55$.

**17.28** Assuming that the assumptions for a normal correlation analysis are met, use the Fisher $Z$ transformation to construct 99% confidence intervals for $\rho$, given that

(a) $n = 20$ and $r = -0.82$;

(b) $n = 25$ and $r = 0.34$;

(c) $n = 75$ and $r = 0.18$.

# *17.4 MULTIPLE AND PARTIAL CORRELATION

In Section 17.1 we introduced the coefficient of correlation as a measure of the goodness of the fit of a least-squares line to a set of paired data. If predictions are to be made with an equation of the form

$$\hat{y} = b_0 + b_1 x_1 + b_2 x_2 + \cdots + b_k x_k$$

obtained by the method of least squares as in Section 16.4, we define the **multiple correlation coefficient** in the same way in which we originally defined $r$. We take the square root of the quantity

$$\frac{\sum (\hat{y} - \bar{y})^2}{\sum (y - \bar{y})^2}$$

which is the proportion of the total variation of the $y$'s that can be attributed to the relationship with the $x$'s. The only difference is that we now calculate $\hat{y}$ by means of the multiple regression equation instead of the equation $\hat{y} = a + bx$.

To give an example, let us refer to Example 16.9, where we based a multiple linear regression equation on the computer printout shown in Figure 16.10. As we pointed out at the time, part of the printout had been deleted, but since that part is needed now, the complete printout is reproduced in Figure 17.5.

**EXAMPLE   17.7**   Use the definition of the multiple correlation coefficient given above to determine its value for the data of Example 16.9, which dealt with the number of bedrooms, the number of baths, and the prices at which eight one-family houses had recently been sold.

**Solution**   In the analysis of variance of Figure 17.5 (in the column headed by SS), we find that the regression sum of squares is 4,877,608,452 and that the total sum of squares is 5,455,580,000. Thus, the multiple correlation coefficient equals the square root of

$$\frac{4,877,608,452}{5,455,580,000} \approx 0.894$$

Denoting it by $R$, we write $R \approx \sqrt{0.894} = 0.95$ rounded to two decimals, which is considered as "essentially nonnegative" according to the Kendall and Buckland dictionary of statistical terms. In actual practice, $R^2$ is used more often than $R$, and it should be observed that its value, denoted by R-Sq, is actually given as 89.4% in the printout of Figure 17.5.

This example also serves to illustrate that adding more independent variables in a correlation study may not be sufficiently productive to justify the extra work.

**Figure 17.5**
Complete printout for the multiple regression example.

**Regression Analysis: y versus x1, x2**

```
The regression equation is
y = 65430 + 16752 x1 + 11235 x2

Predictor    Coef    SE Coef        T        P
Constant    65430      12134     5.39    0.003
x1          16752       6636     2.52    0.053
x2          11235       9885     1.14    0.307

S = 10751    R-Sq = 89.4%   R-Sq(adj) = 85.2%

Analysis of Variance

Source            DF          SS         MS        F        P
Regression         2  4877608452  2438804226   21.10    0.004
Residual Error     5   577971548   115594310
Total              7  5455580000

Source            DF      Seq SS
x1                 1  4728284225
x2                 1   149324227
```

As it can be shown that $r = 0.93$ for $y$ and $x_1$ (number of bedrooms) alone, it is apparent that very little is gained by also considering $x_2$ (number of baths). The situation is quite different, though, in Exercise 17.30, where the two independent variables $x_1$ and $x_2$ together account for a much higher proportion of the total variation in $y$ than does either $x_1$ or $x_2$ alone.

When we discussed the problem of correlation and causation, we showed that a strong correlation between two variables may be due entirely to their dependence on a third variable. We illustrated this with the examples of chewing gum sales and the crime rate, and child births and the number of storks. To give another example, let us consider the two variables $x_1$, the weekly amount of hot chocolate sold by a refreshment stand at a summer resort, and $x_2$, the weekly number of visitors to the resort. If, on the basis of suitable data, we get $r = -0.30$ for these variables, this should come as a surprise—after all, we would expect more sales of hot chocolate when there are more visitors, and vice versa, and hence a positive correlation.

However, if we think for a moment, we may surmise that the negative correlation of $-0.30$ may well be due to the fact that the variables $x_1$ and $x_2$ are both related to a third variable $x_3$, the average weekly temperature at the resort. If the temperature is high, there will be more visitors, but they will prefer cold drinks to hot chocolate; if the temperature is low, there will be fewer visitors, but they will prefer hot chocolate to cold drinks. So, let us suppose that further data yield $r = -0.70$ for $x_1$ and $x_3$, and $r = 0.80$ for $x_2$ and $x_3$. These values seem reasonable since low sales of hot chocolate should go with high temperatures and vice versa, while the number of visitors should be high when the temperature is high, and low when the temperature is low.

In the preceding example, we should really have investigated the relationship between $x_1$ and $x_2$ (hot chocolate sales and the number of visitors to the resort) when all other factors, primarily temperature, are held fixed. As it is seldom possible to control matters to such an extent, it has been found that a statistic called the **partial correlation coefficient** does a fair job of eliminating the effects of other variables. If we write the ordinary correlation coefficients for $x_1$ and $x_2$, $x_1$ and $x_3$, and $x_2$ and $x_3$, as $r_{12}$, $r_{13}$, and $r_{23}$, the partial correlation coefficient for $x_1$ and $x_2$, with $x_3$ fixed, is given by

PARTIAL
CORRELATION
COEFFICIENT

$$r_{12.3} = \frac{r_{12} - r_{13} \cdot r_{23}}{\sqrt{1 - r_{13}^2}\sqrt{1 - r_{23}^2}}$$

EXAMPLE **17.8**    Calculate $r_{12.3}$ for the preceding example, which dealt with the sales of hot chocolate and the number of persons who visit the resort.

**Solution**    Substituting $r_{12} = -0.30$, $r_{13} = -0.70$, and $r_{23} = 0.80$ into the formula for $r_{12.3}$, we get

$$r_{12.3} = \frac{(-0.30) - (-0.70)(0.80)}{\sqrt{1 - (-0.70)^2}\sqrt{1 - (0.80)^2}} \approx 0.607$$

This result shows that, as we expected, there is a positive relationship between the sales of hot chocolate and the number of visitors to the resort when the effect of differences in temperature is eliminated.   ▪▪

We have given this example primarily to illustrate what we mean by a partial correlation coefficient, but it also serves as a reminder that correlation coefficients can be very misleading unless they are interpreted with care.

EXERCISES

**17.29**   *In a multiple regression problem, the regression sum of squares is

$$\sum (\hat{y} - \bar{y})^2 = 45{,}225$$

and the total sum of squares is

$$\sum (y - \bar{y})^2 = 136{,}210$$

Find the value of the multiple correlation coefficient.

**\*17.30**   With reference to Exercise 16.34 on page 420, use the same software as before to obtain the multiple correlation coefficient. Also determine the correlation coefficients for $y$ and $x_1$ (age) alone and for $y$ and $x_2$ (years postgraduate work) alone, and compare them with the multiple correlation coefficient.

**17.31**   With reference to Exercise 16.34 on page 420, use the same software as in that exercise to obtain the multiple correlation coefficient.

**17.32**   A team of research workers conducted an experiment to see if the height of certain rose bushes can be predicted on the basis of the amount of fertilizer and the amount of irrigation that is applied to the soil. For predicting the height on the basis of both variables, they obtained a multiple correlation coefficient of 0.58; for predicting the height on the basis of the fertilizer alone, they obtained $r = 0.66$. Comment on these results.

**17.33**   With reference to Exercise 16.35 on page 421, use a computer or a graphing calculator to determine the necessary correlation coefficients in order to calculate the partial correlation coefficient for hardness and annealing temperature when the copper content is held fixed.

**17.34**   An experiment yielded the following results: $r_{12} = 0.80$, $r_{13} = -0.70$, and $r_{23} = 0.90$. Explain why these figures cannot all be correct.

## CHECKLIST OF KEY TERMS (with page references to their definitions)

Coefficient of correlation, 434
Coefficient of determination, 434
Fisher $Z$ transformation, 444
*Multiple correlation coefficient, 447
Negative correlation, 434
Normal correlation analysis, 444
*Partial correlation coefficient, 449
Population correlation coefficient, 444

Positive correlation, 434
Product moment, 436
Product-moment coefficient
   of correlation, 436
Regression sum of squares, 433
Residual sum of squares, 433
Sample covariance, 437
Total sum of squares, 433

# References

*More detailed information about multiple and partial correlation may be found in*

EZEKIEL, M., and FOX, K. A., *Methods of Correlation and Regression Analysis*, 3rd ed. New York: John Wiley & Sons, Inc., 1959.

HARRIS, R. J., *A Primer of Multivariate Statistics*. New York: Academic Press, Inc., 1975.

*and an advanced theoretical treatment is given in Volume 2 of*

KENDALL, M. G., and STUART, A., *The Advanced Theory of Statistics*, 3rd ed. New York: Hafner Press, 1973.

*Volume 1 of this book provides the theoretical foundation of significance tests for r.*

# 18
# NONPARA-
# METRIC TESTS

**18.1**     The Sign Test  453

**18.2**     The Sign Test (Large Samples)  455

*__18.3__    The Signed-Rank Test  458

*__18.4__    The Signed-Rank Test (Large Samples)  462

**18.5**     The $U$ Test  465

**18.6**     The $U$ Test (Large Samples)  469

**18.7**     The $H$ Test  471

**18.8**     Tests of Randomness: Runs  475

**18.9**     Tests of Randomness: Runs (Large Samples)  476

**18.10**    Tests of Randomness: Runs Above and Below
              the Median  477

**18.11**    Rank Correlation  479

**18.12**    Some Further Considerations  483

**18.13**    Summary  483

             Checklist of Key Terms  484

             References  484

To treat problems in which the various assumptions underlying standard tests cannot be met, statisticians have developed many alternative techniques which have become known as **nonparametric tests**. This name is meant to imply that we are *not* testing hypotheses concerning the parameters of populations *of a given kind*. Many of these tests can also be classified under the heading of "short-cut statistics." They are usually easy to perform and require fewer computations than the corresponding "standard tests." As should be expected, though, by making fewer assumptions we expose ourselves to greater risks. In other words, for a fixed $\alpha$, a nonparametric test is apt to expose us to greater probabilities of committing Type II errors.

In Sections 18.1 and 18.2 we present the sign test as a nonparametric alternative to tests concerning means and to tests concerning differences between means based on paired data. Another nonparametric test that serves the same purpose but is less wasteful of information is given in Sections 18.3 and 18.4. In Sections 18.5 through 18.7 we present a nonparametric alternative to tests concerning the difference between the means of independent samples and a somewhat similar alternative to the one-way analysis of variance. In Sections 18.8 through 18.10 we shall learn how to test the randomness of a sample after the data have actually been collected; and in Section 18.11 we present a nonparametric test of the significance of a relationship between paired data. Finally, in Section 18.12 we mention some of the weaknesses of nonparametric methods, and in Section 18.13 we give a table that lists the various nonparametric tests and the "standard" test that they replace.

# 18.1 THE SIGN TEST

Except for the large-sample tests, all the tests concerning means that we studied in Chapter 12 were based on the assumption that the populations sampled have roughly the shape of normal distributions. When this assumption is untenable in practice, these standard tests can be replaced by any one of several nonparametric alternatives, and these are the subject matter of Sections 18.1 through 18.7. Simplest among these is the **sign test**, which we shall study in this section and in Section 18.2.

The **one-sample sign test** applies when we sample a continuous population, so that the probability of getting a sample value less than the median and the probability of getting a sample value greater than the median are both $\frac{1}{2}$. Of course, when the population is symmetrical, the mean $\mu$ and the median $\tilde{\mu}$ will coincide, and we can phrase the hypotheses in terms of either of these parameters.

To test the null hypothesis $\tilde{\mu} = \tilde{\mu}_0$ against an appropriate alternative on the basis of a random sample of size $n$, we replace each sample value greater than $\tilde{\mu}_0$ with a plus sign and each sample value less than $\tilde{\mu}_0$ with a minus sign. Then we test the null hypothesis that the number of plus signs are the values of a random variable having the binomial distribution with $p = \frac{1}{2}$. If any sample value actually equals $\tilde{\mu}_0$, which can easily happen when we deal with rounded data, we simply discard it.

To perform a one-sample sign test when the sample is fairly small, we refer directly to a table of binomial probabilities, such as Table V at the end of the book or the National Bureau of Standards table referred to among the references on page 205. Alternatively, we may use a computer or a calculator to obtain the required binomial probabilities. When the sample is large, however, we use the normal approximation to the binomial distribution, as is illustrated in Section 18.2.

**EXAMPLE  18.1**   To check a teacher's claim that the published value for the coefficient of friction for well-oiled metals, 0.050, might be too low, a science class made 18 determinations getting 0.054, 0.052, 0.044, 0.056, 0.050, 0.051, 0.055, 0.053, 0.047, 0.053, 0.052, 0.050, 0.051, 0.051, 0.054, 0.046, 0.053, and 0.043. Ordinarily, the one-sample $t$ test would be a logical choice to test the claim, but the skewness of the data suggests the use of a nonparametric alternative. Therefore, the teacher suggests to his science class to use the one-sample sign test to test the null hypothesis $\tilde{\mu} = 0.050$ against the alternative $\tilde{\mu} > 0.050$ at the 0.05 level of significance.

**Solution**   1.  $H_0 : \tilde{\mu} = 0.050$
$H_A : \tilde{\mu} > 0.050$
2.  $\alpha = 0.05$
3′.  The test statistic is the number of plus signs, namely, the number of values exceeding 0.050.

**4'.** Replacing each value greater than 0.050 with a plus sign, each value less than 0.050 with a minus sign, and discarding the two values that equal 0.050, we get

$$+ + - + + + + - + + + + + - + -$$

Thus, $x = 12$, and Table V shows that for $n = 16$ and $p = 0.50$ the probability of $x \geq 12$, the $p$-value, is $0.028 + 0.009 + 0.002 = 0.039$.

**5'.** Since 0.039 is less than 0.05, the null hypothesis must be rejected. The data support the claim that the published value of the coefficient of friction is too low. ◼

Note that we used the alternative method for testing hypotheses as described on page 303. As in Section 14.1, the alternative method simplifies matters when tests are based directly on binomial tables. Although it would not seem necessary, we could have used a computer for Example 18.1. If we had done so, we would have obtained a printout like the one shown in Figure 18.1. The difference between the two $p$-values, 0.039 and 0.0384, is, of course, due to rounding.

**Figure 18.1**
MINITAB printout for
Example 18.1.

```
Sign Test for Median: Data

Sign test of median = 0.05000 versus  >  0.05000

                N    BELOW  EQUAL  ABOVE        P    MEDIAN
Data           18        4      2     12   0.0384   0.05150
```

The sign test can also be used when we deal with paired data as in Section 12.7. In such problems, each pair of sample values is replaced with a plus sign if the first value is greater than the second, with a minus sign if the first value is smaller than the second, and pairs of equal values are discarded. For paired data, the sign test is used to test the null hypothesis that the median of the population of differences is zero. When it is used in this way, it is referred to as the **paired-sample sign test**.

**EXAMPLE 18.2**    In Example 12.9 we gave the following data on the average weekly losses in hours of labor due to accidents in 10 industrial plants before and after the installation of an elaborate safety program:

| | | | | |
|---|---|---|---|---|
| 45 and 36 | 73 and 60 | 46 and 44 | 124 and 119 | 33 and 35 |
| 57 and 51 | 83 and 77 | 34 and 29 | 26 and 24 | 17 and 11 |

Based on the paired-sample $t$ test, we showed at the 0.05 level of significance that the safety program is effective. Use the paired-sample sign test to rework this example.

**Solution**

**1.** $H_0 : \tilde{\mu}_D = 0$, where $\tilde{\mu}_D$ is the median of the population of differences sampled.
$H_A : \tilde{\mu}_D > 0$

**2.** $\alpha = 0.05$

**3'.** The test statistic is the number of plus signs, namely the number of industrial plants in which the average weekly losses in hours of labor has decreased.

**4'.** Replacing each pair of values with a plus sign if the first value is greater than the second, and with a minus sign if the first value is smaller than the second, we get

$$+ + + + - + + + + +$$

Thus, $x = 9$, and Table V shows that for $n = 10$ and $p = 0.50$ the probability of $x \geq 9$, the $p$-value, is $0.010 + 0.001 = 0.011$.

**5'.** Since 0.011 is less than 0.05, the null hypothesis must be rejected. As in Example 12.9, we conclude that the safety program is effective.

# 18.2 THE SIGN TEST (Large Samples)

When $np$ and $n(1 - p)$ are both greater than 5, so that we can use the normal approximation to the binomial distribution, we can base the sign test on the large-sample test of Section 14.2, namely, on the statistic

$$z = \frac{x - np_0}{\sqrt{np_0(1 - p_0)}}$$

with $p_0 = 0.50$, which has approximately the standard normal distribution. When $n$ is small, it may be wise to use the continuity correction suggested on page 328. This is true, especially if, without the continuity correction, we can *barely* reject the null hypothesis. As we have pointed out before, the continuity correction does not have to be considered when *without* it the null hypothesis cannot be rejected.

**EXAMPLE 18.3**   In Exercise 2.44 we gave the following attendance figures for 48 outings sponsored by the alumni association of a university for its single members:

| 28 | 51 | 31 | 38 | 27 | 35 | 33 | 40 | 37 | 28 | 33 | 27 |
|----|----|----|----|----|----|----|----|----|----|----|----|
| 33 | 31 | 41 | 46 | 40 | 36 | 53 | 23 | 33 | 27 | 40 | 30 |
| 33 | 22 | 37 | 38 | 36 | 48 | 22 | 36 | 45 | 34 | 26 | 28 |
| 40 | 42 | 43 | 41 | 35 | 50 | 31 | 48 | 38 | 33 | 39 | 35 |

Use the one-sample sign test to test the null hypothesis $\tilde{\mu} = 32$ against the alternative hypothesis that $\tilde{\mu} \neq 32$ at the 0.01 level of significance.

**Solution**   **1.** $H_0 : \tilde{\mu} = 32$
$H_A : \tilde{\mu} \neq 32$

**2.** $\alpha = 0.01$

**3.** Reject the null hypothesis if $z \leq -2.575$ or $z \geq 2.575$, where

$$z = \frac{x - np_0}{\sqrt{np_0(1 - p_0)}}$$

with $p_0 = 0.50$; otherwise, accept the null hypothesis or reserve judgment.

**4.** Counting the number of values exceeding 32 (plus signs), the number of values less than 32 (minus signs), and the number of values that are equal to 32, we get 34, 14, and 0 (and, hence none have to be discarded). Thus, $x = 34, n = 48$, and

$$z = \frac{34 - 48(0.50)}{\sqrt{48(0.50)(0.50)}} \approx 2.89$$

**5.** Since 2.89 exceeds 2.575, the null hypothesis must be rejected. (Had we used the continuity correction, we would have obtained $z = 2.74$, and the conclusion would have been the same.) ▪■

The next example illustrates the importance of using the continuity correction when, without it, we could barely reject the null hypothesis.

**EXAMPLE 18.4**   Following are two supervisors' ratings of the performance of a random sample of a large company's employees, on a scale from 0 to 100:

| Supervisor 1 | Supervisor 2 |
|:---:|:---:|
| 88 | 73 |
| 69 | 67 |
| 97 | 81 |
| 60 | 73 |
| 82 | 78 |
| 90 | 82 |
| 65 | 62 |
| 77 | 80 |
| 86 | 81 |
| 79 | 79 |
| 65 | 77 |
| 95 | 82 |
| 88 | 84 |
| 91 | 93 |
| 68 | 66 |
| 77 | 76 |
| 74 | 74 |
| 85 | 78 |

Use the paired-sample sign test (based on the normal approximation to the binomial distribution) to test at the 0.05 level of significance whether the differences between the two sets of ratings can be attributed to chance,

**(a)** without using the continuity correction;

**(b)** using the continuity correction.

**Solution (a) 1.** $H_0 : \tilde{\mu}_D = 0$, where $\tilde{\mu}_D$ is the median of the population of differences (between the supervisors' ratings) sampled.

$H_A : \tilde{\mu}_D \neq 0$

**2.** $\alpha = 0.05$

**3.** Reject the null hypothesis if $z \leq -1.96$ or $z \geq 1.96$, where

$$z = \frac{x - np_0}{\sqrt{np_0(1 - p_0)}}$$

with $p_0 = 0.50$; otherwise, accept the null hypothesis or reserve judgment.

**4.** Counting the number of positive differences (plus signs), the number of negative differences (minus signs), and the number of pairs that are equal (and, hence, have to be discarded), we find that there are, respectively, 12, 4, and 2. Thus, $x = 12$ and $n = 16$, and since $np = 16(0.50) = 8$ and $n(1 - p) = 16(0.50) = 8$ are both greater than 5, we are justified in using getting a value less than the median equals the probability of getting a value greater than the median. We shall illustrate it here with measurements of the octane rating of a certain brand of premium gasoline, on the basis of which we will test the null hypothesis $\tilde{\mu} = 98.5$ against the alternative hypothesis $\tilde{\mu} < 98.5$ at the 0.01 level of significance.

The actual data are shown in the left-hand column of the following table, and the middle column contains the differences obtained by subtracting 98.5 from each measurement:

| Measurements | Differences | Ranks |
|---|---|---|
| 97.5 | −1.0 | 4 |
| 95.2 | −3.3 | 12 |
| 97.3 | −1.2 | 6 |
| 96.0 | −2.5 | 10 |
| 96.8 | −1.7 | 7 |
| 100.3 | 1.8 | 8 |
| 97.4 | −1.1 | 5 |
| 95.3 | −3.2 | 11 |
| 93.2 | −5.3 | 14 |
| 99.1 | 0.6 | 2 |
| 96.1 | −2.4 | 9 |
| 97.6 | −0.9 | 3 |
| 98.2 | −0.3 | 1 |
| 98.5 | 0.0 | |
| 94.9 | −3.6 | 13 |

After we discard the zero difference, we find that the smallest numerical difference is 0.3, the next smallest numerical difference is 0.6, the next smallest after that is 0.9, ..., and the largest numerical difference is 5.3. These ranks are shown in the third column, and it follows that

$$T^+ = 8 + 2 = 10$$

$$T^- = 4 + 12 + 6 + 10 + 7 + 5 + 11 + 14 + 9 + 3 + 1 + 13$$

$$= 95$$

and, hence, $T = 10$. Since $T^+ + T^-$ equals the sum of the integers from 1 through $n$, namely, $\frac{n(n+1)}{2}$, we could have obtained $T^-$ more easily by subtracting $T^+ = 10$ from $\frac{14 \cdot 15}{2} = 105$. [There were no ties in rank in this example, but as we pointed out earlier, when there are ties in rank, we assign to each of the tied values (differences) the mean of the ranks that they jointly occupy.]

The close relationship among $T^+$, $T^-$, and $T$ is also reflected by their sampling distributions, an example of which, for $n = 5$, is shown in Figure 18.2. Since there is a fifty-fifty chance for each rank to go to one of the positive differences or one of the negative differences, there are altogether $2^n$ possibilities, each with the probability $(\frac{1}{2})^n$. To get the probabilities associated with the various values of $T^+$, $T^-$, and $T$, we count the number of ways in which these values of $T^+$, $T^-$, and $T$ can be obtained and multiply by $(\frac{1}{2})^n$. For instance, for $n = 5$ and $T^+ = 6$ there

**Figure 18.2**
Distribution of $T^+$, $T^-$,
and $T$ for $n = 5$.

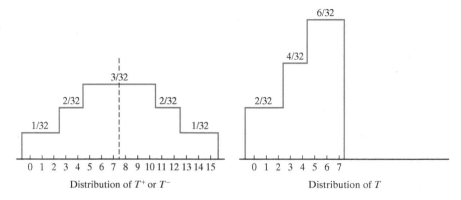

Distribution of $T^+$ or $T^-$          Distribution of $T$

are the three possibilities 1 and 5, 2 and 4, and 1 and 2 and 3, and the probability is $3 \cdot (\frac{1}{2})^5 = \frac{3}{32}$, as shown in Figure 18.2.

To simplify the construction of tables of critical values, we shall base all tests of the null hypothesis $\tilde{\mu} = \tilde{\mu}_0$ on the distribution of $T$ and reject it for values falling into the left-hand tail. We have to be careful, though, to use the right statistic and the right critical value. When $\tilde{\mu} < \tilde{\mu}_0$ then $T^+$ tends to be small, so when the alternative hypothesis is $\tilde{\mu} < \tilde{\mu}_0$ we base the test on $T^+$; when $\tilde{\mu} > \tilde{\mu}_0$ then $T^-$ tends to be small, so when the alternative hypothesis is $\tilde{\mu} > \tilde{\mu}_0$ we base the test on $T^-$; and when $\tilde{\mu} \neq \tilde{\mu}_0$ then either $T^+$ or $T^-$ tends to be small, so when the alternative hypothesis is $\tilde{\mu} \neq \tilde{\mu}_0$ we base the test on $T$. These relationships are summarized in the following table:

| Alternative hypothesis | Reject the null hypothesis if | Accept the null hypothesis or reserve judgment if |
|:---:|:---:|:---:|
| $\tilde{\mu} < \tilde{\mu}_0$ | $T^+ \leq T_{2\alpha}$ | $T^+ > T_{2\alpha}$ |
| $\tilde{\mu} > \tilde{\mu}_0$ | $T^- \leq T_{2\alpha}$ | $T^- > T_{2\alpha}$ |
| $\tilde{\mu} \neq \tilde{\mu}_0$ | $T \leq T_\alpha$ | $T > T_\alpha$ |

The necessary values of $T_\alpha$, which are the largest values of $T$ for which the probability of $T \leq T_\alpha$ does not exceed $\alpha$, may be found in Table VI at the end of the book; the blank spaces in the table indicate that the null hypothesis cannot be rejected regardless of the value we obtain for the test statistic. Note that the same critical values serve for tests at different levels of significance depending on whether the alternative hypothesis is one sided or two sided.

**EXAMPLE 18.5**  With reference to the octane ratings on page 459, use the signed-rank test at the 0.01 level of significance to test the null hypothesis $\tilde{\mu} = 98.5$ against the alternative hypothesis $\tilde{\mu} < 98.5$.

**Solution**   **1.** $H_0 : \tilde{\mu} = 98.5$
$H_A : \tilde{\mu} < 98.5$

2. $\alpha = 0.01$

3. Reject the null hypothesis if $T^+ \leq 16$, where 16 is the value of $T_{0.02}$ for $n = 14$; otherwise, accept it or reserve judgment.

4. As shown on page 459, $T^+ = 10$.

5. Since $T^+ = 10$ is less than 16, the null hypothesis must be rejected. We conclude that the median octane rating of the given premium gasoline is less than 98.5. ◼

Had we wanted to use a computer in this example, we would have obtained a printout like the MINITAB printout shown in Figure 18.3. One advantage of using a computer is that we do not have to refer to a special table; Figure 18.3 gives the *p*-value as 0.004. This would also have led to the rejection of the null hypothesis.

**Figure 18.3**
Computer printout for Example 18.5.

| **Wilcoxon Signed Rank Test: Octanes** | | | | |
|---|---|---|---|---|
| Test of median = 98.50 versus median < 98.50 | | | | |
| | N | N for Test | Wilcoxon Statistic | P | Estimated Median |
| Octanes | 15 | 14 | 10.0 | 0.004 | 96.85 |

The signed-rank test can also be used as an alternative to the paired-sample sign test. The procedure is exactly the same, but when we write the null hypothesis as $\tilde{\mu}_D = 0$, then $\tilde{\mu}_D$ is the median of the population of differences sampled.

**EXAMPLE 18.6**   Use the signed-rank test to rework Example 18.2. The actual data, the average weekly losses in hours of labor due to accidents in 10 industrial plants before and after the installation of the safety program, are shown in the left-hand column of the following table. The middle column contains their differences, and, discarding signs, the ranks of the numerical differences are shown in the column on the right.

| Losses in man-hours before and after | Differences | Ranks |
|---|---|---|
| 45 and 36 | 9 | 9 |
| 73 and 60 | 13 | 10 |
| 46 and 44 | 2 | 2 |
| 124 and 119 | 5 | 4.5 |
| 33 and 35 | −2 | 2 |
| 57 and 51 | 6 | 7 |
| 83 and 77 | 6 | 7 |
| 34 and 29 | 5 | 4.5 |
| 26 and 24 | 2 | 2 |
| 17 and 11 | 6 | 7 |

Thus, $T^- = 2$ and $T^+ = 53$.

**Solution**
1. $H_0 : \tilde{\mu}_D = 0$, where $\tilde{\mu}_D$ is the median of the population of differences (between losses in hours of labor before and after the installation of the safety program) sampled.
   $H_A : \tilde{\mu}_D > 0$
2. $\alpha = 0.05$
3. Reject the null hypothesis if $T^- \leq 11$, where 11 is the value of $T_{0.10}$ for $n = 10$; otherwise, accept the null hypothesis or reserve judgment.
4. As shown previously, $T^- = 2$.
5. Since $T^- = 2$ is less than 11, the null hypothesis must be rejected. We conclude that the safety program is effective. (Had we used a computer for this example, we would have obtained a $p$-value of 0.005, and the conclusion would have been the same.) ∎

# *18.4 THE SIGNED-RANK TEST (Large Samples)

When $n$ is 15 or more, it is considered reasonable to assume that the distributions of $T^+$ and $T^-$ can be approximated closely with normal curves. In that case we can base all tests on either $T^+$ or $T^-$, and as it does not matter which one we choose, we shall use here the statistic $T^+$.

Based on the assumption that each difference is as likely to be positive as negative, it can be shown that the mean and the standard deviation of the sampling distribution of $T^+$ are

**MEAN AND STANDARD DEVIATION OF $T^+$ STATISTIC**

and

$$\mu_{T^+} = \frac{n(n+1)}{4}$$

$$\sigma_{T^+} = \sqrt{\frac{n(n+1)(2n+1)}{24}}$$

Thus, for large samples, in this case, $n \geq 15$, we can base the signed-rank test on the statistic

**STATISTIC FOR LARGE-SAMPLE SIGNED-RANK TEST**

$$z = \frac{T^+ - \mu_{T^+}}{\sigma_{T^+}}$$

which is a value of a random variable having approximately the standard normal distribution. When the alternative hypothesis is $\tilde{\mu} \neq \tilde{\mu}_0$ (or $\tilde{\mu}_D \neq 0$), we reject the null hypothesis if $z \leq -z_{\alpha/2}$ or $z \geq z_{\alpha/2}$; when the alternative hypothesis is $\tilde{\mu} > \tilde{\mu}_0$ (or $\tilde{\mu}_D > 0$), we reject the null hypothesis if $z \geq z_\alpha$; and when the alternative hypothesis is $\tilde{\mu} < \tilde{\mu}_0$ (or $\tilde{\mu}_D < 0$), we reject the null hypothesis if $z \leq -z_\alpha$.

**EXAMPLE 18.7** The following are the weights in pounds, before and after, of 16 persons who stayed on a certain weight-reducing diet for two weeks: 169.0 and 159.9, 188.6 and 181.3, 222.1 and 209.0, 160.1 and 162.3, 187.5 and 183.5, 202.5 and 197.6, 167.8 and

171.4, 214.3 and 202.1, 143.8 and 145.1, 198.2 and 185.5, 166.9 and 158.6, 142.9 and 145.4, 160.5 and 159.5, 198.7 and 190.6, 149.7 and 149.0, and 181.6 and 183.1. Use the large-sample signed-rank test at the 0.05 level of significance to test whether the weight-reducing diet is effective.

**Solution**

1. $H_0 : \tilde{\mu}_D = 0$, where $\tilde{\mu}_D$ is the median of the population of differences (between weights before and after) sampled.
   $H_A : \tilde{\mu}_D > 0$
2. $\alpha = 0.05$
3. Reject the null hypothesis if $z \geq 1.645$, where

$$z = \frac{T^+ - \mu_{T^+}}{\sigma_{T^+}}$$

and otherwise accept the null hypothesis or reserve judgment.
4. The original data, the differences, and the ranks of their absolute values are shown in the following table:

| Weights before and after | Differences | Ranks |
|---|---|---|
| 169.0 and 159.9 | 9.1 | 13 |
| 188.6 and 181.3 | 7.3 | 10 |
| 222.1 and 209.0 | 13.1 | 16 |
| 160.1 and 162.3 | −2.2 | 5 |
| 187.5 and 183.5 | 4.0 | 8 |
| 202.5 and 197.6 | 4.9 | 9 |
| 167.8 and 171.4 | −3.6 | 7 |
| 214.3 and 202.1 | 12.2 | 14 |
| 143.8 and 145.1 | −1.3 | 3 |
| 198.2 and 185.5 | 12.7 | 15 |
| 166.9 and 158.6 | 8.3 | 12 |
| 142.9 and 145.4 | −2.5 | 6 |
| 160.5 and 159.5 | 1.0 | 2 |
| 198.7 and 190.6 | 8.1 | 11 |
| 149.7 and 149.0 | 0.7 | 1 |
| 181.6 and 183.1 | −1.5 | 4 |

It follows that

$$T^+ = 13 + 10 + 16 + 8 + 9 + 14 + 15 + 12 + 2 + 11 + 1 = 111$$

and since

$$\mu_{T^+} = \frac{16 \cdot 17}{4} = 68 \quad \text{and} \quad \sigma_{T^+} = \sqrt{\frac{16 \cdot 17 \cdot 33}{24}} \approx 19.34$$

we finally obtain

$$z = \frac{111 - 68}{19.34} \approx 2.22$$

5. Since $z = 2.22$ exceeds 1.645, the null hypothesis must be rejected; we conclude that the weight-reducing diet is effective.

**\*18.13**  On what statistic do we base our decision and for what values of the statistic do we reject the null hypothesis if we have a random sample of size $n = 10$ and are using the signed-rank test at the 0.05 level of significance to test the null hypothesis $\tilde{\mu} = \tilde{\mu}_0$ against the alternative hypothesis

(a)  $\tilde{\mu} \neq \tilde{\mu}_0$;

(b)  $\tilde{\mu} > \tilde{\mu}_0$;

(c)  $\tilde{\mu} < \tilde{\mu}_0$?

**\*18.14**  Rework Exercise 18.13 with the level of significance changed to 0.01.

**\*18.15**  On what statistic do we base our decision and for what values of the statistic do we reject the null hypothesis if we have a random sample of $n = 12$ pairs of values and we are using the signed-rank test at the 0.01 level of significance to test the null hypothesis $\tilde{\mu}_D = 0$ against the alternative hypothesis

(a)  $\tilde{\mu}_D \neq 0$;

(b)  $\tilde{\mu}_D > 0$;

(c)  $\tilde{\mu}_D < 0$?

**\*18.16**  Rework Exercise 18.15 with the level of significance changed to 0.05.

**\*18.17**  In a random sample of 13 issues, a newspaper listed 40, 52, 43, 27, 35, 36, 57, 39, 41, 34, 46, 32, and 37 apartments for rent. Use the signed-rank test at the 0.05 level of significance to test the null hypothesis $\tilde{\mu} = 45$ against the alternative hypothesis

(a)  $\tilde{\mu} < 45$;

(b)  $\tilde{\mu} \neq 45$.

**\*18.18**  Use the signed-rank test to rework Exercise 18.1.

**\*18.19**  Use the signed-rank test to rework Exercise 18.3.

**\*18.20**  In a random sample taken at a public playground, it took 38, 43, 36, 29, 44, 28, 40, 50, 39, 47, and 33 minutes to play a set of tennis. Use the signed-rank test at the 0.05 level of significance to test whether or not it takes on the average 35 minutes to play a set of tennis at that public playground.

**\*18.21**  In a random sample of 15 summer days, Casa Grande and Gila Bend reported high temperatures of 102 and 106, 103 and 110, 106 and 106, 104 and 107, 105 and 108, 102 and 109, 103 and 102, 104 and 107, 110 and 112, 109 and 110, 100 and 104, 110 and 109, 105 and 108, 111 and 114, and 105 and 106. Use the signed-rank test at the 0.05 level of significance to test the null hypothesis $\tilde{\mu}_D = 0$ against the alternative hypothesis $\tilde{\mu}_D < 0$.

**\*18.22**  Following are the numbers of three-month and six-month certificates of deposit (CDs) that a bank sold on a random sample of 16 business days: 37 and 32, 33 and 22, 29 and 26, 18 and 33, 41 and 25, 42 and 34, 33 and 43, 51 and 31, 36 and 24, 29 and 22, 23 and 30, 28 and 29, 44 and 30, 24 and 26, 27 and 18, and 30 and 35. Test at the 0.05 level of significance whether the bank sells equally many of the two kinds of CDs against the alternative that it sells more of the three-month CDs using

(a)  the signed-rank test based on Table VI;

(b)  the large-sample signed-rank test.

**\*18.23**  Use the large-sample signed-rank test to rework Exercise 18.21.

**\*18.24**  Use the large-sample signed-rank test to rework Exercise 18.7.

**\*18.25**  Use the large-sample signed-rank test to rework Exercise 18.10.

**\*18.26** Following is a random sample of the scores obtained by husbands and their wives on a spatial abilities test:

| Husbands | Wives | Husbands | Wives |
|----------|-------|----------|-------|
| 108 | 103 | 125 | 120 |
| 104 | 116 | 96 | 98 |
| 103 | 106 | 107 | 117 |
| 112 | 104 | 115 | 130 |
| 99 | 99 | 110 | 101 |
| 105 | 94 | 101 | 100 |
| 102 | 110 | 103 | 96 |
| 112 | 128 | 105 | 99 |
| 119 | 106 | 124 | 120 |
| 106 | 103 | 113 | 116 |

Use the large-sample signed-rank test at the 0.05 level of significance to test whether husbands and their wives do equally well on this test.

**\*18.27** Use a computer to rework Exercise 18.20.

**\*18.28** Use a computer to rework Exercise 18.21.

**\*18.29** Use a computer to rework part (a) of Exercise 18.22.

# 18.5 THE *U* TEST

We now consider a nonparametric alternative to the two-sample *t* test concerning the difference between two population means. It is called the ***U* test**, the **Wilcoxon rank-sum test**, or the **Mann-Whitney test**, named after the statisticians who contributed to its development. The different names reflect the manner in which the calculations are organized; logically, they are all equivalent.

With this test we are able to check whether two independent samples come from identical populations. In particular, we can test the null hypothesis $\mu_1 = \mu_2$ without having to assume that the populations sampled have roughly the shape of normal distributions. In fact, the test requires only that the populations be continuous (to avoid ties), and even that requirement is not critical so long as the number of ties is small. Observe, however, that according to the Kendall and Buckland dictionary of statistical terms, the *U* test is a test of the equality of the location parameters of two otherwise identical populations. And of course, there are various kinds of location parameters. For instance, in Figure 18.5 on page 469, the location parameters in question are the population medians. Here they are denoted by ETA1 and ETA2, where previously we had denoted the population medians by $\tilde{\mu}_1$ and $\tilde{\mu}_2$.

To illustrate how the *U* test is performed, suppose that we want to compare the grain size of sand obtained from two different locations on the moon on the basis of the following diameters (in millimeters):

**Location 1**: 0.37  0.70  0.75  0.30  0.45  0.16  0.62  0.73  0.33
**Location 2**: 0.86  0.55  0.80  0.42  0.97  0.84  0.24  0.51  0.92  0.69

The means of these two samples are 0.49 and 0.68, and their difference seems large, but it remains to be seen whether that is significant.

To perform the $U$ test we first arrange the data jointly, as if they comprise one sample, in an increasing order of magnitude. For our data we get

| 0.16 | 0.24 | 0.30 | 0.33 | 0.37 | 0.42 | 0.45 | 0.51 | 0.55 | 0.62 |
|------|------|------|------|------|------|------|------|------|------|
| 1 | 2 | 1 | 1 | 1 | 2 | 1 | 2 | 2 | 1 |

| 0.69 | 0.70 | 0.73 | 0.75 | 0.80 | 0.84 | 0.86 | 0.92 | 0.97 |
|------|------|------|------|------|------|------|------|------|
| 2 | 1 | 1 | 1 | 2 | 2 | 2 | 2 | 2 |

where we indicated for each value whether it came from location 1 or location 2. Assigning the data, in this order, the ranks 1, 2, 3, ..., and 19, we find that the values of the first sample (location 1) occupy ranks 1, 3, 4, 5, 7, 10, 12, 13, and 14, while those of the second sample (location 2) occupy ranks, 2, 6, 8, 9, 11, 15, 16, 17, 18, and 19. There are no ties here between values in different samples, but if there are, we assign each of the tied observations the mean of the ranks that they jointly occupy. For instance, if the third and fourth values are the same, we assign each the rank $\frac{3+4}{2} = 3.5$, and if the ninth, tenth, and eleventh values are the same, we assign each the rank $\frac{9+10+11}{3} = 10$. When there are ties among values belonging to the same sample, it does not matter how they are ranked. For instance, if the third and fourth values are the same but belong to the same sample, it does not matter which one is ranked 3 and which one is ranked 4.

Now, if there is an appreciable difference between the means of the two populations, most of the lower ranks are likely to go to the values of one sample while most of the higher ranks are likely to go to the values of the other sample. The test of the null hypothesis that the two samples come from identical populations may thus be based on $W_1$, the sum of the ranks of the values of the first sample, or on $W_2$, the sum of the ranks of the values of the second sample. In practice, it does not matter which sample we refer to as sample 1 and which sample we refer to as sample 2, and whether we base the test on $W_1$ or $W_2$. (When the sample sizes are unequal, we usually let the smaller of the two samples be sample 1; however, this is not required for the work in this book.)

If the sample sizes are $n_1$ and $n_2$, the sum of $W_1$ and $W_2$ is simply the sum of the first $n_1 + n_2$ positive integers, which is known to be

$$\frac{(n_1 + n_2)(n_1 + n_2 + 1)}{2}$$

This formula enables us to find $W_2$ if we know $W_1$, and vice versa. For our illustration we get

$$W_1 = 1 + 3 + 4 + 5 + 7 + 10 + 12 + 13 + 14 = 69$$

and since the sum of the first 19 positive integers is $\frac{19 \cdot 20}{2} = 190$, it follows that

$$W_2 = 190 - 69 = 121$$

(This value is the sum of the ranks 2, 6, 8, 9, 11, 15, 16, 17, 18, and 19.)

When the use of **rank sums** was first proposed as a nonparametric alternative to the two-sample $t$ test, the decision was based on $W_1$ or $W_2$. Nowadays, it is more common to base the decision on either of the related statistics

$U_1$ AND $U_2$ STATISTICS

or

$$U_1 = W_1 - \frac{n_1(n_1 + 1)}{2}$$

$$U_2 = W_2 - \frac{n_2(n_2 + 1)}{2}$$

or on the statistic $U$, which always equals the smaller of the two. The resulting tests are equivalent to those based on $W_1$ or $W_2$, but they have the advantage that they lend themselves more readily to the construction of tables of critical values. Not only do $U_1$ and $U_2$ take on values on the same interval from 0 to $n_1 n_2$—indeed, their sum is always equal to $n_1 n_2$—but they have identical distributions that are symmetrical about $\frac{n_1 n_2}{2}$. The relationship between the sampling distributions of $U_1$, $U_2$, and $U$ is pictured in Figure 18.4 for the special case where $n_1 = 3$ and $n_2 = 3$.

**Figure 18.4**
Distributions of $U_1$, $U_2$, and $U$ for $n_1 = 3$ and $n_2 = 3$.

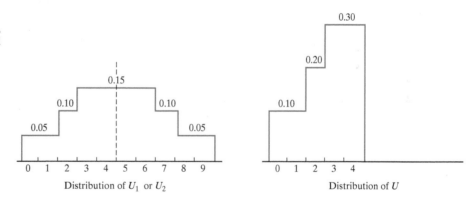

Distribution of $U_1$ or $U_2$          Distribution of $U$

As we have said earlier, it is assumed that we are dealing with independent random samples from identical populations, but we care mainly whether $\mu_1 = \mu_2$. All the tests will be based on the sampling distribution of one and the same statistic, as in Section 18.3. However, here it is the statistic $U$, and the null hypothesis will be rejected for values falling into its left-hand tail. Again, we have to be careful, though, to use the right statistic and the right critical value. When $\mu_1 < \mu_2$ then $U_1$ will tend to be small, so when the alternative hypothesis is $\mu_1 < \mu_2$ we base the test on $U_1$; when $\mu_1 > \mu_2$ then $U_2$ will tend to be small, so when the alternative hypothesis is $\mu_1 > \mu_2$ we base the test on $U_2$; and when $\mu_1 \neq \mu_2$ then either $U_1$ or $U_2$ will tend to be small, so when the alternative hypothesis is $\mu_1 \neq \mu_2$ we base the test on $U$. All this is summarized in the following table:

| Alternative hypothesis | Reject the null hypothesis if | Accept the null hypothesis or reserve judgment if |
|---|---|---|
| $\mu_1 < \mu_2$ | $U_1 \leq U_{2\alpha}$ | $U_1 > U_{2\alpha}$ |
| $\mu_1 > \mu_2$ | $U_2 \leq U_{2\alpha}$ | $U_2 > U_{2\alpha}$ |
| $\mu_1 \neq \mu_2$ | $U \leq U_\alpha$ | $U > U_\alpha$ |

The necessary values of $U_\alpha$, which are the largest values of $U$ for which the probability of $U \leq U_\alpha$ does not exceed $\alpha$, may be found in Table VII at the end of the book; the blank spaces in the table indicate that the null hypothesis cannot be rejected regardless of the value we obtain for the test statistic. Note that the same critical values serve for tests at different levels of significance depending on whether the alternative hypothesis is one sided or two sided.

**EXAMPLE 18.8**   With reference to the grain-size data on page 465, use the $U$ test at the 0.05 level of significance to test whether or not the two samples come from populations with equal means.

**Solution**   **1.** $H_0 : \mu_1 = \mu_2$
$H_A : \mu_1 \neq \mu_2$
**2.** $\alpha = 0.05$
**3.** Reject the null hypothesis if $U \leq 20$, where 20 is the value of $U_{0.05}$ for $n_1 = 9$ and $n_2 = 10$; otherwise, reserve judgment.
**4.** Having already shown on page 466 that $W_1 = 69$ and $W_2 = 121$, we get

$$U_1 = 69 - \frac{9 \cdot 10}{2} = 24$$

$$U_2 = 121 - \frac{10 \cdot 11}{2} = 66$$

and, hence, $U = 24$. Note that $U_1 + U_2 = 24 + 66 = 90$, which equals $n_1 n_2 = 9 \cdot 10$.
**5.** Since $U = 24$ is greater than 20, the null hypothesis cannot be rejected; in other words, we cannot conclude that there is a difference in the mean grain size of sand from the two locations on the moon.

Had we wanted to use a computer in this example, MINITAB would have yielded Figure 18.5. In this printout, ETA (the Greek letter $\eta$) denotes the population median, which we had previously denoted by $\tilde{\mu}$. Also, $W = 69$ is the statistic we had previously called $W_1$, and the $p$-value is again 0.0942. Since 0.0942 exceeds 0.05, we conclude (as before) that the null hypothesis cannot be rejected.

**Figure 18.5**
Computer printout for
Example 18.8.

**Mann-Whitney Test and CI: Loc 1, Loc 2**

```
          N   Median
Loc 1    9   0.4500
Loc 2   10   0.7450

Point estimate for ETA1-ETA2 is -0.1850
95.5 Percent CI for ETA1-ETA2 is (-0.4701,0.0601)
W = 69.0
Test of ETA1 = ETA2 vs ETA1 not = ETA2 is significant at
0.0942
```

**EXAMPLE 18.9**  The following are the burning times (rounded to the nearest tenth of a minute) of random samples of two kinds of emergency flares:

**Brand 1**:  17.2  18.1  19.3  21.1  14.4  13.7  18.8  15.2  20.3  17.5
**Brand 2**:  13.6  19.1  11.8  14.6  14.3  22.5  12.3  13.5  10.9  14.8

Use the *U* test at the 0.05 level of significance to test whether it is reasonable to say that on the average brand 1 flares are better (last longer) than brand 2 flares.

**Solution**
1. $H_0 : \mu_1 = \mu_2$
   $H_A : \mu_1 > \mu_2$
2. $\alpha = 0.05$
3. Reject the null hypothesis if $U_2 \leq 27$, where 27 is the value of $U_{0.10}$ for $n_1 = 10$ and $n_2 = 10$; otherwise, accept it or reserve judgment.
4. Ranking the data jointly according to size, we find that the values of the second sample occupy rank 5, 16, 2, 9, 7, 20, 3, 4, 1, and 10, so that

$$W_2 = 5 + 16 + 2 + 9 + 7 + 20 + 3 + 4 + 1 + 10 = 77$$

and

$$U_2 = 77 - \frac{10 \cdot 11}{2} = 22$$

5. Since $U_2 = 22$ is less than 27, the null hypothesis must be rejected; we conclude that brand 1 flares are, indeed, better than brand 2 flares. ∎

# 18.6 THE *U* TEST (Large Samples)

The large-sample *U* test may be based on either $U_1$ or $U_2$ as defined on page 467, but since the resulting tests are equivalent and it does not matter which sample we denote sample 1 and which sample we denote sample 2, we shall use here the statistic $U_1$.

Based on the assumption that the two samples come from identical continuous populations, it can be shown that the mean and the standard deviation of the sampling distribution of $U_1$ are[†]

---

[†] When there are ties in rank the formula for the standard deviation provides only an approximation, but unless the number of ties is very large, there is rarely any need to make an adjustment.

MEAN AND
STANDARD
DEVIATION OF
$U_1$ STATISTIC

and

$$\mu_{U_1} = \frac{n_1 n_2}{2}$$

$$\sigma_{U_1} = \sqrt{\frac{n_1 n_2 (n_1 + n_2 + 1)}{12}}$$

Observe that these formulas remain the same when we interchange the subscripts 1 and 2, but this should not come as a surprise—as we pointed out on page 467, the distributions of $U_1$ and $U_2$ are the same.

Furthermore, if $n_1$ and $n_2$ are both greater than 8, the sampling distribution of $U_1$ can be approximated closely by a normal distribution. Thus, we base the test of the null hypothesis $\mu_1 = \mu_2$ on the statistic

STATISTIC FOR
LARGE-SAMPLE
$U$ TEST

$$z = \frac{U_1 - \mu_{U_1}}{\sigma_{U_1}}$$

which has approximately the standard normal distribution. When the alternative hypothesis is $\mu_1 \neq \mu_2$, we reject the null hypothesis if $z \leq -z_{\alpha/2}$ or $z \geq z_{\alpha/2}$; when the alternative hypothesis is $\mu_1 > \mu_2$, we reject the null hypothesis if $z \geq z_{\alpha}$; and when the alternative hypothesis is $\mu_1 < \mu_2$, we reject the null hypothesis if $z \leq -z_{\alpha}$.

EXAMPLE 18.10   The following are the weight gains (in pounds) of two random samples of young turkeys fed two different diets but otherwise kept under identical conditions:

**Diet 1:**   16.3   10.1   10.7   13.5   14.9   11.8   14.3   10.2
12.0   14.7   23.6   15.1   14.5   18.4   13.2   14.0

**Diet 2:**   21.3   23.8   15.4   19.6   12.0   13.9   18.8   19.2
15.3   20.1   14.8   18.9   20.7   21.1   15.8   16.2

Use the large-sample $U$ test at the 0.01 level of significance to test the null hypothesis that the two populations sampled are identical against the alternative hypothesis that on the average the second diet produces a greater gain in weight.

**Solution**

1. $H_0 : \mu_1 = \mu_2$ (populations are identical)
   $H_A : \mu_1 < \mu_2$
2. $\alpha = 0.01$
3. Reject the null hypothesis if $z \leq -2.33$, where

$$z = \frac{U_1 - \mu_{U_1}}{\sigma_{U_1}}$$

and otherwise accept the null hypothesis or reserve judgment.

4. Ranking the data jointly according to size, we find that the values of the first sample occupy ranks 21, 1, 3, 8, 15, 4, 11, 2, 5.5, 13, 31, 16, 12, 22, 7, and 10. (The fifth and sixth values are both equal to 12.0, so we assign each the rank 5.5.) Thus

$$W_1 = 1 + 2 + 3 + 4 + 5.5 + 7 + 8 + 10 + 11 + 12 + 13$$
$$+ 15 + 16 + 21 + 22 + 31$$
$$= 181.5$$

and

$$U_1 = 181.5 - \frac{16 \cdot 17}{2} = 45.5$$

Since $\mu_{U_1} = \frac{16 \cdot 16}{2} = 128$ and $\sigma_{U_1} = \sqrt{\frac{16 \cdot 16 \cdot 33}{12}} \approx 26.53$, it follows that

$$z = \frac{45.5 - 128}{26.53} \approx -3.11$$

5. Since $z = -3.11$ is less than $-2.33$, the null hypothesis must be rejected; we conclude that on the average the second diet produces a greater gain in weight.  ◼

# 18.7 THE *H* TEST

The **H test**, or the **Kruskal-Wallis test**, is a rank-sum test that serves to test the assumption that $k$ independent random samples come from identical populations, and in particular the null hypothesis $\mu_1 = \mu_2 = \cdots = \mu_k$, against the alternative hypothesis that these means are not all equal. Unlike the standard test that it replaces, the one-way analysis of variance of Section 15.3, it does not require the assumption that the populations sampled have, at least approximately, normal distributions.

As in the $U$ test, the data are ranked jointly from low to high as though they constitute a single sample. Then, if $R_i$ is the sum of the ranks assigned to the $n_i$ values of the $i$th sample and $n = n_1 + n_2 + \cdots + n_k$, the $H$ test is based on the statistic

**STATISTIC FOR *H* TEST**

$$H = \frac{12}{n(n+1)} \sum_{i=1}^{k} \frac{R_i^2}{n_i} - 3(n+1)$$

If the null hypothesis is true and each sample has at least five observations, it is generally considered reasonable to approximate the sampling distribution of $H$ with a chi-square distribution having $k - 1$ degrees of freedom. Consequently, we reject the null hypothesis $\mu_1 = \mu_2 = \cdots = \mu_k$ and accept the alternative hypothesis that these means are not all equal, when the value we get for $H$ exceeds or equals $\chi_\alpha^2$ for $k - 1$ degrees of freedom.

**EXAMPLE 18.11**    Students are randomly assigned to groups that are taught Spanish by three different methods: (1) classroom instruction and language laboratory, (2) only classroom instruction, and (3) only self-study in language laboratory. Following are the final examination scores of samples of students from the three groups:

| **Method 1:** | 94 | 88 | 91 | 74 | 86 | 97 | |
| **Method 2:** | 85 | 82 | 79 | 84 | 61 | 72 | 80 |
| **Method 3:** | 89 | 67 | 72 | 76 | 69 | | |

Use the $H$ test at the 0.05 level of significance to test the null hypothesis that the populations sampled are identical against the alternative hypothesis that their means are not all equal.

**Solution**

1. $H_0 : \mu_1 = \mu_2 = \mu_3$ (The populations are identical).
   $H_A : \mu_1, \mu_2,$ and $\mu_3$ are not all equal.
2. $\alpha = 0.05$
3. Reject the null hypothesis if $H \geq 5.991$, which is the value of $\chi^2_{0.05}$ for $3 - 1 = 2$ degrees of freedom; otherwise, accept it or reserve judgment.
4. Arranging the data jointly according to size, we get 61, 67, 69, 72, 72, 74, 76, 79, 80, 82, 84, 85, 86, 88, 89, 91, 94, and 97. Assigning the data, in this order, the ranks 1, 2, 3, . . ., and 18, we find that

$$R_1 = 6 + 13 + 14 + 16 + 17 + 18 = 84$$

$$R_2 = 1 + 4.5 + 8 + 9 + 10 + 11 + 12 = 55.5$$

$$R_3 = 2 + 3 + 4.5 + 7 + 15 = 31.5$$

and it follows that

$$H = \frac{12}{18 \cdot 19} \left( \frac{84^2}{6} + \frac{55.5^2}{7} + \frac{31.5^2}{5} \right) - 3 \cdot 19$$

$$\approx 6.67$$

5. Since $H = 6.67$ exceeds 5.991, the null hypothesis must be rejected; we conclude that the three methods of instruction are not all equally effective. ∎

Had we used a computer for this example, we would have found that the $p$-value corresponding to $H = 6.67$ is 0.036, and that the $p$-value adjusted for the tie is also 0.036. Since 0.036 is less than 0.05, we would have concluded, as before, that the null hypothesis must be rejected.

**EXERCISES**

**18.30** On what statistic do we base the decision and for what values of the statistic do we reject the null hypothesis $\mu_1 = \mu_2$ if we have random samples of size $n_1 = 9$ and $n_2 = 9$ and are using the $U$ test based on Table VII and the 0.05 level of significance to test the null hypothesis against the alternative hypothesis
   (a) $\mu_1 > \mu_2$;
   (b) $\mu_1 \neq \mu_2$;
   (c) $\mu_1 < \mu_2$?

**18.31** Rework Exercise 18.30 with the level of significance changed to 0.01.

**18.32** On what statistic do we base the decision and for what values of the statistic do we reject the null hypothesis $\mu_1 = \mu_2$ if we have random samples of size $n_1 = 10$ and $n_2 = 14$ and are using the $U$ test based on Table VII and the

0.01 level of significance to test the given null hypothesis against the alternative hypothesis

(a) $\mu_1 > \mu_2$;

(b) $\mu_1 \neq \mu_2$;

(c) $\mu_1 < \mu_2$?

**18.33** Rework Exercise 18.32 with the level of significance changed to 0.05.

**18.34** On what statistic do we base the decision and for what values of the statistic do we reject the null hypothesis $\mu_1 = \mu_2$ against the alternative hypothesis $\mu_1 \neq \mu_2$ if we are using the *U* test based on Table VII and the 0.05 level of significance, and

(a) $n_1 = 4$ and $n_2 = 6$;

(b) $n_1 = 9$ and $n_2 = 8$;

(c) $n_1 = 5$ and $n_2 = 12$;

(d) $n_1 = 7$ and $n_2 = 3$?

**18.35** Rework Exercise 18.34 with the alternative hypothesis changed to $\mu_1 > \mu_2$.

**18.36** Explain why there is no value in Table VII for $U_{0.05}$ corresponding to $n_1 = 3$ and $n_2 = 3$. (*Hint*: Refer to Figure 18.4.)

**18.37** Following are the scores that random samples of students from two minority groups obtained on a current events test:

**Minority group 1**:  73  82  39  68  91  75
               89  67  50  86  57  65

**Minority group 2**:  51  42  36  53  88  59
               49  66  25  64  18  76

Use the *U* test based on Table VII to test at the 0.05 level of significance whether or not students from the two minority groups can be expected to score equally well on this test.

 **18.38** Use a computer to rework Exercise 18.37.

**18.39** The following are the numbers of minutes taken by random samples of 15 men and 12 women to complete a written test for the renewal of their driver's licenses:

**Men**:    9.9  7.4   8.9   9.1   7.7   9.7  11.8  7.5
       9.2  10.0  10.2   9.5  10.8  8.0  11.0

**Women**:  8.6  10.9   9.8  10.7  9.4  10.3
       7.3  11.5   7.6   9.3  8.8   9.6

Use the *U* test based on Table VII to test at the 0.05 level of significance whether or not $\mu_1 = \mu_2$, where $\mu_1$ and $\mu_2$ are the average amounts of time it takes men and women to complete the test.

**18.40** Use the large-sample *U* test to rework Exercise 18.39.

**18.41** The following are the Rockwell hardness numbers obtained for six aluminum die castings randomly selected from production lot A and eight from production lot B:

**Production lot A**:  75  56  63  70  58  74
**Production lot B**:  63  85  77  80  86  76  72  82

Use the *U* test based on Table VII to test at the 0.05 level of significance whether the castings of production lot *B* are on the average harder than those of production lot *A*.

 **18.42** Use a computer to rework Exercise 18.41.

**18.43** Use the large-sample $U$ test to rework Example 18.8.

**18.44** Use the large-sample $U$ test to rework Example 18.9.

**18.45** Use a computer to rework Example 18.10.

**18.46** Following are data for the breaking strength (in pounds) of random samples of two kinds of 2-inch cotton ribbons:

|  |  |
|---|---|
| **Type I ribbon:** | 133  144  165  169  171  176  180  181 |
|  | 182  183  186  187  194  197  198  200 |
| **Type II ribbon:** | 134  154  159  161  164  164  164  169 |
|  | 170  172  175  176  185  189  194  198 |

(For convenience, the values have been arranged in ascending order.) Use the large-sample $U$ test at the 0.05 level of significance to test the claim that Type I ribbon is, on the average, stronger than Type II ribbon.

**18.47** Use a computer to rework Exercise 18.46.

**18.48** Following are the miles per gallon that a test driver got in random samples of six tankfuls of each of three kinds of gasoline:

**Gasoline 1**: 15  24  27  29  30  32
**Gasoline 2**: 17  20  22  28  32  33
**Gasoline 3**: 18  19  22  23  25  32

(For convenience, the values have been arranged in ascending order.) Use the $H$ test at the 0.05 level of significance to test the claim that there is no difference in the true average mileage yield of the three kinds of gasoline.

**18.49** Use the $H$ test at the 0.01 level of significance to rework Exercise 15.15.

**18.50** Use the $H$ test at the 0.01 level of significance to rework Exercise 15.17.

**18.51** To compare four bowling balls, a professional bowler bowls five games with each ball and gets the following scores:

| | | | | | |
|---|---|---|---|---|---|
| **Bowling ball D**: | 221 | 232 | 207 | 198 | 212 |
| **Bowling ball E**: | 202 | 225 | 252 | 218 | 226 |
| **Bowling ball F** | 210 | 205 | 189 | 196 | 216 |
| **Bowling ball G**: | 229 | 192 | 247 | 220 | 208 |

Use the $H$ test at the 0.05 level of significance to test the null hypothesis that on the average the bowler performs equally well with the four bowling balls.

**18.52** Use a computer to rework Exercise 18.51.

**18.53** Three groups of guinea pigs, injected with 0.5, 1.0, and 1.5 mg, respectively, of a tranquilizer, fell asleep in the following number of seconds:

| | | | | | |
|---|---|---|---|---|---|
| **0.5 mg dose**: | 7.8 | 8.2 | 10.0 | 10.2 | |
|  | 10.9 | 12.7 | 13.7 | 14.0 | |
| **1.0 mg dose**: | 7.5 | 7.9 | 8.8 | 9.7 | 10.5 |
|  | 11.0 | 12.5 | 12.9 | 13.1 | 13.3 |
| **1.5 mg dose**: | 7.2 | 8.0 | 8.5 | 9.0 | |
|  | 9.4 | 11.3 | 11.5 | 12.0 | |

(For convenience, the values have been arranged in increasing order.) Use the $H$ test at the 0.01 level of significance to test the null hypothesis that the differences in dosage have no effect on the length of time it takes guinea pigs to fall asleep.

**18.54** Use a computer to rework Exercise 18.53.

# 18.8 TESTS OF RANDOMNESS: RUNS

All the methods of inference studied in this book are based on the assumption that our samples are random; yet there are many applications where it is difficult to determine whether this assumption is justifiable. This is true, particularly, when we have little or no control over the selection of the data, as is the case, for example, when we rely on whatever records are available to make long-range predictions of the weather, when we use whatever data are available to estimate the mortality rate of a disease, or when we use sales records for past months to make predictions of a department store's future sales. None of this information constitutes a random sample in the strict sense.

There are several methods of judging the randomness of a sample on the basis of the order in which the observations are obtained; they enable us to decide, after the data have been collected, whether patterns that look suspiciously nonrandom may be attributed to chance. The technique we describe here and in the next two sections, the **u test**, is based on the **theory of runs**.

A **run** is a succession of identical letters (or other kinds of symbols) that is followed and preceded by different letters or no letters at all. To illustrate, consider the following arrangement of healthy, $H$, and diseased, $D$, elm trees that were planted many years ago along a country road:

$$\underline{H\,H\,H\,H}\ \ \underline{D\,D\,D}\ \ \underline{H\,H\,H\,H\,H\,H\,H}\ \ \underline{D\,D}\ \ \underline{H\,H}\ \ \underline{D\,D\,D\,D}$$

Using underlines to combine the letters that constitute the runs, we find that there is first a run of four $H$'s, then a run of three $D$'s, then a run of seven $H$'s, then a run of two $D$'s, then a run of two $H$'s, and finally a run of four $D$'s.

The **total number of runs** appearing in an arrangement of this kind is often a good indication of a possible lack of randomness. If there are too few runs, we might suspect a definite grouping or clustering, or perhaps a trend; if there are too many runs, we might suspect some sort of repeated alternating, or cyclical pattern. In the preceding example there seems to be a definite clustering—the diseased trees seem to come in groups—but it remains to be seen whether this is significant or whether it can be attributed to chance.

If there are $n_1$ letters of one kind, $n_2$ letters of another kind, and $u$ runs, we base this kind of decision on the following criterion:

---

**Reject the null hypothesis of randomness if**

$$u \leqslant u'_{\alpha/2} \quad \text{or} \quad u \geqslant u_{\alpha/2}$$

**where $u'_{\alpha/2}$ and $u_{\alpha/2}$ are given in Table VIII for values of $n_1$ and $n_2$ through 15, and $\alpha = 0.05$ and $\alpha = 0.01$.**

---

In the construction of Table VIII, $u'_{\alpha/2}$ is the largest value of $u$ for which the probability of $u \leq u'_{\alpha/2}$ does not exceed $\alpha/2$ and $u_{\alpha/2}$ is the smallest value of $u$ for which the probability of $u \geq u_{\alpha/2}$ does not exceed $\alpha/2$; the blank spaces in the table indicate that the null hypothesis cannot be rejected for values in that tail of the sampling distribution regardless of the value we obtain for $u$.

**EXAMPLE  18.12**   With reference to the arrangement of healthy and diseased elm trees cited previously, use the $u$ test at the 0.05 level of significance to test the null hypothesis of randomness against the alternative hypothesis that the arrangement is not random.

**Solution**   1. $H_0$: Arrangement is random.
   $H_A$: Arrangement is not random.
   2. $\alpha = 0.05$.
   3. Reject the null hypothesis if $u \leq 6$ or $u \geq 17$, where 6 and 17 are the values of $u'_{0.025}$ and $u_{0.025}$ for $n_1 = 13$ and $n_2 = 9$; otherwise, accept it or reserve judgment.
   4. $u = 6$ by inspection of the data.
   5. Since $u = 6$ equals the value of $u'_{0.025}$, the null hypothesis must be rejected; we conclude that the arrangement of healthy and diseased elm trees is not random. There are fewer runs than might have been expected, and it appears that the diseased trees come in clusters.

# 18.9 TESTS OF RANDOMNESS: RUNS (Large Samples)

Under the null hypothesis that $n_1$ letters of one kind and $n_2$ letters of another kind are arranged at random, it can be shown that the mean and the standard deviation of $u$, the total number of runs, are

**MEAN AND STANDARD DEVIATION OF $u$**

and

$$\mu_u = \frac{2n_1 n_2}{n_1 + n_2} + 1$$

$$\sigma_u = \sqrt{\frac{2n_1 n_2(2n_1 n_2 - n_1 - n_2)}{(n_1 + n_2)^2(n_1 + n_2 - 1)}}$$

Furthermore, if neither $n_1$ nor $n_2$ is less than 10, the sampling distribution of $u$ can be approximated closely by a normal distribution. Thus, we base the test of the null hypothesis of randomness on the statistic

**STATISTIC FOR LARGE-SAMPLE $u$ TEST**

$$z = \frac{u - \mu_u}{\sigma_u}$$

which has approximately the standard normal distribution. If the alternative hypothesis is that the arrangement is not random, we reject the null hypothesis for $z \leq -z_{\alpha/2}$ or $z \geq z_{\alpha/2}$; if the alternative hypothesis is that there is a clustering or a trend, we reject the null hypothesis for $z \leq -z_{\alpha}$; and if the alternative hypothesis is that there is an alternating, or cyclical, pattern, we reject the null hypothesis for $z \geq z_{\alpha}$.

EXAMPLE **18.13**   The following is an arrangement of men, $M$, and women, $W$, lined up single file to purchase tickets for a rock concert:

M W M W M M M W M W M M M W W M M M M W W M W M

M M W M M M W W W M W M M M W M W M M M M W W M

Test for randomness at the 0.05 level of significance.

**Solution**   1. $H_0$: Arrangement is random.

$H_A$: Arrangement is not random.

2. $\alpha = 0.05$

3. Reject the null hypothesis if $z \leq -1.96$ or $z \geq 1.96$, where

$$z = \frac{u - \mu_u}{\sigma_u}$$

and otherwise accept the null hypothesis or reserve judgment.

4. Since $n_1 = 30$, $n_2 = 18$, and $u = 27$, we get

$$\mu_u = \frac{2 \cdot 30 \cdot 18}{30 + 18} + 1 = 23.5$$

$$\sigma_u = \sqrt{\frac{2 \cdot 30 \cdot 18(2 \cdot 30 \cdot 18 - 30 - 18)}{(30 + 18)^2(30 + 18 - 1)}} = 3.21$$

and, hence,

$$z = \frac{27 - 23.5}{3.21} \approx 1.09$$

5. Since $z = 1.09$ falls between $-1.96$ and $1.96$, the null hypothesis cannot be rejected; in other words, there is no real evidence to suggest that the arrangement is not random.  ■

# 18.10 TESTS OF RANDOMNESS: RUNS ABOVE AND BELOW THE MEDIAN

The $u$ test is not limited to testing the randomness of sequences of attributes, such as the $H$'s and $D$'s, or $M$'s and $W$'s, of our examples. Any sample consisting of numerical measurements or observations can be treated similarly by using the letters $a$ and $b$ to denote values falling above and below the median of the sample. Numbers equal to the median are omitted. The resulting sequence of $a$'s and $b$'s (representing the data in their original order) can then be tested for randomness on the basis of the total number of runs of $a$'s and $b$'s, namely, the total number of **runs above and below the median**. Depending on the size of $n_1$ and $n_2$, we use Table VIII or the large-sample test of Section 18.9.

EXAMPLE **18.14**   On 24 successive runs between two cities, a bus carried

24   19   32   28   21   23   26   17   20   28   30   24

13   35   26   21   19   29   27   18   26   14   21   23

passengers. Use the total number of runs above and below the median to test at the 0.01 level of significance whether it is reasonable to treat these data as if they constitute a random sample.

**Solution** Since the median of the data is 23.5, we get the following arrangement of values above and below the median:

$$a\,b\,a\,a\,b\,b\,a\,b\,b\,a\,a\,a\,b\,a\,a\,b\,b\,a\,a\,b\,a\,b\,b\,b$$

1. $H_0$: Arrangement is random.
   $H_A$: Arrangement is not random.
2. $\alpha = 0.01$
3. Reject the null hypothesis if $u \leq 6$ or $u \geq 20$, where 6 and 20 are the values of $u'_{0.005}$ and $u_{0.005}$ for $n_1 = 12$ and $n_2 = 12$; otherwise, accept the null hypothesis or reserve judgment.
4. $u = 14$ by inspection of the preceding arrangement of $a$'s and $b$'s.
5. Since $u = 14$ falls between 6 and 20, the null hypothesis cannot be rejected; in other words, there is no real evidence to indicate that the data do not constitute a random sample. ∎

## EXERCISES

**18.55** Following is the order in which a broker received 25 orders to buy, $B$, or sell, $S$, shares of a certain stock:

$$S\,S\,S\,B\,B\,B\,B\,B\,B\,B\,B\,S\,B\,S\,S\,S\,S\,S\,S\,B\,B\,B\,B\,S\,S$$

Use Table VIII to test for randomness at the 0.05 level of significance.

**18.56** Use the large-sample test to rework Exercise 18.55.

**18.57** A driver buys gasoline either at a Chevron station, $C$, or an Arco station, $A$, and the following arrangement shows the order of the stations from which he made 29 purchases of gasoline over a recent period of time:

$$A\,C\,A\,C\,C\,A\,C\,A\,C\,A\,C\,C\,A\,A\,A\,C\,A\,C\,C\,C\,A\,C\,A\,A\,A\,C\,C\,A\,C$$

Use Table VIII to test for randomness at the 0.01 level of significance.

**18.58** Use the large-sample test to rework Exercise 18.57.

**18.59** Test at the 0.05 level of significance whether the following arrangement of defective, $D$, and nondefective, $N$, engines coming off an assembly line may be regarded as random:

$$N\,N\,N\,N\,N\,N\,D\,N\,N\,N\,N\,N\,D\,D\,D\,N\,N\,N\,D\,D\,D\,N\,N$$

**18.60** The following arrangement indicates whether 60 consecutive persons interviewed by a pollster are for, $F$, or against, $A$, an increase of half a cent in the state sales tax to build a new football stadium:

$$A\,F\,F\,F\,A\,A\,A\,A\,F\,A\,A\,F\,F\,A\,A\,A\,A\,F\,F\,F\,A\,F\,A\,A\,A\,F\,F\,F\,A\,A$$
$$A\,A\,F\,F\,F\,F\,A\,A\,F\,A\,A\,F\,F\,F\,F\,F\,A\,F\,A\,A\,A\,A\,F\,F\,A\,F\,F\,F\,F\,A$$

Test for randomness at the 0.01 level of significance.

**18.61** To test whether a radio signal contains a message or constitutes random noise, an interval of time is subdivided into a number of very short intervals and for each of these it is determined whether the signal strength exceeds, $E$, or does not exceed, $N$, a certain level of background noise. Test at the 0.05 level of significance whether the following arrangement, thus obtained, may be regarded as random, and hence that the signal contains no message and may be regarded as random noise:

$$E\,N\,N\,N\,N\,E\,N\,E\,N\,N\,N\,E\,E\,N\,N\,N\,E\,E\,N\,E\,N\,N\,N\,E\,E\,N\,N\,N$$
$$N\,N\,E\,E\,N\,E\,N\,N\,E\,N\,N\,N\,E\,E\,E\,N\,N\,N\,E\,N\,E\,N\,N\,N\,N\,E\,N$$

**18.62** Flip a coin 50 times and test at the 0.05 level of significance whether the resulting sequence of $H$'s and $T$'s (heads and tails) may be regarded as random.

**18.63** Record whether 60 consecutive cars arriving at an intersection from the north have local license plates, $L$, or out-of-state plates, $O$. Test for randomness at the 0.05 level of significance.

**18.64** In the beginning of Section 11.1 we gave the following data on the number of minutes it took 36 persons to assemble an "easy to assemble" toy: 17, 13, 18, 19, 17, 21, 29, 22, 16, 28, 21, 15, 26, 23, 24, 20, 8, 17, 17, 21, 32, 18, 25, 22, 16, 10, 20, 22, 19, 14, 30, 22, 12, 24, 28, and 11. Test for randomness at the 0.05 level of significance.

**18.65** Following are the weights of 40 mature dogs of a certain breed: 66.2, 59.2, 70.8, 58.0, 64.3, 50.7, 62.5, 58.4, 48.7, 52.4, 51.0, 35.7, 62.6, 52.3, 41.2, 61.1, 52.9, 58.8, 64.1, 48.9, 74.3, 50.3, 55.7, 55.5, 51.8, 55.8, 48.9, 51.8, 63.1, 44.6, 47.0, 49.0, 62.5, 45.0, 78.6, 54.2, 72.2, 52.4, 60.5, and 46.8 ounces. Given that the median of these weights is 54.85 ounces, test for randomness at the 0.05 level of significance.

**18.66** Following are the examination grades of 42 students in the order in which they finished an examination:

75  95  77  93  89  83  69  77  92  88  62  64  91  72
76  83  50  65  84  67  63  54  58  76  70  62  65  41
63  55  32  58  61  68  54  28  35  49  82  60  66  57

Test for randomness at the 0.05 level of significance.

**18.67** The total number of retail stores opening for business and also quitting business within the calendar years 1976–2005 in a large city were

107  125  142  147  122  116  153  144  106  138
126  125  129  134  137  143  150  148  152  145
112  162  139  132  122  143  148  155  146  158

Making use of the fact that the median is 140.5, test at the 0.05 level of significance whether there is a significant trend.

**18.68** The following are six years' quarterly sales (in millions of dollars) of a manufacturer of heavy machinery:

83.8   102.5  121.0  90.5  106.6  104.8  114.7  93.6
98.9   96.9   122.6  85.6  103.2  96.9   118.0  92.1
100.5  92.9   125.6  79.2  110.8  95.1   125.6  86.7

Making use of the fact that the median is 99.7, test at the 0.05 level of significance whether there is a real cyclical pattern.

# 18.11 RANK CORRELATION

Since the significance test for $r$ of Section 17.3 is based on very stringent assumptions, we sometimes use a nonparametric alternative that can be applied under much more general conditions. This test of the null hypothesis of no correlation is based on the **rank-correlation coefficient**, often called **Spearman's rank-correlation coefficient** and denoted by $r_s$.

To calculate the rank-correlation coefficient for a given set of paired data, we first rank the $x$'s among themselves from low to high or high to low; then we rank the $y$'s in the same way, find the sum of the squares of the differences, $d$, between the ranks of the $x$'s and the $y$'s, and substitute into the formula

RANK-
CORRELATION
COEFFICIENT

$$r_S = 1 - \frac{6\left(\sum d^2\right)}{n(n^2 - 1)}$$

where $n$ is the number of pairs of $x$'s and $y$'s. When there are ties in rank, we proceed as before and assign to each of the tied observations the mean of the ranks that they jointly occupy.

EXAMPLE **18.15**   The following are the numbers of hours that ten students studied for an examination and the scores that they obtained:

| Number of hours studied | Scores in examination |
|:---:|:---:|
| x | y |
| 9 | 56 |
| 5 | 44 |
| 11 | 79 |
| 13 | 72 |
| 10 | 70 |
| 5 | 54 |
| 18 | 94 |
| 15 | 85 |
| 2 | 33 |
| 8 | 65 |

Calculate $r_S$.

**Solution**   Ranking the $x$'s among themselves from low to high and also the $y$'s, we get the ranks shown in the first two columns in the following table:

| Rank of x | Rank of y | d | d² |
|:---:|:---:|:---:|:---:|
| 5 | 4 | 1.0 | 1.00 |
| 2.5 | 2 | 0.5 | 0.25 |
| 7 | 8 | −1.0 | 1.00 |
| 8 | 7 | 1.0 | 1.00 |
| 6 | 6 | 0.0 | 0.00 |
| 2.5 | 3 | −0.5 | 0.25 |
| 10 | 10 | 0.0 | 0.00 |
| 9 | 9 | 0.0 | 0.00 |
| 1 | 1 | 0.0 | 0.00 |
| 4 | 5 | −1.0 | 1.00 |
| | | | 4.50 |

Note that the second and third smallest values among the $x$'s are both equal to 5, so we assign each of them the rank $\frac{2+3}{2} = 2.5$. Then, determining the $d$'s (differences between the ranks) and their squares, and substituting $n = 10$ and $\sum d^2 = 4.50$ into the formula for $r_S$, we get

$$r_S = 1 - \frac{6(4.50)}{10(10^2 - 1)} \approx 0.97$$

As can be seen from this example, $r_S$ is easy to compute manually, and this is why it is sometimes used instead of $r$ when no calculator is available. When there are no ties, $r_S$ actually equals the correlation coefficient $r$ calculated for the two sets of ranks; when ties exist there may be a small (but usually negligible) difference. Of course, by using ranks instead of the original data we lose some information, but this is usually offset by the rank-correlation coefficient's computational ease. It is of interest to note that if we had calculated $r$ for the original $x$'s and $y$'s in the preceding example, we would have obtained 0.96 instead of 0.97; at least in this case, the difference between $r$ and $r_S$ is very small.

The main advantage in using $r_S$ is that we can test the null hypothesis of no correlation without having to make any assumptions about the populations sampled. Under the null hypothesis of no correlation—indeed, the null hypothesis that the $x$'s and the $y$'s are randomly matched—the sampling distribution of $r_S$ has the mean 0 and the standard deviation

$$\sigma_{r_S} = \frac{1}{\sqrt{n-1}}$$

Since this sampling distribution can be approximated with a normal distribution even for relatively small values of $n$, we base the test of the null hypothesis on the statistic

**S**TATISTIC FOR TESTING SIGNIFICANCE OF $r_S$

$$z = \frac{r_S - 0}{1/\sqrt{n-1}} = r_S\sqrt{n-1}$$

which has approximately the standard normal distribution.

**EXAMPLE 18.16** With reference to Example 18.15, where we had $n = 10$ and $r_S = 0.97$, test the null hypothesis of no correlation at the 0.01 level of significance.

**Solution**
1. $H_0 : \rho = 0$ (no correlation)
   $H_A : \rho \neq 0$
2. $\alpha = 0.01$
3. Reject the null hypothesis if $z \leq -2.575$ or $z \geq 2.575$, where

$$z = r_S\sqrt{n-1}$$

and otherwise, accept it or reserve judgment.
4. For $n = 10$ and $r_S = 0.97$ we get

$$z = 0.97\sqrt{10-1} = 2.91$$

5. Since $z = 2.91$ exceeds 2.575, the null hypothesis must be rejected; we conclude that there is a relationship between study time and scores in the population sampled.

**18.69** Calculate $r_S$ for the following sample data representing the number of minutes it took 12 mechanics to assemble a piece of machinery in the morning, $x$, and in the late afternoon, $y$:

| $x$ | $y$ |
|------|------|
| 10.8 | 15.1 |
| 16.6 | 16.8 |
| 11.1 | 10.9 |
| 10.3 | 14.2 |
| 12.0 | 13.8 |
| 15.1 | 21.5 |
| 13.7 | 13.2 |
| 18.5 | 21.1 |
| 17.3 | 16.4 |
| 14.2 | 19.3 |
| 14.8 | 17.4 |
| 15.3 | 19.0 |

**18.70** If a sample of $n = 37$ pairs of data yielded $r_S = 0.39$, is this rank-correlation coefficient significant at the 0.01 level of significance?

**18.71** If a sample of $n = 50$ pairs of data yielded $r_S = 0.31$, is this rank-correlation coefficient significant at the 0.05 level of significance?

**18.72** The following table shows how a panel of nutrition experts and a panel of heads of household ranked 15 breakfast foods on their palatability:

| Breakfast food | Nutrition experts | Heads of household |
|:---:|:---:|:---:|
| I | 7 | 5 |
| II | 3 | 4 |
| III | 11 | 8 |
| IV | 9 | 14 |
| V | 1 | 2 |
| VI | 4 | 6 |
| VII | 10 | 12 |
| VIII | 8 | 7 |
| IX | 5 | 1 |
| X | 13 | 9 |
| XI | 12 | 15 |
| XII | 2 | 3 |
| XIII | 15 | 10 |
| XIV | 6 | 11 |
| XV | 14 | 13 |

Calculate $r_S$ as a measure of the consistency of the two ratings.

**18.73** The following are the rankings which three judges gave to the work of ten artists:

| **Judge A**: | 5 | 8 | 4 | 2 | 3 | 1 | 10 | 7 | 9 | 6 |
| **Judge B**: | 3 | 10 | 1 | 4 | 2 | 5 | 6 | 7 | 8 | 9 |
| **Judge C**: | 8 | 5 | 6 | 4 | 10 | 2 | 3 | 1 | 7 | 9 |

Calculate $r_S$ for each pair of rankings and decide

(a) which two judges are most alike in their opinions about these artists;

(b) which two judges differ the most in their opinions about these artists.

# 18.12 SOME FURTHER CONSIDERATIONS

Although nonparametric tests have a great deal of intuitive appeal and are widely applicable, it should not be overlooked that they are usually **less efficient** than the standard tests that they replace. To illustrate what we mean here by "less efficient," let us refer to Example 10.11, where we showed that the mean of a random sample of size $n = 128$ is as reliable an estimate of the mean of a symmetrical population as a median of a random sample of size $n = 200$. Thus, the median requires a larger sample than the mean, and this is what we mean when we say that it is "less efficient."

Put another way, nonparametric tests tend to be wasteful of information. The one-sample sign test and the paired-sample sign test are especially wasteful, whereas the other procedures introduced in this chapter are wasteful to a lesser degree. Above all, nonparametric tests should not be used indiscriminately when the assumptions underlying the corresponding standard tests are satisfied.

In actual practice, nonparametric procedures are often used to confirm conclusions based on standard tests when there is some uncertainty about the validity of the assumptions that underly the standard tests. Nonparametric tests are indispensable when sample sizes are too small to form an opinion one way or the other about the validity of assumptions.

# 18.13 SUMMARY

The table that follows summarizes the various nonparametric tests we have discussed (except for the tests of randomness based on runs) and the corresponding standard tests that they replace. In each case we list the section or sections of the book where they are discussed.

| *Null hypothesis* | *Standard tests* | *Nonparametric tests* |
| --- | --- | --- |
| $\mu = \mu_0$ | One-sample $t$ test (Section 12.4) or one-sample $z$ test (Section 12.3) | One-sample sign test (Sections 18.1 and 18.2) or signed-rank test (Sections 18.3 and 18.4) |
| $\mu_1 = \mu_2$ (independent samples) | Two-sample $t$ test (Section 12.6) or two-sample $z$ test (Section 12.5) | $U$ test (Sections 18.5 and 18.6) |

| Null hypothesis | Standard tests | Nonparametric tests |
|---|---|---|
| $\mu_1 = \mu_2$ (paired data) | Paired-sample $t$ test or paired-sample $z$ test (Section 12.7) | Paired-sample sign test (Sections 18.1 and 18.2) or signed-rank test (Sections 18.3 and 18.4) |
| $\mu_1 = \mu_2 = \cdots = \mu_k$ | One-way analysis of variance (Section 15.3) | $H$ test (Section 18.7) |
| $\rho = 0$ | Test based on Fisher $Z$ transformation (Section 17.3) | Test based on rank-correlation coefficient (Section 18.11) |

Tests of randomness are discussed in Sections 18.8, 18.9, and 18.10, but there are no corresponding standard tests.

# CHECKLIST OF KEY TERMS (with page references to their definitions)

Efficiency, 483
$H$ test, 471
Kruskal-Wallis test, 471
Mann-Whitney test, 465
Nonparametric tests, 452
One-sample sign test, 453
Paired-sample sign test, 454
Rank sums, 467
Rank-correlation coefficient, 479
Runs, 475

Runs above and below the median, 477
Sign test, 453
Signed-rank test, 458
Spearman's rank-correlation coefficient, 479
Theory of runs, 475
Total number of runs, 475
$u$ test, 475
$U$ test, 465
Wilcoxon signed-rank test, 458
Wilcoxon rank-sum test, 465

# REFERENCES

*Further information about the nonparametric tests discussed in this chapter and many others may be found in*

CONOVER, W. J., *Practical Nonparametric Statistics*. New York: John Wiley & Sons, Inc., 1971.

DANIEL, W. W., *Applied Nonparametric Statistics*. Boston: Houghton-Mifflin Company, 1978.

GIBBONS, J. D., *Nonparametric Statistical Inference*. New York: Marcel Dekker, 1985.

LEHMANN, E. L., *Nonparametrics: Statistical Methods Based on Ranks*. San Francisco: Holden-Day, Inc., 1975.

MOSTELLER, F., and ROURKE, R. E. K., *Sturdy Statistics, Nonparametrics and Order Statistics*. Reading, Mass.: Addison-Wesley Publishing Company, Inc., 1973.

NOETHER, G. E., *Introduction to Statistics: The Nonparametric Way*. New York: Springer-Verlag, 1990.

RANDLES, R., and WOLFE, D., *Introduction to the Theory of Nonparametric Statistics*. New York: John Wiley & Sons, Inc., 1979.

SIEGEL, S., *Nonparametric Statistics for the Behavioral Sciences*. New York: McGraw-Hill Book Company, 1956.

**R.169** Recent government statistics show that for couples with 0, 1, 2, 3, or 4 children, the relationship between the number of children, $x$, and family income in dollars, $y$, is fairly well described by the least-squares line $\hat{y} = 38,600 + 3,500x$. If a childless couple has twins, will this increase their income by $2(3,500) = \$7,000$?

**R.170** Following are the numbers of hours that 10 persons (interviewed as part of a sample survey) spent watching television, $x$, and reading books or magazines, $y$, per week:

| $x$ | $y$ |
|-----|-----|
| 18 | 7 |
| 25 | 5 |
| 19 | 1 |
| 12 | 5 |
| 12 | 10 |
| 27 | 2 |
| 15 | 3 |
| 9 | 9 |
| 12 | 8 |
| 18 | 4 |

For these data, $\sum x = 167$, $\sum x^2 = 3,101$, $\sum y = 54$, $\sum y^2 = 374$, and $\sum xy = 798$.

(a) Fit a least-squares line that will enable us to predict $y$ in terms of $x$.

(b) If a person spends 22 hours watching television per week, predict how many hours he or she will spend reading books or magazines.

**R.171** Calculate $r$ for the data of Exercise R.170.

**R.172** Following are the numbers of minutes that patients had to wait for their appointments with four doctors:

| | | | | | |
|---|---|---|---|---|---|
| **Doctor A**: | 18 | 26 | 29 | 22 | 16 |
| **Doctor B**: | 9 | 11 | 28 | 26 | 15 |
| **Doctor C**: | 20 | 13 | 22 | 25 | 10 |
| **Doctor D**: | 21 | 26 | 39 | 32 | 24 |

Use the $H$ test at the 0.05 level of significance to test the null hypothesis that the four samples come from identical populations against the alternative hypothesis that the means of the four populations are not all equal.

**R.173** The following sequence shows whether a certain senator was present, $P$, or absent, $A$, at 30 consecutive meetings of an appropriations committee:

$P\ P\ P\ P\ P\ P\ P\ A\ A\ P\ P\ P\ P\ P\ P\ A\ A\ P\ P\ P\ P\ P\ A\ P\ P\ P\ P\ A\ A\ P$

At the 0.01 level of significance, is there any real indication of a lack of randomness?

**R.174** If $k = 6$ and $n = 9$ in a two-way analysis of variance without interaction, what are the degrees of freedom for treatments, blocks, and error?

**R.175** The following data pertain to a study of the effects of environmental pollution on wildlife; in particular, the relationship between DDT and the thickness of the eggshells of certain birds:

| DDT residue in yolk lipids (parts per million) | Thickness of eggshells (millimeters) |
|---|---|
| 117 | 0.49 |
| 65 | 0.52 |
| 303 | 0.37 |
| 98 | 0.53 |
| 122 | 0.49 |
| 150 | 0.42 |

If $x$ denotes the DDT residue and $y$ denotes the eggshell thickness, then $S_{xx} = 34{,}873.50$, $S_{xy} = -23.89$, and $S_{yy} = 0.0194$. Calculate the coefficient of correlation.

**R.176** With reference to Exercise R.175, use the 0.05 level of significance to test whether the value obtained for $r$ is significant.

**R.177** Following are the numbers of computer modem cards produced by four assembly lines on 12 workdays:

| Line 1 | Line 2 | Line 3 | Line 4 |
|---|---|---|---|
| 904 | 835 | 873 | 839 |
| 852 | 857 | 803 | 849 |
| 861 | 822 | 855 | 913 |
| 770 | 796 | 851 | 840 |
| 877 | 808 | 856 | 843 |
| 929 | 832 | 857 | 892 |
| 955 | 777 | 873 | 841 |
| 836 | 830 | 830 | 807 |
| 870 | 808 | 921 | 875 |
| 843 | 862 | 886 | 898 |
| 847 | 843 | 834 | 976 |
| 864 | 802 | 939 | 822 |

(a) Perform an analysis of variance and, assuming that the required assumptions can be met, test at the 0.05 level of significance whether the differences obtained for the means of the four samples, 867.33, 822.67, 864.83, and 866.25, can be attributed to chance.

(b) Use the studentized-range method at the 0.05 level of significance to analyze the performance of the four assembly lines.

**R.178** An experiment yielded $r_{12} = 0.40$, $r_{13} = -0.90$, and $r_{23} = 0.90$. Explain why these figures cannot all be correct.

**R.179** Working a homework assignment, a student obtained $S_{xx} = 145.22$, $S_{xy} = -210.58$, and $S_{yy} = 287.45$ for a given set of paired data. Explain why there must be an error in these calculations.

**R.180** Following are the numbers of burglaries committed in two cities on 22 days: 87 and 81, 83 and 80, 98 and 87, 114 and 86, 112 and 120, 77 and 102, 103 and 94, 116 and 81, 136 and 95, 156 and 158, 83 and 127, 105 and 104, 117 and 102, 86 and 100, 150 and 108, 119 and 124, 111 and 91, 137 and 103, 160 and 153, 121 and 140, 143 and 105, and 129 and 129. Use the large-sample sign test at the 0.05 level of significance to test whether or not $\tilde{\mu}_D = 0$, where $\tilde{\mu}_D$ is the median of the population of differences sampled.

**R.181** Use the large-sample signed-rank test to rework the preceding exercise.

**R.182** To study the earnings of faculty members in statistics and economics from speeches, writing, and consulting, a research worker interviewed four male assistant professors of economics, four male professors of economics, four female professors of statistics, four male associate professors of economics, four female assistant professors of statistics, and four female associate professors of statistics. If he combines the first and fifth groups, the second and third groups, and the fourth and sixth groups, performs an analysis of variance with $k = 3$ and $n = 8$, and gets a significant value of $F$, to what source of variation (sex, rank, or subject) can this be attributed?

**R.183** With reference to Exercise R.182, explain why there is no way in which the research worker can use the data to test whether there is a significant difference that can be attributed to sex.

**R.184** Following are the batting averages, $x$, and home runs hit, $y$, by a random sample of 15 major league baseball players during the first half of the season:

| $x$ | $y$ |
|-------|-----|
| 0.252 | 12 |
| 0.305 | 6 |
| 0.299 | 4 |
| 0.303 | 15 |
| 0.285 | 2 |
| 0.191 | 2 |
| 0.283 | 16 |
| 0.272 | 6 |
| 0.310 | 8 |
| 0.266 | 10 |
| 0.215 | 0 |
| 0.211 | 3 |
| 0.272 | 14 |
| 0.244 | 6 |
| 0.320 | 7 |

Calculate the rank-correlation coefficient and test at the 0.01 level of significance whether it is statistically significant.

**R.185** In the years 1999–2005 the profits of stockholders in petroleum and coal products corporations were 6.1, 7.7, 14.9, 14.6, 12.8, 7.7, and 2.4 percent of equity. Code the years $-3, -2, -1, 0, 1, 2$, and 3, and fit a parabola by the method of least squares.

**R.186** The following are the closing prices of a commodity (in dollars) on 20 consecutive trading days: 378, 379, 379, 378, 377, 376, 374, 374, 373, 373, 374, 375, 376, 376, 376, 375, 374, 374, 373, and 374. Test for randomness at the 0.01 level of significance.

**R.187** The following are the scores of 16 golfers on the first two days of a tournament: 68 and 71, 73 and 76, 70 and 73, 74 and 71, 69 and 72, 72 and 74, 67 and 70, 72 and 68, 71 and 72, 73 and 74, 68 and 69, 70 and 72, 73 and 70, 71 and 75, 67 and 69, and 73 and 71. Use the sign test at the 0.05 level of significance to test whether on the average the hundreds of golfers participating in the tournament scored equally well on the first two days or whether they tended to score lower on the first day.

**R.188** If $r = 0.28$ for the ages of a group of college students and their knowledge of foreign affairs, what percentage of the variation of their knowledge of foreign affairs can be attributed to differences in age?

**R.189** State in each case whether you would expect a positive correlation, a negative correlation, or no correlation:

(a) Family expenditures on restaurant meals and family expenditures on property taxes.

(b) Daily low temperatures and closing prices of shares of an electronics firm.

(c) The number of hours that basketball players practice and their free-throw percentages.

(d) The number of passengers on a cruise ship and the number of empty cabins.

**R.190** In a multiple regression problem, the residual sum of squares is 926 and the total sum of squares is 1,702. Find the value of the multiple correlation coefficient.

**R.191** Following are the number of inquiries a car dealer received in eight weeks about cars for sale, $x$, and cars for lease, $y$:

| $x$ | $y$ |
|-----|-----|
| 325 | 29 |
| 212 | 20 |
| 278 | 22 |
| 167 | 14 |
| 201 | 17 |
| 265 | 23 |
| 305 | 26 |
| 259 | 19 |

Calculate $r$.

**R.192** With reference to Exercise R.191, calculate 95% confidence limits for $\rho$.

**R.193** If $r = 0.41$ for one set of data and $r = -0.92$ for another, compare the strengths of the two relationships.

**R.194** The following sequence shows whether a television news program had at least 25% of a city's viewing audience, $A$, or less than 25%, $L$, on 36 consecutive weekday evenings:

$$L\,L\,L\,L\,A\,A\,L\,L\,L\,A\,L\,L\,L\,A\,L\,L\,L\,A\,A\,A\,A\,L$$

$$A\,L\,L\,L\,A\,A\,L\,L\,L\,L\,L\,A\,L\,L\,L\,L\,L\,A$$

Test for randomness at the 0.05 level of significance.

**R.195** To find the best arrangement of instruments on a control panel of an airplane, three different arrangements were tested by simulating emergency conditions and observing the reaction time required to correct the condition. The reaction times (in tenths of a second) of 12 pilots (randomly assigned to the different arrangements) were as follows:

| **Arrangement 1:** | 8 | 15 | 10 | 11 |
|---|---|---|---|---|
| **Arrangement 2:** | 16 | 11 | 14 | 19 |
| **Arrangement 3:** | 12 | 7 | 13 | 8 |

(a) Calculate $n \cdot s_{\bar{x}}^2$ for these data, and also the mean of the variances of the three samples and the value of $F$.

(b) Assuming that the necessary assumptions can be met, test at the 0.01 level of significance whether the differences among the three sample means can be attributed to chance.

**R.196** Following are the prices (in dollars) charged for a certain camera in a random sample of 15 discount stores: 57.25, 58.14, 54.19, 56.17, 57.21, 55.38, 54.75, 57.29, 57.80, 54.50, 55.00, 56.35, 54.26, 60.23, and 53.99. Use the sign test based on Table V to test at the 0.05 level of significance whether or not the median price charged for such cameras in the population sampled is $55.00.

**R.197** A school has seven department heads who are assigned to seven different committees, as shown in the following table:

| Committee | Department heads | | | |
|---|---|---|---|---|
| Textbooks | Dodge, | Fleming, | Griffith, | Anderson |
| Athletics | Bowman, | Evans, | Griffith, | Anderson |
| Band | Bowman, | Carlson, | Fleming, | Anderson |
| Dramatics | Bowman, | Carlson, | Dodge, | Griffith |
| Tenure | Carlson, | Evans, | Fleming, | Griffith |
| Salaries | Bowman, | Dodge, | Evans, | Fleming |
| Discipline | Carlson, | Dodge, | Evans, | Anderson |

(a) Verify that this arrangement is a balanced incomplete block design.

(b) If Dodge, Bowman, and Carlson are (in that order) appointed chairpersons of the first three committees, how will the chairpersons of the other four committees have to be chosen so that each of the department heads is chairperson of one of the committees?

**R.198** The sample data in the following table are the grades in a statistics test obtained by nine college students from three majors who were taught by three different instructors:

| | Instructor A | Instructor B | Instructor C |
|---|---|---|---|
| Marketing | 77 | 88 | 71 |
| Finance | 88 | 97 | 81 |
| Insurance | 85 | 95 | 72 |

Assuming that the necessary assumptions can be met, use the 0.05 level of significance to analyze this two-factor experiment.

**R.199** On what statistic do we base our decision and for what values of the statistic do we reject the null hypothesis $\tilde{\mu}_1 = \tilde{\mu}_2$ if we have random samples of size $n_1 = 8$ and $n_2 = 11$ and are using the $U$ test based on Table VII at the 0.05 level of significance to test the null hypothesis against the alternative hypothesis

(a) $\tilde{\mu}_1 \neq \tilde{\mu}_2$;

(b) $\tilde{\mu}_1 < \tilde{\mu}_2$;

(c) $\tilde{\mu}_1 > \tilde{\mu}_2$?

**R.200** Assuming that the conditions underlying normal correlation analysis are met, use the Fisher $Z$ transformation to construct approximate 99% confidence intervals for $\rho$ when

(a) $r = 0.45$ and $n = 18$;

(b) $r = -0.32$ and $n = 38$.

**R.201** The figures in the following $5 \times 5$ Latin square are the numbers of minutes engines $E_1, E_2, E_3, E_4,$ and $E_5$, tuned up by mechanics $M_1, M_2, M_3, M_4,$ and $M_5$, ran with a gallon of fuel brand $A, B, C, D,$ and $E$:

|  | $E_1$ | $E_2$ | $E_3$ | $E_4$ | $E_5$ |
|---|---|---|---|---|---|
| $M_1$ | A<br>31 | B<br>24 | C<br>20 | D<br>20 | E<br>18 |
| $M_2$ | B<br>21 | C<br>27 | D<br>23 | E<br>25 | A<br>31 |
| $M_3$ | C<br>21 | D<br>27 | E<br>25 | A<br>29 | B<br>21 |
| $M_4$ | D<br>21 | E<br>25 | A<br>33 | B<br>25 | C<br>22 |
| $M_5$ | E<br>21 | A<br>37 | B<br>24 | C<br>24 | D<br>20 |

Analyze this Latin square, using the 0.01 level of significance for each of the tests.

**R.202** Following are data on the percentage kill of two kinds of insecticides used against mosquitos:

**Insecticide X:**    41.9    46.9    44.6    43.9    42.0    44.0
                      41.0    43.1    39.0    45.2    44.6    42.0

**Insecticide Y:**    45.7    39.8    42.8    41.2    45.0    40.2
                      40.2    41.7    37.4    38.8    41.7    38.7

Use the $U$ test based on Table VII to test at the 0.05 level of significance whether or not the two insecticides are on the average equally effective.

**R.203** Use the large-sample $U$ test to rework Exercise R.202.

**R.204** Following are the numbers of persons who attended a "singles only" dance on 12 Saturdays: 172, 208, 169, 232, 123, 165, 197, 178, 221, 195, 209, and 182. Use the sign test based on Table V to test at the 0.05 level of significance whether or not the median of the population sampled is $\tilde{\mu} = 169$.

**R.205** Use the signed-rank test based on Table VI to rework Exercise R.204.

**R.206** Use the large-sample sign test to rework Exercise R.204.

**\*R.207** The following data pertain to the cosmic-ray doses measured at various altitudes:

| Altitude<br>(hundreds of feet)<br>$x$ | Dose rate<br>(mrem/year)<br>$y$ |
|---|---|
| 0.5 | 28 |
| 4.5 | 30 |
| 7.8 | 32 |
| 12.0 | 36 |
| 48.0 | 58 |
| 53.0 | 69 |

Use a computer or a graphing calculator to fit an exponential curve and use it to estimate the cosmic radiation dose rate at 6,000 feet.

**R.208** The manager of a restaurant wants to determine whether the sales of chicken dinners depend on how this entree is described on the menu. He has three kinds of menus printed, listing chicken dinners among the other entrees or featuring them as "Chef's Special," or as "Gourmet's Delight," and he intends to use each kind of menu on six different Sundays. Actually, the manager collects only the following data showing the number of chicken dinners sold on 12 Sundays:

| | |
|---|---|
| **Listed among**<br>**other entrees:** | 76   94   85   77 |
| **Featured as**<br>**Chef's Special:** | 109  117  102  92  115 |
| **Featured as**<br>**Gourmet's Delight:** | 100   83  102 |

Perform a one-way analysis of variance at the 0.05 level of significance.

**R.209** Following are the high school grade-point averages, $x$, and the first-year college grade-point averages, $y$, of seven students:

| $x$ | $y$ |
|-----|-----|
| 2.7 | 2.5 |
| 3.6 | 3.8 |
| 3.0 | 2.8 |
| 2.4 | 2.1 |
| 2.4 | 2.5 |
| 3.1 | 3.2 |
| 3.5 | 2.9 |

Fit a least-squares line that will enable us to predict $y$ in terms of $x$, and use it to predict $y$ for a student with $x = 2.8$.

**R.210** With reference to Exercise R.209, construct a 95% confidence interval for the regression coefficient $\beta$.

**R.211** Market research shows that weekly sales of a new candy bar will be related to its price as follows:

| Price<br>(cents) | Weekly sales<br>(number of bars) |
|------------------|----------------------------------|
| 50 | 232,000 |
| 55 | 194,000 |
| 60 | 169,000 |
| 65 | 157,000 |

Finding that the parabola $\hat{y} = 1{,}130{,}000 - 28{,}000x + 200x^2$ provides an excellent fit, the person conducting the study substitutes $x = 85$, gets $\hat{y} = 195{,}000$, and predicts that weekly sales will total 195,000 bars if the candy is priced at 85 cents. Comment on this argument.

# STATISTICAL TABLES

**I**     Normal Curve Areas   495

**II**    Critical Values of $t$   497

**III**   Critical Values of $\chi^2$   499

**IV**   Critical Values of $F$   501

**V**    Binomial Probabilities   503

**VI**   Critical Values of $T$   508

**VII**   Critical Values of $U$   509

**VIII**   Critical Values of $u$   511

**IX**   Critical Values of $q_\alpha$   513

**X**    Values of $Z = \dfrac{1}{2} \cdot \ln \dfrac{1+r}{1-r}$   515

**XI**   Binomial Coefficients   516

**XII**   Values of $e^{-x}$   517

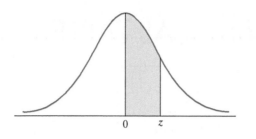

The entries in Table I are the probabilities that a random variable having the standard normal distribution will take on a value between 0 and $z$; they are given by the area of the tinted region in the figure shown above.

| TABLE I | Normal-Curve Areas | | | | | | | | |
|---|---|---|---|---|---|---|---|---|---|
| $z$ | .00 | .01 | .02 | .03 | .04 | .05 | .06 | .07 | .08 | .09 |
| 0.0 | .0000 | .0040 | .0080 | .0120 | .0160 | .0199 | .0239 | .0279 | .0319 | .0359 |
| 0.1 | .0398 | .0438 | .0478 | .0517 | .0557 | .0596 | .0636 | .0675 | .0714 | .0753 |
| 0.2 | .0793 | .0832 | .0871 | .0910 | .0948 | .0987 | .1026 | .1064 | .1103 | .1141 |
| 0.3 | .1179 | .1217 | .1255 | .1293 | .1331 | .1368 | .1406 | .1443 | .1480 | .1517 |
| 0.4 | .1554 | .1591 | .1628 | .1664 | .1700 | .1736 | .1772 | .1808 | .1844 | .1879 |
| 0.5 | .1915 | .1950 | .1985 | .2019 | .2054 | .2088 | .2123 | .2157 | .2190 | .2224 |
| 0.6 | .2257 | .2291 | .2324 | .2357 | .2389 | .2422 | .2454 | .2486 | .2517 | .2549 |
| 0.7 | .2580 | .2611 | .2642 | .2673 | .2704 | .2734 | .2764 | .2794 | .2823 | .2852 |
| 0.8 | .2881 | .2910 | .2939 | .2967 | .2995 | .3023 | .3051 | .3078 | .3106 | .3133 |
| 0.9 | .3159 | .3186 | .3212 | .3238 | .3264 | .3289 | .3315 | .3340 | .3365 | .3389 |
| 1.0 | .3413 | .3438 | .3461 | .3485 | .3508 | .3531 | .3554 | .3577 | .3599 | .3621 |
| 1.1 | .3643 | .3665 | .3686 | .3708 | .3729 | .3749 | .3770 | .3790 | .3810 | .3830 |
| 1.2 | .3849 | .3869 | .3888 | .3907 | .3925 | .3944 | .3962 | .3980 | .3997 | .4015 |
| 1.3 | .4032 | .4049 | .4066 | .4082 | .4099 | .4115 | .4131 | .4147 | .4162 | .4177 |
| 1.4 | .4192 | .4207 | .4222 | .4236 | .4251 | .4265 | .4279 | .4292 | .4306 | .4319 |
| 1.5 | .4332 | .4345 | .4357 | .4370 | .4382 | .4394 | .4406 | .4418 | .4429 | .4441 |
| 1.6 | .4452 | .4463 | .4474 | .4484 | .4495 | .4505 | .4515 | .4525 | .4535 | .4545 |
| 1.7 | .4554 | .4564 | .4573 | .4582 | .4591 | .4599 | .4608 | .4616 | .4625 | .4633 |
| 1.8 | .4641 | .4649 | .4656 | .4664 | .4671 | .4678 | .4686 | .4693 | .4699 | .4706 |
| 1.9 | .4713 | .4719 | .4726 | .4732 | .4738 | .4744 | .4750 | .4756 | .4761 | .4767 |
| 2.0 | .4772 | .4778 | .4783 | .4788 | .4793 | .4798 | .4803 | .4808 | .4812 | .4817 |
| 2.1 | .4821 | .4826 | .4830 | .4834 | .4838 | .4842 | .4846 | .4850 | .4854 | .4857 |
| 2.2 | .4861 | .4864 | .4868 | .4871 | .4875 | .4878 | .4881 | .4884 | .4887 | .4890 |
| 2.3 | .4893 | .4896 | .4898 | .4901 | .4904 | .4906 | .4909 | .4911 | .4913 | .4916 |
| 2.4 | .4918 | .4920 | .4922 | .4925 | .4927 | .4929 | .4931 | .4932 | .4934 | .4936 |
| 2.5 | .4938 | .4940 | .4941 | .4943 | .4945 | .4946 | .4948 | .4949 | .4951 | .4952 |
| 2.6 | .4953 | .4955 | .4956 | .4957 | .4959 | .4960 | .4961 | .4962 | .4963 | .4964 |
| 2.7 | .4965 | .4966 | .4967 | .4968 | .4969 | .4970 | .4971 | .4972 | .4973 | .4974 |
| 2.8 | .4974 | .4975 | .4976 | .4977 | .4977 | .4978 | .4979 | .4979 | .4980 | .4981 |
| 2.9 | .4981 | .4982 | .4982 | .4983 | .4984 | .4984 | .4985 | .4985 | .4986 | .4986 |
| 3.0 | .4987 | .4987 | .4987 | .4988 | .4988 | .4989 | .4989 | .4989 | .4990 | .4990 |

Also, for $z = 4.0$, 5.0, and 6.0, the areas are 0.49997, 0.4999997, and 0.499999999.

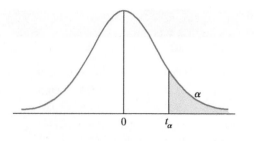

The entries in Table II are values for which the area to their right under the $t$ distribution with given degrees of freedom (the area of the tinted region under the curve shown above) is equal to $\alpha$.

## TABLE II   Critical Values of $t$[†]

| d.f. | $t_{.100}$ | $t_{.050}$ | $t_{.025}$ | $t_{.010}$ | $t_{.005}$ | d.f. |
|------|-----------|-----------|-----------|-----------|-----------|------|
| 1 | 3.078 | 6.314 | 12.706 | 31.821 | 63.657 | 1 |
| 2 | 1.886 | 2.920 | 4.303 | 6.965 | 9.925 | 2 |
| 3 | 1.638 | 2.353 | 3.182 | 4.541 | 5.841 | 3 |
| 4 | 1.533 | 2.132 | 2.776 | 3.747 | 4.604 | 4 |
| 5 | 1.476 | 2.015 | 2.571 | 3.365 | 4.032 | 5 |
| 6 | 1.440 | 1.943 | 2.447 | 3.143 | 3.707 | 6 |
| 7 | 1.415 | 1.895 | 2.365 | 2.998 | 3.499 | 7 |
| 8 | 1.397 | 1.860 | 2.306 | 2.896 | 3.355 | 8 |
| 9 | 1.383 | 1.833 | 2.262 | 2.821 | 3.250 | 9 |
| 10 | 1.372 | 1.812 | 2.228 | 2.764 | 3.169 | 10 |
| 11 | 1.363 | 1.796 | 2.201 | 2.718 | 3.106 | 11 |
| 12 | 1.356 | 1.782 | 2.179 | 2.681 | 3.055 | 12 |
| 13 | 1.350 | 1.771 | 2.160 | 2.650 | 3.012 | 13 |
| 14 | 1.345 | 1.761 | 2.145 | 2.624 | 2.977 | 14 |
| 15 | 1.341 | 1.753 | 2.131 | 2.602 | 2.947 | 15 |
| 16 | 1.337 | 1.746 | 2.120 | 2.583 | 2.921 | 16 |
| 17 | 1.333 | 1.740 | 2.110 | 2.567 | 2.898 | 17 |
| 18 | 1.330 | 1.734 | 2.101 | 2.552 | 2.878 | 18 |
| 19 | 1.328 | 1.729 | 2.093 | 2.539 | 2.861 | 19 |
| 20 | 1.325 | 1.725 | 2.086 | 2.528 | 2.845 | 20 |
| 21 | 1.323 | 1.721 | 2.080 | 2.518 | 2.831 | 21 |
| 22 | 1.321 | 1.717 | 2.074 | 2.508 | 2.819 | 22 |
| 23 | 1.319 | 1.714 | 2.069 | 2.500 | 2.807 | 23 |
| 24 | 1.318 | 1.711 | 2.064 | 2.492 | 2.797 | 24 |
| 25 | 1.316 | 1.708 | 2.060 | 2.485 | 2.787 | 25 |
| 26 | 1.315 | 1.706 | 2.056 | 2.479 | 2.779 | 26 |
| 27 | 1.314 | 1.703 | 2.052 | 2.473 | 2.771 | 27 |
| 28 | 1.313 | 1.701 | 2.048 | 2.467 | 2.763 | 28 |
| 29 | 1.311 | 1.699 | 2.045 | 2.462 | 2.756 | 29 |
| inf. | 1.282 | 1.645 | 1.960 | 2.326 | 2.576 | inf. |

[†]From Richard A. Johnson and Dean W. Wichern, *Applied Multivariate Statistical Analysis*, © 1982, p. 582. Adapted by permission of Pearson Education, Inc., Upper Saddle River, N.J.

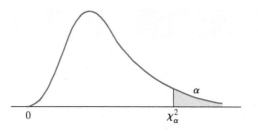

The entries in Table III are values for which the area to their right under the chi-square distribution with given degrees of freedom (the area of the tinted region under the curve shown above) is equal to $\alpha$.

**TABLE III**  Critical Values of $\chi^2$†

| d.f. | $\chi^2_{.995}$ | $\chi^2_{.99}$ | $\chi^2_{.975}$ | $\chi^2_{.95}$ | $\chi^2_{.05}$ | $\chi^2_{.025}$ | $\chi^2_{.01}$ | $\chi^2_{.005}$ | d.f. |
|---|---|---|---|---|---|---|---|---|---|
| 1 | .0000393 | .000157 | .000982 | .00393 | 3.841 | 5.024 | 6.635 | 7.879 | 1 |
| 2 | .0100 | .0201 | .0506 | .103 | 5.991 | 7.378 | 9.210 | 10.597 | 2 |
| 3 | .0717 | .115 | .216 | .352 | 7.815 | 9.348 | 11.345 | 12.838 | 3 |
| 4 | .207 | .297 | .484 | .711 | 9.488 | 11.143 | 13.277 | 14.860 | 4 |
| 5 | .412 | .554 | .831 | 1.145 | 11.070 | 12.832 | 15.086 | 16.750 | 5 |
| 6 | .676 | .872 | 1.237 | 1.635 | 12.592 | 14.449 | 16.812 | 18.548 | 6 |
| 7 | .989 | 1.239 | 1.690 | 2.167 | 14.067 | 16.013 | 18.475 | 20.278 | 7 |
| 8 | 1.344 | 1.646 | 2.180 | 2.733 | 15.507 | 17.535 | 20.090 | 21.955 | 8 |
| 9 | 1.735 | 2.088 | 2.700 | 3.325 | 16.919 | 19.023 | 21.666 | 23.589 | 9 |
| 10 | 2.156 | 2.558 | 3.247 | 3.940 | 18.307 | 20.483 | 23.209 | 25.188 | 10 |
| 11 | 2.603 | 3.053 | 3.816 | 4.575 | 19.675 | 21.920 | 24.725 | 26.757 | 11 |
| 12 | 3.074 | 3.571 | 4.404 | 5.226 | 21.026 | 23.337 | 26.217 | 28.300 | 12 |
| 13 | 3.565 | 4.107 | 5.009 | 5.892 | 22.362 | 24.736 | 27.688 | 29.819 | 13 |
| 14 | 4.075 | 4.660 | 5.629 | 6.571 | 23.685 | 26.119 | 29.141 | 31.319 | 14 |
| 15 | 4.601 | 5.229 | 6.262 | 7.261 | 24.996 | 27.488 | 30.578 | 32.801 | 15 |
| 16 | 5.142 | 5.812 | 6.908 | 7.962 | 26.296 | 28.845 | 32.000 | 34.267 | 16 |
| 17 | 5.697 | 6.408 | 7.564 | 8.672 | 27.587 | 30.191 | 33.409 | 35.718 | 17 |
| 18 | 6.265 | 7.015 | 8.231 | 9.390 | 28.869 | 31.526 | 34.805 | 37.156 | 18 |
| 19 | 6.844 | 7.633 | 8.907 | 10.117 | 30.144 | 32.852 | 36.191 | 38.582 | 19 |
| 20 | 7.434 | 8.260 | 9.591 | 10.851 | 31.410 | 34.170 | 37.566 | 39.997 | 20 |
| 21 | 8.034 | 8.897 | 10.283 | 11.591 | 32.671 | 35.479 | 38.932 | 41.401 | 21 |
| 22 | 8.643 | 9.542 | 10.982 | 12.338 | 33.924 | 36.781 | 40.289 | 42.796 | 22 |
| 23 | 9.260 | 10.196 | 11.689 | 13.091 | 35.172 | 38.076 | 41.638 | 44.181 | 23 |
| 24 | 9.886 | 10.856 | 12.401 | 13.848 | 36.415 | 39.364 | 42.980 | 45.558 | 24 |
| 25 | 10.520 | 11.524 | 13.120 | 14.611 | 37.652 | 40.646 | 44.314 | 46.928 | 25 |
| 26 | 11.160 | 12.198 | 13.844 | 15.379 | 38.885 | 41.923 | 45.642 | 48.290 | 26 |
| 27 | 11.808 | 12.879 | 14.573 | 16.151 | 40.113 | 43.194 | 46.963 | 49.645 | 27 |
| 28 | 12.461 | 13.565 | 15.308 | 16.928 | 41.337 | 44.461 | 48.278 | 50.993 | 28 |
| 29 | 13.121 | 14.256 | 16.047 | 17.708 | 42.557 | 45.722 | 49.588 | 52.336 | 29 |
| 30 | 13.787 | 14.953 | 16.791 | 18.493 | 43.773 | 46.979 | 50.892 | 53.672 | 30 |

†Based on Table 8 of *Biometrika Tables for Statisticians, Volume I* (Cambridge University Press, 1954), by permission of the *Biometrika* trustees.

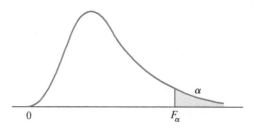

The entries in Table IV are values for which the area to their right under the $F$ distribution with given degrees of freedom (the area of the tinted region under the curve shown above) is equal to $\alpha$.

**TABLE IV**   Critical Values of $F$[†]

**Values of $F_{0.05}$**

| | Degrees of freedom for numerator | | | | | | | | | | | | | | | | | | |
|---|---|---|---|---|---|---|---|---|---|---|---|---|---|---|---|---|---|---|---|
| Degrees of freedom for denominator | 1 | 2 | 3 | 4 | 5 | 6 | 7 | 8 | 9 | 10 | 12 | 15 | 20 | 24 | 30 | 40 | 60 | 120 | ∞ |
| 1 | 161 | 200 | 216 | 225 | 230 | 234 | 237 | 239 | 241 | 242 | 244 | 246 | 248 | 249 | 250 | 251 | 252 | 253 | 254 |
| 2 | 18.5 | 19.0 | 19.2 | 19.2 | 19.3 | 19.3 | 19.4 | 19.4 | 19.4 | 19.4 | 19.4 | 19.4 | 19.4 | 19.5 | 19.5 | 19.5 | 19.5 | 19.5 | 19.5 |
| 3 | 10.1 | 9.55 | 9.28 | 9.12 | 9.01 | 8.94 | 8.89 | 8.85 | 8.81 | 8.79 | 8.74 | 8.70 | 8.66 | 8.64 | 8.62 | 8.59 | 8.57 | 8.55 | 8.53 |
| 4 | 7.71 | 6.94 | 6.59 | 6.39 | 6.26 | 6.16 | 6.09 | 6.04 | 6.00 | 5.96 | 5.91 | 5.86 | 5.80 | 5.77 | 5.75 | 5.72 | 5.69 | 5.66 | 5.63 |
| 5 | 6.61 | 5.79 | 5.41 | 5.19 | 5.05 | 4.95 | 4.88 | 4.82 | 4.77 | 4.74 | 4.68 | 4.62 | 4.56 | 4.53 | 4.50 | 4.46 | 4.43 | 4.40 | 4.37 |
| 6 | 5.99 | 5.14 | 4.76 | 4.53 | 4.39 | 4.28 | 4.21 | 4.15 | 4.10 | 4.06 | 4.00 | 3.94 | 3.87 | 3.84 | 3.81 | 3.77 | 3.74 | 3.70 | 3.67 |
| 7 | 5.59 | 4.74 | 4.35 | 4.12 | 3.97 | 3.87 | 3.79 | 3.73 | 3.68 | 3.64 | 3.57 | 3.51 | 3.44 | 3.41 | 3.38 | 3.34 | 3.30 | 3.27 | 3.23 |
| 8 | 5.32 | 4.46 | 4.07 | 3.84 | 3.69 | 3.58 | 3.50 | 3.44 | 3.39 | 3.35 | 3.28 | 3.22 | 3.15 | 3.12 | 3.08 | 3.04 | 3.01 | 2.97 | 2.93 |
| 9 | 5.12 | 4.26 | 3.86 | 3.63 | 3.48 | 3.37 | 3.29 | 3.23 | 3.18 | 3.14 | 3.07 | 3.01 | 2.94 | 2.90 | 2.86 | 2.83 | 2.79 | 2.75 | 2.71 |
| 10 | 4.96 | 4.10 | 3.71 | 3.48 | 3.33 | 3.22 | 3.14 | 3.07 | 3.02 | 2.98 | 2.91 | 2.85 | 2.77 | 2.74 | 2.70 | 2.66 | 2.62 | 2.58 | 2.54 |
| 11 | 4.84 | 3.98 | 3.59 | 3.36 | 3.20 | 3.09 | 3.01 | 2.95 | 2.90 | 2.85 | 2.79 | 2.72 | 2.65 | 2.61 | 2.57 | 2.53 | 2.49 | 2.45 | 2.40 |
| 12 | 4.75 | 3.89 | 3.49 | 3.26 | 3.11 | 3.00 | 2.91 | 2.85 | 2.80 | 2.75 | 2.69 | 2.62 | 2.54 | 2.51 | 2.47 | 2.43 | 2.38 | 2.34 | 2.30 |
| 13 | 4.67 | 3.81 | 3.41 | 3.18 | 3.03 | 2.92 | 2.83 | 2.77 | 2.71 | 2.67 | 2.60 | 2.53 | 2.46 | 2.42 | 2.38 | 2.34 | 2.30 | 2.25 | 2.21 |
| 14 | 4.60 | 3.74 | 3.34 | 3.11 | 2.96 | 2.85 | 2.76 | 2.70 | 2.65 | 2.60 | 2.53 | 2.46 | 2.39 | 2.35 | 2.31 | 2.27 | 2.22 | 2.18 | 2.13 |
| 15 | 4.54 | 3.68 | 3.29 | 3.06 | 2.90 | 2.79 | 2.71 | 2.64 | 2.59 | 2.54 | 2.48 | 2.40 | 2.33 | 2.29 | 2.25 | 2.20 | 2.16 | 2.11 | 2.07 |
| 16 | 4.49 | 3.63 | 3.24 | 3.01 | 2.85 | 2.74 | 2.66 | 2.59 | 2.54 | 2.49 | 2.42 | 2.35 | 2.28 | 2.24 | 2.19 | 2.15 | 2.11 | 2.06 | 2.01 |
| 17 | 4.45 | 3.59 | 3.20 | 2.96 | 2.81 | 2.70 | 2.61 | 2.55 | 2.49 | 2.45 | 2.38 | 2.31 | 2.23 | 2.19 | 2.15 | 2.10 | 2.06 | 2.01 | 1.96 |
| 18 | 4.41 | 3.55 | 3.16 | 2.93 | 2.77 | 2.66 | 2.58 | 2.51 | 2.46 | 2.41 | 2.34 | 2.27 | 2.19 | 2.15 | 2.11 | 2.06 | 2.02 | 1.97 | 1.92 |
| 19 | 4.38 | 3.52 | 3.13 | 2.90 | 2.74 | 2.63 | 2.54 | 2.48 | 2.42 | 2.38 | 2.31 | 2.23 | 2.16 | 2.11 | 2.07 | 2.03 | 1.98 | 1.93 | 1.88 |
| 20 | 4.35 | 3.49 | 3.10 | 2.87 | 2.71 | 2.60 | 2.51 | 2.45 | 2.39 | 2.35 | 2.28 | 2.20 | 2.12 | 2.08 | 2.04 | 1.99 | 1.95 | 1.90 | 1.84 |
| 21 | 4.32 | 3.47 | 3.07 | 2.84 | 2.68 | 2.57 | 2.49 | 2.42 | 2.37 | 2.32 | 2.25 | 2.18 | 2.10 | 2.05 | 2.01 | 1.96 | 1.92 | 1.87 | 1.81 |
| 22 | 4.30 | 3.44 | 3.05 | 2.82 | 2.66 | 2.55 | 2.46 | 2.40 | 2.34 | 2.30 | 2.23 | 2.15 | 2.07 | 2.03 | 1.98 | 1.94 | 1.89 | 1.84 | 1.78 |
| 23 | 4.28 | 3.42 | 3.03 | 2.80 | 2.64 | 2.53 | 2.44 | 2.37 | 2.32 | 2.27 | 2.20 | 2.13 | 2.05 | 2.01 | 1.96 | 1.91 | 1.86 | 1.81 | 1.76 |
| 24 | 4.26 | 3.40 | 3.01 | 2.78 | 2.62 | 2.51 | 2.42 | 2.36 | 2.30 | 2.25 | 2.18 | 2.11 | 2.03 | 1.98 | 1.94 | 1.89 | 1.84 | 1.79 | 1.73 |
| 25 | 4.24 | 3.39 | 2.99 | 2.76 | 2.60 | 2.49 | 2.40 | 2.34 | 2.28 | 2.24 | 2.16 | 2.09 | 2.01 | 1.96 | 1.92 | 1.87 | 1.82 | 1.77 | 1.71 |
| 30 | 4.17 | 3.32 | 2.92 | 2.69 | 2.53 | 2.42 | 2.33 | 2.27 | 2.21 | 2.16 | 2.09 | 2.01 | 1.93 | 1.89 | 1.84 | 1.79 | 1.74 | 1.68 | 1.62 |
| 40 | 4.08 | 3.23 | 2.84 | 2.61 | 2.45 | 2.34 | 2.25 | 2.18 | 2.12 | 2.08 | 2.00 | 1.92 | 1.84 | 1.79 | 1.74 | 1.69 | 1.64 | 1.58 | 1.51 |
| 60 | 4.00 | 3.15 | 2.76 | 2.53 | 2.37 | 2.25 | 2.17 | 2.10 | 2.04 | 1.99 | 1.92 | 1.84 | 1.75 | 1.70 | 1.65 | 1.59 | 1.53 | 1.47 | 1.39 |
| 120 | 3.92 | 3.07 | 2.68 | 2.45 | 2.29 | 2.18 | 2.09 | 2.02 | 1.96 | 1.91 | 1.83 | 1.75 | 1.66 | 1.61 | 1.55 | 1.50 | 1.43 | 1.35 | 1.25 |
| ∞ | 3.84 | 3.00 | 2.60 | 2.37 | 2.21 | 2.10 | 2.01 | 1.94 | 1.88 | 1.83 | 1.75 | 1.67 | 1.57 | 1.52 | 1.46 | 1.39 | 1.32 | 1.22 | 1.00 |

[†]Reproduced from M. Merrington and C. M. Thompson, "Tables of percentage points of the inverted beta (F) distribution," *Biometrika*, vol. 33 (1943), by permission of the *Biometrika* trustees.

**TABLE IV** Critical Values of $F$ (Continued)

Values of $F_{0.01}$

Degrees of freedom for numerator

| Denom. df | 1 | 2 | 3 | 4 | 5 | 6 | 7 | 8 | 9 | 10 | 12 | 15 | 20 | 24 | 30 | 40 | 60 | 120 | ∞ |
|---|---|---|---|---|---|---|---|---|---|---|---|---|---|---|---|---|---|---|---|
| 1 | 4,052 | 5,000 | 5,403 | 5,625 | 5,764 | 5,859 | 5,928 | 5,982 | 6,023 | 6,056 | 6,106 | 6,157 | 6,209 | 6,235 | 6,261 | 6,287 | 6,313 | 6,339 | 6,366 |
| 2 | 98.5 | 99.0 | 99.2 | 99.2 | 99.3 | 99.3 | 99.4 | 99.4 | 99.4 | 99.4 | 99.4 | 99.4 | 99.4 | 99.5 | 99.5 | 99.5 | 99.5 | 99.5 | 99.5 |
| 3 | 34.1 | 30.8 | 29.5 | 28.7 | 28.2 | 27.9 | 27.7 | 27.5 | 27.3 | 27.2 | 27.1 | 26.9 | 26.7 | 26.6 | 26.5 | 26.4 | 26.3 | 26.2 | 26.1 |
| 4 | 21.2 | 18.0 | 16.7 | 16.0 | 15.5 | 15.2 | 15.0 | 14.8 | 14.7 | 14.5 | 14.4 | 14.2 | 14.0 | 13.9 | 13.8 | 13.7 | 13.7 | 13.6 | 13.5 |
| 5 | 16.3 | 13.3 | 12.1 | 11.4 | 11.0 | 10.7 | 10.5 | 10.3 | 10.2 | 10.1 | 9.89 | 9.72 | 9.55 | 9.47 | 9.38 | 9.29 | 9.20 | 9.11 | 9.02 |
| 6 | 13.7 | 10.9 | 9.78 | 9.15 | 8.75 | 8.47 | 8.26 | 8.10 | 7.98 | 7.87 | 7.72 | 7.56 | 7.40 | 7.31 | 7.23 | 7.14 | 7.05 | 6.97 | 6.88 |
| 7 | 12.2 | 9.55 | 8.45 | 7.85 | 7.46 | 7.19 | 6.99 | 6.84 | 6.72 | 6.62 | 6.47 | 6.31 | 6.16 | 6.07 | 5.99 | 5.91 | 5.82 | 5.74 | 5.65 |
| 8 | 11.3 | 8.65 | 7.59 | 7.01 | 6.63 | 6.37 | 6.18 | 6.03 | 5.91 | 5.81 | 5.67 | 5.52 | 5.36 | 5.28 | 5.20 | 5.12 | 5.03 | 4.95 | 4.86 |
| 9 | 10.6 | 8.02 | 6.99 | 6.42 | 6.06 | 5.80 | 5.61 | 5.47 | 5.35 | 5.26 | 5.11 | 4.96 | 4.81 | 4.73 | 4.65 | 4.57 | 4.48 | 4.40 | 4.31 |
| 10 | 10.0 | 7.56 | 6.55 | 5.99 | 5.64 | 5.39 | 5.20 | 5.06 | 4.94 | 4.85 | 4.71 | 4.56 | 4.41 | 4.33 | 4.25 | 4.17 | 4.08 | 4.00 | 3.91 |
| 11 | 9.65 | 7.21 | 6.22 | 5.67 | 5.32 | 5.07 | 4.89 | 4.74 | 4.63 | 4.54 | 4.40 | 4.25 | 4.10 | 4.02 | 3.94 | 3.86 | 3.78 | 3.69 | 3.60 |
| 12 | 9.33 | 6.93 | 5.95 | 5.41 | 5.06 | 4.82 | 4.64 | 4.50 | 4.39 | 4.30 | 4.16 | 4.01 | 3.86 | 3.78 | 3.70 | 3.62 | 3.54 | 3.45 | 3.36 |
| 13 | 9.07 | 6.70 | 5.74 | 5.21 | 4.86 | 4.62 | 4.44 | 4.30 | 4.19 | 4.10 | 3.96 | 3.82 | 3.66 | 3.59 | 3.51 | 3.43 | 3.34 | 3.25 | 3.17 |
| 14 | 8.86 | 6.51 | 5.56 | 5.04 | 4.70 | 4.46 | 4.28 | 4.14 | 4.03 | 3.94 | 3.80 | 3.66 | 3.51 | 3.43 | 3.35 | 3.27 | 3.18 | 3.09 | 3.00 |
| 15 | 8.68 | 6.36 | 5.42 | 4.89 | 4.56 | 4.32 | 4.14 | 4.00 | 3.89 | 3.80 | 3.67 | 3.52 | 3.37 | 3.29 | 3.21 | 3.13 | 3.05 | 2.96 | 2.87 |
| 16 | 8.53 | 6.23 | 5.29 | 4.77 | 4.44 | 4.20 | 4.03 | 3.89 | 3.78 | 3.69 | 3.55 | 3.41 | 3.26 | 3.18 | 3.10 | 3.02 | 2.93 | 2.84 | 2.75 |
| 17 | 8.40 | 6.11 | 5.19 | 4.67 | 4.34 | 4.10 | 3.93 | 3.79 | 3.68 | 3.59 | 3.46 | 3.31 | 3.16 | 3.08 | 3.00 | 2.92 | 2.83 | 2.75 | 2.65 |
| 18 | 8.29 | 6.01 | 5.09 | 4.58 | 4.25 | 4.01 | 3.84 | 3.71 | 3.60 | 3.51 | 3.37 | 3.23 | 3.08 | 3.00 | 2.92 | 2.84 | 2.75 | 2.66 | 2.57 |
| 19 | 8.19 | 5.93 | 5.01 | 4.50 | 4.17 | 3.94 | 3.77 | 3.63 | 3.52 | 3.43 | 3.30 | 3.15 | 3.00 | 2.92 | 2.84 | 2.76 | 2.67 | 2.58 | 2.49 |
| 20 | 8.10 | 5.85 | 4.94 | 4.43 | 4.10 | 3.87 | 3.70 | 3.56 | 3.46 | 3.37 | 3.23 | 3.09 | 2.94 | 2.86 | 2.78 | 2.69 | 2.61 | 2.52 | 2.42 |
| 21 | 8.02 | 5.78 | 4.87 | 4.37 | 4.04 | 3.81 | 3.64 | 3.51 | 3.40 | 3.31 | 3.17 | 3.03 | 2.88 | 2.80 | 2.72 | 2.64 | 2.55 | 2.46 | 2.36 |
| 22 | 7.95 | 5.72 | 4.82 | 4.31 | 3.99 | 3.76 | 3.59 | 3.45 | 3.35 | 3.26 | 3.12 | 2.98 | 2.83 | 2.75 | 2.67 | 2.58 | 2.50 | 2.40 | 2.31 |
| 23 | 7.88 | 5.66 | 4.76 | 4.26 | 3.94 | 3.71 | 3.54 | 3.41 | 3.30 | 3.21 | 3.07 | 2.93 | 2.78 | 2.70 | 2.62 | 2.54 | 2.45 | 2.35 | 2.26 |
| 24 | 7.82 | 5.61 | 4.72 | 4.22 | 3.90 | 3.67 | 3.50 | 3.36 | 3.26 | 3.17 | 3.03 | 2.89 | 2.74 | 2.66 | 2.58 | 2.49 | 2.40 | 2.31 | 2.21 |
| 25 | 7.77 | 5.57 | 4.68 | 4.18 | 3.86 | 3.63 | 3.46 | 3.32 | 3.22 | 3.13 | 2.99 | 2.85 | 2.70 | 2.62 | 2.53 | 2.45 | 2.36 | 2.27 | 2.17 |
| 30 | 7.56 | 5.39 | 4.51 | 4.02 | 3.70 | 3.47 | 3.30 | 3.17 | 3.07 | 2.98 | 2.84 | 2.70 | 2.55 | 2.47 | 2.39 | 2.30 | 2.21 | 2.11 | 2.01 |
| 40 | 7.31 | 5.18 | 4.31 | 3.83 | 3.51 | 3.29 | 3.12 | 2.99 | 2.89 | 2.80 | 2.66 | 2.52 | 2.37 | 2.29 | 2.20 | 2.11 | 2.02 | 1.92 | 1.80 |
| 60 | 7.08 | 4.98 | 4.13 | 3.65 | 3.34 | 3.12 | 2.95 | 2.82 | 2.72 | 2.63 | 2.50 | 2.35 | 2.20 | 2.12 | 2.03 | 1.94 | 1.84 | 1.73 | 1.60 |
| 120 | 6.85 | 4.79 | 3.95 | 3.48 | 3.17 | 2.96 | 2.79 | 2.66 | 2.56 | 2.47 | 2.34 | 2.19 | 2.03 | 1.95 | 1.86 | 1.76 | 1.66 | 1.53 | 1.38 |
| ∞ | 6.63 | 4.61 | 3.78 | 3.32 | 3.02 | 2.80 | 2.64 | 2.51 | 2.41 | 2.32 | 2.18 | 2.04 | 1.88 | 1.79 | 1.70 | 1.59 | 1.47 | 1.32 | 1.00 |

Degrees of freedom for denominator

## TABLE V   Binomial Probabilities

| | | | | | | | $p$ | | | | | |
|---|---|---|---|---|---|---|---|---|---|---|---|---|
| $n$ | $x$ | 0.05 | 0.1 | 0.2 | 0.3 | 0.4 | 0.5 | 0.6 | 0.7 | 0.8 | 0.9 | 0.95 |
| 2 | 0 | 0.902 | 0.810 | 0.640 | 0.490 | 0.360 | 0.250 | 0.160 | 0.090 | 0.040 | 0.010 | 0.002 |
| | 1 | 0.095 | 0.180 | 0.320 | 0.420 | 0.480 | 0.500 | 0.480 | 0.420 | 0.320 | 0.180 | 0.095 |
| | 2 | 0.002 | 0.010 | 0.040 | 0.090 | 0.160 | 0.250 | 0.360 | 0.490 | 0.640 | 0.810 | 0.902 |
| 3 | 0 | 0.857 | 0.729 | 0.512 | 0.343 | 0.216 | 0.125 | 0.064 | 0.027 | 0.008 | 0.001 | |
| | 1 | 0.135 | 0.243 | 0.384 | 0.441 | 0.432 | 0.375 | 0.288 | 0.189 | 0.096 | 0.027 | 0.007 |
| | 2 | 0.007 | 0.027 | 0.096 | 0.189 | 0.288 | 0.375 | 0.432 | 0.441 | 0.384 | 0.243 | 0.135 |
| | 3 | | 0.001 | 0.008 | 0.027 | 0.064 | 0.125 | 0.216 | 0.343 | 0.512 | 0.729 | 0.857 |
| 4 | 0 | 0.815 | 0.656 | 0.410 | 0.240 | 0.130 | 0.062 | 0.026 | 0.008 | 0.002 | | |
| | 1 | 0.171 | 0.292 | 0.410 | 0.412 | 0.346 | 0.250 | 0.154 | 0.076 | 0.026 | 0.004 | |
| | 2 | 0.014 | 0.049 | 0.154 | 0.265 | 0.346 | 0.375 | 0.346 | 0.265 | 0.154 | 0.049 | 0.014 |
| | 3 | | 0.004 | 0.026 | 0.076 | 0.154 | 0.250 | 0.346 | 0.412 | 0.410 | 0.292 | 0.171 |
| | 4 | | | 0.002 | 0.008 | 0.026 | 0.062 | 0.130 | 0.240 | 0.410 | 0.656 | 0.815 |
| 5 | 0 | 0.774 | 0.590 | 0.328 | 0.168 | 0.078 | 0.031 | 0.010 | 0.002 | | | |
| | 1 | 0.204 | 0.328 | 0.410 | 0.360 | 0.259 | 0.156 | 0.077 | 0.028 | 0.006 | | |
| | 2 | 0.021 | 0.073 | 0.205 | 0.309 | 0.346 | 0.312 | 0.230 | 0.132 | 0.051 | 0.008 | 0.001 |
| | 3 | 0.001 | 0.008 | 0.051 | 0.132 | 0.230 | 0.312 | 0.346 | 0.309 | 0.205 | 0.073 | 0.021 |
| | 4 | | | 0.006 | 0.028 | 0.077 | 0.156 | 0.259 | 0.360 | 0.410 | 0.328 | 0.204 |
| | 5 | | | | 0.002 | 0.010 | 0.031 | 0.078 | 0.168 | 0.328 | 0.590 | 0.774 |
| 6 | 0 | 0.735 | 0.531 | 0.262 | 0.118 | 0.047 | 0.016 | 0.004 | 0.001 | | | |
| | 1 | 0.232 | 0.354 | 0.393 | 0.303 | 0.187 | 0.094 | 0.037 | 0.010 | 0.002 | | |
| | 2 | 0.031 | 0.098 | 0.246 | 0.324 | 0.311 | 0.234 | 0.138 | 0.060 | 0.015 | 0.001 | |
| | 3 | 0.002 | 0.015 | 0.082 | 0.185 | 0.276 | 0.312 | 0.276 | 0.185 | 0.082 | 0.015 | 0.002 |
| | 4 | | 0.001 | 0.015 | 0.060 | 0.138 | 0.234 | 0.311 | 0.324 | 0.246 | 0.098 | 0.031 |
| | 5 | | | 0.002 | 0.010 | 0.037 | 0.094 | 0.187 | 0.303 | 0.393 | 0.354 | 0.232 |
| | 6 | | | | 0.001 | 0.004 | 0.016 | 0.047 | 0.118 | 0.262 | 0.531 | 0.735 |
| 7 | 0 | 0.698 | 0.478 | 0.210 | 0.082 | 0.028 | 0.008 | 0.002 | | | | |
| | 1 | 0.257 | 0.372 | 0.367 | 0.247 | 0.131 | 0.055 | 0.017 | 0.004 | | | |
| | 2 | 0.041 | 0.124 | 0.275 | 0.318 | 0.261 | 0.164 | 0.077 | 0.025 | 0.004 | | |
| | 3 | 0.004 | 0.023 | 0.115 | 0.227 | 0.290 | 0.273 | 0.194 | 0.097 | 0.029 | 0.003 | |
| | 4 | | 0.003 | 0.029 | 0.097 | 0.194 | 0.273 | 0.290 | 0.227 | 0.115 | 0.023 | 0.004 |
| | 5 | | | 0.004 | 0.025 | 0.077 | 0.164 | 0.261 | 0.318 | 0.275 | 0.124 | 0.041 |
| | 6 | | | | 0.004 | 0.017 | 0.055 | 0.131 | 0.247 | 0.367 | 0.372 | 0.257 |
| | 7 | | | | | 0.002 | 0.008 | 0.028 | 0.082 | 0.210 | 0.478 | 0.698 |
| 8 | 0 | 0.663 | 0.430 | 0.168 | 0.058 | 0.017 | 0.004 | 0.001 | | | | |
| | 1 | 0.279 | 0.383 | 0.336 | 0.198 | 0.090 | 0.031 | 0.008 | 0.001 | | | |
| | 2 | 0.051 | 0.149 | 0.294 | 0.296 | 0.209 | 0.109 | 0.041 | 0.010 | 0.001 | | |
| | 3 | 0.005 | 0.033 | 0.147 | 0.254 | 0.279 | 0.219 | 0.124 | 0.047 | 0.009 | | |
| | 4 | | 0.005 | 0.046 | 0.136 | 0.232 | 0.273 | 0.232 | 0.136 | 0.046 | 0.005 | |
| | 5 | | | 0.009 | 0.047 | 0.124 | 0.219 | 0.279 | 0.254 | 0.147 | 0.033 | 0.005 |
| | 6 | | | 0.001 | 0.010 | 0.041 | 0.109 | 0.209 | 0.296 | 0.294 | 0.149 | 0.051 |
| | 7 | | | | 0.001 | 0.008 | 0.031 | 0.090 | 0.198 | 0.336 | 0.383 | 0.279 |
| | 8 | | | | | 0.001 | 0.004 | 0.017 | 0.058 | 0.168 | 0.430 | 0.663 |

All values omitted in this table are 0.0005 or less.

## TABLE V    Binomial Probabilities (*Continued*)

| n | x | 0.05 | 0.1 | 0.2 | 0.3 | 0.4 | 0.5 | 0.6 | 0.7 | 0.8 | 0.9 | 0.95 |
|---|---|------|-----|-----|-----|-----|-----|-----|-----|-----|-----|------|
| 9 | 0 | 0.630 | 0.387 | 0.134 | 0.040 | 0.010 | 0.002 | | | | | |
| | 1 | 0.299 | 0.387 | 0.302 | 0.156 | 0.060 | 0.018 | 0.004 | | | | |
| | 2 | 0.063 | 0.172 | 0.302 | 0.267 | 0.161 | 0.070 | 0.021 | 0.004 | | | |
| | 3 | 0.008 | 0.045 | 0.176 | 0.267 | 0.251 | 0.164 | 0.074 | 0.021 | 0.003 | | |
| | 4 | 0.001 | 0.007 | 0.066 | 0.172 | 0.251 | 0.246 | 0.167 | 0.074 | 0.017 | 0.001 | |
| | 5 | | 0.001 | 0.017 | 0.074 | 0.167 | 0.246 | 0.251 | 0.172 | 0.066 | 0.007 | 0.001 |
| | 6 | | | 0.003 | 0.021 | 0.074 | 0.164 | 0.251 | 0.267 | 0.176 | 0.045 | 0.008 |
| | 7 | | | | 0.004 | 0.021 | 0.070 | 0.161 | 0.267 | 0.302 | 0.172 | 0.063 |
| | 8 | | | | | 0.004 | 0.018 | 0.060 | 0.156 | 0.302 | 0.387 | 0.299 |
| | 9 | | | | | | 0.002 | 0.010 | 0.040 | 0.134 | 0.387 | 0.630 |
| 10 | 0 | 0.599 | 0.349 | 0.107 | 0.028 | 0.006 | 0.001 | | | | | |
| | 1 | 0.315 | 0.387 | 0.268 | 0.121 | 0.040 | 0.010 | 0.002 | | | | |
| | 2 | 0.075 | 0.194 | 0.302 | 0.233 | 0.121 | 0.044 | 0.011 | 0.001 | | | |
| | 3 | 0.010 | 0.057 | 0.201 | 0.267 | 0.215 | 0.117 | 0.042 | 0.009 | 0.001 | | |
| | 4 | 0.001 | 0.011 | 0.088 | 0.200 | 0.251 | 0.205 | 0.111 | 0.037 | 0.006 | | |
| | 5 | | 0.001 | 0.026 | 0.103 | 0.201 | 0.246 | 0.201 | 0.103 | 0.026 | 0.001 | |
| | 6 | | | 0.006 | 0.037 | 0.111 | 0.205 | 0.251 | 0.200 | 0.088 | 0.011 | 0.001 |
| | 7 | | | 0.001 | 0.009 | 0.042 | 0.117 | 0.215 | 0.267 | 0.201 | 0.057 | 0.010 |
| | 8 | | | | 0.001 | 0.011 | 0.044 | 0.121 | 0.233 | 0.302 | 0.194 | 0.075 |
| | 9 | | | | | 0.002 | 0.010 | 0.040 | 0.121 | 0.268 | 0.387 | 0.315 |
| | 10 | | | | | | 0.001 | 0.006 | 0.028 | 0.107 | 0.349 | 0.599 |
| 11 | 0 | 0.569 | 0.314 | 0.086 | 0.020 | 0.004 | | | | | | |
| | 1 | 0.329 | 0.384 | 0.236 | 0.093 | 0.027 | 0.005 | 0.001 | | | | |
| | 2 | 0.087 | 0.213 | 0.295 | 0.200 | 0.089 | 0.027 | 0.005 | 0.001 | | | |
| | 3 | 0.014 | 0.071 | 0.221 | 0.257 | 0.177 | 0.081 | 0.023 | 0.004 | | | |
| | 4 | 0.001 | 0.016 | 0.111 | 0.220 | 0.236 | 0.161 | 0.070 | 0.017 | 0.002 | | |
| | 5 | | 0.002 | 0.039 | 0.132 | 0.221 | 0.226 | 0.147 | 0.057 | 0.010 | | |
| | 6 | | | 0.010 | 0.057 | 0.147 | 0.226 | 0.221 | 0.132 | 0.039 | 0.002 | |
| | 7 | | | 0.002 | 0.017 | 0.070 | 0.161 | 0.236 | 0.220 | 0.111 | 0.016 | 0.001 |
| | 8 | | | | 0.004 | 0.023 | 0.081 | 0.177 | 0.257 | 0.221 | 0.071 | 0.014 |
| | 9 | | | | 0.001 | 0.005 | 0.027 | 0.089 | 0.200 | 0.295 | 0.213 | 0.087 |
| | 10 | | | | | 0.001 | 0.005 | 0.027 | 0.093 | 0.236 | 0.384 | 0.329 |
| | 11 | | | | | | | 0.004 | 0.020 | 0.086 | 0.314 | 0.569 |
| 12 | 0 | 0.540 | 0.282 | 0.069 | 0.014 | 0.002 | | | | | | |
| | 1 | 0.341 | 0.377 | 0.206 | 0.071 | 0.017 | 0.003 | | | | | |
| | 2 | 0.099 | 0.230 | 0.283 | 0.168 | 0.064 | 0.016 | 0.002 | | | | |
| | 3 | 0.017 | 0.085 | 0.236 | 0.240 | 0.142 | 0.054 | 0.012 | 0.001 | | | |
| | 4 | 0.002 | 0.021 | 0.133 | 0.231 | 0.213 | 0.121 | 0.042 | 0.008 | 0.001 | | |
| | 5 | | 0.004 | 0.053 | 0.158 | 0.227 | 0.193 | 0.101 | 0.029 | 0.003 | | |
| | 6 | | | 0.016 | 0.079 | 0.177 | 0.226 | 0.177 | 0.079 | 0.016 | | |
| | 7 | | | 0.003 | 0.029 | 0.101 | 0.193 | 0.227 | 0.158 | 0.053 | 0.004 | |
| | 8 | | | 0.001 | 0.008 | 0.042 | 0.121 | 0.213 | 0.231 | 0.133 | 0.021 | 0.002 |
| | 9 | | | | 0.001 | 0.012 | 0.054 | 0.142 | 0.240 | 0.236 | 0.085 | 0.017 |
| | 10 | | | | | 0.002 | 0.016 | 0.064 | 0.168 | 0.283 | 0.230 | 0.099 |
| | 11 | | | | | | 0.003 | 0.017 | 0.071 | 0.206 | 0.377 | 0.341 |
| | 12 | | | | | | | 0.002 | 0.014 | 0.069 | 0.282 | 0.540 |

## TABLE V    Binomial Probabilities (*Continued*)

| n | x | 0.05 | 0.1 | 0.2 | 0.3 | 0.4 | 0.5 | 0.6 | 0.7 | 0.8 | 0.9 | 0.95 |
|---|---|------|-----|-----|-----|-----|-----|-----|-----|-----|-----|------|
| **13** | 0 | 0.513 | 0.254 | 0.055 | 0.010 | 0.001 | | | | | | |
| | 1 | 0.351 | 0.367 | 0.179 | 0.054 | 0.011 | 0.002 | | | | | |
| | 2 | 0.111 | 0.245 | 0.268 | 0.139 | 0.045 | 0.010 | 0.001 | | | | |
| | 3 | 0.021 | 0.100 | 0.246 | 0.218 | 0.111 | 0.035 | 0.006 | 0.001 | | | |
| | 4 | 0.003 | 0.028 | 0.154 | 0.234 | 0.184 | 0.087 | 0.024 | 0.003 | | | |
| | 5 | | 0.006 | 0.069 | 0.180 | 0.221 | 0.157 | 0.066 | 0.014 | 0.001 | | |
| | 6 | | 0.001 | 0.023 | 0.103 | 0.197 | 0.209 | 0.131 | 0.044 | 0.006 | | |
| | 7 | | | 0.006 | 0.044 | 0.131 | 0.209 | 0.197 | 0.103 | 0.023 | 0.001 | |
| | 8 | | | 0.001 | 0.014 | 0.066 | 0.157 | 0.221 | 0.180 | 0.069 | 0.006 | |
| | 9 | | | | 0.003 | 0.024 | 0.087 | 0.184 | 0.234 | 0.154 | 0.028 | 0.003 |
| | 10 | | | | 0.001 | 0.006 | 0.035 | 0.111 | 0.218 | 0.246 | 0.100 | 0.021 |
| | 11 | | | | | 0.001 | 0.010 | 0.045 | 0.139 | 0.268 | 0.245 | 0.111 |
| | 12 | | | | | | 0.002 | 0.011 | 0.054 | 0.179 | 0.367 | 0.351 |
| | 13 | | | | | | | 0.001 | 0.010 | 0.055 | 0.254 | 0.513 |
| **14** | 0 | 0.488 | 0.229 | 0.044 | 0.007 | 0.001 | | | | | | |
| | 1 | 0.359 | 0.356 | 0.154 | 0.041 | 0.007 | 0.001 | | | | | |
| | 2 | 0.123 | 0.257 | 0.250 | 0.113 | 0.032 | 0.006 | 0.001 | | | | |
| | 3 | 0.026 | 0.114 | 0.250 | 0.194 | 0.085 | 0.022 | 0.003 | | | | |
| | 4 | 0.004 | 0.035 | 0.172 | 0.229 | 0.155 | 0.061 | 0.014 | 0.001 | | | |
| | 5 | | 0.008 | 0.086 | 0.196 | 0.207 | 0.122 | 0.041 | 0.007 | | | |
| | 6 | | 0.001 | 0.032 | 0.126 | 0.207 | 0.183 | 0.092 | 0.023 | 0.002 | | |
| | 7 | | | 0.009 | 0.062 | 0.157 | 0.209 | 0.157 | 0.062 | 0.009 | | |
| | 8 | | | 0.002 | 0.023 | 0.092 | 0.183 | 0.207 | 0.126 | 0.032 | 0.001 | |
| | 9 | | | | 0.007 | 0.041 | 0.122 | 0.207 | 0.196 | 0.086 | 0.008 | |
| | 10 | | | | 0.001 | 0.014 | 0.061 | 0.155 | 0.229 | 0.172 | 0.035 | 0.004 |
| | 11 | | | | | 0.003 | 0.022 | 0.085 | 0.194 | 0.250 | 0.114 | 0.026 |
| | 12 | | | | | 0.001 | 0.006 | 0.032 | 0.113 | 0.250 | 0.257 | 0.123 |
| | 13 | | | | | | 0.001 | 0.007 | 0.041 | 0.154 | 0.356 | 0.359 |
| | 14 | | | | | | | 0.001 | 0.007 | 0.044 | 0.229 | 0.488 |
| **15** | 0 | 0.463 | 0.206 | 0.035 | 0.005 | | | | | | | |
| | 1 | 0.366 | 0.343 | 0.132 | 0.031 | 0.005 | | | | | | |
| | 2 | 0.135 | 0.267 | 0.231 | 0.092 | 0.022 | 0.003 | | | | | |
| | 3 | 0.031 | 0.129 | 0.250 | 0.170 | 0.063 | 0.014 | 0.002 | | | | |
| | 4 | 0.005 | 0.043 | 0.188 | 0.219 | 0.127 | 0.042 | 0.007 | 0.001 | | | |
| | 5 | 0.001 | 0.010 | 0.103 | 0.206 | 0.186 | 0.092 | 0.024 | 0.003 | | | |
| | 6 | | 0.002 | 0.043 | 0.147 | 0.207 | 0.153 | 0.061 | 0.012 | 0.001 | | |
| | 7 | | | 0.014 | 0.081 | 0.177 | 0.196 | 0.118 | 0.035 | 0.003 | | |
| | 8 | | | 0.003 | 0.035 | 0.118 | 0.196 | 0.177 | 0.081 | 0.014 | | |
| | 9 | | | 0.001 | 0.012 | 0.061 | 0.153 | 0.207 | 0.147 | 0.043 | 0.002 | |
| | 10 | | | | 0.003 | 0.024 | 0.092 | 0.186 | 0.206 | 0.103 | 0.010 | 0.001 |
| | 11 | | | | 0.001 | 0.007 | 0.042 | 0.127 | 0.219 | 0.188 | 0.043 | 0.005 |
| | 12 | | | | | 0.002 | 0.014 | 0.063 | 0.170 | 0.250 | 0.129 | 0.031 |
| | 13 | | | | | | 0.003 | 0.022 | 0.092 | 0.231 | 0.267 | 0.135 |
| | 14 | | | | | | | 0.005 | 0.031 | 0.132 | 0.343 | 0.366 |
| | 15 | | | | | | | | 0.005 | 0.035 | 0.206 | 0.463 |

## TABLE V   Binomial Probabilities (*Continued*)

| n | x | 0.05 | 0.1 | 0.2 | 0.3 | 0.4 | *p* 0.5 | 0.6 | 0.7 | 0.8 | 0.9 | 0.95 |
|---|---|------|-----|-----|-----|-----|-----|-----|-----|-----|-----|------|
| **16** | 0 | 0.440 | 0.185 | 0.028 | 0.003 | | | | | | | |
| | 1 | 0.371 | 0.329 | 0.113 | 0.023 | 0.003 | | | | | | |
| | 2 | 0.146 | 0.275 | 0.211 | 0.073 | 0.015 | 0.002 | | | | | |
| | 3 | 0.036 | 0.142 | 0.246 | 0.146 | 0.047 | 0.009 | 0.001 | | | | |
| | 4 | 0.006 | 0.051 | 0.200 | 0.204 | 0.101 | 0.028 | 0.004 | | | | |
| | 5 | 0.001 | 0.014 | 0.120 | 0.210 | 0.162 | 0.067 | 0.014 | 0.001 | | | |
| | 6 | | 0.003 | 0.055 | 0.165 | 0.198 | 0.122 | 0.039 | 0.006 | | | |
| | 7 | | | 0.020 | 0.101 | 0.189 | 0.175 | 0.084 | 0.019 | 0.001 | | |
| | 8 | | | 0.006 | 0.049 | 0.142 | 0.196 | 0.142 | 0.049 | 0.006 | | |
| | 9 | | | 0.001 | 0.019 | 0.084 | 0.175 | 0.189 | 0.101 | 0.020 | | |
| | 10 | | | | 0.006 | 0.039 | 0.122 | 0.198 | 0.165 | 0.055 | 0.003 | |
| | 11 | | | | 0.001 | 0.014 | 0.067 | 0.162 | 0.210 | 0.120 | 0.014 | 0.001 |
| | 12 | | | | | 0.004 | 0.028 | 0.101 | 0.204 | 0.200 | 0.051 | 0.006 |
| | 13 | | | | | 0.001 | 0.009 | 0.047 | 0.146 | 0.246 | 0.142 | 0.036 |
| | 14 | | | | | | 0.002 | 0.015 | 0.073 | 0.211 | 0.275 | 0.146 |
| | 15 | | | | | | | 0.003 | 0.023 | 0.113 | 0.329 | 0.371 |
| | 16 | | | | | | | | 0.003 | 0.028 | 0.185 | 0.440 |
| **17** | 0 | 0.418 | 0.167 | 0.023 | 0.002 | | | | | | | |
| | 1 | 0.374 | 0.315 | 0.096 | 0.017 | 0.002 | | | | | | |
| | 2 | 0.158 | 0.280 | 0.191 | 0.058 | 0.010 | 0.001 | | | | | |
| | 3 | 0.041 | 0.156 | 0.239 | 0.125 | 0.034 | 0.005 | | | | | |
| | 4 | 0.008 | 0.060 | 0.209 | 0.187 | 0.080 | 0.018 | 0.002 | | | | |
| | 5 | 0.001 | 0.017 | 0.136 | 0.208 | 0.138 | 0.047 | 0.008 | 0.001 | | | |
| | 6 | | 0.004 | 0.068 | 0.178 | 0.184 | 0.094 | 0.024 | 0.003 | | | |
| | 7 | | 0.001 | 0.027 | 0.120 | 0.193 | 0.148 | 0.057 | 0.009 | | | |
| | 8 | | | 0.008 | 0.064 | 0.161 | 0.185 | 0.107 | 0.028 | 0.002 | | |
| | 9 | | | 0.002 | 0.028 | 0.107 | 0.185 | 0.161 | 0.064 | 0.008 | | |
| | 10 | | | | 0.009 | 0.057 | 0.148 | 0.193 | 0.120 | 0.027 | 0.001 | |
| | 11 | | | | 0.003 | 0.024 | 0.094 | 0.184 | 0.178 | 0.068 | 0.004 | |
| | 12 | | | | 0.001 | 0.008 | 0.047 | 0.138 | 0.208 | 0.136 | 0.017 | 0.001 |
| | 13 | | | | | 0.002 | 0.018 | 0.080 | 0.187 | 0.209 | 0.060 | 0.008 |
| | 14 | | | | | | 0.005 | 0.034 | 0.125 | 0.239 | 0.156 | 0.041 |
| | 15 | | | | | | 0.001 | 0.010 | 0.058 | 0.191 | 0.280 | 0.158 |
| | 16 | | | | | | | 0.002 | 0.017 | 0.096 | 0.315 | 0.374 |
| | 17 | | | | | | | | 0.002 | 0.023 | 0.167 | 0.418 |
| **18** | 0 | 0.397 | 0.150 | 0.018 | 0.002 | | | | | | | |
| | 1 | 0.376 | 0.300 | 0.081 | 0.013 | 0.001 | | | | | | |
| | 2 | 0.168 | 0.284 | 0.172 | 0.046 | 0.007 | 0.001 | | | | | |
| | 3 | 0.047 | 0.168 | 0.230 | 0.105 | 0.025 | 0.003 | | | | | |
| | 4 | 0.009 | 0.070 | 0.215 | 0.168 | 0.061 | 0.012 | 0.001 | | | | |
| | 5 | 0.001 | 0.022 | 0.151 | 0.202 | 0.115 | 0.033 | 0.004 | | | | |
| | 6 | | 0.005 | 0.082 | 0.187 | 0.166 | 0.071 | 0.015 | 0.001 | | | |
| | 7 | | 0.001 | 0.035 | 0.138 | 0.189 | 0.121 | 0.037 | 0.005 | | | |
| | 8 | | | 0.012 | 0.081 | 0.173 | 0.167 | 0.077 | 0.015 | 0.001 | | |
| | 9 | | | 0.003 | 0.039 | 0.128 | 0.185 | 0.128 | 0.039 | 0.003 | | |
| | 10 | | | 0.001 | 0.015 | 0.077 | 0.167 | 0.173 | 0.081 | 0.012 | | |
| | 11 | | | | 0.005 | 0.037 | 0.121 | 0.189 | 0.138 | 0.035 | 0.001 | |
| | 12 | | | | 0.001 | 0.015 | 0.071 | 0.166 | 0.187 | 0.082 | 0.005 | |

## TABLE V    Binomial Probabilities (*Continued*)

| n | x | 0.05 | 0.1 | 0.2 | 0.3 | 0.4 | *p* 0.5 | 0.6 | 0.7 | 0.8 | 0.9 | 0.95 |
|---|---|------|-----|-----|-----|-----|---------|-----|-----|-----|-----|------|
|   | 13 |      |       |       |       | 0.004 | 0.033 | 0.115 | 0.202 | 0.151 | 0.022 | 0.001 |
|   | 14 |      |       |       |       | 0.001 | 0.012 | 0.061 | 0.168 | 0.215 | 0.070 | 0.009 |
|   | 15 |      |       |       |       |       | 0.003 | 0.025 | 0.105 | 0.230 | 0.168 | 0.047 |
|   | 16 |      |       |       |       |       | 0.001 | 0.007 | 0.046 | 0.172 | 0.284 | 0.168 |
|   | 17 |      |       |       |       |       |       | 0.001 | 0.013 | 0.081 | 0.300 | 0.376 |
|   | 18 |      |       |       |       |       |       |       | 0.002 | 0.018 | 0.150 | 0.397 |
| 19 | 0 | 0.377 | 0.135 | 0.014 | 0.001 |       |       |       |       |       |       |       |
|   | 1 | 0.377 | 0.285 | 0.068 | 0.009 | 0.001 |       |       |       |       |       |       |
|   | 2 | 0.179 | 0.285 | 0.154 | 0.036 | 0.005 |       |       |       |       |       |       |
|   | 3 | 0.053 | 0.180 | 0.218 | 0.087 | 0.017 | 0.002 |       |       |       |       |       |
|   | 4 | 0.011 | 0.080 | 0.218 | 0.149 | 0.047 | 0.007 | 0.001 |       |       |       |       |
|   | 5 | 0.002 | 0.027 | 0.164 | 0.192 | 0.093 | 0.022 | 0.002 |       |       |       |       |
|   | 6 |      | 0.007 | 0.095 | 0.192 | 0.145 | 0.052 | 0.008 | 0.001 |       |       |       |
|   | 7 |      | 0.001 | 0.044 | 0.153 | 0.180 | 0.096 | 0.024 | 0.002 |       |       |       |
|   | 8 |      |       | 0.017 | 0.098 | 0.180 | 0.144 | 0.053 | 0.008 |       |       |       |
|   | 9 |      |       | 0.005 | 0.051 | 0.146 | 0.176 | 0.098 | 0.022 | 0.001 |       |       |
|   | 10 |      |       | 0.001 | 0.022 | 0.098 | 0.176 | 0.146 | 0.051 | 0.005 |       |       |
|   | 11 |      |       |       | 0.008 | 0.053 | 0.144 | 0.180 | 0.098 | 0.017 |       |       |
|   | 12 |      |       |       | 0.002 | 0.024 | 0.096 | 0.180 | 0.153 | 0.044 | 0.001 |       |
|   | 13 |      |       |       | 0.001 | 0.008 | 0.052 | 0.145 | 0.192 | 0.095 | 0.007 |       |
|   | 14 |      |       |       |       | 0.002 | 0.022 | 0.093 | 0.192 | 0.164 | 0.027 | 0.002 |
|   | 15 |      |       |       |       | 0.001 | 0.007 | 0.047 | 0.149 | 0.218 | 0.080 | 0.011 |
|   | 16 |      |       |       |       |       | 0.002 | 0.017 | 0.087 | 0.218 | 0.180 | 0.053 |
|   | 17 |      |       |       |       |       |       | 0.005 | 0.036 | 0.154 | 0.285 | 0.179 |
|   | 18 |      |       |       |       |       |       | 0.001 | 0.009 | 0.068 | 0.285 | 0.377 |
|   | 19 |      |       |       |       |       |       |       | 0.001 | 0.014 | 0.135 | 0.377 |
| 20 | 0 | 0.358 | 0.122 | 0.012 | 0.001 |       |       |       |       |       |       |       |
|   | 1 | 0.377 | 0.270 | 0.058 | 0.007 |       |       |       |       |       |       |       |
|   | 2 | 0.189 | 0.285 | 0.137 | 0.028 | 0.003 |       |       |       |       |       |       |
|   | 3 | 0.060 | 0.190 | 0.205 | 0.072 | 0.012 | 0.001 |       |       |       |       |       |
|   | 4 | 0.013 | 0.090 | 0.218 | 0.130 | 0.035 | 0.005 |       |       |       |       |       |
|   | 5 | 0.002 | 0.032 | 0.175 | 0.179 | 0.075 | 0.015 | 0.001 |       |       |       |       |
|   | 6 |      | 0.009 | 0.109 | 0.192 | 0.124 | 0.037 | 0.005 |       |       |       |       |
|   | 7 |      | 0.002 | 0.055 | 0.164 | 0.166 | 0.074 | 0.015 | 0.001 |       |       |       |
|   | 8 |      |       | 0.022 | 0.114 | 0.180 | 0.120 | 0.035 | 0.004 |       |       |       |
|   | 9 |      |       | 0.007 | 0.065 | 0.160 | 0.160 | 0.071 | 0.012 |       |       |       |
|   | 10 |      |       | 0.002 | 0.031 | 0.117 | 0.176 | 0.117 | 0.031 | 0.002 |       |       |
|   | 11 |      |       |       | 0.012 | 0.071 | 0.160 | 0.160 | 0.065 | 0.007 |       |       |
|   | 12 |      |       |       | 0.004 | 0.035 | 0.120 | 0.180 | 0.114 | 0.022 |       |       |
|   | 13 |      |       |       | 0.001 | 0.015 | 0.074 | 0.166 | 0.164 | 0.055 | 0.002 |       |
|   | 14 |      |       |       |       | 0.005 | 0.037 | 0.124 | 0.192 | 0.109 | 0.009 |       |
|   | 15 |      |       |       |       | 0.001 | 0.015 | 0.075 | 0.179 | 0.175 | 0.032 | 0.002 |
|   | 16 |      |       |       |       |       | 0.005 | 0.035 | 0.130 | 0.218 | 0.090 | 0.013 |
|   | 17 |      |       |       |       |       | 0.001 | 0.012 | 0.072 | 0.205 | 0.190 | 0.060 |
|   | 18 |      |       |       |       |       |       | 0.003 | 0.028 | 0.137 | 0.285 | 0.189 |
|   | 19 |      |       |       |       |       |       |       | 0.007 | 0.058 | 0.270 | 0.377 |
|   | 20 |      |       |       |       |       |       |       | 0.001 | 0.012 | 0.122 | 0.358 |

## TABLE VI   Critical Values of $T^{\dagger}$

| $n$ | $T_{0.10}$ | $T_{0.05}$ | $T_{0.02}$ | $T_{0.01}$ |
|---|---|---|---|---|
| 4 | | | | |
| 5 | 1 | | | |
| 6 | 2 | 1 | | |
| 7 | 4 | 2 | 0 | |
| 8 | 6 | 4 | 2 | 0 |
| 9 | 8 | 6 | 3 | 2 |
| 10 | 11 | 8 | 5 | 3 |
| 11 | 14 | 11 | 7 | 5 |
| 12 | 17 | 14 | 10 | 7 |
| 13 | 21 | 17 | 13 | 10 |
| 14 | 26 | 21 | 16 | 13 |
| 15 | 30 | 25 | 20 | 16 |
| 16 | 36 | 30 | 24 | 19 |
| 17 | 41 | 35 | 28 | 23 |
| 18 | 47 | 40 | 33 | 28 |
| 19 | 54 | 46 | 38 | 32 |
| 20 | 60 | 52 | 43 | 37 |
| 21 | 68 | 59 | 49 | 43 |
| 22 | 75 | 66 | 56 | 49 |
| 23 | 83 | 73 | 62 | 55 |
| 24 | 92 | 81 | 69 | 61 |
| 25 | 101 | 90 | 77 | 68 |

$\dagger$From F. Wilcoxon and R. A. Wilcox, *Some Rapid Approximate Statistical Procedures*, American Cyanamid Company, Pearl River, N.Y., 1964. Reproduced with permission of American Cyanamid Company.

## TABLE VII   Critical Values of $U^{\dagger}$

| $n_1$ \ $n_2$ | 2 | 3 | 4 | 5 | 6 | 7 | 8 | 9 | 10 | 11 | 12 | 13 | 14 | 15 |
|---|---|---|---|---|---|---|---|---|---|---|---|---|---|---|
| **Values of $U_{0.10}$** | | | | | | | | | | | | | | |
| 2 |  |  |  | 0 | 0 | 0 | 1 | 1 | 1 | 1 | 2 | 2 | 3 | 3 |
| 3 |  | 0 | 0 | 1 | 2 | 2 | 3 | 4 | 4 | 5 | 5 | 6 | 7 | 7 |
| 4 |  | 0 | 1 | 2 | 3 | 4 | 5 | 6 | 7 | 8 | 9 | 10 | 11 | 12 |
| 5 | 0 | 1 | 2 | 4 | 5 | 6 | 8 | 9 | 11 | 12 | 13 | 15 | 16 | 18 |
| 6 | 0 | 2 | 3 | 5 | 7 | 8 | 10 | 12 | 14 | 16 | 17 | 19 | 21 | 23 |
| 7 | 0 | 2 | 4 | 6 | 8 | 11 | 13 | 15 | 17 | 19 | 21 | 24 | 26 | 28 |
| 8 | 1 | 3 | 5 | 8 | 10 | 13 | 15 | 18 | 20 | 23 | 26 | 28 | 31 | 33 |
| 9 | 1 | 4 | 6 | 9 | 12 | 15 | 18 | 21 | 24 | 27 | 30 | 33 | 36 | 39 |
| 10 | 1 | 4 | 7 | 11 | 14 | 17 | 20 | 24 | 27 | 31 | 34 | 37 | 41 | 44 |
| 11 | 1 | 5 | 8 | 12 | 16 | 19 | 23 | 27 | 31 | 34 | 38 | 42 | 46 | 50 |
| 12 | 2 | 5 | 9 | 13 | 17 | 21 | 26 | 30 | 34 | 38 | 42 | 47 | 51 | 55 |
| 13 | 2 | 6 | 10 | 15 | 19 | 24 | 28 | 33 | 37 | 42 | 47 | 51 | 56 | 61 |
| 14 | 3 | 7 | 11 | 16 | 21 | 26 | 31 | 36 | 41 | 46 | 51 | 56 | 61 | 66 |
| 15 | 3 | 7 | 12 | 18 | 23 | 28 | 33 | 39 | 44 | 50 | 55 | 61 | 66 | 72 |

| $n_1$ \ $n_2$ | 2 | 3 | 4 | 5 | 6 | 7 | 8 | 9 | 10 | 11 | 12 | 13 | 14 | 15 |
|---|---|---|---|---|---|---|---|---|---|---|---|---|---|---|
| **Values of $U_{0.05}$** | | | | | | | | | | | | | | |
| 2 |  |  |  |  |  |  | 0 | 0 | 0 | 0 | 1 | 1 | 1 | 1 |
| 3 |  |  | 0 | 1 | 1 | 2 | 2 | 3 | 3 | 4 | 4 | 5 | 5 |
| 4 |  |  | 0 | 1 | 2 | 3 | 4 | 4 | 5 | 6 | 7 | 8 | 9 | 10 |
| 5 |  | 0 | 1 | 2 | 3 | 5 | 6 | 7 | 8 | 9 | 11 | 12 | 13 | 14 |
| 6 |  | 1 | 2 | 3 | 5 | 6 | 8 | 10 | 11 | 13 | 14 | 16 | 17 | 19 |
| 7 |  | 1 | 3 | 5 | 6 | 8 | 10 | 12 | 14 | 16 | 18 | 20 | 22 | 24 |
| 8 | 0 | 2 | 4 | 6 | 8 | 10 | 13 | 15 | 17 | 19 | 22 | 24 | 26 | 29 |
| 9 | 0 | 2 | 4 | 7 | 10 | 12 | 15 | 17 | 20 | 23 | 26 | 28 | 31 | 34 |
| 10 | 0 | 3 | 5 | 8 | 11 | 14 | 17 | 20 | 23 | 26 | 29 | 30 | 36 | 39 |
| 11 | 0 | 3 | 6 | 9 | 13 | 16 | 19 | 23 | 26 | 30 | 33 | 37 | 40 | 44 |
| 12 | 1 | 4 | 7 | 11 | 14 | 18 | 22 | 26 | 29 | 33 | 37 | 41 | 45 | 49 |
| 13 | 1 | 4 | 8 | 12 | 16 | 20 | 24 | 28 | 30 | 37 | 41 | 45 | 50 | 54 |
| 14 | 1 | 5 | 9 | 13 | 17 | 22 | 26 | 31 | 36 | 40 | 45 | 50 | 55 | 59 |
| 15 | 1 | 5 | 10 | 14 | 19 | 24 | 29 | 34 | 39 | 44 | 49 | 54 | 59 | 64 |

$^{\dagger}$This table is based on Table 11.4 of D. B. Owen, *Handbook of Statistical Tables*, © 1962, U.S. Department of Energy. Published by Addison-Wesley Publishing Company, Inc., Reading Mass. Reprinted with permission of the publisher.

## TABLE VII   Critical Values of $U$ (*Continued*)

### Values of $U_{0.02}$

| $n_1$ \ $n_2$ | 2 | 3 | 4 | 5 | 6 | 7 | 8 | 9 | 10 | 11 | 12 | 13 | 14 | 15 |
|---|---|---|---|---|---|---|---|---|---|---|---|---|---|---|
| 2  |   |   |   |   |   |   |   |   |    |    |    | 0  | 0  | 0  |
| 3  |   |   |   |   |   | 0 | 0 | 1 | 1  | 1  | 2  | 2  | 2  | 3  |
| 4  |   |   |   | 0 | 1 | 1 | 2 | 3 | 3  | 4  | 5  | 5  | 6  | 7  |
| 5  |   |   | 0 | 1 | 2 | 3 | 4 | 5 | 6  | 7  | 8  | 9  | 10 | 11 |
| 6  |   |   | 1 | 2 | 3 | 4 | 6 | 7 | 8  | 9  | 11 | 12 | 13 | 15 |
| 7  |   | 0 | 1 | 3 | 4 | 6 | 7 | 9 | 11 | 12 | 14 | 16 | 17 | 19 |
| 8  |   | 0 | 2 | 4 | 6 | 7 | 9 | 11| 13 | 15 | 17 | 20 | 22 | 24 |
| 9  |   | 1 | 3 | 5 | 7 | 9 | 11| 14| 16 | 18 | 21 | 23 | 26 | 28 |
| 10 |   | 1 | 3 | 6 | 8 | 11| 13| 16| 19 | 22 | 24 | 27 | 30 | 33 |
| 11 |   | 1 | 4 | 7 | 9 | 12| 15| 18| 22 | 25 | 28 | 31 | 34 | 37 |
| 12 |   | 2 | 5 | 8 | 11| 14| 17| 21| 24 | 28 | 31 | 35 | 38 | 42 |
| 13 | 0 | 2 | 5 | 9 | 12| 16| 20| 23| 27 | 31 | 35 | 39 | 43 | 47 |
| 14 | 0 | 2 | 6 | 10| 13| 17| 22| 26| 30 | 34 | 38 | 43 | 47 | 51 |
| 15 | 0 | 3 | 7 | 11| 15| 19| 24| 28| 33 | 37 | 42 | 47 | 51 | 56 |

### Values of $U_{0.01}$

| $n_1$ \ $n_2$ | 3 | 4 | 5 | 6 | 7 | 8 | 9 | 10 | 11 | 12 | 13 | 14 | 15 |
|---|---|---|---|---|---|---|---|---|---|---|---|---|---|
| 3  |   |   |   |   |   |   | 0 | 0  | 0  | 1  | 1  | 1  | 2  |
| 4  |   |   |   | 0 | 0 | 1 | 1 | 2  | 2  | 3  | 3  | 4  | 5  |
| 5  |   |   | 0 | 1 | 1 | 2 | 3 | 4  | 5  | 6  | 7  | 7  | 8  |
| 6  |   | 0 | 1 | 2 | 3 | 4 | 5 | 6  | 7  | 9  | 10 | 11 | 12 |
| 7  |   | 0 | 1 | 3 | 4 | 6 | 7 | 9  | 10 | 12 | 13 | 15 | 16 |
| 8  |   | 1 | 2 | 4 | 6 | 7 | 9 | 11 | 13 | 15 | 17 | 18 | 20 |
| 9  | 0 | 1 | 3 | 5 | 7 | 9 | 11| 13 | 16 | 18 | 20 | 22 | 24 |
| 10 | 0 | 2 | 4 | 6 | 9 | 11| 13| 16 | 18 | 21 | 24 | 26 | 29 |
| 11 | 0 | 2 | 5 | 7 | 10| 13| 16| 18 | 21 | 24 | 27 | 30 | 33 |
| 12 | 1 | 3 | 6 | 9 | 12| 15| 18| 21 | 24 | 27 | 31 | 34 | 37 |
| 13 | 1 | 3 | 7 | 10| 13| 17| 20| 24 | 27 | 31 | 34 | 38 | 42 |
| 14 | 1 | 4 | 7 | 11| 15| 18| 22| 26 | 30 | 34 | 38 | 42 | 46 |
| 15 | 2 | 5 | 8 | 12| 16| 20| 24| 29 | 33 | 37 | 42 | 46 | 51 |

## TABLE VIII   Critical Values of $u$[†]

**Values of $u_{0.025}$**

| $n_1$ \ $n_2$ | 4 | 5 | 6 | 7 | 8 | 9 | 10 | 11 | 12 | 13 | 14 | 15 |
|---|---|---|---|---|---|---|---|---|---|---|---|---|
| 4 |  | 9 | 9 |  |  |  |  |  |  |  |  |  |
| 5 | 9 | 10 | 10 | 11 | 11 |  |  |  |  |  |  |  |
| 6 | 9 | 10 | 11 | 12 | 12 | 13 | 13 | 13 | 13 |  |  |  |
| 7 |  | 11 | 12 | 13 | 13 | 14 | 14 | 14 | 14 | 15 | 15 | 15 |
| 8 |  | 11 | 12 | 13 | 14 | 14 | 15 | 15 | 16 | 16 | 16 | 16 |
| 9 |  |  | 13 | 14 | 14 | 15 | 16 | 16 | 16 | 17 | 17 | 18 |
| 10 |  |  | 13 | 14 | 15 | 16 | 16 | 17 | 17 | 18 | 18 | 18 |
| 11 |  |  | 13 | 14 | 15 | 16 | 17 | 17 | 18 | 19 | 19 | 19 |
| 12 |  |  | 13 | 14 | 16 | 16 | 17 | 18 | 19 | 19 | 20 | 20 |
| 13 |  |  |  | 15 | 16 | 17 | 18 | 19 | 19 | 20 | 20 | 21 |
| 14 |  |  |  | 15 | 16 | 17 | 18 | 19 | 20 | 20 | 21 | 22 |
| 15 |  |  |  | 15 | 16 | 18 | 18 | 19 | 20 | 21 | 22 | 22 |

**Values of $u'_{0.025}$**

| $n_1$ \ $n_2$ | 2 | 3 | 4 | 5 | 6 | 7 | 8 | 9 | 10 | 11 | 12 | 13 | 14 | 15 |
|---|---|---|---|---|---|---|---|---|---|---|---|---|---|---|
| 2 |  |  |  |  |  |  |  |  |  |  | 2 | 2 | 2 | 2 |
| 3 |  |  |  |  |  | 2 | 2 | 2 | 2 | 2 | 2 | 2 | 2 | 3 |
| 4 |  |  |  | 2 | 2 | 2 | 3 | 3 | 3 | 3 | 3 | 3 | 3 | 3 |
| 5 |  |  | 2 | 2 | 3 | 3 | 3 | 3 | 3 | 4 | 4 | 4 | 4 | 4 |
| 6 |  | 2 | 2 | 3 | 3 | 3 | 3 | 4 | 4 | 4 | 4 | 5 | 5 | 5 |
| 7 |  | 2 | 2 | 3 | 3 | 3 | 4 | 4 | 5 | 5 | 5 | 5 | 5 | 6 |
| 8 |  | 2 | 3 | 3 | 3 | 4 | 4 | 5 | 5 | 5 | 6 | 6 | 6 | 6 |
| 9 |  | 2 | 3 | 3 | 4 | 4 | 5 | 5 | 5 | 6 | 6 | 6 | 7 | 7 |
| 10 |  | 2 | 3 | 3 | 4 | 5 | 5 | 5 | 6 | 6 | 7 | 7 | 7 | 7 |
| 11 |  | 2 | 3 | 4 | 4 | 5 | 5 | 6 | 6 | 7 | 7 | 7 | 8 | 8 |
| 12 | 2 | 2 | 3 | 4 | 4 | 5 | 6 | 6 | 7 | 7 | 7 | 8 | 8 | 8 |
| 13 | 2 | 2 | 3 | 4 | 5 | 5 | 6 | 6 | 7 | 7 | 8 | 8 | 9 | 9 |
| 14 | 2 | 2 | 3 | 4 | 5 | 5 | 6 | 7 | 7 | 8 | 8 | 9 | 9 | 9 |
| 15 | 2 | 3 | 3 | 4 | 5 | 6 | 6 | 7 | 7 | 8 | 8 | 9 | 9 | 10 |

[†]This table is adapted, by permission, from F. S. Swed and C. Eisenhart, "Tables for testing randomness of grouping in a sequence of alternatives," *Annals of Mathematical Statistics*, Vol. 14.

## TABLE VIII   Critical Values of *u* (*Continued*)

| $n_1$ \ $n_2$ | 5 | 6 | 7 | 8 | 9 | 10 | 11 | 12 | 13 | 14 | 15 |
|---|---|---|---|---|---|---|---|---|---|---|---|
| | | | | **Values of $u_{0.005}$** | | | | | | | |
| 5 | | 11 | | | | | | | | | |
| 6 | 11 | 12 | 13 | 13 | | | | | | | |
| 7 | | 13 | 13 | 14 | 15 | 15 | 15 | | | | |
| 8 | | 13 | 14 | 15 | 15 | 16 | 16 | 17 | 17 | 17 | |
| 9 | | | 15 | 15 | 16 | 17 | 17 | 18 | 18 | 18 | 19 |
| 10 | | | 15 | 16 | 17 | 17 | 18 | 19 | 19 | 19 | 20 |
| 11 | | | 15 | 16 | 17 | 18 | 19 | 19 | 20 | 20 | 21 |
| 12 | | | | 17 | 18 | 19 | 19 | 20 | 21 | 21 | 22 |
| 13 | | | | 17 | 18 | 19 | 20 | 21 | 21 | 22 | 22 |
| 14 | | | | 17 | 18 | 19 | 20 | 21 | 22 | 23 | 23 |
| 15 | | | | | 19 | 20 | 21 | 22 | 22 | 23 | 24 |

| $n_1$ \ $n_2$ | 3 | 4 | 5 | 6 | 7 | 8 | 9 | 10 | 11 | 12 | 13 | 14 | 15 |
|---|---|---|---|---|---|---|---|---|---|---|---|---|---|
| | | | | | | **Values of $u'_{0.005}$** | | | | | | | |
| 3 | | | | | | | | | | 2 | 2 | 2 | 2 |
| 4 | | | | | | 2 | 2 | 2 | 2 | 2 | 2 | 2 | 3 |
| 5 | | | | 2 | 2 | 2 | 2 | 3 | 3 | 3 | 3 | 3 | 3 |
| 6 | | | 2 | 2 | 2 | 3 | 3 | 3 | 3 | 3 | 3 | 4 | 4 |
| 7 | | | 2 | 2 | 3 | 3 | 3 | 3 | 4 | 4 | 4 | 4 | 4 |
| 8 | | 2 | 2 | 3 | 3 | 3 | 3 | 4 | 4 | 4 | 5 | 5 | 5 |
| 9 | | 2 | 2 | 3 | 3 | 3 | 4 | 4 | 5 | 5 | 5 | 5 | 6 |
| 10 | | 2 | 3 | 3 | 3 | 4 | 4 | 5 | 5 | 5 | 5 | 6 | 6 |
| 11 | | 2 | 3 | 3 | 4 | 4 | 5 | 5 | 5 | 6 | 6 | 6 | 7 |
| 12 | 2 | 2 | 3 | 3 | 4 | 4 | 5 | 5 | 6 | 6 | 6 | 7 | 7 |
| 13 | 2 | 2 | 3 | 3 | 4 | 5 | 5 | 5 | 6 | 6 | 7 | 7 | 7 |
| 14 | 2 | 2 | 3 | 4 | 4 | 5 | 5 | 6 | 6 | 7 | 7 | 7 | 8 |
| 15 | 2 | 3 | 3 | 4 | 4 | 5 | 6 | 6 | 7 | 7 | 7 | 8 | 8 |

**TABLE IX   Critical Values of $q_\alpha^\dagger$**

$\alpha = 0.05$

| Degrees of freedom | Number of treatments | | | | | | | | | | | | | | | | | | | |
|---|---|---|---|---|---|---|---|---|---|---|---|---|---|---|---|---|---|---|---|---|
| | 3 | 4 | 5 | 6 | 7 | 8 | 9 | 10 | 11 | 12 | 13 | 14 | 15 | 16 | 20 | 24 | 28 | 32 | 36 | 40 |
| 1 | 27.0 | 32.8 | 37.1 | 40.4 | 43.1 | 45.4 | 47.4 | 49.1 | 50.6 | 52.0 | 53.2 | 54.3 | 55.4 | 56.3 | 59.6 | 62.1 | 64.2 | 66.0 | 67.6 | 68.9 |
| 2 | 8.33 | 9.80 | 10.9 | 11.7 | 12.4 | 13.0 | 13.5 | 14.0 | 14.4 | 14.8 | 15.1 | 15.4 | 15.7 | 15.9 | 16.8 | 17.5 | 18.0 | 18.5 | 18.9 | 19.3 |
| 3 | 5.91 | 6.83 | 7.50 | 8.04 | 8.48 | 8.85 | 9.18 | 9.46 | 9.72 | 9.95 | 10.2 | 10.4 | 10.5 | 10.7 | 11.2 | 11.7 | 12.1 | 12.4 | 12.6 | 12.9 |
| 4 | 5.04 | 5.76 | 6.29 | 6.71 | 7.05 | 7.35 | 7.60 | 7.83 | 8.03 | 8.21 | 8.37 | 8.53 | 8.66 | 8.79 | 9.23 | 9.58 | 9.88 | 10.1 | 10.3 | 10.5 |
| 5 | 4.60 | 5.22 | 5.67 | 6.03 | 6.33 | 6.58 | 6.80 | 7.00 | 7.17 | 7.32 | 7.47 | 7.60 | 7.72 | 7.83 | 8.21 | 8.51 | 8.76 | 8.98 | 9.17 | 9.33 |
| 6 | 4.34 | 4.90 | 5.31 | 5.63 | 5.90 | 6.12 | 6.32 | 6.49 | 6.65 | 6.79 | 6.92 | 7.03 | 7.14 | 7.24 | 7.59 | 7.86 | 8.09 | 8.28 | 8.45 | 8.60 |
| 7 | 4.17 | 4.68 | 5.06 | 5.36 | 5.61 | 5.82 | 6.00 | 6.16 | 6.30 | 6.43 | 6.55 | 6.66 | 6.76 | 6.85 | 7.17 | 7.42 | 7.63 | 7.81 | 7.97 | 8.11 |
| 8 | 4.04 | 4.53 | 4.89 | 5.17 | 5.40 | 5.60 | 5.77 | 5.92 | 6.05 | 6.18 | 6.29 | 6.39 | 6.48 | 6.57 | 6.87 | 7.11 | 7.31 | 7.48 | 7.63 | 7.76 |
| 9 | 3.95 | 4.42 | 4.76 | 5.02 | 5.24 | 5.43 | 5.60 | 5.74 | 5.87 | 5.98 | 6.09 | 6.19 | 6.28 | 6.36 | 6.64 | 6.87 | 7.06 | 7.22 | 7.36 | 7.49 |
| 10 | 3.88 | 4.33 | 4.65 | 4.91 | 5.12 | 5.31 | 5.46 | 5.60 | 5.72 | 5.83 | 5.94 | 6.03 | 6.11 | 6.19 | 6.47 | 6.69 | 6.87 | 7.02 | 7.16 | 7.28 |
| 11 | 3.82 | 4.26 | 4.57 | 4.82 | 5.03 | 5.20 | 5.35 | 5.49 | 5.61 | 5.71 | 5.81 | 5.90 | 5.98 | 6.06 | 6.33 | 6.54 | 6.71 | 6.86 | 6.99 | 7.11 |
| 12 | 3.77 | 4.20 | 4.51 | 4.75 | 4.95 | 5.12 | 5.27 | 5.40 | 5.51 | 5.62 | 5.71 | 5.80 | 5.88 | 5.95 | 6.21 | 6.41 | 6.59 | 6.73 | 6.86 | 6.97 |
| 13 | 3.74 | 4.15 | 4.45 | 4.69 | 4.89 | 5.05 | 5.19 | 5.32 | 5.43 | 5.53 | 5.63 | 5.71 | 5.79 | 5.86 | 6.11 | 6.31 | 6.48 | 6.62 | 6.74 | 6.85 |
| 14 | 3.70 | 4.11 | 4.41 | 4.64 | 4.83 | 4.99 | 5.13 | 5.25 | 5.36 | 5.46 | 5.55 | 5.64 | 5.71 | 5.79 | 6.03 | 6.22 | 6.39 | 6.53 | 6.65 | 6.75 |
| 15 | 3.67 | 4.08 | 4.37 | 4.60 | 4.78 | 4.94 | 5.08 | 5.20 | 5.31 | 5.40 | 5.49 | 5.57 | 5.65 | 5.72 | 5.96 | 6.15 | 6.31 | 6.45 | 6.56 | 6.67 |
| 16 | 3.65 | 4.05 | 4.33 | 4.56 | 4.74 | 4.90 | 5.03 | 5.15 | 5.26 | 5.35 | 5.44 | 5.52 | 5.59 | 5.66 | 5.90 | 6.08 | 6.24 | 6.37 | 6.49 | 6.59 |
| 17 | 3.63 | 4.02 | 4.30 | 4.52 | 4.71 | 4.86 | 4.99 | 5.11 | 5.21 | 5.31 | 5.39 | 5.47 | 5.54 | 5.61 | 5.84 | 6.03 | 6.18 | 6.31 | 6.43 | 6.53 |
| 18 | 3.61 | 4.00 | 4.28 | 4.50 | 4.67 | 4.82 | 4.96 | 5.07 | 5.17 | 5.27 | 5.35 | 5.43 | 5.50 | 5.57 | 5.79 | 5.98 | 6.13 | 6.26 | 6.37 | 6.47 |
| 19 | 3.59 | 3.98 | 4.25 | 4.47 | 4.65 | 4.79 | 4.92 | 5.04 | 5.14 | 5.23 | 5.32 | 5.39 | 5.46 | 5.53 | 5.75 | 5.93 | 6.08 | 6.21 | 6.32 | 6.42 |
| 20 | 3.58 | 3.96 | 4.23 | 4.45 | 4.62 | 4.77 | 4.90 | 5.01 | 5.11 | 5.20 | 5.28 | 5.36 | 5.43 | 5.49 | 5.71 | 5.89 | 6.04 | 6.17 | 6.28 | 6.37 |
| 24 | 3.53 | 3.90 | 4.17 | 4.37 | 4.54 | 4.68 | 4.81 | 4.92 | 5.01 | 5.10 | 5.18 | 5.25 | 5.32 | 5.38 | 5.59 | 5.76 | 5.91 | 6.03 | 6.13 | 6.23 |
| 30 | 3.49 | 3.85 | 4.10 | 4.30 | 4.46 | 4.60 | 4.72 | 4.82 | 4.92 | 5.00 | 5.08 | 5.15 | 5.21 | 5.27 | 5.48 | 5.64 | 5.77 | 5.89 | 5.99 | 6.08 |
| 40 | 3.44 | 3.79 | 4.04 | 4.23 | 4.39 | 4.52 | 4.64 | 4.74 | 4.82 | 4.90 | 4.98 | 5.04 | 5.11 | 5.16 | 5.36 | 5.51 | 5.64 | 5.75 | 5.85 | 5.93 |
| 60 | 3.40 | 3.74 | 3.98 | 4.16 | 4.31 | 4.44 | 4.55 | 4.65 | 4.73 | 4.81 | 4.88 | 4.94 | 5.00 | 5.06 | 5.24 | 5.39 | 5.51 | 5.62 | 5.71 | 5.79 |
| 120 | 3.36 | 3.69 | 3.92 | 4.10 | 4.24 | 4.36 | 4.47 | 4.56 | 4.64 | 4.71 | 4.78 | 4.84 | 4.90 | 4.95 | 5.13 | 5.27 | 5.38 | 5.48 | 5.57 | 5.64 |
| ∞ | 3.31 | 3.63 | 3.86 | 4.03 | 4.17 | 4.29 | 4.39 | 4.47 | 4.55 | 4.62 | 4.69 | 4.74 | 4.80 | 4.85 | 5.01 | 5.14 | 5.25 | 5.35 | 5.43 | 5.50 |

$\dagger$SOURCE: Adapted from H. L. Harter (1969), *Order Statistics and Their Use in Testing and Estimation. Vol 1: Tests Based on Range and Studentized Range of Samples From a Normal Population*, Aerospace Research Laboratories, U.S. Air Force.

**TABLE IX** Critical Values of $q_\alpha$ (Continued)

$\alpha = 0.01$

| Degrees of freedom | Number of treatments | | | | | | | | | | | | | | | | | | | |
|---|---|---|---|---|---|---|---|---|---|---|---|---|---|---|---|---|---|---|---|---|
| | 3 | 4 | 5 | 6 | 7 | 8 | 9 | 10 | 11 | 12 | 13 | 14 | 15 | 16 | 20 | 24 | 28 | 32 | 36 | 40 |
| 1 | 135 | 164 | 186 | 202 | 216 | 227 | 237 | 246 | 253 | 260 | 266 | 272 | 277 | 282 | 298 | 311 | 321 | 330 | 338 | 345 |
| 2 | 19.0 | 22.3 | 24.7 | 26.6 | 28.2 | 29.5 | 30.7 | 31.7 | 32.6 | 33.4 | 34.1 | 34.9 | 35.4 | 36.0 | 38.0 | 39.5 | 40.8 | 41.8 | 42.8 | 43.6 |
| 3 | 10.6 | 12.2 | 13.3 | 14.2 | 15.0 | 15.6 | 16.2 | 16.7 | 17.1 | 17.5 | 17.9 | 18.3 | 18.5 | 18.8 | 19.8 | 20.6 | 21.2 | 21.7 | 22.2 | 22.6 |
| 4 | 8.12 | 9.17 | 9.96 | 10.6 | 11.1 | 11.6 | 11.9 | 12.3 | 12.6 | 12.8 | 13.1 | 13.4 | 13.5 | 13.7 | 14.4 | 15.0 | 15.4 | 15.8 | 16.1 | 16.4 |
| 5 | 6.98 | 7.80 | 8.42 | 8.91 | 9.32 | 9.67 | 9.97 | 10.2 | 10.5 | 10.7 | 10.9 | 11.1 | 11.2 | 11.4 | 11.9 | 12.4 | 12.7 | 13.0 | 13.3 | 13.5 |
| 6 | 6.33 | 7.03 | 7.56 | 7.97 | 8.32 | 8.61 | 8.87 | 9.10 | 9.30 | 9.49 | 9.65 | 9.81 | 9.95 | 10.1 | 10.5 | 11.0 | 11.2 | 11.5 | 11.7 | 11.9 |
| 7 | 5.92 | 6.54 | 7.01 | 7.37 | 7.68 | 7.94 | 8.17 | 8.37 | 8.55 | 8.71 | 8.86 | 9.00 | 9.12 | 9.24 | 9.65 | 9.97 | 10.2 | 10.5 | 10.7 | 10.9 |
| 8 | 5.64 | 6.20 | 6.63 | 6.96 | 7.24 | 7.47 | 7.68 | 7.86 | 8.03 | 8.18 | 8.31 | 8.44 | 8.55 | 8.66 | 9.03 | 9.33 | 9.57 | 9.78 | 9.96 | 10.1 |
| 9 | 5.43 | 5.96 | 6.35 | 6.66 | 6.92 | 7.13 | 7.33 | 7.50 | 7.65 | 7.78 | 7.91 | 8.03 | 8.13 | 8.23 | 8.57 | 8.85 | 9.08 | 9.27 | 9.44 | 9.59 |
| 10 | 5.27 | 5.77 | 6.14 | 6.43 | 6.67 | 6.88 | 7.06 | 7.21 | 7.36 | 7.49 | 7.60 | 7.72 | 7.81 | 7.91 | 8.23 | 8.49 | 8.70 | 8.88 | 9.04 | 9.19 |
| 11 | 5.15 | 5.62 | 5.97 | 6.25 | 6.48 | 6.67 | 6.84 | 6.99 | 7.13 | 7.25 | 7.36 | 7.47 | 7.56 | 7.65 | 7.95 | 8.20 | 8.40 | 8.58 | 8.73 | 8.86 |
| 12 | 5.05 | 5.50 | 5.84 | 6.10 | 6.32 | 6.51 | 6.67 | 6.81 | 6.94 | 7.06 | 7.17 | 7.27 | 7.36 | 7.44 | 7.73 | 7.97 | 8.16 | 8.33 | 8.47 | 8.60 |
| 13 | 4.96 | 5.40 | 5.73 | 5.98 | 6.19 | 6.37 | 6.53 | 6.67 | 6.79 | 6.90 | 7.01 | 7.11 | 7.19 | 7.27 | 7.55 | 7.78 | 7.96 | 8.12 | 8.26 | 8.39 |
| 14 | 4.90 | 5.32 | 5.63 | 5.88 | 6.09 | 6.26 | 6.41 | 6.54 | 6.66 | 6.77 | 6.87 | 6.97 | 7.05 | 7.13 | 7.40 | 7.62 | 7.79 | 7.95 | 8.08 | 8.20 |
| 15 | 4.84 | 5.25 | 5.56 | 5.80 | 5.99 | 6.16 | 6.31 | 6.44 | 6.56 | 6.66 | 6.76 | 6.85 | 6.93 | 7.00 | 7.26 | 7.48 | 7.65 | 7.80 | 7.93 | 8.05 |
| 16 | 4.79 | 5.19 | 5.49 | 5.72 | 5.92 | 6.08 | 6.22 | 6.35 | 6.46 | 6.56 | 6.66 | 6.75 | 6.82 | 6.90 | 7.15 | 7.36 | 7.53 | 7.67 | 7.80 | 7.92 |
| 17 | 4.74 | 5.14 | 5.43 | 5.66 | 5.85 | 6.01 | 6.15 | 6.27 | 6.38 | 6.48 | 6.57 | 6.66 | 6.73 | 6.81 | 7.05 | 7.26 | 7.42 | 7.56 | 7.69 | 7.80 |
| 18 | 4.70 | 5.09 | 5.38 | 5.60 | 5.79 | 5.94 | 6.08 | 6.20 | 6.31 | 6.41 | 6.50 | 6.58 | 6.66 | 6.73 | 6.97 | 7.17 | 7.33 | 7.47 | 7.59 | 7.70 |
| 19 | 4.67 | 5.05 | 5.33 | 5.55 | 5.74 | 5.89 | 6.02 | 6.14 | 6.25 | 6.34 | 6.43 | 6.51 | 6.59 | 6.65 | 6.89 | 7.09 | 7.24 | 7.38 | 7.50 | 7.61 |
| 20 | 4.64 | 5.02 | 5.29 | 5.51 | 5.69 | 5.84 | 5.97 | 6.09 | 6.19 | 6.29 | 6.37 | 6.45 | 6.52 | 6.59 | 6.82 | 7.02 | 7.17 | 7.30 | 7.42 | 7.52 |
| 24 | 4.55 | 4.91 | 5.17 | 5.37 | 5.54 | 5.69 | 5.81 | 5.92 | 6.02 | 6.11 | 6.19 | 6.27 | 6.33 | 6.39 | 6.61 | 6.79 | 6.94 | 7.06 | 7.17 | 7.27 |
| 30 | 4.46 | 4.80 | 5.05 | 5.24 | 5.40 | 5.54 | 5.65 | 5.76 | 5.85 | 5.93 | 6.01 | 6.08 | 6.14 | 6.20 | 6.41 | 6.58 | 6.71 | 6.83 | 6.93 | 7.02 |
| 40 | 4.37 | 4.70 | 4.93 | 5.11 | 5.27 | 5.39 | 5.50 | 5.60 | 5.69 | 5.76 | 5.84 | 5.90 | 5.96 | 6.02 | 6.21 | 6.37 | 6.49 | 6.60 | 6.70 | 6.78 |
| 60 | 4.28 | 4.60 | 4.82 | 4.99 | 5.13 | 5.25 | 5.36 | 5.45 | 5.53 | 5.60 | 5.67 | 5.73 | 5.79 | 5.84 | 6.02 | 6.16 | 6.28 | 6.38 | 6.47 | 6.55 |
| 120 | 4.20 | 4.50 | 4.71 | 4.87 | 5.01 | 5.12 | 5.21 | 5.30 | 5.38 | 5.44 | 5.51 | 5.57 | 5.61 | 5.66 | 5.83 | 5.96 | 6.07 | 6.16 | 6.24 | 6.32 |
| ∞ | 4.12 | 4.40 | 4.60 | 4.76 | 4.88 | 4.99 | 5.08 | 5.16 | 5.23 | 5.29 | 5.35 | 5.40 | 5.45 | 5.49 | 5.65 | 5.77 | 5.87 | 5.95 | 6.03 | 6.09 |

**TABLE X**   Values of $Z = \frac{1}{2} \cdot \ln \frac{1+r}{1-r}$

| r | .00 | .01 | .02 | .03 | .04 | .05 | .06 | .07 | .08 | .09 |
|---|---|---|---|---|---|---|---|---|---|---|
| 0.0 | 0.000 | 0.010 | 0.020 | 0.030 | 0.040 | 0.050 | 0.060 | 0.070 | 0.080 | 0.090 |
| 0.1 | 0.100 | 0.110 | 0.121 | 0.131 | 0.141 | 0.151 | 0.161 | 0.172 | 0.182 | 0.192 |
| 0.2 | 0.203 | 0.213 | 0.224 | 0.234 | 0.245 | 0.255 | 0.266 | 0.277 | 0.288 | 0.299 |
| 0.3 | 0.310 | 0.321 | 0.332 | 0.343 | 0.354 | 0.365 | 0.377 | 0.388 | 0.400 | 0.412 |
| 0.4 | 0.424 | 0.436 | 0.448 | 0.460 | 0.472 | 0.485 | 0.497 | 0.510 | 0.523 | 0.536 |
| 0.5 | 0.549 | 0.563 | 0.576 | 0.590 | 0.604 | 0.618 | 0.633 | 0.648 | 0.662 | 0.678 |
| 0.6 | 0.693 | 0.709 | 0.725 | 0.741 | 0.758 | 0.775 | 0.793 | 0.811 | 0.829 | 0.848 |
| 0.7 | 0.867 | 0.887 | 0.908 | 0.929 | 0.950 | 0.973 | 0.996 | 1.020 | 1.045 | 1.071 |
| 0.8 | 1.099 | 1.127 | 1.157 | 1.188 | 1.221 | 1.256 | 1.293 | 1.333 | 1.376 | 1.422 |
| 0.9 | 1.472 | 1.528 | 1.589 | 1.658 | 1.738 | 1.832 | 1.946 | 2.092 | 2.298 | 2.647 |

For negative values of $r$ put a minus sign in front of the corresponding $Z$'s, and vice versa.

## TABLE XI　Binomial Coefficients

| $n$ | $\binom{n}{0}$ | $\binom{n}{1}$ | $\binom{n}{2}$ | $\binom{n}{3}$ | $\binom{n}{4}$ | $\binom{n}{5}$ | $\binom{n}{6}$ | $\binom{n}{7}$ | $\binom{n}{8}$ | $\binom{n}{9}$ | $\binom{n}{10}$ |
|---|---|---|---|---|---|---|---|---|---|---|---|
| 0 | 1 | | | | | | | | | | |
| 1 | 1 | 1 | | | | | | | | | |
| 2 | 1 | 2 | 1 | | | | | | | | |
| 3 | 1 | 3 | 3 | 1 | | | | | | | |
| 4 | 1 | 4 | 6 | 4 | 1 | | | | | | |
| 5 | 1 | 5 | 10 | 10 | 5 | 1 | | | | | |
| 6 | 1 | 6 | 15 | 20 | 15 | 6 | 1 | | | | |
| 7 | 1 | 7 | 21 | 35 | 35 | 21 | 7 | 1 | | | |
| 8 | 1 | 8 | 28 | 56 | 70 | 56 | 28 | 8 | 1 | | |
| 9 | 1 | 9 | 36 | 84 | 126 | 126 | 84 | 36 | 9 | 1 | |
| 10 | 1 | 10 | 45 | 120 | 210 | 252 | 210 | 120 | 45 | 10 | 1 |
| 11 | 1 | 11 | 55 | 165 | 330 | 462 | 462 | 330 | 165 | 55 | 11 |
| 12 | 1 | 12 | 66 | 220 | 495 | 792 | 924 | 792 | 495 | 220 | 66 |
| 13 | 1 | 13 | 78 | 286 | 715 | 1287 | 1716 | 1716 | 1287 | 715 | 286 |
| 14 | 1 | 14 | 91 | 364 | 1001 | 2002 | 3003 | 3432 | 3003 | 2002 | 1001 |
| 15 | 1 | 15 | 105 | 455 | 1365 | 3003 | 5005 | 6435 | 6435 | 5005 | 3003 |
| 16 | 1 | 16 | 120 | 560 | 1820 | 4368 | 8008 | 11440 | 12870 | 11440 | 8008 |
| 17 | 1 | 17 | 136 | 680 | 2380 | 6188 | 12376 | 19448 | 24310 | 24310 | 19448 |
| 18 | 1 | 18 | 153 | 816 | 3060 | 8568 | 18564 | 31824 | 43758 | 48620 | 43758 |
| 19 | 1 | 19 | 171 | 969 | 3876 | 11628 | 27132 | 50388 | 75582 | 92378 | 92378 |
| 20 | 1 | 20 | 190 | 1140 | 4845 | 15504 | 38760 | 77520 | 125970 | 167960 | 184756 |

For $r > 10$ it may be necessary to make use of the identity $\binom{n}{r} = \binom{n}{n-r}$.

**TABLE XII**   Values of $e^{-x}$

| $x$ | $e^{-x}$ | $x$ | $e^{-x}$ | $x$ | $e^{-x}$ | $x$ | $e^{-x}$ |
|-----|----------|-----|----------|-----|----------|-----|----------|
| 0.0 | 1.0000 | 2.5 | 0.082085 | 5.0 | 0.006738 | 7.5 | 0.00055308 |
| 0.1 | 0.9048 | 2.6 | 0.074274 | 5.1 | 0.006097 | 7.6 | 0.00050045 |
| 0.2 | 0.8187 | 2.7 | 0.067206 | 5.2 | 0.005517 | 7.7 | 0.00045283 |
| 0.3 | 0.7408 | 2.8 | 0.060810 | 5.3 | 0.004992 | 7.8 | 0.00040973 |
| 0.4 | 0.6703 | 2.9 | 0.055023 | 5.4 | 0.004517 | 7.9 | 0.00037074 |
| 0.5 | 0.6065 | 3.0 | 0.049787 | 5.5 | 0.004087 | 8.0 | 0.00033546 |
| 0.6 | 0.5488 | 3.1 | 0.045049 | 5.6 | 0.003698 | 8.1 | 0.00030354 |
| 0.7 | 0.4966 | 3.2 | 0.040762 | 5.7 | 0.003346 | 8.2 | 0.00027465 |
| 0.8 | 0.4493 | 3.3 | 0.036883 | 5.8 | 0.003028 | 8.3 | 0.00024852 |
| 0.9 | 0.4066 | 3.4 | 0.033373 | 5.9 | 0.002739 | 8.4 | 0.00022487 |
| 1.0 | 0.3679 | 3.5 | 0.030197 | 6.0 | 0.002479 | 8.5 | 0.00020347 |
| 1.1 | 0.3329 | 3.6 | 0.027324 | 6.1 | 0.002243 | 8.6 | 0.00018411 |
| 1.2 | 0.3012 | 3.7 | 0.024724 | 6.2 | 0.002029 | 8.7 | 0.00016659 |
| 1.3 | 0.2725 | 3.8 | 0.022371 | 6.3 | 0.001836 | 8.8 | 0.00015073 |
| 1.4 | 0.2466 | 3.9 | 0.020242 | 6.4 | 0.001662 | 8.9 | 0.00013639 |
| 1.5 | 0.2231 | 4.0 | 0.018316 | 6.5 | 0.001503 | 9.0 | 0.00012341 |
| 1.6 | 0.2019 | 4.1 | 0.016573 | 6.6 | 0.001360 | 9.1 | 0.00011167 |
| 1.7 | 0.1827 | 4.2 | 0.014996 | 6.7 | 0.001231 | 9.2 | 0.00010104 |
| 1.8 | 0.1653 | 4.3 | 0.013569 | 6.8 | 0.001114 | 9.3 | 0.00009142 |
| 1.9 | 0.1496 | 4.4 | 0.012277 | 6.9 | 0.001008 | 9.4 | 0.00008272 |
| 2.0 | 0.1353 | 4.5 | 0.011109 | 7.0 | 0.000912 | 9.5 | 0.00007485 |
| 2.1 | 0.1225 | 4.6 | 0.010052 | 7.1 | 0.000825 | 9.6 | 0.00006773 |
| 2.2 | 0.1108 | 4.7 | 0.009095 | 7.2 | 0.000747 | 9.7 | 0.00006128 |
| 2.3 | 0.1003 | 4.8 | 0.008230 | 7.3 | 0.000676 | 9.8 | 0.00005545 |
| 2.4 | 0.0907 | 4.9 | 0.007447 | 7.4 | 0.000611 | 9.9 | 0.00005017 |

# ANSWERS TO ODD-NUMBERED EXERCISES

## CHAPTER 1

**1.1** (a) The results can be misleading because "Xerox copiers" is often used as a generic term for photocopiers.

(b) Since Rolex watches are very expensive, persons wearing them can hardly be described as average individuals.

(c) The cost of such cruises is greater than the cost of a typical vacation.

**1.3** (a) Many persons are reluctant to give honest answers about their personal health habits.

(b) Successful graduates are more likely to return the questionnaire than graduates that have not done so well.

**1.5** (a) Since $4 + 2 = 6$ and $3 + 3 = 6$, the statement is descriptive.

(b) The data relate to a given day, so that "always" requires a generalization.

(c) The data relate to a given day, so that a statement about what happens over a week requires a generalization.

(d) Since the statement does not tell us anything about the time required to treat a patient of either sex, the statement requires a generalization.

**1.7** (a) The statement is a generalization based upon the misconception that trucks necessarily get better mileage on rural roads.

(b) The statement is a generalization based on the idea that higher speeds lead to poorer mileage.

(c) Since 15.5 occurs twice while each of the other figures occurs only once, the statement is purely descriptive.

(d) Since none of the values exceeds 16.0, the statement is purely descriptive.

**1.9** (a) The conclusion is nonsense. The same number of elevators go up and down.

(b) When operating, most elevators in the building will be on the upper floors; and few elevators will be on the first and second floors. Thus, the first elevator is more likely to be coming down from above.

**1.11** Nominal

**1.13** (a) Interval

(b) Ordinal

(c) Ratio

## CHAPTER 2

**2.1** The dots are arranged in vertical columns for each of the eleven years. The columns contain 12, 11, 6, 7, 12, 11, 14, 8, 7, 8, and 7 dots, respectively.

**2.3** a.

| Number of days | Number of Prescriptions |
|---|---|
| 4 | 2 |
| 5 | 3 |
| 6 | 7 |
| 7 | 11 |
| 8 | 9 |
| 9 | 5 |
| 10 | 3 |
| Total | 40 |

b. The asterisks are arranged in seven vertical columns, which are labeled 4, 5, 6, 7, 8, 9, and 10. These columns contain 2, 3, 7, 11, 9, 5, and 3 asterisks, respectively.

**2.5** The dot diagram explained here is a horizontal dot diagram (the dots are arranged horizontally). If we elected to draw a vertical dot diagram, the dots would be arranged vertically. The breeds of the dogs named by 30 persons are alphabetized (Afghan, Basset, Beagle, Bloodhound, Dachshund, and Greyhound) and are used to identify the corresponding rows of dots. These rows contain 5, 2, 8, 1, 8, and 6 dots.

**2.7** If we elect to use small circles instead of dots, and use rows instead of columns, we can label the 5 rows as A, B, C, D, and E. These rows contain 7, 5, 4, 2, and 1 small circles.

**2.9** Presenting these rows in descending order of numbers of defects, they contain 16, 9, 5, 3, and 2 dots. The rows can be labeled by the given codes as 3, 2, 0, 1, and 4, respectively.

**2.11** a. 36, 31, 37, 35, 32

b. 415, 438, 450, 477

c. 254, 254, 250, 253, 259.

**2.13**
```
5 | 8  6
6 | 5  6  4  0  7
7 | 9  7  8  1  2  1  3  5
8 | 6  4  3  8  1  1  5  9  0
9 | 5
```

**2.15** The frequencies are 1, 2, 9, 22, 15, 8, 2, 1.

**2.17** The frequencies are 3, 5, 11, 6.

**2.19** The frequencies are 1, 4, 9, 5, 1.

**2.21** The frequencies are 2, 6, 10, 3, 3.

**2.23** A convenient choice would be $220 - 239, 240 - 259, \ldots$, and $360 - 379$.

**2.25** (a) $0 - 49.99, 50.00 - 99.99, 100.00 - 149.99, 150.00 - 199.99$.

(b) $20.00 - 49.99, 50.00 - 79.99, 80.00 - 109.99, \ldots, 170 - 199.99$.

(c) $30.00 - 49.99, 50.00 - 69.99, 70.00 - 89.99, \ldots, 170.00 - 189.99$.

**2.27** (a) 5.0, 20.0, 35.0, 50.0, 65.0, 80.0.

(b) 19.9, 34.9, 49.9, 64.9, 79.9, 94.9.

(c) 4.95, 19.95, 34.95, 49.95, 64.95, 79.95, 94.95.

(d) 15.

**2.29** There is no provision for values from 50.00 to 59.99, and values from 70.00 to 79.99 go into two classes.

**2.31**    There is no provision for some items and confusion about items that could go into several classes.

**2.33**    (a)  $20 - 24, 25 - 29, 30 - 34, 35 - 39, 40 - 44$.

      (b)  $22, 27, 32, 37, 42$.

      (c)  All 5s.

**2.35**    (a)  $60.0 - 74.9, 75.0 - 89.9, 90.0 - 104.9, 105.0 - 119.9, 120.0 - 134.9$.

      (b)  $67.45, 82.45, 97.45, 112.45, 127.45$.

**2.37**    2.5, 5.0, 37.5, 40.0, 10.0, 5.0 percent.

**2.39**    13, 14, 16, 12, 4, 1.

**2.41**    0, 21.67, 45.0, 71.67, 91.67, 98.33, 100.00.

**2.43**    120, 118, 112, 100, 62, 36, 23, 16, 8, 3, 0.

**2.45**    100%, 93.75%, 79.17%, 56.25%, 31.25%, 14.58%, 6.25%, 0%.

**2.55**    0, 3, 16, 42, 62, 72, 79, 80.

**2.57**    It might easily give a misleading impression because we tend to compare the areas of rectangles rather than their heights.

**2.59**    The central angles are $110.2°, 56.0°, 52.9°, 41.6°, 27.3°, 20.9°, 18.6°, 32.5°$.

**2.63**    The central angles are 28.8, 79.2, 172.8, 64.8, 14.4 degrees.

**2.65**    1, 5, 8, 33, 40, 30, 20, 11, 2.

**2.67**    There is an upward trend, but the points are fairly widely scattered.

## CHAPTER 3

**3.1**    a.  If we are interested only in the blood pressures of patients in this specific cardiac ward, the data constitute a population.

      b.  If we want to generalize about patients in other cardiac wards, or in other cardiac patients, the data constitute a sample.

**3.3**    The information would be a sample if it is to be used for planning future tournaments. It would be a population if it is to be used to pay the club's employees who were to receive a bonus for each day it rained.

**3.5**    $\bar{x} = 97.5$

**3.7**    $\bar{x} = 9.96$. On the average, the calibration is off by 0.04.

**3.9**    The total weight, 2,988 pounds, does not exceed 3,200 pounds.

**3.11**    $\bar{x} = 5.25$.

**3.13**    $\dfrac{(32)(78) + (48)(84)}{32 + 48} = \dfrac{2,496 + 4,032}{80} = 82$ points.

**3.15**    (a) 0.67;   (b) 0.86.

**3.17**    (a) 18;   (b) 6;   (c) The predictions are 96 and 192.

**3.19**    3.755%

**3.21**    $\bar{x}_W = \dfrac{382(33,373) + (450)(31,684) + 113(40,329)}{382 + 450 + 113} = \dfrac{31,563,463}{945} = \$33,400.49$

**3.23**    $\bar{x} = 78.27$ minutes.

**3.25**    (a)  The median is the 28th value.

      (b)  The median is the mean of the 17th and 18th values.

**3.27**    The median is 55.

**3.29**    The median is 142 minutes.

**3.31**   The error is only 37.5.

**3.33**   The median is 118.5 grams.

**3.37**   The manufacturers of car C can use the midrange to substantiate the claim that their car performed best.

**3.39**   a. Since $\dfrac{41+1}{2} = 21$, the median is the 21st value. Since $\dfrac{20+1}{2} = 10.5$, $Q_1$ is the mean of the 10th and 11th values, and $Q_3$ is the mean of the 10th and 11th values from the other end.

b. Since $\dfrac{50+1}{2} = 25.5$, the median is the mean of the 25th and 26th values. Since $\dfrac{25+1}{2} = 13$, $Q_1$ is the 13th value and $Q_3$ is the 13th value from the other end.

**3.41**   Since $\dfrac{34+1}{2} = 17.5$, the median is the mean of the 17th and 18th values. Since $\dfrac{17+1}{2} = 9$, $Q_1$ is the 9th value and $Q_3$ is the 9th value from the other end. There are eight values to the left of the $Q_1$ position, eight values between the $Q_1$ position and the median position, eight values between the median position and the $Q_3$ position, and eight values to the right of the $Q_3$ position.

**3.43**   The smallest value is 41 and the largest value is 66. Also, from Exercise 3.42, $Q_1 = 47$, the median is 56, $Q_3 = 62$, so that the boxplot is

**3.45**   The smallest value is 405 and the largest value is 440. Also, from Exercise 3.44, $Q_1 = 411$, the median is 417, and $Q_3 = 432$, so that the boxplot is

**3.47**   The smallest value is 33 and the largest value is 118. Also, from Exercise 3.46, $Q_1 = 71$, the median is 80, and $Q_3 = 87$, so that the boxplot is

**3.49**   The smallest value is 82 and the largest value is 148. From Exercise 3.48, $Q_1 = 109$, the median is 118.5, and $Q_3 = 126.5$.

**3.51**   The mode is 48.

**3.53**   The mode is 0, which occurs six times. There seems to be a cyclical (up and down) pattern, which does not follow by just giving the mode.

**3.55**   Occasionally is the mode.

**3.57**   a. The mean and the median can both be determined.

b. The mean cannot be determined because of the open class; the median can be determined because it does not fall into one of the open classes.

c. The mean cannot be determined because of the open class; the median cannot be determined because it falls into the open class.

**3.59**   The mean is 4.88 and the median is 4.89, both rounded to two decimal places.

**3.61**   The mean is 47.64 and the median is 46.20, both rounded to two decimal places.

**3.63**   Since $P_{95}$ would have fallen into the open class, it could not have been determined.

**3.65**   $Q_1 = 0.82$, the median is 0.90, the mean is 0.94, and $Q_3 = 1.04$, all rounded to two decimal places.

**3.67**

a. $\sum\limits_{i=1}^{5} z_i$          e. $\sum\limits_{i=1}^{7} 2x_i$

b. $\sum\limits_{i=5}^{12} x_i$          f. $\sum\limits_{i=2}^{4} (x_i - y_i)$

c. $\sum\limits_{i=1}^{6} x_i f_i$          g. $\sum\limits_{i=2}^{5} (z_i + 3)$

d. $\sum\limits_{i=1}^{3} y_i^2$          h. $\sum\limits_{i=1}^{4} x_i y_i f_i$

**3.69**   a. $2 + 3 + 4 + 5 + 6 = 20$
b. $2 + 8 + 9 + 3 + 2 = 24$
c. $4 + 24 + 36 + 15 + 12 = 91$
d. $8 + 72 + 144 + 75 + 72 = 371$

**3.71**   a. $10, 6, 1$, and $13$
b. $8, 12$, and $10$

**3.73**   $\left(\sum\limits_{i=1}^{2} x_i\right)^2 = (x_1 + x_2)^2 = x_1^2 + 2x_1 x_2 + x_2^2$

$\sum\limits_{i=1}^{2} x_i^2 = x_1^2 + x_2^2$

It is not a true statement.

## CHAPTER 4

**4.1**   (a) The range is 0.07;   (b) $s = 0.032$.

**4.3**   (a) The range is 11;   (b) $s = 3.13$.

**4.5**   The range is 11 and twice the interquartile range is 8.

**4.7**   2.3452, or 2.3 rounded to one decimal.

**4.9**   $\sigma = 3.84$.

**4.11**   The range is $8.98 - 8.92 = 0.06$ for part (a); and $0.08 - 0.02 = 0.06$ for part (b). The range was not affected by the subtraction of the constant value.

**4.13**   (a) 10 claims; (b) $s = 4.0$; (c) 4.0

**4.15**   Verify.

**4.17**   $s = 20.68$

**4.19**   (a) 84%; (b) 99.6%.

**4.21**   (a) Between 94.8 and 128.4 minutes; (b) Between 83.6 and 139.6 minutes.

**4.23** The percentages are 65%, 97.5%, and 100%, which are close to 68%, 95%, and 99.7%.

**4.25** $z = 1.68$ for stock A and $z = 3.00$ for stock B. Stock B is relatively more overpriced.

**4.27** $V = \dfrac{2{,}236}{52} \cdot 100 = 4.3\%$; $V = \dfrac{0.0365}{0.20} \cdot 100 = 18.25\%$. Since 18.25% is greater than 4.3%, the rainfall data are relatively more variable.

**4.29** The coefficient of quartile variation is 12.0%.

**4.31** (a) This is like comparing apples with oranges.

(b) The data of Exercise 4.8 are relatively more variable than those of Exercise 4.11.

**4.33** $s = 5.66$.

**4.35** $s = 0.277$.

**4.37** (a) The mean is 56.45 and the median is 58.99;

(b) $s = 20.62$.

**4.39** $SK = -0.16$.

**4.41** The data are negatively skewed.

**4.43** The data are positively skewed.

**4.45** The distribution is J-shaped and positively skewed.

**4.47** The distribution is U-shaped.

## REVIEW EXERCISES FOR CHAPTERS 1, 2, 3, AND 4

**R.1** Two numbers can go into two classes and some numbers cannot be accommodated.

**R.3** 123, 125, 130, 134, 137, 138, 141, 143, 144, 146, 146, 149, 150, 152, 152, 155, 158, 161, and 167.

**R.5**

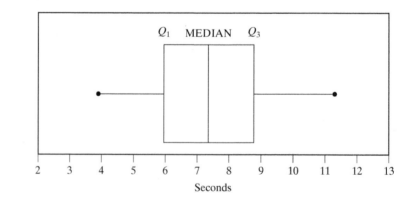

**R.7** (a) 7.31;  (b) 6;  (c) 5.70;  (d) 0.69.

**R.9** a.
```
0.04 | 5  6  7  8  9  9
0.05 | 0  2  2  4  4  4  5  5  6  7  7  8  8  8
0.06 | 0  1  2  2  3  3  5  6  7  8
0.07 | 2  2
```
b. The median is 0.057. $Q_1$ is 0.052 and $Q_3$ is 0.0625.
The data are slightly positively skewed.

**R.11**   a. The data would constitute a population if the meteorologist is interested only in the given ten years.

b. The data would constitute a sample if the meteorologist is interested in making predictions for future years.

**R.13**   The two coefficients of variation are, respectively, $V = \dfrac{s}{\bar{x}} \cdot 100 = \dfrac{2}{10} \cdot 100 = 20\%$ for the first company; and $\dfrac{3}{25} \cdot 100 \approx 12\%$ for the other company. There is relatively less variability in the firm where $V$ is smaller ($12\% < 20\%$).

**R.15**   At least $\left(1 - \dfrac{1}{3^2}\right) \cdot 100 = 88.9\%$ (rounded to one decimal) have diameters between 23.91 and 24.09 mm.

**R.17**   (a) 11;   (b) 44;   (c) cannot be determined;   (d) cannot be determined.

**R.19**   Mean, 12.58; median, 12.33; standard deviation, 5.38.

**R.21**

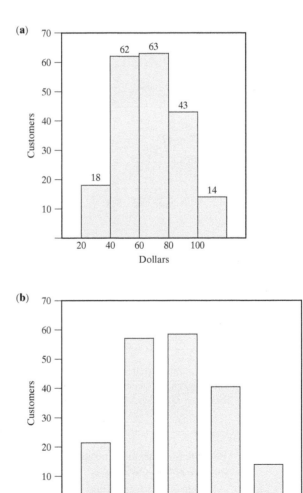

**R.23**  (a) 9.5, 29.5, 49.5, 69.5, 89.5, and 109.5;  (b) 19.5, 39.5, 59.5, 79.5, and 99.5;  (c) 20.

**R.25**  (a) 17.1;  (b) 87.45;  (c) 292.41.

**R.27**  $s = 6.24$.

**R.29**  There are other kinds of fibers and also shirts made of combinations of fibers.

**R.31**  (a) Cannot be determined;  (b) Yes the number in the fourth class;  (c) Yes the sum of the numbers in the 2nd and 3rd classes;  (d) Cannot be determined.

**R.33**  (a) This is begging the question. (b) Whether or not a person has a telephone may affect the results.

**R.35**  a.  14.5, 29.5, 44.5, 59.5, 74.5, 89.5, 104.5, and 119.5.

     b.  22, 37, 52, 67, 82, 97, and 112.

     c.  15.

**R.37**  The cumulative "less than" frequencies are 0, 3, 17, 35, 61, 81, 93, and 100.

**R.39**
```
12  4
13  0  0  5
14  2  6  9
15  1  3  4  5  6  8  9
16  2  2  2  5
17  2  3
18  2
19
20  4
```

**R.41**  $57 - 47 = 47 - 37 = 10 = 2.5k, k = 4$, and the percentage is at least $\left(1 - \dfrac{1}{4^2}\right)(100)$ $= 93.75\%$.

**R.43**

**R.45**  $V = \dfrac{0.537}{2.1} \cdot 100 = 25.57\%$.

**R.47**  It is assumed that the difference between $A$ and $B$ counts for as much as the differences between $B$ and $C$, the difference between $C$ and $D$, and the difference between $D$ and $E$.

## CHAPTER 5

**5.1**  Possible Monday and Tuesday sales of 0 and 0, 0 and 1, 0 and 2, 1 and 0, 1 and 1, 2 and 0, 2 and 1, and 2 and 2.

**5.3**  AL team wins 5th game and the series, AL team loses 5th game, wins 6th game and the series, AL team loses 5th game, loses 6th game, wins 7th game and series, NL team wins 5th game, wins 6th game, and 7th game and series.

**5.5**  $50 \times 49 \times 48 \times 47 = 5,527,200$

**5.7**  $6! = 6 \cdot 5 \cdot 4 \cdot 3 \cdot 2 \cdot 1 = 720$
$5! = 5 \cdot 4 \cdot 3 \cdot 2 \cdot 1 = 120$
$4! = 4 \cdot 3 \cdot 2 \cdot 1 = 24$
$3! = 3 \cdot 2 \cdot 1 = 6$

**5.9**  $_{12}P_4 = 12 \cdot 11 \cdot 10 \cdot 9 = 11,880$

**5.11**  a. In three cases.

b. In two cases.

**5.13**  a. 0 and 0, 0 and 1, 0 and 2, 1 and 0, 1 and 1, and 2 and 0.

b. Label the paintings Q and R, and let N denote none. The possibilities are N and N, N and Q, N and R, N and Q and R, Q and N, Q and R, R and N, R and Q, Q and R, and N.

**5.15**  $6 \cdot 4 = 24$

**5.17**  $4 \cdot 32 = 128$

**5.19**  a. 4

b. $4 \cdot 4 = 16$

c. $4 \cdot 3 = 12$

**5.21**  $5 \cdot 5 \cdot 2 \cdot 3 = 150$

**5.23**  a. $3^{10} = 59,049$

b. $2^{10} = 1,024$

**5.25**  a. True

b. False

c. $3! + 0! = 6 + 1 = 7$, true

d. $6! + 3! = 720 + 6 \neq 362,880$; false

e. $\dfrac{9!}{7!2!} = \dfrac{9 \cdot 8}{2} = 36$; true

f. $17! = 17 \cdot 16 \cdot 15! \neq 15! \cdot 2$; false

**5.27**  $6! = 720$

**5.29**  $\dfrac{32!}{5!27!} = \dfrac{32 \cdot 31 \cdot 30 \cdot 29 \cdot 28}{120} = 201,376$

**5.31**  $8 \cdot 7 \cdot 6 \cdot 5 \cdot 4 = 6,720$

**5.33**  $9! = 362,880$

**5.35**  a. $4 \cdot 3 \cdot 3 \cdot 2 = 72$

b. $3 \cdot 3 \cdot 2 = 18$

c. $3 \cdot 3 = 9$

**5.37**  a. $\dfrac{7!}{2! \cdot 2!} = 1,260$

b. $\dfrac{6!}{3! \cdot 3!} = 20$

c. $\dfrac{6!}{2! \cdot 2! \cdot 2!} = 90$

**5.39**  $\dfrac{15 \cdot 14 \cdot 13}{6} = 455$

**5.41**　$\dfrac{18 \cdot 17 \cdot 16}{6} = 816$

**5.43**　a. $\dfrac{1}{52}$　b. $\dfrac{6}{52} = \dfrac{3}{26}$　c. $\dfrac{12}{52} = \dfrac{3}{13}$　d. $\dfrac{13}{52} = \dfrac{1}{4}$

**5.45**　$s = 4 \cdot 4 \cdot 4,\ n = \dfrac{52 \cdot 51 \cdot 50}{3 \cdot 2} = 22{,}100$

　　　and $\dfrac{s}{n} = \dfrac{4 \cdot 4 \cdot 4}{22{,}100} = \dfrac{16}{5{,}525}$

**5.47**　The 36 possible outcomes are 1 and 1, 1 and 2, 1 and 3, 1 and 4, 1 and 5, 1 and 6, 2 and 1, 2 and 2, 2 and 3, 2 and 4, 2 and 5, 2 and 6, 3 and 1, 3 and 2, 3 and 3, 3 and 4, 3 and 5, 3 and 6, 4 and 1, 4 and 2, 4 and 3, 4 and 4, 4 and 5, 4 and 6, 5 and 1, 5 and 2, 5 and 3, 5 and 4, 5 and 5, 5 and 6, 6 and 1, 6 and 2, 6 and 3, 6 and 4, 6 and 5, 6 and 6.

　a. $\dfrac{3}{36} = \dfrac{1}{12}$

　b. $\dfrac{4}{36} = \dfrac{1}{9}$

　c. $\dfrac{8}{36} = \dfrac{2}{9}$

**5.49**　a. $\dfrac{15}{72} = \dfrac{5}{24}$　b. $\dfrac{50}{72} = \dfrac{25}{36}$　c. $\dfrac{7}{72}$　d. $\dfrac{35}{72}$

**5.51**　a. $\dfrac{37}{75}$　b. $\dfrac{15}{75} = \dfrac{1}{5}$　c. $\dfrac{16}{75}$

**5.53**　a. $\dfrac{56}{220} = \dfrac{14}{55}$　b. $\dfrac{48}{220} = \dfrac{12}{55}$

**5.55**　a. $\dfrac{475}{1{,}683}$　b. $\dfrac{96}{462}$

**5.57**　$\dfrac{424}{954} = \dfrac{4}{9}$

**5.59**　$\dfrac{678}{904} = \dfrac{3}{4}$

**5.61**　$\dfrac{28}{52} = \dfrac{7}{13}$

## CHAPTER 6

**6.1**　a. $U' = \{a, c, d, f, g\}$, the scholarship is awarded to Ms. Adam, Miss Clark, Mrs. Daly, Ms. Fuentes, or Ms. Gardner. $U \cap V = \{e, h\}$; the scholarship is awarded to Mr. Earl or Mr. Hall. $U \cup V' = \{a, b, c, d, e, h\}$; the scholarship is not awarded to Ms. Fuentes or Ms. Gardner.

**6.3**　a. $\{(0, 2), (1, 1), (2, 0)\}$　　b. $\{(0, 0), (1, 1)\}$　　c. $\{(1, 1), (2, 1), (1, 2)\}$

**6.5**　a. There is one professor less than there are assistants.

　　b. Altogether there are four professors and assistants.

　　c. There are two assistants.
　　　$K$ and $L$ are mutually exclusive; $K$ and $M$ are not mutually exclusive; and $L$ and $M$ are not mutually exclusive.

**6.7**　a. $\{(4, 1), (3, 2)\}$　　b. $\{(4, 3)\}$　　c. $\{(3, 3), (4, 3)\}$

**6.9**　a. $K' = \{(0, 0), (1, 0), (2, 0), (3, 0), (0, 1), (1, 1), (2, 1)\}$; at most one boat is rented out for the day.

　　b. $L \cap M = \{(2, 1), (3, 0)\}$

**6.11**　a. $\{A, D\}$　　b. $\{C, E\}$　　c. $\{B\}$

**6.13**  a. Not mutually exclusive; there can be sunshine and rain on the same day.

b. Not mutually exclusive

c. Mutually exclusive; when it is 11 P.M. in Los Angeles it is already the next day in New York.

d. Not mutually exclusive; one could be a bachelor's degree and the other could be a master's degree.

**6.15**  a. $98 - 50 = 48$    b. $224 - 50 = 174$    c. $360 - 224 - 48 = 88$

**6.17**  a. The car needs an engine overhaul, transmission repairs, and new tires.

b. The car needs transmission repairs and new tires, but no engine overhaul.

c. The car needs an engine overhaul, but no transmission repairs and no new tires.

d. The car needs an engine overhaul and new tires.

e. The car needs transmission repairs but no new tires.

f. The car does not need transmission repairs.

**6.19**  a. $8 + 8 + 5 + 3 = 24$    b. $3 + 8 + 3 + 2 = 16$

**6.21**  $P(C')$ is the probability that there will not be enough capital for the planned expansion.

$P(E')$ is the probability that the planned expansion will not provide enough parking.

$P(C' \cap E)$ is the probability that there will not be enough capital for the planned expansion and that the planned expansion will provide enough parking.

$P(C \cap E')$ is the probability that there will be enough capital for the planned expansion but that the planned expansion will not provide enough parking.

**6.23**  $P(A')$ is the probability that the attendance at the concert will not be good. $P(A' \cup W)$ is the probability that the attendance will not be good and/or more than half the crowd will walk out during the intermission. $P(A \cap W)$ is the probability that there will be a good attendance at the concert and at most half the crowd will walk out during the intermission.

**6.25**  a. Postulate 1    b. Postulate 2    c. Postulate 2    d. Postulate 3.

**6.27**  The corresponding probabilities are $\dfrac{2}{2+1} = \dfrac{2}{3}$ and $\dfrac{3}{3+1} = \dfrac{3}{4}$, and since $\dfrac{2}{3} + \dfrac{3}{4} > 1$, the odds cannot be right.

**6.29**  a. Since $A$ is contained in $A \cup B$, $P(A)$ cannot exceed $P(A \cup B)$.

b. Since $A \cap B$ is contained in $A$, $P(A \cap B)$ cannot exceed $P(A)$.

**6.31**  If $P(A) = 0$.

**6.33**  (a) The odds are 11 to 5 for getting at least two heads in four flips.

(b) The probability that at least one of the tiles will have a blemish is $\frac{34}{55}$.

(c) The odds are 19 to 5 that any particular household will not be included.

(d) The probability that at least one of the envelopes will end up in a wrong envelope is $\dfrac{719}{720}$.

**6.35**  The probability is greater than or equal to $\frac{6}{11}$, but less than $\frac{3}{5}$.

**6.37**  The probability for the \$1,000 raise is $\dfrac{5}{12}$, the probability for the \$2,000 raise is $\dfrac{1}{12}$, and the probability for either raise is $\dfrac{1}{2}$. Since $\dfrac{5}{12} + \dfrac{1}{12} = \dfrac{1}{2}$, the probabilities are consistent.

**6.39**  $\dfrac{a}{b} = \dfrac{p}{1-p}$ yields $a(1-p) = bp$, $a - ap = bp$, $a = ap + bp$, $a = p(a+b)$, and $p = \dfrac{a}{a+b}$.

**6.41**  $1 - (0.19 + 0.26 + 0.25 + 0.20 + 0.07) = 0.03$

**6.43**  a. $0.23 + 0.15 = 0.38$;    b. $0.31 + 0.24 + 0.07 = 0.62$;    c. $0.23 + 0.24 = 0.47$;
d. $1 - 0.07 = 0.93$

**6.45**  1/32, 5/32, 10/32, 10/32, 5/32, and 1/32.

**6.47**  $0.33 + 0.27 - 0.19 = 0.41$

**6.49**  $0.39 + 0.46 - 0.31 = 0.54$

**6.51**  (a) $P(A|T)$;    (b) $P(W|A)$;    (c) $P(T|W')$;    (d) $P(W|A' \cap T')$.

**6.53**  (a) $P(N|I)$;    (b) $P(I'|A')$;    (c) $P(I' \cap A'|N)$.

**6.55**  $\dfrac{1}{3} = \dfrac{0.2}{0.6}$

**6.57**  $\dfrac{3}{7} = \dfrac{0.3}{0.7}$

**6.59**  $\dfrac{0.44}{0.80} = 0.55$

**6.61**  $\dfrac{\dbinom{30}{2}}{\dbinom{40}{2}} = \dfrac{29}{52}$

**6.63**  Since $(0.80)(0.95) = 0.76$; the two events are independent.

**6.65**  0.42, 0.18, and 0.12

**6.67**  $(0.25)(0.40)^2(0.60) = 0.024$

**6.69**  $(0.70)(0.70)(0.30)(0.60)(0.40) = 0.03528$

**6.71**  $\dfrac{(0.40)(0.66)}{0.372} = 0.71$

**6.73**  $\dfrac{(0.50)(0.68)}{0.76} = 0.447$

**6.75**  $\dfrac{(0.10)(0.95)}{(0.10)(0.95) + (0.90)(0.05)} = 0.679$

**6.77**  a. $\dfrac{3}{4} \cdot \dfrac{1}{3} + \dfrac{1}{4} \cdot \dfrac{3}{4} = \dfrac{7}{16}$;    b. $\dfrac{\frac{3}{16}}{\frac{7}{16}} = \dfrac{3}{7}$

**6.79**  The probabilities for the respective causes are 0.229, 0.244, 0.183, and 0.344 all rounded to three decimals. On the basis of this information, the most likely cause is purposeful action.

## CHAPTER 7

**7.1**  $750 \cdot \dfrac{1}{3,000} = \$0.25$

**7.3**  $\dfrac{3,000 + 1,000}{15,000} = \$0.27$ rounded up to the nearest cent.

**7.5**  (a) $E = 300,000 \cdot \dfrac{1}{2} + 120,000 \cdot \dfrac{1}{2} = \$210,000$
Each golfer has the expectation of $210,000.

(b) If the younger golfer is favored by odds of 3 to 2. his probability of winning is $\dfrac{3}{5}$.

$E$ younger $= \dfrac{3}{5}(300,000) + \dfrac{2}{5}(120,000) = \$228,000$. $E$ older $= \dfrac{2}{5}(300,000) + \dfrac{3}{5}(120,000) = \$192,000$.

**7.7**  $16,000(0.25) + 13,000(0.46) + 12,000(0.19) + 10,000(0.10) = \$13,260$, so that the expected gross profit is $\$13,260 - \$12,000 = \$1,260$.

**7.9**  $7,500 > (30,000)p, \quad \dfrac{7,500}{30,000} > p, \quad 0.25 > p.$

**7.11**  $50,000 - 12,500p < 45,000$, so that $5,000 < 12,500p$ and $p > \dfrac{5,000}{12,500} = 0.40.$

**7.13**  $x \cdot \dfrac{1}{2} = 1,000 \cdot \dfrac{1}{2} + 400$, so that $\dfrac{x}{2} = 900$ and $x = 2 \cdot 900 = \$1,800.$

**7.15**  The two expectations are $120,000 \cdot \dfrac{3}{4} - 30,000 \cdot \dfrac{1}{4} = \$82,500$ and $180,000 \cdot \dfrac{1}{2} - 45,000 \cdot \dfrac{1}{2} = \$67,500$; the contractor should take the first job.

**7.17**  If the driver goes to the barn first, the expected distance is $(18 + 18) \cdot \dfrac{1}{6} + (18 + 8 + 22) \cdot \dfrac{5}{6} = 46$ miles. If the driver goes to the shopping center first, the expected distance is $(22 + 22) \cdot \dfrac{5}{6} + (22 + 8 + 18) \cdot \dfrac{1}{6} = 44\dfrac{2}{3}$ miles. He should go first to the shopping center.

**7.19**  If the driver goes to the barn first, the expected distance is $(18 + 18) \cdot \dfrac{1}{4} + (18 + 8 + 22) \cdot \dfrac{3}{4} = 45$ miles; if the driver goes to the shopping center first, the expected distance is $(22 + 22) \cdot \dfrac{3}{4} + (22 + 8 + 18) \cdot \dfrac{1}{4} = 45$ miles. It does not matter where he goes first.

**7.21**  If they continue, the expected profit is $-\$300,000$; if they do not continue, the expected profit is $-\$300,000$. It does not matter whether or not they continue the operation.

**7.23**  a.  The maximum losses would be \$600,000 if the tests are continued and \$500,000 if the tests are discontinued. To minimize the maximum loss, the tests should be discontinued.

b.  If he first goes to the barn, the possible distances are 36 and 48 miles, and if he first goes to the shopping center, the possible distances are 44 and 48 miles. In either case, the maximum distance is 48 miles, so it does not matter where he goes first.

**7.25**  (a)  The maximum profit would be maximized if the operation is continued.

(b)  The worst that can happen is minimized if the operation is discontinued.

**7.27**  a.  The errors are 0, 1, 4, or 5, and correspondingly, the consultant will get 600, 580, 280, or 100 dollars. He can expect to get $600 \cdot \dfrac{2}{5} + 580 \cdot \dfrac{1}{5} + 280 \cdot \dfrac{1}{5} + 100 \cdot \dfrac{1}{5} = \$432.$

b.  The errors are 1, 0, 3, or 4, and correspondingly, the consultant will get \$580, \$600, \$420, or \$280 dollars. He can expect to get

$$580 \cdot \dfrac{2}{5} + 600 \cdot \dfrac{1}{5} + 420 \cdot \dfrac{1}{5} + 280 \cdot \dfrac{1}{5}$$
$$= \$492.$$

**7.29**  a.  The median, 18.     b.  The mean, 19.

## REVIEW EXERCISES FOR CHAPTERS 5, 6, AND 7

**R.49**  a.  $P(C) = 0.12 + 0.48 = 0.60$    b.  $P(D') = 0.12 + 0.08 = 0.20$
c.  $P(C \cup D) = 0.92$    d.  $P(C \cap D') = 0.12$

**R.51**  a.  $0.01 + 0.02 + 0.05 + 0.14 + 0.16 = 0.38$    b.  $0.18 + 0.15 + 0.09 = 0.42$
c.  $0.14 + 0.16 + 0.20 = 0.50$

**R.53**

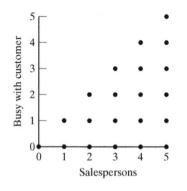

**R.55**  $\dfrac{3}{4} \le p < \dfrac{4}{5}$

**R.57**  $\dbinom{15}{4} = 1{,}365$

**R.59**  $\dfrac{1{,}134}{1{,}800} = 0.63$

**R.61**  a.  If the mortgage manager accepts or rejects the application, the expected profits are, respectively, $8{,}000(0.9) - 20.000(0.1) = 5.200$ and 0. To maximize the
expected profit, the mortgage manager should accept the application.

b.  If the mortgage manager accepts or rejects the application, the expected profits are, respectively, $8{,}000(0.7) - 20{,}000(0.3) = -400$ and 0. To maximize the
expected profit, the mortgage manager should reject the application.

c.  If the mortgage manager accepts or rejects the application, the maximum losses
are, respectively, 20,000 and 0. To minimize the maximum loss, the mortgage
manager should reject the application.

**R.63**  $2^8 \cdot 4^4 = 65{,}536$

**R.65**  a.  $11 \cdot 15 = 165$    b.  $12 \cdot 14 = 168$

**R.67**  $(0.24)^3 = 0.014$ rounded to three decimals.

**R.69**  a.  The probability is  $\dfrac{3}{21 + 3} = \dfrac{1}{8}$;    b.  The probability is  $\dfrac{11}{11 + 5} = \dfrac{11}{16}$.

**R.71**  a.  Since $P(A \cap B) = 0.62 - (0.37 + 0.25) = 0$, events $A$ and $B$ are mutually exclusive.

b.  Since $(0.37)(0.25) \ne 0$, events $A$ and $B$ are not independent.

**R.73**  $30 = 3 \cdot \dfrac{15}{60} + V \cdot \dfrac{45}{60}$. Simplification yields $120 = 3 + 3V$, and $V = \$39$.

**R.75**  a.  The accountant should use the mode, 30.
b.  The accountant should use the mean, 31.

**R.77**  a.  $5! = 120$;    b.  $\dfrac{6!}{2!} = 360$;    c.  $\dfrac{6!}{2!2!} = 180$;    d.  $\dfrac{7!}{3!} = 840$

**R.79**  a. $\dfrac{\binom{3}{1}\binom{2}{1}\binom{5}{1}}{\binom{10}{3}} = \dfrac{1}{4}$;   b. $\dfrac{\binom{5}{3}}{120} = \dfrac{1}{12}$;   c. $\dfrac{\binom{3}{1}\binom{5}{2}}{120} = \dfrac{1}{4}$.

**R.81**  a. $\dfrac{(0.02)(0.90)}{(0.02)(0.90) + (0.98)(0.08)} = \dfrac{0.0180}{0.0964} = 0.187$

b. $\dfrac{(0.98)(0.92)}{(0.02)(0.90) + (0.98)(0.92)} = \dfrac{0.9016}{0.9196} = 0.980$

**R.83**  Many persons would prefer a guaranteed 4.5% to a potentially risky 6.2%.

**R.85**  There are 12 outcomes in event A, and 8 outcomes in the event outside A. If $p$ is the probability of each outcome outside $A$, then $12 \cdot 2p + 8p = 1$ and $p = \dfrac{1}{32}$.

Therefore, $P(A) = 12 \cdot \dfrac{2}{32} = \dfrac{24}{32} = \dfrac{3}{4}$.

**R.87**  $\dfrac{\binom{18}{10}}{2^{18}} = \dfrac{43{,}758}{262{,}144} = 0.167$

**R.89**  a. $4! = 24$;   b. $5 \cdot 4! = 120$.

## CHAPTER 8

**8.1**  a. No; $0.52 + 0.26 + 0.32 = 1.10 > 1.00$

b. No; $0.18 + 0.02 + 1.00 = 1.20 > 1.00$

c. Yes; the values are all non-negative and their sum equals 1.

**8.3**  a. Yes; the values are all non-negative and $7 \cdot \dfrac{1}{7} = 1$

b. No; the sum of the values is $10 \cdot \dfrac{1}{9} > 1$.

c. Yes; the values are all non-negative and $\dfrac{3}{18} + \dfrac{4}{18} + \dfrac{5}{18} + \dfrac{6}{18} = 1$.

**8.5**  $\binom{3}{2}\left(\dfrac{3}{4}\right)^2\left(\dfrac{1}{4}\right) = \dfrac{27}{64} = 0.42$

**8.7**  $\binom{4}{0}(0.10)^0(0.90)^4 = 0.6561$; the value in Table V is 0.656 rounded to three decimals.

**8.9**  a. $0.028 + 0.121 + 0.233 + 0.267 = 0.649$;   b. $0.037 + 0.009 + 0.001 = 0.047$.

**8.11**  a. $0.002 + 0.007 + 0.024 = 0.033$

b. $0.177 + 0.207 + 0.186 + 0.127 = 0.697$

c. $0.127 + 0.063 + 0.022 + 0.005 = 0.217$

**8.13**  a. $1 - 0.282 = 0.718$;   b. $0.069$;   c. $0.014$.

**8.15**  a. $0.0000 + 0.0008 + 0.0063 + 0.0285 + 0.0849 = 0.1205$   b. $0.1205$

**8.17**  a. $(0.40)(0.60)^3 = 0.0864$

b. $(0.25)(0.75)^4 = 0.079$ rounded to three decimal places

c. $(0.70)(0.30)^2 = 0.063$

**8.19**  a. $\dfrac{\binom{10}{3}\binom{4}{0}}{\binom{14}{3}} = \dfrac{30}{91}$   b. $\dfrac{\binom{10}{2}\binom{4}{1}}{\binom{14}{3}} = \dfrac{45}{91}$

**8.21**  a. $\dfrac{\binom{3}{2}\binom{9}{0}}{\binom{12}{2}} = \dfrac{1}{22}$;   b. $\dfrac{\binom{3}{1}\binom{9}{1}}{\binom{12}{2}} = \dfrac{9}{22}$;   c. $\dfrac{\binom{3}{0}\binom{9}{2}}{\binom{12}{2}} = \dfrac{12}{22}$.

**8.23**  a. $(0.05)(140+60) = 10$. Since $n = 12 > 10$, condition is not satisfied.

b. $(0.05)(220+280) = 25$. Since $n = 20 < 25$, condition is satisfied.

c. $(0.05)(250+390) = 32$. Since $n = 30 < 32$, condition is satisfied.

d. $(0.05)(220+220) = 22$. Since $n = 25 > 22$, condition is not satisfied.

**8.25**  The binomial approximation is 0.0750. Since the hypergeometric probability is 0.0762, the error of the binomial approximation is $0.0750 - 0.0762 = 0.0012$.

**8.27**  a. Since $np = 12.5 > 10$, conditions are not satisfied.

b. Since $n = 400 > 100$ and $np = 8 < 10$, the conditions are satisfied.

c. Since $n = 90 < 100$, the conditions are not satisfied.

**8.29**  $f(3) = \dfrac{6^3(0.002479)}{3!} = 0.089$ rounded to three decimal places.

**8.31**  $np = 150(0.05) = 7.5$.   $\dfrac{7.5^0 \cdot e^{-7.5}}{0!} = 0.00055.$

$\dfrac{7.5^1 e^{-7.5}}{1!} = 0.00415$, and   $\dfrac{7.5^2 e^{-7.5}}{2!} = 0.01555$

The probability for at most two will be involved in an accident is $0.00055 + 0.00415 + 0.01555 = 0.02025$.

**8.33**  Since $n = 120 < 0.05(3{,}200) = 160$, the given hypergeometric distribution can be approximated with the binomial distribution with $n = 120$ and $p = \dfrac{50}{3{,}200} = 0.0156$. Then since $n = 120 > 100$ and $120(0.0156) = 1.87 < 10$, this binomial distribution can be approximated with the Poisson distribution with $np = 1.87$.

**8.35**  a. $\dfrac{1.6^0 \cdot e^{-1.6}}{0!} = 0.2019$;   b. $\dfrac{1.6^1 \cdot e^{-1.6}}{1!} = 0.3230$;   c. $\dfrac{1.6^2 \cdot e^{-1.6}}{2!} = 0.2584.$

**8.37**  $\dfrac{10!}{6!3!1!}(0.70)^6(0.20)^3(0.10)^1 = 0.0791$

**8.39**  $\dfrac{10!}{7!1!1!1!}(0.60)^7(0.20)(0.10)(0.10) = 0.0403$

**8.41**  $\sigma^2 = \Sigma\, x^2 f(x) - \mu^2$

$\sigma^2 = 1^2(0.4) + 2^2(0.3) + 3^2(0.2) + 4^2(0.1) - 2^2$

$\sigma^2 = 0.4 + 1.2 + 1.8 + 1.6 - 4.0$

$\sigma^2 = 5.0 - 4.0 = 1.0$

**8.43**  $\mu = 0(0.0035) + 1(0.0231) + 2(0.0725) + \cdots + 12(0.0004) = 4.8587$

$\sigma^2 = 0^2(0.0035) + 1^2(0.0231) + 2^2(0.0725) + \cdots + 12^2(0.0004) - (4.8587)^2$

$= 3.5461$ rounded to four decimal places.

$\sigma = \sqrt{3.5461} = 1.883$ rounded to three decimal places.

**8.45**  $\mu = 4 \cdot \dfrac{1}{2} = 2, \sigma^2 = 4 \cdot \dfrac{1}{2} \cdot \dfrac{1}{2} = 1$, and $\sigma = 1$.

**8.47**  a. $\mu = 484 \cdot \dfrac{1}{2} = 242, \sigma^2 = 484 \cdot \dfrac{1}{2} \cdot \dfrac{1}{2} = 121$, and $\sigma = 11$.

b. $\mu = 120$ and $\sigma = 10$.

c. $\mu = 180$ and $\sigma = 11.225$ rounded to three decimal places.

d. $\mu = 24$, and $\sigma = 4.8$.

e. $\mu = 520$, and $\sigma = 13.491$ rounded to three decimals.

**8.49** $\mu = 0(0.013) + 1(0.128) + 2(0.359) + 3(0.359)$

$\qquad + 4(0.128) + 5(0.013) = 2.5$

$\qquad \mu = \dfrac{8.5}{5+11} = \dfrac{40}{16} = 2.5$

**8.51** $\mu = 0(0.082) + 1(0.205) + 2(0.256) + \cdots + 9(0.001)$

$\qquad = 2.501$, which is very close to $\lambda = 2.5$.

**8.53** a. The proportion of heads will be between 0.475 and 0.525;

b. same as (a) with value of $n$ changed.

## CHAPTER 9

**9.1** a. First area is bigger.     b. First area is bigger.     c. Areas are equal.

**9.3** a. First area is bigger.     b. Second area is bigger.     c. Areas are equal.

**9.5** a. $0.1064 + 0.5000 = 0.6064$.     b. $0.5000 + 0.4032 = 0.9032$.

c. $0.5000 - 0.2852 = 0.2148$.

**9.7** a. $2(0.2794) = 0.5588$.     b. $0.4177 + 0.5000 = 0.9177$.

**9.9** a. $1.63 = z$.     b. $0.9868 - 0.5000 = 0.4868$.     $z = 2.22$.

**9.11** a. $0.4678 - 0.3023 = 0.1655$.     b. $0.4772 - 0.2157 = 0.2615$.

**9.13** a. $z = 2.03$.     b. $z = 0.98$.     c. $z = \pm 1.47$.     d. $z = -0.41$.

**9.15** a. $2(0.3413) = 0.6826$     b. $2(0.4772) = 0.9544$     c. $2(0.4987) = 0.9974$

d. $2(0.49997) = 0.99994$     e. $2(0.4999997) = 0.9999994$

**9.17** a. 0.9332     b. 0.7734     c. 0.2957     d. 0.9198

**9.19** Since the entry in Table I closest to $0.5000 - 0.2000 = 0.3000$ is 0.2995 corresponding to $z = 0.84$, $\dfrac{79.2 - 62.4}{\sigma} = 0.84$ and $\sigma = \dfrac{16.8}{0.84} = 20$.

**9.21** a. $1 - e^{-0.4} = 0.3297$     b. $(1 - e^{-0.9}) - (1 - e^{-0.5}) = 0.1999$

**9.23** a. $e^{-2} = 0.1353$     b. $e^{-3} = 0.049787$     c. $1 - e^{-0.5} = 0.3935$

**9.25** a. $z = \dfrac{19.0 - 17.4}{2.2} = 0.73$; the normal curve area is 0.2678 and the probability is 0.77.

b. $z = \dfrac{12.0 - 17.4}{2.2} = -2.45$ and $z = \dfrac{15.0 - 17.4}{2.2} = -1.09$; the normal curve areas are 0.4930 and 0.3621, and the probability is 0.13.

**9.27** a. $z = \dfrac{33.4 - 38.6}{6.5} = -0.80$; the normal curve area is 0.2881 and the probability is 0.7881.

b. $z = \dfrac{34.7 - 38.6}{6.5} = 0.60$; the normal curve area is 0.2257 and the probability is

$0.5000 - 0.2257 = 0.2743$.

**9.29** $\dfrac{x - 18.2}{1.2} = -0.84$, $x = 17.2$ ounces.

**9.31** $z = \dfrac{2.0 - 1.96}{0.08} = 0.50$; the normal curve area is 0.1915 and the percentage is

$(0.5000 - 0.1915)100 = 30.85\%$.

**9.33** The entry corresponding to $z = \dfrac{79.0 - 63}{20} = 0.80$ is 0.2881 and $(0.5000 - 0.2881)8{,}000{,}000 = 1{,}695{,}200$ or approximately 1.7 million men would be discharged.

**9.35** a. $\dfrac{18 - 25.8}{\sigma} = -0.84$, so that $\sigma = \dfrac{7.8}{0.84} = 9.3$.

b. $z = \dfrac{30 - 25.8}{9.3} = 0.45$; the normal curve area is 0.1736 and the probability is $0.5000 - 0.1736 = 0.33$.

**9.37** $\dfrac{400 - 338}{\sigma} = 1.556$, so that $\sigma = \dfrac{62}{1.556} = 39.8$; $z = \dfrac{300 - 338}{39.8} = -0.955$ and the probability is $0.5000 - 0.3302 = 0.17$.

**9.39** $\mu = 10$ and $\sigma = \sqrt{5} = 2.236$; $z = \dfrac{11.5 - 10}{2.236} = 0.67$

a. probability is $0.500 - 0.2486 = 0.2514$;

b. probability $= 0.120 + 0.074 + 0.037 + 0.015 + 0.005 + 0.001 = 0.252$.

**9.41** a. $np = 32 \cdot \dfrac{1}{7} = 4.57$; conditions not satisfied;

b. $np = 75(0.10) = 7.5$ and $n(1 - p) = 67.5$; conditions satisfied;

c. $np = 50(0.08) = 4$; conditions not satisfied.

**9.43** $\mu = 200(0.80) = 160$ and $\sigma = \sqrt{32} = 5.66$; $z = \dfrac{169.5 - 160}{5.66} = 1.68$ and the probability is $0.5000 - 0.4535 = 0.0465$

**9.45** a. Using Figure 9.24 and normal curve approximation the probabilities are 0.6261 and 0.6141; error is $0.6261 - 0.6141 = 0.0120$ and percentage error is $\dfrac{0.0120}{0.6261} \cdot 100 = 1.9\%$.

b. Using Figure 9.24 and normal curve approximation the probabilities are 0.1623 and 0.1639; error is $0.1639 - 0.1623 = 0.0016$ and percentage error is $\dfrac{0.0016}{0.1623} \cdot 100 = 0.99\%$.

**9.47** $\mu = 100(0.18) = 18$; $\sigma = \sqrt{100(0.18)(0.82)} = 3.84$; $z = \dfrac{11.5 - 18}{3.84} = -1.69$. The probability is $0.4545 + 0.5000 = 0.9545$.

## CHAPTER 10

**10.1** a. $\dbinom{6}{2} = 15$   b. $\dbinom{20}{2} = 190$   c. $\dbinom{32}{2} = 496$   d. $\dbinom{75}{2} = 2{,}775$

**10.3** a. $\dfrac{1}{\dbinom{12}{4}} = \dfrac{1}{495}$   b. $\dfrac{1}{\dbinom{20}{4}} = \dfrac{1}{4{,}845}$

**10.5** $uvw,\ uvx,\ uvy,\ uvz,\ uwx,\ uwy,\ uwz,\ uxy,\ uxz,\ uyz,\ vwx,\ vwy,\ vwz,\ vxy,\ vxz,\ vyz,\ wxy,\ wxz,\ wyz,\ xyz$

**10.7** $\dfrac{4}{20} = \dfrac{1}{5}$

**10.9** a. $\dfrac{1}{\dbinom{5}{3}} = \dfrac{1}{10}$   b. $\dfrac{\dbinom{4}{2}}{10} = \dfrac{6}{10} = \dfrac{3}{5}$   c. $\dfrac{\dbinom{3}{1}}{10} = \dfrac{3}{10}$

**10.11** 3406, 3591, 3383, 3554, 3513, 3439, 3707, 3416, 3795, and 3329.

**10.13**   264, 429, 437, 419, 418, 252, 326, 443, 410, 472, 446, and 318.

**10.15**   6094, 2749, 0160, 0081, 0662, 5676, 6441, 6356, 2269, 4341, 0922, 6762, 5454, 7323, 1522, 1615, 4363, 3019, 3743, 5173, 5186, 4030, 0276, 7845, 5025, 0792, 0903, 5667, 4814, 3676, 1435, 5552, 7885, 1186, 6769, 5006, 0165, 1380, 0831, 3327, 0279, 7607, 3231, 5015, 4909, 6100, 0633, 6299, 3350, 3597.

**10.17**   $\dfrac{3}{5} \cdot \dfrac{2}{4} \cdot \dfrac{1}{3} = \dfrac{1}{10}$

**10.19**   $\dfrac{n}{N} \cdot \dfrac{n-1}{N-1} \cdot \ldots \cdot \dfrac{1}{N-n+1} = \dfrac{n}{N} \cdot \dfrac{n-1}{N-1} \cdot \ldots \cdot \dfrac{1}{N-n+1} \cdot \dfrac{(N-n)!}{(N-n)!}$

$= \dfrac{n!(N-n)}{N!} = \dfrac{1}{\dbinom{N}{n}}$

**10.21**   16.8, 24.0, 20.1, 21.9, 15.8, 22.1, 20.9, 21.3, 18.8, and 18.5;
27.6, 15.9, 24.2, 15.2, 20.4, 21.1, 15.7, 25.0, 16.9, and 25.0;
25.4, 16.9, 19.4, 17.2, 19.0, 25.8, 16.8, 12.9, 21.1, and 13.2;
16.6, 19.4, 16.6, 16.1, 16.9, 18.0, 20.5, 15.0, 27.9, and 16.7;
22.6, 17.0, 17.0, 18.6, 18.4, 15.5, 17.0, 15.8, 14.7, and 24.2

**10.23**   All the December figures that are, of course, much higher than the others, go into the same (sixth) sample.

**10.25**   a. $\dbinom{9}{2}\dbinom{3}{2} = 36 \cdot 3 = 108$     b. $\dbinom{9}{3}\dbinom{3}{1} = 84 \cdot 3 = 252$

**10.27**   $n_1 = \dfrac{250}{1,000} \cdot 40 = 10, \ n_2 = \dfrac{600}{1,000} \cdot 40 = 24, \ n_3 = \dfrac{100}{1,000} \cdot 40 = 4,$ and

$n_4 = \dfrac{50}{1,000} \cdot 40 = 2.$

**10.29**   a. $n_1 = \dfrac{100 \cdot 10,000 \cdot 45}{10,000 \cdot 45 + 30,000 \cdot 60} = 20,$ and $n_2 = 100 - 20 = 80.$

b. $n_1 = \dfrac{84 \cdot 5,000 \cdot 15}{5,000 \cdot 15 + 2,000 \cdot 18 + 3,000 \cdot 5}$

$= 50$

$n_2 = \dfrac{84 \cdot 2,000 \cdot 18}{126,000} = 24,$ and

$n_3 = 84 - (50 + 24) = 10.$

**10.31**   a. $\dfrac{4+5+4}{25} = \dfrac{13}{25} = 0.52$

b. $\dfrac{3+4+5+4+3}{25} = \dfrac{19}{25} = 0.76$

**10.33**   a. Finite,     b. Infinite,     c. Infinite,     d. Finite,     e. Infinite.

**10.35**   a. It is divided by $\sqrt{\dfrac{120}{30}} = 2$     b. Multiplied by $\sqrt{\dfrac{245}{5}} = 7$

**10.37**   a. $\sqrt{\dfrac{90}{99}} = 0.954$     b. $\sqrt{\dfrac{275}{299}} = 0.958$     c. $\sqrt{\dfrac{4,900}{4,999}} = 0.9900$

**10.39**   $0.6745 \cdot \dfrac{12.8}{\sqrt{60}} = 1.115$

**10.41**   $\sigma_{\bar{x}} = \dfrac{2.4}{\sqrt{25}} = 0.48$

a. Since $k = \dfrac{1.2}{0.48} = 2.5,$ the probability is at least $1 - \dfrac{1}{2.5^2} = 0.84.$

b. Since $z = 2.50$, the probability is $2(0.4938) = 0.9876$.

**10.43**  $\sigma_{\bar{x}} = \dfrac{0.025}{\sqrt{16}} = 0.00625,\quad z = \dfrac{0.01}{0.00625} = 1.6$ and the probability is $2(0.4452) = 0.8904$.

**10.45**  $\dfrac{\sigma}{\sqrt{144}} = 1.25 \cdot \dfrac{\sigma}{\sqrt{n}}$, so that $\sqrt{n} = 12(1.25) = 15$ and $n = 225$.

**10.47**  The medians are 11, 15, 19, 17, 14, 13, 14, 17, 18, 14, 14, 19, 16, 18, 17, 18, 19, 17, 14, 12, 11, 16, 16, 19, 18, 17, 17, 15, 13, 14, 15, 16, 14, 15, 12, 16, 16, 17, 14, and 17. Their standard deviation is 2.20 and the corresponding standard error formula yields $1.25 \cdot \dfrac{4}{\sqrt{5}} = 2.24$.

**10.49**  The sum of the forty sample variances is 615.9, and their mean is $\dfrac{615.9}{40} = 15.40$ and the percentage error is $\dfrac{16 - 15.40}{16} \cdot 100 = 3.75\%$.

## REVIEW EXERCISES FOR CHAPTERS 8, 9, AND 10

**R.91**  a. $\dfrac{\binom{10}{4}}{\binom{18}{4}} = \dfrac{210}{3,060} = 0.069$   b. $\dfrac{\binom{10}{2}\binom{8}{2}}{3,060} = \dfrac{45 \cdot 28}{3,060} = 0.412$

**R.93**  $f(2) = \binom{5}{2}(0.70)^2(1 - 0.70)^{5-2} \approx 0.132$

**R.95**  a. Since $8 < 0.05(40 + 160) = 10$, the condition is satisfied.
b. Since $10 > 0.05(100 + 60) = 8$, the condition is not satisfied.
c. Since $12 > 0.05(68 + 82) = 7.5$, the condition is not satisfied.

**R.97**  $\dfrac{1}{\binom{45}{6}} = \dfrac{720}{45 \cdot 44 \cdot 43 \cdot 42 \cdot 41 \cdot 40} = \dfrac{1}{8,145,060}$

**R.99**  a. $0(0.017) + 1(0.090) + 2(0.209) + 3(0.279) + 4(0.232) + 5(0.124) + 6(0.041) + 7(0.008) + 8(0.001) = 3.203$
b. $np = 8(0.40) = 3.20$

**R.101**  a. $np = 180 \cdot \dfrac{1}{9} = 20 > 10$, conditions not satisfied.
b. $n = 480 > 100$ and $np = 480 \cdot \dfrac{1}{60} = 8 < 10$, conditions are satisfied.
c. $n = 575 > 100$ and $np = 575 \cdot \dfrac{1}{100} = 5.75 < 10$, conditions are satisfied.

**R.103**  Since 0.4713 corresponds to $z = 1.90$, $\dfrac{82.6 - \mu}{4} = 190$ and $\mu = 75$.
Since $\dfrac{80 - 75}{4} = 1.25$, and $\dfrac{70 - 75}{4} = -1.25$, the probability is $2(0.3944) = 0.7888$.

**R.105**  $\dfrac{9!}{4!4!1!}(0.30)^4(0.60)^4(0.10) = 0.066$.

**R.107**  a. $\dfrac{\binom{5}{3}\binom{4}{0}}{\binom{9}{3}} = \dfrac{5}{42}$   b. $\dfrac{\binom{5}{1}\binom{4}{2}}{\binom{9}{3}} = \dfrac{5}{14}$

**R.109**  Use formula for the area of a triangle.

   a. $\dfrac{1}{2} \cdot 3 \cdot \dfrac{2}{3} = 1$    b. $\dfrac{1}{2} \cdot \dfrac{3}{2} \cdot \dfrac{1}{3} = \dfrac{1}{4}$

**R.111**  $z = \dfrac{20 - 24.55}{3.16} = -1.44.$  Area  corresponding  to  $z = -1.44 = 0.4251.$  $z =$ $\dfrac{30 - 24.55}{3.16} = 1.72.$  Area  corresponding  to  $z = 1.72 = 0.4573.$  Probability  is $0.4251 + 0.4573 = 0.8824.$

**R.113**  a.  Ratio is $\dfrac{\binom{30}{1}\binom{270}{11}}{\binom{300}{12}}$ to $\dfrac{\binom{30}{0}\binom{270}{12}}{\binom{300}{12}}$ or 360 to 259.

   b. $\binom{12}{1}(0.1)^1(0.9)^{11}$ to $\binom{12}{0}(0.1)^0(0.9)^{12}$ or 4 to 3.

**R.115**  a.  $0.2019 + 0.3230 + 0.2584 = 0.7833$

   b.  $0.1378 + 0.0551 + 0.0176 = 0.2105$

   c.  $0.0176 + 0.0047 + 0.0011 + 0.0002 + 0.0000 = 0.0236.$

**R.117**  $\mu = 0(0.2019) + 1(0.3230) + 2(0.2584) + 3(0.1378) + 4(0.0551) + 5(0.0176) +$ $6(0.0047) + 7(0.0011) + 8(0.0002) + 9(0.0000) = 1.627.$

**R.119**  a.  $\sqrt{\dfrac{120 - 30}{120 - 1}} = 0.8697$    b.  $\sqrt{\dfrac{450 - 50}{400 - 1}} = 0.9366$

**R.121**  a.  $f(0) = \dfrac{(2.3)^0 \cdot e^{-2.3}}{0!} = 0.1003$    b.  $f(1) = \dfrac{(2.3)^1 \cdot e^{-2.3}}{1!} = 0.2307$

**R.123**  a.  $\binom{40}{2}\binom{20}{2}\binom{10}{2}\binom{10}{2} = 780 \cdot 190 \cdot 45 \cdot 45 = 300{,}105{,}000$

   b.  $\binom{40}{4}\binom{20}{2}\binom{10}{1}\binom{10}{1} = 91{,}390 \cdot 190 \cdot 10 \cdot 10 = 1{,}736{,}410{,}000$

**R.125**  a.  $0.2171 + 0.1585 + 0.0844 = 0.4600$

   b.  $0.0319 + 0.0082 + 0.0013 + 0.0001 = 0.0415$

**R.127**  a.  Since $np = 55 \cdot \dfrac{1}{5} = 11 > 5$ and $n(1 - p) = 55 \cdot \dfrac{4}{5} = 44$ are both greater than 5, the conditions are satisfied;

   b.  since $np = 105 \cdot \dfrac{1}{35} = 3$ is less than 5, the conditions are not satisfied;

   c.  since $np = 210 \cdot \dfrac{1}{30} = 7$ and $n(1 - p) = 210 \cdot \dfrac{29}{30} = 203$ are both greater than 5, the conditions are satisfied;

   d.  since $n(1 - p) = 40(0.05) = 2$ is less than 5, the conditions are not satisfied.

**R.129**  $\dfrac{\binom{7}{1}\binom{3}{1}\binom{2}{1}}{\binom{12}{3}} = \dfrac{7 \cdot 3 \cdot 2}{220} = \dfrac{21}{110}.$

## CHAPTER 11

**11.1**  Maximum error is $1.96 \cdot \dfrac{135}{\sqrt{40}} \approx 41.84.$

**11.3**  Maximum error is $2.575 \cdot \dfrac{3.2}{\sqrt{40}} \approx 1.30$ mm

**11.5**  Maximum error is $2.33 \cdot \dfrac{269}{\sqrt{35}} = \$106.$

**11.7**  $z = \dfrac{24 - 23.5}{\dfrac{3.3}{8}} = 1.21$ and the probability is $2(0.3869) = 0.77.$

**11.9**  $n = \left(\dfrac{2.575 \cdot 138}{40}\right)^2 = 78.92$ and $n = 79.$

**11.11**  $n = \left(\dfrac{2.575 \cdot 0.77}{0.25}\right)^2 = 62.90$ and $n = 63.$

**11.13**  a.  $30(0.90) = 27$

b.  26; this is within one of what we could have expected.

**11.15**  a.  $2.34 \pm 2.306 \cdot \dfrac{0.48}{3}, 2.34 - 0.37 = 1.97 < \mu < 2.34 + 0.37 = 2.71$ micrograms.

b.  Maximum error is $E = 3.355 \cdot \dfrac{0.48}{3} = 0.54$ microgram.

**11.17**  Maximum error is $E = 2.306 \cdot \dfrac{1{,}527}{\sqrt{9}} = 1{,}174$ pound.

**11.19**  a. 1.771     b. 2.101     c. 2.508     d. 2.947

**11.21**  a.  It is reasonable to treat the data as a sample from a normal population.

b.  $31.693 \pm 2.977 \cdot \dfrac{2.156}{\sqrt{15}}$, which yields $30.04 < \mu < 33.35.$

**11.23**  $12 \pm 2.821 \cdot \dfrac{2.75}{\sqrt{10}}$, which yields $9.55 < \mu < 14.45.$

**11.25**  $0.15 \pm 2.447 \cdot \dfrac{0.03}{\sqrt{7}}$ which yields $0.122 < \mu < 0.178.$

**11.27**  $E = 3.106 \cdot \dfrac{1.859}{\sqrt{12}} = 1.67$ fillings.

**11.29**  $n = 12$ and $s = 2.75$; $\dfrac{11(2.75)^2}{26.757} < \sigma^2 < \dfrac{11(2.75)^2}{2.603}$ and $1.76 < \sigma < 5.65.$

**11.31**  $n = 5$ and $s = 0.381.$ $\dfrac{4(0.381)^2}{11.143} < \sigma^2 < \dfrac{4(0.381)^2}{0.484}$ and $0.052 < \sigma^2 < 1.200.$

**11.33**  $n = 12$ and $s = 1.859.$ $\dfrac{(11)(1.859)^2}{26.757} < \sigma^2 < \dfrac{(11)(1.859)^2}{2.603}$ and $1.19 < \sigma < 3.82.$

**11.35**  a.  $\dfrac{81.0 - 70.2}{3.26} = 3.31.$ This is not too close to 2.75.

b.  $\dfrac{14.34 - 14.26}{2.06} = 0.039.$ This is quite close to $s = 0.0365.$

**11.37**  a.  $400p > 5$ and $400(1 - p) > 5$ yields $0.0125 < p < 0.9875.$

b.  $500p > 5$ and $500(1 - p) > 5$ yields $0.01 < p < 0.99.$

**11.39**  $\dfrac{x}{n} = 0.570$, so that $0.570 \pm 1.96\sqrt{\dfrac{(0.570)(0.430)}{400}}$ and $0.570 \pm 0.0485$ and $0.52 < p < 0.62.$

**11.41**  $\dfrac{x}{n} = \dfrac{56}{400} = 0.140$, so that $0.140 \pm 2.575\sqrt{\dfrac{(0.140)(0.860)}{400}}$ and $0.095 < p < 0.185.$

**11.43**  $\dfrac{x}{n} = \dfrac{54}{120} = 0.45; E = 1.645\sqrt{\dfrac{(0.45)(0.55)}{120}} = 0.075$

**11.45**  $\dfrac{x}{n} = \dfrac{412}{1,600} = 0.2575$ and $0.2575 \pm 1.96\sqrt{\dfrac{(0.2575)(0.7425)}{1,600}}$. This yields $23.6 < 100p < 27.89$ percent.

**11.47**  $\dfrac{x}{n} = \dfrac{119}{140} = 0.85$  and  $0.85 \pm 2.575\sqrt{\dfrac{(0.85)(0.15)}{140}}$.  This  yields  $0.85 \pm 0.078$  and $0.772 < p < 0.928$.

**11.49**  a.  $\dfrac{x}{n} = \dfrac{34}{100} = 0.34$  and  $0.34 \pm 1.96\sqrt{\dfrac{(0.34)(0.66)}{100} \cdot \dfrac{(360 - 100)}{(360 - 1)}}$.  This yields $0.261 < p < 0.419$.

b.  Continuing with Exercise 11.47, we get $0.85 \pm 0.078\sqrt{\dfrac{350 - 140}{350 - 1}}$. This yields $0.85 \pm 0.061$ and $0.789 < p < 0.911$.

**11.51**  a.  $\dfrac{1}{4}\left(\dfrac{1.645}{0.05}\right)^2 = 271$ rounded up to the nearest integer.

b.  $n = \dfrac{1}{4}\left(\dfrac{1.96}{0.05}\right)^2 = 385$ rounded up to the nearest integer.

c.  $n = \dfrac{1}{4}\left(\dfrac{2.575}{0.05}\right)^2 = 664$ rounded up to the nearest integer.

**11.53**  a. $n = 2.172$       b. $n = 1.824$

# CHAPTER 12

**12.1**  (a) $\mu < \mu_0$ and buy the new van only if the null hypothesis can be rejected;

(b) $\mu > \mu_0$ and buy the new van unless the null hypothesis can be rejected.

**12.3**  (a) Since the null hypothesis is true and accepted, the psychologist will not be making an error.

(b) Since the null hypothesis is false but accepted, the psychologist will be making a Type II error.

**12.5**  If the testing service erroneously rejects the null hypothesis, it will be committing a Type I error. If the testing service erroneously accepts the null hypothesis, it will be committing a Type II error.

**12.7**  Use the null hypothesis that the antipollution device is not effective.

**12.9**  (a) The probability of a Type I error is 0.0478; (b) the probability of a Type II error is 0.3446, increased from 0.21.

**12.11**  Since we are not dealing with sample data, there is no question here of statistical significance.

**12.13**  To reject the null hypothesis that there is no such thing as extrasensory perception, at least 2.8 persons would have to get high scores.

**12.15**  The null hypothesis is $\mu = 2.6$. As it is of concern that there may be more absences than that the alternative hypothesis should be $\mu > 2.6$.

**12.17**  Use the null hypothesis $\mu = 20$ and the alternative hypothesis $\mu > 20$.

**12.19**  a. 0.05       b. $1 - (0.95)^2 = 0.0975$       c. $1 - (0.95)^{32} = 0.8063$

**12.21**  1. $H_0: \mu = 12.3$ and $H_A: \mu \neq 12.3$       2. $\alpha = 0.05$
3. Reject $H_0$ if $z \leq -1.96$ or $z \geq 1.96$.       4. $z = -1.25$.
5. Null hypothesis cannot be rejected.

**12.23**  1. $H_0: \mu = 3.52$ and $H_A: \mu \neq 3.52$     2. $\alpha = 0.05$
3. Reject $H_0$ if $z \leq -1.96$ or $z \geq 1.96$.    4. $z = \dfrac{3.55 - 3.52}{\dfrac{0.07}{\sqrt{32}}} \approx 2.42$

5. Null hypothesis must be rejected since $2.42 > 1.96$.

**12.25**  1. $H_0: \mu = 83.2$ and $H_A: \mu > 83.2$     2. $\alpha = 0.01$     3. Reject $H_0$ if $z \geq 2.33$.
4. $z = \dfrac{86.7 - 83.2}{\dfrac{8.6}{\sqrt{45}}} \approx 2.73$     5. Null hypothesis must be rejected.

**12.27**  1. $H_0: \mu = 80$ and $H_A: \mu < 80$     2. $\alpha = 0.05$     3. Reject $H_0$ if $t \leq -1.796$.
4. $t = \dfrac{78.2 - 80}{\dfrac{7.9}{\sqrt{12}}} \approx -0.79$     5. The null hypothesis cannot be rejected.

**12.29**  1. $H_0: \mu = 0.125$ and $H_A: \mu > 0.125$     2. $\alpha = 0.01$     3. Reject $H_0$ if $t \geq 3.747$
4. $t = \dfrac{13.1 - 12.5}{\dfrac{0.51}{\sqrt{5}}} = 2.63$     5. The null hypotheses cannot be rejected.

**12.31**  1. $H_0: \mu = 14$ and $H_A: \mu > 14$     2. $\alpha = 0.05$     3. Reject $H_0$ if $t \geq 1.753$
4. $t = \dfrac{15.25 - 14}{\dfrac{2.70}{\sqrt{16}}} = 1.85$     5. The null hypothesis must be rejected.

**12.33**  The normal probability plot indicates that the population is not normal.

**12.37**  $s_p = 3.084$ and $t = 2.29$. The null hypothesis must be rejected.

**12.39**  $s_p = 19.10$ and $t = -2.12$. The null hypothesis cannot be rejected.

**12.41**  The $p$-value is 0.0744. Since this is less than 0.10, the null hypothesis could have been rejected.

**12.43**  The $p$-value is 0.246, and this is the lowest level of significance at which the null hypothesis could have been rejected.

**12.45**  $z = -2.53$. Since $z = -2.53$ is less than $-1.96$, it follows that the null hypothesis must be rejected.

## CHAPTER 13

**13.1**  1. $H_0: \sigma = 0.0100$ and $H_A: \sigma < 0.0100$     2. $\alpha = 0.05$
3. Reject $H_0$ if $\chi^2 \leq 3.325$

4. $\chi^2 = \dfrac{9(0.0086)^2}{(0.010)^2} = 6.66$     5. Null hypothesis cannot be rejected.

**13.3**  The $p$-value is $2(0.0092) = 0.0184$. Since $0.0184 \leq 0.03$, the null hypothesis must be rejected.

**13.5**  1. $H_0: \sigma = 0.80$ and $H_A: \sigma < 0.80$     2. $\sigma = 0.01$     3. Reject $H_0$ if $z \leq -2.33$.
4. $z = \dfrac{0.74 - 0.80}{\dfrac{0.80}{\sqrt{80}}} = -0.674$     5. Null hypothesis cannot be rejected.

**13.7**  1. $H_0: \sigma_1 = \sigma_2$ and $H_A: \sigma_1 < \sigma_2$     2. $\alpha = 0.05$     3. Reject $H_0$ if $F \geq 2.72$.
4. $F = \dfrac{(4.4)^2}{(2.6)^2} = 2.86$     5. The null hypothesis must be rejected.

## CHAPTER 14

**14.1** 1. $H_0$: $p = 0.05$ and $H_A$: $p > 0.05$      2. $\alpha = 0.01$

3. The test statistic is the observed number of persons who take such a cruise within a year's time.

4. $x = 3$ and the probability of 3 or more successes is 0.043.

5. Since 0.043 is more than 0.01 the null hypothesis cannot be rejected.

**14.3** 1. $H_0$: $p = 0.50$ and $H_A$: $p \neq 0.50$.      2. $\alpha = 0.10$

3. Test statistic is $x$, the number of persons in the sample who are opposed to capital punishment.

4. From Table V, with $n = 20$, $x = 14$, $p = 0.5$. The $p$-value of this two-tailed test is $2(0.058) = 0.116$.

5. The null hypothesis cannot be rejected.

**14.5** a. 1. $H_0$: $p = 0.36$ and $H_A$: $p < 0.36$      2. $\alpha = 0.05$      3. Reject $H_0$ if $z \leq -1.645$.

4. $z = \dfrac{94 - 300(0.36)}{\sqrt{300(0.36)(0.64)}} \approx -1.68$      5. Null hypothesis must be rejected.

b. Using continuity correction, steps 1, 2, and 3 are the same.

4. $z = \dfrac{94.5 - 300(0.36)}{\sqrt{300(0.36)(0.64)}} \approx -1.62$      5. Null hypothesis cannot be rejected.

**14.7** 1. $H_0$: $p = 0.95$ and $H_A$: $p \neq 0.95$      2. $\alpha = 0.01$

3. Reject $H_0$ if $z \leq -2.575$ or $z = \geq 2.575$      4. $z = \dfrac{464.5 - 500(0.95)}{\sqrt{500(0.95)(0.05)}} = -2.15$

5. The null hypothesis cannot be rejected.

**14.9** 1. $H_0$: $p_1 = p_2$ and $H_A$: $p_1 > p_2$      2. $\alpha = 0.05$      3. Reject $H_0$ if $z \geq 1.645$

4. $\hat{p} = \dfrac{54 + 33}{250} = 0.348$ and $z = \dfrac{0.36 - 0.33}{\sqrt{(0.348)(0.652)\left(\dfrac{1}{150} + \dfrac{1}{100}\right)}} = 0.49$

5. Null hypothesis cannot be rejected.

**14.11** 1. $H_0$: $p_1 = p_2$ and $H_A$: $p_1 \neq p_2$      2. $\alpha = 0.05$

3. Reject $H_0$ if $z \leq -1.96$ or $z \geq 1.96$.

4. $z = \dfrac{0.22 - 0.275}{\sqrt{(0.25)(0.75)\left(\dfrac{1}{100} + \dfrac{1}{120}\right)}} = -0.94$

5. Null hypothesis cannot be rejected.

**14.13** 1. $H_0$: $p_1 = p_2$ and $H_A$: $p_1 \neq p_2$      2. $\alpha = 0.05$

3. Reject $H_0$ if $z \leq -1.96$ or $z \geq 1.96$.

4. $\dfrac{x_1}{n_1} = \dfrac{62}{100} = 0.62$,      $\dfrac{x_2}{n_2} = \dfrac{44}{100} = 0.44$,      and      $\hat{p} = \dfrac{62 + 44}{100 + 100} = 0.53$

$z = \dfrac{0.62 - 0.44}{\sqrt{(0.53)(0.47)\left(\dfrac{1}{100} + \dfrac{1}{100}\right)}} \approx 2.55$

5. The null hypothesis must be rejected. The difference between the sample proportions is not significant.

**14.15** $\chi^2 = \dfrac{67^2}{46.4} + \dfrac{64^2}{63.6} + \cdots + \dfrac{37^2}{20.7} - 400 = 40.89$

**14.17** $H_0$: The probabilities about the three response categories (part-time employment, full-time employment, no employment) are all equal regardless of the number of children.

$H_A$: The probabilities for at least one of the response categories are not all the same.

**14.21**  1. $H_0$: The probabilities for the three response categories are the same for all four ranks.

$H_A$: The probabilities for at least one response category are not the same for all four ranks.

2. $\alpha = 0.01$

3. Reject $H_0$ if $\chi^2 \geq 16.812$.

4. The expected frequencies are 10.8, 18.9, 13.5, 10.8 for the first row, 24.2, 42.35, 30.25, and 24.2 for the second row, and 45, 78.75, 56.25, and 45 for the third row.

$$\chi^2 = \frac{(8 - 10.8)^2}{10.8} + \cdots + \frac{(52 - 45)^2}{45}$$

$$= 20.72$$

5. The null hypothesis must be rejected.

**14.23**  1. $H_0$: Students' interest and ability in studying a foreign language are independent.

$H_A$: The two variables are not independent.

2. $\alpha = 0.05$    3. Reject $H_0$ if $\chi^2 \geq 9.488$.

4. The expected values for the first row are 16.364, 23.182, and 20.455, for the second row are 21.818, 30.909, and 27.273 and for the third row are 21.818, 30.909, and 27.273.

$$\chi^2 = 26.77.$$

5. The null hypothesis must be rejected.

**14.27**  The new first column becomes 28, 74, and 18.

1. $H_0$: The handicaps do not affect the performance.

$H_A$: The handicaps do affect the performance.

2. $\alpha = 0.05$    3. Reject $H_0$ if $\chi^2 \geq 5.991$, which is the value of $\chi^2_{0.05}$ for 2 degrees of freedom.

4. The expected values for the first row are 24 and 40, those for the second row are 78 and 130, and those for the third row are 18 and 30.

$$\chi^2 = 1.39$$

5. The null hypothesis cannot be rejected.

**14.29**  1. $H_0$: There is no relationship between the fidelity and the selectivity of the radios.

$H_A$: There is a relationship between the two variables

2. $\alpha = 0.01$

3. Reject the null hypothesis if

$$\chi^2 \geq 13.277$$

4. The expected frequencies are 15.00, 22.11, and 12.89 for the first row, 33.6, 49.52, and 28.88 second row, and 8.40, 12.38, and 7.22 for the third row.

$$\chi^2 = 52.72$$

5. The null hypothesis must be rejected.

**14.31**  $\chi^2 = 1.23$ and $z = -1.11$ from Exercise 14.12. $z^2 = 1.23$, which equals $\chi^2$.

**14.33**  1. $H_0$: The probability of a response favoring the candidate is the same for all five unions.

$H_A$: The probabilities are not all the same.

2. $\alpha = 0.01$    3. Reject $H_0$ if $\chi^2 \geq 13.277$.

4. The expected frequencies for the first row are all 78, and those for the second row are all 22.

$$\chi^2 = 16.55$$

5. The null hypothesis must be rejected.

**14.35**  Let $r_i$ be the total of the observed frequencies for the $i$th row, $c_j$ the total of the observed frequencies for the $j$th column, $e_{ij}$ the expected frequency for the $i$th row and the $j$th column, and $n$ the grand total for the entire table.

$$\sum_i e_{ij} = \sum_j \frac{r_i \cdot c_j}{n} = \frac{c_j}{n} \cdot \sum r_i = \frac{c_j}{n} \cdot n = c_j$$

**14.37**  1. $H_0$: Data constitute sample from a binomial population with $n = 4$ and $p = 0.50$.
$H_A$: Data do not constitute sample from a binomial population with $n = 4$ and $p = 0.50$.

2. $\alpha = 0.01$      3. Reject $H_0$ if $\chi^2 \geq 13.277$

4. The expected frequencies are 10, 40, 60, 40, and 10. $\chi^2 = 2.37$

5. The null hypothesis cannot be rejected.

**14.39**  $\mu = 0 \cdot \dfrac{2}{300} + 1 \cdot \dfrac{10}{300} + \ldots + 4 \cdot \dfrac{119}{300} = 3.2 = 4p$, so that $p = \dfrac{3.2}{4} = 0.8$

1. $H_0$: Data constitute sample from a binomial population with $n = 4$ and $p = 0.8$.
$H_A$: Data do not constitute sample from binomial population with $n = 4$ and $p = 0.8$.

2. $\alpha = 0.05$      3. Reject $H_0$ if $x^2 \geq 5.991$

4. The probabilities are 0.002, 0.026, 0.154, 0.410, and 0.410, so that the expected frequencies for 0, 1, 2, 3 and 4 are 0.6, 7.8, 46.2, 123, and 123. Combine 0 and 1.

$$\chi^2 = 6.82$$

5. The null hypothesis must be rejected.

## SOLUTIONS OF REVIEW EXERCISES FOR CHAPTERS 11, 12, 13, 14

**R.131**  $n = (0.22)(0.78) \left( \dfrac{1.96}{0.035} \right)^2 = 539$

**R.135**  1. $H_0$: $\mu = 78$ and $H_A$: $\mu > 78$.      2. $\alpha = 0.01$      3. Reject $H_0$ if $t \geq 3.747$.

4. $t = \dfrac{82.2 - 78}{1.351\sqrt{5}} = 6.95$      5. The null hypothesis must be rejected.

**R.137**  1. $H_0$: The probabilities of the responses are the same regardless of the number of children.
$H_A$: The probabilities of the responses are not all the same at least for one of the numbers of children.

2. $\alpha = 0.05$      3. Reject the null hypothesis if $\chi^2 \geq 9.488$.

4. The expected frequencies for the first row are 44.4, 38.6, and 16.9, those for the second row are 60.9, 52.9, and 23.2, and those for the third row are 54.7, 47.5, and 20.8.

$$\chi^2 = \frac{(48 - 44.4)^2}{44.4} + \cdots + \frac{(20 - 20.8)^2}{20.8} = 3.97$$

5. The null hypothesis cannot be rejected.

**R.139**  1. $H_0$: $p_1 = p_3$ and $H_A$: $p_1 \neq p_2$

2. $\alpha = 0.01$      3. Reject $H_0$ if $\chi_2 > 6.635$

4. The expected frequencies for the first two rows are 67.8 and 45.2. Those for the second row are 2 and 154.8

$$\chi^2 = \frac{(81-67.8)^2}{67.8} + \ldots + \frac{(168-154.8)}{154.8} = 8.30$$

5. The null hypothesis must be rejected.

**R.141**   $\mu = 1.4 = 7p$, so that $p = 0.20$. Combine 3, 4 and 5.

1. $H_0: p = 0.20$ and $H_A: p \neq 0.20$.    2. $\alpha = 0.05$.

3. Reject $H_0$ if $\chi^2 \geq 25.991$.    4. $\chi^2 = \frac{(12-10.5)^2}{10.5} + \ldots + \frac{(9-7.4)^2}{7.4} = 0.91$.

5. The null hypothesis cannot be rejected.

**R.143**   1. $H_0$: The three programs are equally effective.
   $H_A$: The three programs are not equally effective.

2. $\alpha = 0.05$    3. Reject $H_0$ if $\chi^2 \geq 2.920$.

4. $\chi^2 = \frac{(86-82.9)^2}{82.9} + \ldots + \frac{(38-33.1)^2}{33.1} = 1.659$

5. The null hypothesis cannot be rejected.

**R.145**   The normal probability plot reveals a linear pattern.

**R.147**   Since we are not dealing with samples, there is no question of statistical significance.

**R.149**   $\dfrac{10.4}{1 + \dfrac{2.575}{\sqrt{120}}} < \sigma < \dfrac{10.4}{1 - \dfrac{2.575}{\sqrt{120}}}$   yields $8.42 < \sigma < 13.59$.

**R.151**   $\hat{p}_1 = \dfrac{23}{80} = 0.2875$,    $\hat{p}_2 = \dfrac{19}{80} = 0.2375$, and    $\hat{p} = \dfrac{19+23}{80+80} = 0.2625$

1. $H_0: p_1 = p_2$ and $H_A: p_1 \neq p_2$    2. $\alpha = 0.05$

3. Reject $H_0$ if $z \leq -1.96$ or $z \geq 1.96$.

4. $z = \dfrac{0.2875 - 0.2375}{\sqrt{(0.2625)(0.7375)\left(\dfrac{1}{80} + \dfrac{1}{80}\right)}} = 0.72$.

5. The null hypothesis cannot be rejected.

**R.153**   1. $H_0: p_1 = p_2 = p_3$ and $H_A: p_1, p_2,$ and $p_3$ are not all equal.    2. $\alpha = 0.01$

3. Reject $H_0$ if $\chi^2 \geq 9.210$.

4. The expected frequencies for the first row are all 72; those for the second row are all 28.

$$\chi^2 = \frac{(63-72)^2}{72} + \ldots + \frac{(31-28)^2}{28} = 11.61$$

5. The null hypothesis must be rejected.

**R.155**   $n = \left[\dfrac{2.575(3.4)}{1.2}\right]^2 = 54$

**R.157**   a. 0.0375, 0.1071, 0.2223, 0.2811, 0.2163, 0.1013, and 0.0344

b. 3, 8.57, 17.78, 22.49, 17.30, 8.10, and 2.75

c. 1. $H_0$: population sampled is normal and $H_A$: population sampled is not normal.
   2. $\alpha = 0.05$
   3. Combining first two classes and last two classes, reject $H_0$ if $\chi^2 \geq 5.991$,
   4. $\chi^2 = \dfrac{(13-11.57)^2}{11.57} + \ldots + \dfrac{(11-10.85)^2}{10.85} = 1.27$
   5. The null hypothesis cannot be rejected.

**R.159**   $n_1 = 10, n_2 = 8, s_1 = 4.395$, and $s_2 = 1.637$.

1. $H_0: \sigma_1 = \sigma_2$ and $H_A: \sigma_1 \neq \sigma_2$.    2. $\alpha = 0.02$    3. Reject $H_0$ if $F \geq 6.72$.

4. $F = \dfrac{(4.395)^2}{(1.637)^2} = 7.21$    5. The null hypothesis must be rejected.

**R.161**   a. $s = 5.10$    b. $\dfrac{14}{2.85} = 4.91$

**R.163**   1. $H_0: \sigma_1 = \sigma_2$ and $H_A: \sigma_1 \neq \sigma_2$    2. $\alpha = 0.10$

        3. Reject $H_0$ if $F \geq 5.05$.

        4. $F = \dfrac{(3.3)^2}{(2.1)^2} = 2.47$    5. The null hypothesis cannot be rejected.

**R.165**   1. 1. $H_0: \sigma = 1.0$ and $H_A: \sigma > 1.0$    2. $\alpha = 0.01$

        3. Reject $H_0$ if $\chi^2 \geq 21.666$, the value of $\chi^2_{0.01}$ for 9 degrees of freedom

        4. $\chi^2 = \dfrac{9(1.28)^2}{1.0^2} = 14.75$    5. The null hypothesis cannot be rejected.

**R.167**   a. $\mu > \mu_0$ and replace the old machines only if the null hypothesis can be rejected.

        b. $\mu < \mu_0$ and replace the old machines unless the null hypothesis can be rejected.

## CHAPTER 15

**15.1**   a. $ns^2_{\bar{x}} = 6 \cdot \dfrac{(3.3 - 2.7)^2 + (2.6 - 2.7)^2 + (2.2 - 2.7)^2}{2} = 1.86$

        $\dfrac{1}{3}(s_1^2 + s_2^2 + s_3^2) = \dfrac{4.796 + 2.784 + 2.064}{3} = 3.215$   $F = \dfrac{1.86}{3.215} = 0.58.$

        b. 1. $H_0: \mu_1 = \mu_2 = \mu_3$ and $H_A:$ the $\mu$'s are not all equal.

        2. $\alpha = 0.05$    3. Reject $H_0$ if $F \geq 3.68$.    4. $F = 0.58$

        5. The null hypothesis cannot be rejected.

**15.3**   a. $ns^2_{\bar{x}} = 3 \cdot \dfrac{(63 - 60)^2 + (58 - 60)^2 + (58 - 60)^2 + (61 - 60)^2}{3} = 18$

        $\dfrac{1}{4}(s_1^2 + s_2^2 + s_3^2 + s_4^2) = \dfrac{1}{4}(7 + 19 + 3 + 3) = 8$, and $F = \dfrac{18}{8} = 2.25.$

        b. 1. $H_0: \mu_1 = \mu_2 = \mu_3 = \mu_4$ and $H_A:$ the $\mu$'s are not all equal.

        2. $\alpha = 0.01$    3. Reject $H_0$ if $F \geq 7.59$    4. $F = 2.25$

        5. The null hypothesis cannot be rejected.

**15.5**   The scores for School 2 are much more variable than those for the other two schools.

**15.7**   The three kinds of tulips should have been assigned at random to the twelve locations in the flower bed.

**15.9**   This is controversial, and statisticians argue about the appropriateness of discarding a proper randomization because it happens to possess some undesirable property. For the situation described here it is likely that the analysis of variance will not be performed as is.

**15.15**   The degrees of freedom for treatments and error are 3 and 20, the sums of squares for treatments and error are 32.34 and 20.03, the mean squares for treatments and error are 10.78 and 1.00, and $F$ equals 10.78.

        1. $H_0: \mu_1 = \mu_2 = \mu_3 = \mu_4$ and $H_A:$ the $\mu$'s are not all equal.    2. $\alpha = 0.01$

        3. Reject $H_0$ if $F \geq 4.94$, the value of $F_{0.01}$ for 3 and 20 degrees of freedom.

        4. $F = 10.78$

        5. The null hypothesis must be rejected. The differences among the sample means cannot be attributed to chance.

**15.17**   The degrees of freedom for treatments and error are 3 and 19, the sums of squares for treatments and error are 7,669.19 and 4,152.55, the mean squares for treatments and error are 2,556.40 and 218.56, and the value of $F$ is 11.70.

1. $H_0 : \mu_1 = \mu_2 = \mu_3 = \mu_4$ and $H_A$: the $\mu$'s are not all equal.     2. $\alpha = 0.01$

3. Reject $H_0$ if $F \geq 5.01$     4. $F = 11.70$

5. The null hypothesis must be rejected. The differences among the sample means cannot be attributed to chance.

**15.19**  The degrees of freedom for treatments and error are 2 and 12, the sums of squares for treatments and error are 79.65 and 149.95, the mean squares for treatments and error are 39.82 and 12.50, and the value of $F$ is 3.19.

1. $H_0 : \mu_1 = \mu_2 = \mu_3$ and $H_A$: the $\mu$'s are not all equal.     2. $\alpha = 0.05$

3. Reject $H_0$ if $F \geq 3.89$, the value of $F_{0.05}$ for 2 and 12 degrees of freedom.

4. $F = 3.19$

5. The null hypothesis cannot be rejected. The differences among the three sample means are not significant.

**15.21**  The degrees of freedom for treatments and error are 7 and 40, the sums of squares for treatments and error are 12,696.20 and 7,818.70, the mean squares for treatments and error are 1,813.74 and 195.47, and the value of $F$ is 9.28.

1. $H_0 : \mu_1 = \mu_2 = \cdots = \mu_8$ and $H_A$: the $\mu$'s are not all equal.     2. $\alpha = 0.05$

3. Reject $H_0$ if $F \geq 2.25$     4. $F = 9.28$

5. The null hypothesis must be rejected.

**15.25**  There are six possible pairs of means. The three underlined pairs are not significant. Only the differences between Mr. Brown and Mr. Black, Mr. Brown and Mrs. Smith, and Ms. Jones and Mrs. Smith are significant.

**15.29**  She might consider only programs of the same length, or she might use the program lengths as blocks and perform a two-way analysis of variance.

**15.31**  The degrees of freedom for treatments (diet foods), blocks (laboratories), and error are 2, 3, and 6. The sums of squares for treatments, blocks, and error are 0.49, 0.54, and 0.22, the mean squares for treatments, blocks and error are 0.245, 0.18, and 0.037, and the values of $F$ for treatments and blocks are 6.62 and 4.86.

1. $H_0 : \alpha_1 = \alpha_2 = \alpha_3 = 0; \beta_1 = \beta_2 = \beta_3 = \beta_4 = 0$. $H_A$: The $\alpha$'s are not all equal to zero and the $\beta$'s are not all equal to zero.

2. $\alpha = 0.01$ for both tests.     3. Reject $H_0$ for treatments if $F \geq 10.9$ and for blocks of $F \geq 9.78$.

4. $F = 6.62$ for treatments and $F = 4.86$ for blocks.     5. Neither null hypothesis can be rejected.

**15.33**  The degrees of freedom for treatments (threads), blocks (measuring instruments), and error are 4, 3, and 12. The sums of squares for treatments, blocks, and error are 70.18, 3.69, and 25.31. The mean squares for treatments, blocks, and error are 17.54, 1.23, and 2.11. The values of $F$ for treatments and blocks are 8.31 and 0.58.

1. $H_0$: The $\alpha$'s are all equal to zero and the B's are equal to zero. $H_A$: The $\alpha$'s are not all equal to zero and the $\beta$'s are all equal to zero.

2. $\alpha = 0.05$ for both tests.

3. Reject $H_0$ for treatments if $F \geq 3.26$ and for blocks if $F \geq 3.49$.

4. $F = 8.31$ for treatments and $F = 0.58$ for blocks.

5. The null hypothesis for treatments must be rejected; the null hypothesis for blocks cannot be rejected.

**15.35**  The degrees of freedom for water temperature, detergents, interaction, and error are 2, 2, 4, and 18. The sums of squares for water temperature, detergents, interaction, and error are 34.3, 442.7, 146.6, and 246.7. The mean squares for water temperature, detergents, interaction, and error are 17.1, 221.4, 36.6, and 13.7. The values of $F$ for water temperature, detergents, and interaction are 1.25, 16.16, and 2.67.

1. $H_0$: The water temperature effects are all equal to zero. The detergent effects are all equal to zero. The interaction effects are all equal to zero.
   $H_A$: The water temperature effects are not all equal to zero. The detergent effects are not all equal to zero. The interaction effects are not all equal, to zero.

2. $\alpha = 0.01$ for each test.

3. Reject $H_0$ for water temperatures, if $F \geq 6.01$ for detergents if $F \geq 6.01$, and for interaction if $F \geq 4.58$.

4. $F = 1.25$ for water temperature, $F = 16.6$; for detergents, and $F = 2.67$ for interaction.

5. The null hypothesis for detergents must be rejected; the other two hypotheses cannot be rejected.

**15.37**   $15 \cdot 4 \cdot 3 = 180$

**15.39**   $A_L B_L C_L D_L$, $A_L B_L C_L D_H$, $A_L B_L C_H D_L$, $A_L B_L C_H D_H$, $A_L B_H C_L D_L$, $A_L B_H C_L D_H$, $A_L B_H C_H D_L$, $A_L B_H C_H D_H$, $A_H B_L C_L D_L$, $A_H B_L C_L D_H$, $A_H B_L C_H D_L$, $A_H B_L C_H D_H$, $A_H B_H C_L D_L$, $A_H B_H C_L D_H$, $A_H B_H C_H D_L$, and $A_H B_H C_H D_H$.

**15.43**   Completing the Latin Square leads to the result that the doctor who is a Westerner is a Republican.

**15.45**   Since 2 already appears with 1, 3, 4, and 6, it must appear together with 5 and 7 on Thursday. Since 4 already appears with 1, 2, 5, and 6, it must appear together with 3 and 7 on Tuesday. Since 5 already appears together with 1, 2, 4, and 7, it must appear together with 3 and 6 on Saturday.

## CHAPTER 16

**16.1**   The second line provides a better fit.

**16.3**   a. The points are fairly dispersed, but the overall pattern is that of a straight line.
   c. The estimate is about 12 or 13.

**16.5**   The two normal equations are $100 = 10a + 525b$ and $5,980 = 525a + 32,085b$ Their solution is $a = 1.526$ and $b = 0.161$.

**16.7**   $\hat{y} = 19.6$.

**16.9**   a. $\hat{y} = 2.039 - 0.102x$ b. $\hat{y} = 1.529$
   c. On a very hot day the chlorine would dissipate much faster.

**16.11**   The sum of the squares of the errors is 20.94, which is less than either 44 or 26.

**16.13**   $\hat{y} = 0.4911 + 0.2724x$.

**16.15**   $\hat{y} = 10.83$.

**16.17**   $\hat{y} = 2.66 + 0.6\mu$; $\hat{y} = \$4.46$ million.

**16.19**   a. $a = 12.447$ and $b = 0.898$;

**16.21**   a. $S_{xx} = 88$, $S_{yy} = 92.83$, $S_{xy} = 79$, and $s_e = 2.34$.
   b. 1. $H_0: \beta = 1.5$ and $H_A : \beta < 1.5$       2. $\alpha = 0.05$
   3. Reject $H_0$ if $t \geq -2.132$. where 2.132 is the value of $t_{0.05}$ for 4 degrees of freedom.
   4. $t = -2.41$       5. The null hypothesis must be rejected.

**16.23**   1. $H_0 : \beta = 3.5$ and $H_A : \beta > 3.5$
   2. $\alpha = 0.01$       3. Reject $H_0$ if $t \geq 3.143$.
   4. $t = 2.68$       5. The null hypothesis cannot be rejected.

**16.25**   1. $H_0 : \beta = -0.15$ and $H_A : \beta \neq -0.15$       2. $\alpha = 0.01$
   3. Reject $H_0$ if $t \leq -4.032$ or $t \geq 4.032$.
   4. $t = 4.17$       5. The null hypothesis must be rejected.

**16.27**  a. $S_{yy} = 74$, and $s_e = 1.070$.
  b.  1. $H_0 : \beta = 0.40$ and $H_A : \beta < 0.40$      2. $\alpha = 0.05$
  3. Reject $H_0$ if $t \le -1.812$.      4. $t = -3.48$
  5. The null hypothesis must be rejected.

**16.29**  $a + bx_0 = 14.09$      $13.16 < \mu_{y|50} < 15.02$

**16.31**  a.  $7.78 < \mu_{y|60} < 14.64$      b.  $0.43 - 21.99$

**16.33**  a.  $0.553 < \mu_{y|5} < 1.575$      b. $0.152 - 1.976$

**16.35**  a.  $\hat{y} = 198 + 37.2x_1 - 0.120x_2$      b. $\hat{y} = 198 + 37.2(0.14) - 0.120(1,100) = 71.2$

**16.37**  a.  $\hat{y} = -2.33 + 0.900x_1 + 1.27x_2 + 0.900x \cdot 3$
  b. $\hat{y} = -2.33 + 0.900(12.5) + 1.27(25) + 0.900(15) = 54.17\%$

**16.39**  a.  $11.9286 = 5(\log \alpha) + 30(\log b)$  and  the  solution  is  $75.228 = 30(\log \alpha) + 220(\log b)$. $\alpha = 68.9$ and $b = 1.234$. The equation is $\log \hat{y} = 1.8379 + 0.0913x$.
  b. $\hat{y} = 68.9(1.234)^x$
  c. $\log \hat{y} = 1.8379 + 0.0913(5) = 2.2944$ and $y = 197.0$.

**16.41**  $\hat{y} = (101.17)(0.9575)^x$

**16.43**  $\hat{y} = (1.178)(2.855)^x$

**16.45**  $\hat{y} = (18.99)(1.152)^x$

**16.47**  $\hat{y} = 384.4 - 36.0x + 0.896x^2$
  $\hat{y} = 384.4 - 36.0(12) + 0.896(12)^2 = 81.4$

## CHAPTER 17

**17.1**  $r = \dfrac{23.6}{\sqrt{344(1.915)}} = 0.92$; the printout yields $\sqrt{0.845} = 0.919$.

**17.3**  $(0.78)^2 100 = 60.8\%$

**17.5**  $r = -0.01$

**17.7**  $r = -0.99$
  $(-0.99)^2 100 = 98.01\%$

**17.9**  No correction is needed. The correlation coefficient does not depend on the units of measurement.

**17.11**  a. Positive correlation      b. negative correlation      c. negative correlation
  d. no correlation      e. positive correlation

**17.13**  $\left(\dfrac{0.41}{0.29}\right)^2 = 1.999$. The first relationship is just about twice as strong as the second.

**17.15**  Correlation does not necessarily imply causation. Actually, both variables (foreign language degrees and railroad track) depend on other variables, such as population growth and economic conditions in general.

**17.17**  Labeling the rows $x = -1$, 0, and 1, and the columns $y = -1$, 0, and 1 we get $\sum x = -22$, $\sum y = -8$, $\sum x^2 = 78$, $\sum y^2 = 106$, and $\sum xy = -39$, so that $S_{xx} = 75.45$, $S_{yy} = 105.66$, $S_{xy} = 39.93$ and $r = -0.45$.

**17.19**  a. $z = 2.35$; the null hypothesis must be rejected.
  b. $z = 1.80$; the null hypothesis cannot be rejected.
  c. $z = 2.29$; the null hypothesis must be rejected.

**17.21**  a. $z = 1.22$; the null hypothesis cannot be rejected.
  b. $z = 0.50$; the null hypothesis cannot be rejected.

**17.23**  $n = 12$ and $r = 0.77$

(a)  1. $H_0: \rho = 0$ and $H_A: \rho \neq 0$      2. $\alpha = 0.01$
     3. Reject $H_0$ if $t \leq -3.169$ or $t \geq 3.169$
     4. Use $t = \dfrac{r\sqrt{n-2}}{\sqrt{1-r^2}} = 3.82$.
     5. $3.82 \geq 3.169$. The null hypothesis must be rejected.

$n = 16$ and $r = 0.49$

(b)  1. $H_0: \rho = 0$ and $H_A: \rho \neq 0$      2. $\alpha = 0.01$
     3. Reject $H_0$ if $t \leq -2.977$ or $t \geq 2.977$
     4. Use $t = \dfrac{r\sqrt{n-2}}{\sqrt{1-r^2}} = 2.10$
     5. $2.10 \leq 2.977$. The null hypothesis cannot be rejected.

**17.25**  1. $H_0: \rho = 0.50$ and $H_A: \rho > 0.50$      2. $\alpha = 0.05$
3. Reject $H_0$ if $z \geq 1.645$.      4. $z = 0.93$
5. The null hypothesis cannot be rejected.

**17.27**  a. $0.533 < \mu_Z < 1.665$ and $0.49 < \rho < 0.93$;
b. $-0.637 < \mu_Z < 0.147$ and $-0.56 < \rho < 0.15$
c. $0.365 < \mu_Z < 0.871$ and $0.35 < \rho < 0.70$.

**17.29**  $R^2 = 0.3320$ and $R = 0.576$.

**17.31**  $R^2 = 0.984$ and $R = 0.992$.

**17.33**  $r_{12} = 0$, $r_{13} = 0.20$, $r_{23} = -0.98$, and $r_{12.3} = -1.00$.

## CHAPTER 18

**18.1**  1. $H_0: \mu = 9.00$, and $H_A: \mu > 9.00$
2. $\alpha = 0.05$

3. The criterion may be based on the number of plus signs or the number of minus signs denoted by $x$. Reject the null hypothesis if the probability of getting $x$ or more plus signs is less than or equal to 0.05.

4. Replacing each value greater than 9.00 with a plus sign, we get $+ - + +$ $- + + + + + + - - + +$, where there are 11 plus signs. Table V shows that for $n = 15$ and $p = 0.50$ the probability of 11 or more plus signs is equal to $0.042 + 0.014 + 0.003 = 0.059$.

5. Since 0.059 exceeds 0.05, the null hypothesis cannot be rejected.

**18.3**  1. $H_0: \tilde{\mu} = 110$ and $H_A: \tilde{\mu} > 110$
2. $\alpha = 0.01$
3. The test statistic, $x$, is the number of packages weighing more than 110 grams.
4. $x = 14$ (out of $n = 18$) and the $p$-value is 0.016.
5. The null hypothesis cannot be rejected.

**18.5**  1. $H_0: \tilde{\mu} = 278$ and $H_A: \tilde{\mu} > 278$      2. $\alpha = 0.05$

a. 3. The test statistic, $x$, is the number of scores greater than 278.
   4. $x = 10$ and the $p$-value is 0.047.      5. The null hypothesis must be rejected.

b. 3. Reject $H_0$ if $z \geq 1.645$. Since the two scores that equal 278 must be discarded, $n = (15 - 2) = 13$; $np = 13(0.5) = 6.5$; $\sigma = \sqrt{13(0.5)(0.5)} = 1.803$.
   4. $z = \dfrac{10 - 6.5}{1.803} = 1.94$.
   5. The null hypothesis must be rejected.

**18.7**  1. $H_0: \tilde{\mu} = 24.2$ and $H_A: \tilde{\mu} > 24.2$
2. $\alpha = 0.01$

3. Reject $H_0$ if $z \geq 2.33$.

4. Without continuity correction $z = 2.333$ and with continuity correction $z = 2.17$.

5. Null hypothesis cannot be rejected.

**18.9** The $p$-value is $0.01758$.

**18.11** 1. $H_0: \tilde{\mu}_D = 0$ and $H_A: \tilde{\mu}_D > 0$

2. $\alpha = 0.05$

3. Reject $H_0$ if $z \leq -1.645$

4. $z = \dfrac{14 - 19(0.5)}{\sqrt{19(0.5)(0.5)}} = -2.06$

5. The null hypothesis must be rejected.

**18.13** a. Reject $H_0$ if $T \leq 8$.   b. Reject $H_0$ if $T^- \leq 11$.   c. Reject $H_0$ if $T^+ \leq 11$.

**18.15** a. Reject $H_0$ if $T \leq 7$.   b. Reject $H_0$ if $T^- \leq 10$.   c. Reject $H_0$ if $T^+ \leq 10$.

**18.17** a. 1. $H_0: \mu = 45$ and $H_A: \mu < 45$   2. $\alpha = 0.05$   3. Reject $H_0$ if $T^+ \leq 21$.
4. $T^+ = 18$   5. The null hypothesis must be rejected.

b. 1. $H_0: \mu = 0$ and $H_A: \mu \neq 0$   2. $\alpha = 0.05$   3. Reject $H_0$ if $T^+ \leq 17$
4. $T^+ = 18$   5. The null hypothesis cannot be rejected.

**18.19** 1. $H_0: \mu = 110$ and $H_A: \mu > 110$   2. $\alpha = 0.01$   3. Reject $H_0$ if $T^- \leq 33$
4. $T^- = 18$   5. The null hypothesis must be rejected.

**18.21** 1. $H_0: \mu_D = 0$ and $H_A: \mu_D < 0$   2. $\alpha = 0.05$   3. Reject $H_0$ if $T^+ \leq 26$.
4. $T^+ = 5$   5. The null hypothesis must be rejected.

**18.23** 1. and 2. as in Exercise 18.21   3. Reject $H_0$ if $z \leq -1.645$.

4. Without continuity correction $z = \dfrac{5 - 52.5}{15.93} = -2.98$ and with continuity correction $z = -2.95$.   5. The null hypothesis must be rejected.

**18.25** 1. $H_0: \overline{\mu}_D = 0$ and $H_A: \overline{\mu}_D > 0$   2. $a = 0.05$
3. Reject $H_0$ if $z > 1.645$   4. $z = 1.75$
5. The null hypothesis must be rejected.

**18.31** a. Reject $H_0$ if $U_2 \leq 14$.   b. Reject if $U \leq 11$.   c. Reject if $U_1 \leq 14$.

**18.33** a. Reject $H_0$ if $U_2 \leq 41$.   b. Reject if $U \leq 36$.   c. Reject if $U_1 \leq 41$.

**18.35** a. Reject $H_0$ if $U_2 \leq 3$.   b. Reject $H_0$ if $U_2 \leq 18$.   c. Reject $H_0$ if $U_2 \leq 13$.
d. Reject $H_0$ if $U_2 \leq 2$.

**18.37** 1. $H_0: \mu_1 = \mu_2$ and $H_A: \mu_1 \neq \mu_2$   2. $\alpha = 0.05$   3. Reject $H_0$ if $U \leq 37$.
4. $W_1 = 188$, $W_2 = 112$, $U_1 = 110$, $U_2 = 34$ and $U = 34$.
5. The null hypothesis must be rejected.

**18.39** 1. $H_0: \mu_1 = \mu_2$ and $H_A: \mu_1 \neq \mu_2$   2. $\alpha = 0.05$   3. Reject $H_0$ if $U < 49$.
4. $W_1 = 208$, $W_2 = 170$, $U_1 = 88$, $U_2 = 92$ and $U = 88$.
5. The null hypothesis cannot be rejected.

**18.41** 1. $H_0: \mu_1 = \mu_2$ and $H_A: \mu_1 < \mu_2$   2. $\alpha = 0.05$   3. Reject $H_0$ if $U_1 \leq 10$.
4. $W_1 = 26.5$ and $U_1 = 5.5$   5. The null hypothesis must be rejected.

**18.43** 1. $H_0: \mu_1 = \mu_2$ and $H_A: \mu_1 \neq \mu_2$   2. $\alpha = 0.05$

3. Reject $H_0$ if $z \leq -1.96$ or $z \geq 1.96$.   4. $z = \dfrac{24 - 45}{12.25} = -1.71$.
5. The null hypothesis cannot be rejected.

**18.51** 1. $H_0: \mu_1 = \mu_2 = \mu_3 = \mu_4$ and $H_A:$ the $\mu$'s are not all equal.   2. $\alpha = 0.05$
3. Reject $H_0$ if $H \geq 7.815$.
4. $R_1 = 53$, $R_2 = 68$, $R_3 = 30$, and $R_4 = 59$, and $H = 4.51$.
5. The null hypothesis cannot be rejected.

**18.53** 1. $H_0: \mu_1 = \mu_2 = \mu_3$ and $H_A:$ the $\mu$'s are not all equal.   2. $\alpha = 0.01$
3. Reject $H_0$ if $H \geq 9.210$.   4. $R_1 = 121$, $R_2 = 144$, and $R_3 = 86$, and $H = 1.53$.
5. The null hypothesis cannot be rejected.

**18.55** 1. $H_0:$ Arrangement is random and $H_A:$ arrangement is not random.   2. $\alpha = 0.05$
3. Reject $H_0$ if $u \leq 8$ or $u \geq 19$.   4. $n_1 = 12$, $n_2 = 13$, and $u = 7$

5. The null hypothesis must be rejected.

**18.57** 1. $H_0$: Arrangement is random and $H_A$: arrangement is not random.　　2. $\alpha = 0.01$
3. Reject $H_0$ if $u \le 8$ or $u \ge 23$.　　4. $n_1 = 15$, $n_2 = 14$, and $u = 20$.
5. The null hypothesis cannot be rejected.

**18.59** 1. $H_0$: Arrangement is random and $H_A$: arrangement is not random.　　2. $\alpha = 0.05$
3. Reject $H_0$ if $z \le -1.96$ or $z \ge 1.96$.　　4. $z = \dfrac{7.5 - 10.92}{1.96} = -1.74$.
5. The null hypothesis cannot be rejected.

**18.61** 1. $H_0$: Arrangement is random and $H_A$: arrangement is not random.　　2. $\alpha = 0.05$
3. Reject $H_0$ if $z \le -1.96$ or $z \ge 1.96$.　　4. $z = \dfrac{28 - 26.71}{3.40} = 0.38$.
5. The null hypothesis cannot be rejected.

**18.65** The median is 54.85 and the arrangement of values above and below the median is *aaaaabaabbbbabbabaabaabaabaababbabbbabab abab*, so that $n_1 = 20$, $n_2 = 20$, and $u = 26$.
1. $H_0$: Arrangement is random and $H_A$: arrangement is not random.　　2. $\alpha = 0.05$
3. Reject $H_0$ if $z \le -1.96$ or $z \ge 1.96$.　　4. $z = \dfrac{26 - 21}{3.12} = 1.60$.
5. The null hypothesis cannot be rejected.

**18.67** The arrangement is 140 and the arrangement of values above and below the median is *bbaabbaabbbbbbbaaaaababbbaaaaa*, so that $n_1 = 15$, $n_2 = 15$, and $u = 10$.
1. $H_0$: Arrangement is random and $H_A$: arrangement is not random.　　2. $\alpha = 0.05$
3. Reject $H_0$ if $z \le -1.645$　　4. $z = \dfrac{10.5 - 16}{2.69} = -2.04$
5. The null hypothesis must be rejected.

**18.69** $\sum d^2 = 100$, so that $r_S = 1 - \dfrac{6 \cdot 100}{12 \cdot 143} = 0.65$.

**18.71** $z = 0.31\sqrt{49} = 2.17$. Since $z = 2.17$ exceeds 1.96, the null hypothesis must be rejected.

**18.73** $r_s = 0.61$ for $A$ and $B$, $r_s = -0.05$ for $A$ and $C$, and $r_s = -0.18$ for $B$ and $C$.
a. $A$ and $B$ are most alike.　　b. $B$ and $C$ are least alike.

## REVIEW EXERCISES FOR CHAPTERS 15, 16, 17, AND 18

**R.169** They will belong to a group of families with a higher mean income, but no guaranteed increase.

**R.171** $r = \dfrac{-103.8}{\sqrt{(312.1)(82.4)}} = -0.65$.

**R.173** $n_1 = 23$, $n_2 = 7$, and $u = 9$
1. $H_0$: Arrangement is random and $H_A$: arrangement is not random.　　2. $\alpha = 0.01$
3. Reject $H_0$ if $z \le -2.575$ or $z \ge 2.575$.　　4. $z = \dfrac{9 - 11.73}{1.90} = -1.44$.
5. The null hypothesis cannot be rejected.

**R.175** $r = \dfrac{-23.89}{\sqrt{(34,873.50)(0.0194)}} = -0.92$

**R.177** 1. $H_0$: $\mu_1 = \mu_2 = \mu_3 = \mu_4$ $H_A$: the $\mu$'s are not all equal.　　2. $\alpha = 0.05$
3. Reject $H_0$ if $F \ge 2.83$.　　4. $F = 3.48$
5. The null hypothesis must be rejected.

**R.179** $r = -1.03$, which is an impossible value.

**R.181** 1. $H_0$: $\overline{\mu} = 0$ and $H_A$: $\overline{\mu} = 0$　　2. $a = 0.05$
3. Reject　　4. $z = \dfrac{154 - 115.5}{28.77} = 1.34$
5. The null hypothesis cannot be rejected.

**R.183**  Since the males are all economists and the females are all statisticians, sex and field of specialization are confounded. There is no way in which we can distinguish between these two sources of variation on the basis of the experiment.

**R.185**  The three normal equations are $66.2 = 7a + 28c$, $-13.2 = 28b$, and $165.8 = 28a + 196c$, so that $a = 14.2$, $b = -0.471$, and $c = -1.18$. The equation of the parabola is $\hat{y} = 14.2 - 0.471x - 1.18x^2$.

**R.187**  The data yield $- - - + - - - - + - - - - - + - - +$,
where $n = 16$ and $x = 4$.
1. $H_0: \rho = 0.50$ and $H_A: \rho < 0.50$.      2. $\alpha = 0.05$
3. $x$ is the number of $+$ signs.
4. $x = 4$ and the $p$-value is 0.039.
5. The null hypothesis must be rejected.

**R.189**  a. Positive correlation      b. no correlation      c. positive correlation      d. negative correlation

**R.191**  $r = \dfrac{1{,}727}{\sqrt{(20{,}456)(163.5)}} = 0.94$

**R.193**  $\left(\dfrac{-0.92}{0.41}\right)^2 = 5.04$. The second relationship is just about 5 times as strong as the first relationship.

**R.195**  $\bar{x}_1 = 11$, $\bar{x}_2 = 15$, $\bar{x}_3 = 10$, $\bar{x} = 12$, $s_1^2 = \dfrac{26}{3}$, $s_2^2 = \dfrac{34}{3}$, and $s_3^2 = \dfrac{26}{3}$.

a.  $ns_{\bar{x}}^2 = 28$      $\dfrac{1}{3}(s_1^2 + s_2^2 + s_3^2) = \dfrac{86}{9}$, and      $F = \dfrac{28}{\dfrac{86}{9}} = 2.93$.

b. 1. $H_0: \mu_1 = \mu_2 = \mu_3$ and $H_A$: the $\mu$'s are not all equal.      2. $\alpha = 0.01$
3. Reject $H_0$ if $F \geq 8.02$      4. $F = 2.93$
5. The null hypothesis cannot be rejected.

**R.197**  a. It is a balanced incomplete block design because the seven department heads are not all serving together on a committee, but each department head serves together with each other department head on two committees. Griffith—Dramatics, Anderson—Discipline, Evans—Tenure or Salaries, Fleming—Salaries or Tenure.

b. There are two solutions. Griffith—Dramatics, Griffith—Dramatics, Anderson—Discipline, Anderson—Evans—Tenure, Evans—Salaries, Fleming—Salaries, Fleming—Tenure.

**R.199**  a. Reject $H_0$ if $U \leq 19$.      b. Reject $H_0$ if $U_1 \leq 23$.      c. Reject $H_0$ if $U_2 \leq 23$.

**R.201**  1. $H_0$: The row effects are all equal to zero; the column effects are all equal to zero; the treatment effects are all equal to zero.
$H_A$: The row effects are not all equal to zero. The column effects are not all equal to zero. The treatment effects are not all equal to zero.
2. $\alpha = 0.01$ for each test.
3. For each test, reject $H_0$ if $F \geq 5.41$
4. $F = 2.31$ for rows, $F = 8.24$ for columns, and $F = 31.28$ for treatments.
5. The null hypothesis for rows cannot be rejected. The null hypothesis for columns and treatments must both be rejected.

**R.203**  1. $H_0: \mu_1 = \mu_2$ and $H_A: \mu_1 \neq \mu_2$.
2. $\alpha = 0.05$
3. Reject $H_0$ if $z \leq -1.96$ or $z \geq 1.96$.
4. $z = -2.14$
5. The null hypothesis must be rejected.

**R.205**   1. $H_0$: $\tilde{\mu} = 169$ and $\tilde{H}_A$: $\mu \neq 169$.
2. $\alpha = 0.05$
3. Reject $H_0$ if $T \leq 11$.
4. $T = 11$ so that $T = 11$.
5. The null hypothesis must be rejected.

**R.207**   $\log \hat{y} = 1.45157 + 0.00698x$; $\log \hat{y} = 1.45157 + 0.00698(60) = 1.8704$ $\hat{y} = 74.20$

**R.211**   The parabola provides an excellent fit for the given range of values of $x$, 50 to 65. However, the parabola will go up for greater values of $x$, and this does not make any sense for the given kind of data on price and demand.

# INDEX

## A

Addition rules, 139–143
    special addition rule, 142
Alternative hypothesis, 288
American Statistics Index (ASI), 6
Analysis of variance (ANOVA),
        357–395
    differences among $k$ means,
        358–361
    experimental design, 357, 362–363
        blocking, 376–377
        replication, 382
    multiple comparisons, 370–373
    one-way, 358, 363–370
    randomization, 358, 362–363
        defined, 363
    two-way, 377
        with interaction, 382–384
        without interaction, 378–382
Area sampling, 240
Arithmetic mean, *See* Mean(s)
Arithmetic probability paper, 218
Astragali, 3
Averages, law of, 117

## B

Balanced incomplete block design, 390
Bar charts, 15–16, 31
Bayes' theorem, 152–155
Bell-shaped distribution, 88
Betting odds, 136
Biased data, 4
Biased estimators, 77
Biases:
    hidden, 4–5
    intentional, 5
Binomial coefficients, 108–109
Binomial distribution, 181–186, 280,
        333–334
    normal approximation to, 222–225
    variance of, 201
Block effects, 378
Block sum of squares (SSB), 378–379
    computing formula for, 378
Blocking, 358, 376–377
Blocks, 377
Box-and-whisker plot, 59
Boxplots, 58–59
Bureau of Labor Statistics,
        Consumer Price Index, 5
Business statistics, 5

## C

Categorical data, 7–8
Categorical distributions, 22
    pie charts, 32
Cell, 335–336
Censuses, 2
Central limit theorem, 248–249
Central location, measures of, 43
    fractiles, 57–59
    mean, 45–48
    median, 54–57
    mode, 59–60
    weighted mean, 49–51
Character plot, 39
Charts:
    bar, 15–16, 31
        horizontal, 15
        vertical, 15
    pie, 32
Chebyshev, P. L., 79
Chebyshev's theorem, 79–80, 202–203,
        248–249
Chi-square distribution, 275
Chi-square statistic, 275, 318
    alternative formula for, 338
Class, 22
    boundaries, 25
    frequencies, 24–25
    interval of a distribution, 25
    intervals, 25
    limits, 25
    lower class limits, 25
    marks, 25
    open, 22
    real, 25
    upper class limits, 25
Classical probability concept, 114–115
Cluster sampling, 239–240
Coefficient of correlation, 432–438
        computing formula for, 435
Coefficient of determination, 434
Coefficient of quartile variation, 86
Coefficient of variation, 82
Coefficients:
    binomial, 108–109
    estimated regression, 108–109
    regression, 413
Combinations, 107–110
Complements, 126
Complete blocks, 377
Completely randomized design, 363
Composite hypothesis, 293
Computer revolution, and growth of
        statistics, 4
Computer simulation, 117, 244
Conditional probability, 145–147
    defined, 146
Confidence, use of term, 265
Confidence interval, 267
Confidence limits, 267
    for regression coefficients, 413
Consistency criterion, 137
Consumer Price Index, 5
Contingency coefficient, 344
Contingency tables, 335
Continuity correction, 221
    test with, 328
Continuous distributions, 207–209
Continuous random variables, 206
Controlled experiment, 362
Correlation, 431–451
    coefficient of, 432–438
    coefficient of determination, 434
    correlation analysis, 443–446
    Fisher Z transformation, 444
    interpretation of $r$, 438–439
    multiple, 447–450
    multiple correlation coefficient, 447
    negative, 434
    normal correlation analysis, 444
    partial, 447–450
    partial correlation coefficient, 449
    population correlation coefficient,
        444
    positive, 434
    product-moment coefficient of
        correlation, 436
Count data, 279–280
    defined, 326
    tests of hypotheses based on,
        326–350
Countably infinite random variables,
        179
Counting, 101–104
Critical values, 300
Cross stratification, 239
Cumulative distribution, 26
Cumulative percentage distribution, 27
Curve:
    exponential, 422
    normal curve of errors, 210
    normal-curve areas, 210
    OC, 292
Curve fitting, 397–399
    polynomial, 422

## D

Data, 2
    biased, 4
    categorical, 7–8
    external, 5
    internal, 5
    interval, 8
    listing/grouping, 12–42
    nominal, 8
    numerical, 7–8
    ordinal, 8
    primary, 5
    qualitative, 8
    quantitative, 7
    ratio, 8–9
    raw, 21
    secondary, 5
    summarizing:
        frequency distribution, 21–27
        graphical presentations, 30–34
        measures of location, 43–73
        measures of variation, 74–93
        two-variable data, 37–41
Data collection, examples of, 3–4
Data points, 37, 399, 401
Deciles, 57
Decision making, 164–166, *See also*
        Expectations; Statistical
        decision problems
    Bayesian analysis, 165
    payoff table, 165
    states of nature, 165
Decision theory, 168

Degree of confidence, 267
Degrees of freedom, 268–270
  defined, 269
  denominator, 324
  number of, 268–269, 275
  numerator, 321
Denominator degrees of freedom, 324
Dependent events, 147
Descriptive statistics, 2–3
*Design of Experiments, The* (Fisher), 393
Determinants, 402
Deviations from the mean, 76
Diagrams:
  dot, 13–14
  Pareto, 15–16
  scatter, 37
  tree, 126–127
  Venn, 126–127, 135, 143
Discrete random variable, 179
Discrete uniform distribution, 244
Distributions:
  bell-shaped, 88
  binomial distribution, 181–186, 280,
    333–334
    normal approximation to,
      222–225
    variance of, 201
  categorical, 22, 32
  chi-square, 275
  continuous, 207–209
  cumulative, 26, 26–27
  cumulative percentage, 27
  exponential, 217
  *F* distribution, 321
  frequency, 21–27, *See also*
    Graphical presentations
    defined, 22
    and loss of information, 22
    open classes, 22
    steps in construction of, 22
  geometric, 189
  hypergeometric, 189–192
    mean of, 200
  integer, 244
  multinomial, 197–198
  negatively skewed, 88
  normal, 206–228, 207
    applications of, 219–222
    continuous distributions,
      207–209
    defined, 209
    normal approximation to the
      binomial distribution,
      222–225
    normality, check for, 218–219
  numerical, 22
  percentage, 25
  Poisson, 193–196, 210, 345
  positively skewed, 88
  probability, 177–205
    binomial distribution, 181–186
    defined, 179
    hypergeometric distribution,
      189–192
    mean of, 198–200
    multinomial distribution,
      197–198
    Poisson distribution, 193–196
    random variables, 177, 178–179
    standard deviation of, 201–203
    values of, 180

  variance of, 200, 201
  qualitative, 22
  quantitative, 22
  reverse J-shaped, 91
  sampling, 242–245
    computer simulation, 244
    defined, 230
    discrete uniform distribution,
      244
    integer distribution, 244
    skewed, 88–89
    standard normal, 210
    student *t*-distribution, 268
    U-shaped, 91
*Domesday Book* (William of
    Normandy), 2
Dot diagrams, 13–14
Double summations, 71
Double-stem display, 18
Downward slope, least-squares line, 403

**E**

Efficiency, 483
Elimination, method of, 402
Empirical rule, 80–81
Empty set, 125
Equations:
  linear, 398
  normal, 402
  polynomial, 425
  regression, 404
Equitable game, 162
Error mean square, 365
Error sum of squares (SSE), 364,
    376–379, 383
  for Latin square, 388
Estimated regression coefficients, 410
Estimated regression lines, 405
Estimation:
  defined, 262
  of means, 263–268
    σ unknown, 268–271
  problems of, 262–286
  of proportions, 279–284
  of standard deviations, 274–278
Events, 125–128
  dependent, 147
  independent, 147
  mutually exclusive, 126
  rule for calculating the probability
    of, 140
Expectations, 159–171
  and decision making, 164–166
  mathematical, 159, 160–163
Expected cell frequencies, 336–337,
    339–341
Expected value of a random variable,
    199
Experimental design, 357
Experiments, 125
Exploratory data analysis, 18, 57
Exponential curve, 422
Exponential distribution, 217
External data, 5

**F**

*F* distribution, 321
*F* statistic, 321
Factor A, 382
Factor B, 382
Factorial notation, 105–106, 108

Factors, 382
Fair game, 162
Fair odds, 136
*Federal Reserve Bulletin*, 5
Fermat, Pierre de, 3
Finite population correction factor, 247
Finite sample space, 125
First postulate of probability, 132
Fisher, R. A., 393
Fisher Z transformation, 444
Fractiles, 57–59
Frequency distributions, 21–27, *See also*
    Graphical presentations
    defined, 22
    and loss of information, 22
    open classes, 22
    steps in construction of, 22
Frequency interpretation, of
    probability, 116
Frequency polygon, 31

**G**

Galton, Francis, 396
Games of chance, 3
General multiplication rule, 148–150
Generalizations, 2–3
Generalized multiplication of choices,
    103–104
Generalized (special) addition rule,
    142–143
Geometric distribution, 189
Geometric mean, 45
Goodness of fit, 345–349
Gosset, W. S., 268, 371
Grand mean, 50, 359
Grand total, 335
Graphical presentations, 21, 30–34
  bar charts, 31
  frequency polygon, 31
  histograms, 30–31
  ogives, 32
  pictogram, 32
  pie charts, 32
Grouped data, description of, 63–68,
    86–90
Grouping data, 12
Grouping error, 65

**H**

*H* test, 471–472
Harmonic mean, 45
Hidden biases, 4–5
Histograms, 30–31
Horizontal bar charts, 15
Hypergeometric distribution, 189–192
  mean of, 200
Hypotheses:
  alternative, 288
  composite, 293
  null, 288
  simple, 293
  statistical, 287
  tests of, 262, 287–293
    means, 287–316
    significance tests, 293–297
    tests concerning means,
      300–306

**I**

Incomplete block designs, 390
Independent events, 147

Index to International Statistics (IS), 6
Inductive statistics, 3
Infinite sample space, 125
Integer distribution, 244
Intentional biases, 5
Interaction, 377–378
Interaction sums of squares, 384
Internal data, 5
Interquartile range, 75, 279
intersections, 126
Interval data, 8
Interval estimates, 267

### J

Judgment samples, 239

### K

Kruskal-Wallis test, *See H* test

### L

Large-sample confidence interval, 267
    for $\mu$, 278
$(1 - \alpha)$ 100% large-sample confidence
    interval for $\rho$, 281
Large-sample test, 300
    concerning proportion, statistic for,
        328
        test with continuity correction,
            328
    concerning standard deviation,
        statistic for, 320
Latin squares, 387–389
Law of averages, 117
Law of large numbers, 117
Leaf, 17
Least squares, method of, 170, 397,
        399–406
    defined, 400
Least-squares line, 401
Legendre, Adrien, 400
Level of significance, 296
Limits of prediction, 415
Linear equations, 398
Linear regression analysis, 410
Listing data, 12
    numerical data, 13–16
        bar charts, 15–16
        dot diagrams, 13–14
        Pareto diagrams, 15–16
Location, measures of, 43–73
Log-log paper, 424
Lower boundary, 25
Lower class limits, 25

### M

Mann-Whitney test, *See U* test
Mathematical expectations, 159,
    160–163
Mathematics of chance, 100
Maximax criterion, 168
Maximin criterion, 167
Maximum error of estimate, 264
Mean deviation, 76
Mean(s), 45–48, 311
    deviations from, 76
    differences between, 307–311
        paired data, 313
        $\sigma$'s unknown, 311–312
    estimation of, 263–268
        $\sigma$ unknown, 268–271

geometric, 45
grand, 50
harmonic, 45
of probability distributions, 198–200
small-sample test concerning, 304
standard error of, 245–246
tests concerning, 300–304
weighted, 49–51
Measures of central location, 43
    fractiles, 57–59
    mean, 45–48
    median, 54–57
    mode, 59–60
    weighted mean, 49–51
Measures of location, 43–73
Measures of relative variation, 82
Measures of skewness, 88–89
Measures of variation, 43, 74–93
    defined, 74–75
    grouped data, description of, 86–87
    range, 75
    standard deviation, 75–79
        applications of, 79–82
Median, 48, 54–57, 65
    standard error of, 250
Median position, 55
Method of elimination, 402
Method of least squares, 170, 397,
        399–406
    defined, 400
Midquartile, 57, 58
Midrange, 57
Minimax criterion, 167
Minimin criterion, 168
MINITAB, 25, 219, 232
Mode, 57, 59–60
Mohs scale, 8
*Monthly Labor Review* (Bureau of
    Labor Statistics), 5–6
Multinomial distribution, 197–198
Multiple comparisons, 358, 370–373
Multiple correlation coefficient, 447
Multiple regression, 418–421
    normal equations, 419
Multiplication of choices, 102–104
    generalized, 103–104
Multiplication rules, 147–150
    general, 148–150
Mutually exclusive events, 126

### N

$n$ factorial, 105, 151
Negative correlation, 434
Negatively skewed distribution, 88
Nominal data, 8
Nonlinear regression, 422–430
Nonparametric tests, 452–465
    defined, 452
    $H$ test, 471–472
    one-sample sign test, 453
    paired-sample sign test, 454–455
    rank correlation, 479–481
    sign test, 453
        large samples, 455–457
    signed-rank test, 458–461
        large samples, 461–463
    tests of randomness:
        runs, 475–476
        runs, above and below the
            median, 477–478
        runs, large samples, 476–477

$U$ test, 465–469
    large samples, 469–471
Non–U.S. governmental sources of
    statistical data, 6
Normal correlation analysis, 444
Normal curve of errors, 210
Normal distribution, 206–228, 207
    applications of, 219–222
    continuous distributions, 207–209
    defined, 209
    normal approximation to the
        binomial distribution, 222–225
    normality, check for, 218–219
Normal equations, 402
Normal population, 218
    sampling, 222
Normal probability paper, 218
Normal probability plot, 218–219
Normal regression analysis, 410
Normal-curve areas, 210
Not statistically significant, use of term,
    294
Null hypothesis, 288
Number of degrees of freedom,
    268–269, 275
Numerator degrees of freedom, 321
Numerical data, 7–8, 13–16
    bar charts, 15–16
    dot diagrams, 13–14
    Pareto diagrams, 15–16
    stem-and-leaf displays, 16–17
Numerical distribution, 22

### O

OC curve, 292
Odds, 114
    betting, 136
    defined, 135
    expressing in terms of probabilities,
        135
    expressing probabilities in terms of,
        136
    fair, 136
    and probability, 135–137
Ogives, 32
One-sample sign test, 453
One-sample $z$ test, 300
One-sided alternative, 295
One-sided criterion, 297
One-sided (tailed) test, 297
One-way analysis of variance, 358,
        363–370, *See also* Analysis of
        variance (ANOVA)
    identity for, 364
Open classes, 22
Operating characteristic curve, 292
Optimum allocation, 239, 242
Ordinal data, 8
Outcomes, 125
Outlier, 48

### P

Paired-sample sign test, 454–455
Paired-sample $t$ test, 313
Parabola, 425
Parameters, 46
Pareto diagrams, 15–16
Pareto, Vilfredo, 15
Partial correlation coefficient, 449
Pascal, Blaise, 3

Pearsonian coefficient of skewness, 88–90
Percentage distribution, 25
Percentage error, 224
Percentiles, 57
Permutations, 104–107
Personal probabilities, 119
Pictogram, 32
Pie charts, 32
Point estimate, 263
Poisson distribution, 193–196, 210, 345
Poisson, S. D., 193
Polynomial curve fitting, 422
Polynomial equations, 425
Pooled standard deviation, 311
Pooling, 330
Population correlation coefficient, 444
Population size, 46
Population standard deviation, 77
Population standard variation, 201
Population variance, 77
Populations, 44–45
Positive correlation, 434
Positively skewed distribution, 88
Postulates of, 132–135
Power function, 292, 424
Prediction:
    limits of, 415
    problems of, 262
Primary data, 5
Probability, 3, 100–123
    addition rules, 139–143
    Bayes' theorem, 152–155
    classical probability concept, 114–115
    combinations, 107–110
    computer simulation, 117
    conditional, 145–147
    counting, 101–104
    first postulate of, 132
    frequency interpretation of, 116
    law of large numbers, 117
    multiplication rules, 147–150
    normal probability paper, 218
    normal probability plot, 218–219
    and odds, 114, 135–137
    permutations, 104–107
    personal, 119
    postulates of, 132–135
    random selections, 115
    rules of, 124–158
    sample spaces and events, 125–128
    second postulate of, 5
    subjective, 119, 136
    third postulate of, 133, 139
    total, rule of, 154
Probability density, 207–209, 227
Probability distributions, 177–205
    binomial distribution, 181–186
    defined, 179
    hypergeometric distribution, 189–192
    mean of, 198–200
    multinomial distribution, 197–198
    Poisson distribution, 193–196
    random variables, 177, 178–179
    standard deviation of, 201–203
    values of, 180
    variance of, 200, 201
Probability theory, 3, 100, 124
Problems of estimation, 262–286
Problems of prediction, 262

Product moment, 436
Product-moment coefficient of correlation, 436
Proportional allocation, 238
    sample sizes for, 238
Proportions:
    estimation of, 279–284
    sample, 280
    tests concerning, 327–328
        analysis of an r × c table, 333–342
        differences between proportions, 329–331
        large samples, 328–329
$p$-value, 300

## Q

Qualitative data, 8
Qualitative distribution, 22
Quantitative data, 7
Quantitative distribution, 22
Quantum statistics, 1
Quartile deviation, 75
Quartiles, 57
Quota sampling, 239

## R

$r × c$ table, 334
    analysis of, statistic for, 336
Random, defined, 234
Random numbers, 231
Random samples, 229–234
    defined, 231
    from a finite population,, 230–231
    finite populations, defined, 230
    infinite populations, defined, 230
Random selections, 115
Random variables, 177, 178–179
    discrete, 179
    expected value of, 198
    variance of, 201
Randomization, 358, 362–363
    defined, 363
Randomized block design, 377
Randomized response technique, 156
Rank correlation, 479–481
Rank sums, 467
Rank-correlation coefficient, 479
Ratio data, 8–9
Raw data, 21
Real class, 25
Regression, 396–430
    curve fitting, 397–399
    defined, 396
    estimated regression coefficients, 410
    estimated regression lines, 405
    limits of prediction, 415
    method of least squares, 170, 397, 399–406
    multiple, 418–421
    nonlinear, 422–430
    regression analysis, 410–416
        linear, 410
        normal, 410
    regression coefficients, 410
    regression equation, 404
    regression lines, 405
    standard error of estimate, 411
Regression analysis, 410–416
Regression coefficients, 410

    confidence limits for, 413
Regression sum of squares, 433
Relative importance weight, 49
Relative variation, measures of, 82
Replication, 382
Residual sum of squares, 433
Reverse J-shaped distribution, 91
Robust, use of term, 324
Root-mean-square deviation, 76
Rule of elimination, 154
Rule of total probability, 154
Runs:
    above and below the median, 477–478
    defined, 475
    large samples, 476–477
    theory of, 475
    total number of, 475

## S

Sample covariance, 437, 449
Sample proportion, 280
Sample spaces, 125–128
    finite, 125
    infinite, 125
    subset of, 125
Sample standard deviation, 76
Sample variance, 77
Samples, 2–3, 44–45
Sampling, 222, 229–255
    area, 240
    central limit theorem, 248–249
    cluster, 239–240
    quota, 239
    sample designs, 236–237
    standard error of the mean, 245–246
    standard error of the median, 250
    stratified, 237–238
    survey, 237
    systematic, 237
Sampling a binomial population, 186
Sampling a normal population, 222
Sampling distributions, 242–245
    computer simulation, 244
    defined, 230
    discrete uniform distribution, 244
    integer distribution, 244
Sampling frame., 235–236
Sampling with replacement, 151
Sampling without replacement, 151
Scatter diagram, 37
Scatter plot, 37
Scattergram, 37, 399
Second postulate of probability, 132
Secondary data, 5
Semi-interquartile range, 75
Semilog paper, 422
Sigma, 46, 77
Sign test, 453
    large samples, 455–457
Signed-rank test, 458–461
    large samples, 461–463
Significance tests, 293–297
    defined, 293–294
    level of significance, 296
    one-sided alternative, 295
    one-sided criterion, 297
    one-sided (tailed) test, 297
    reserving judgment in, 294
    test statistic, 297
    two-sided alternative, 295

two-sided criterion, 297
two-sided (tailed) test, 297
Simple hypothesis, 293
Simple random sample, 231
Simple random sampling, 237–238
Simulation, technical note, 252–254
Skewed distribution, 88–89
Skewness, measures of, 88–89
Small-sample confidence interval for $\mu$, 269, 276
Small-sample test:
  concerning means, 304
  concerning standard deviations, 320
  concerning the difference between means, 311
Spearman's rank-correlation coefficient, 479
Special addition rule, 142, 142–143
Standard deviation, 75–79, 317–325
  applications of, 79–82
  estimation of, 274–278
  pooled, 311
  of probability distributions, 201–203
  sample, 76
  tests concerning, 317–318
    equality of two standard deviations, 321–324
Standard error:
  of a proportion, 281
  of estimate, 411
  of the difference between two means, 308
  of the difference between two proportions, 330
  of the mean, 245–246
  of the median, 250
Standard normal distribution, 210
Standard scores, 210
Standard units, 210–211
**STAT CALC ExpReg** command, 424
**STAT CALC QuadReg** command, 427
Statistic for large-sample test concerning proportion, 328
*Statistical Abstract of the United States* (Bureau of Labor Statistics), 5
Statistical data:
  nature of, 7–10
  sources of, 5–6
  summarizing, 12–42
*Statistical Decision Functions* (Wald), 168
Statistical decision problems, 168–170
Statistical hypothesis, defined, 287
Statistical inference, defined, 3
*Statistical Masterfile* (Congressional Information Service), 6
Statistical model, 184
Statistical Reference Index (SRI), 6
Statistician, use of term, 1
Statistics, 46
  arbitrariness in statistical descriptions, 60
  descriptive, 2–3
  growth of, 2–4
  inductive, 3
  use of term, 1
Stem, 17
Stem label, 17
Stem-and-leaf displays, 16–18
  as hybrid arrangement, 17
Strata, 237
Stratification, 237

cross, 239
Stratified sampling, 237–238
  optimum allocation, 239, 242
  proportional allocation, 238
  cross stratification, 239
Student $t$-distribution, 268
Studentized range, 371
Studentizing, 371
Subjective probabilities, 119, 136
Subset, of sample spaces, 125
Summations, 68–69
Sums of squares, computing formulas for:
  equal sample sizes, 367
  unequal sample sizes, 369
*Survey of Current Business* (Department of Commerce), 5
Survey sampling, 237
Systematic sampling, 237

**T**

Tables:
  contingency, 335
  $r \times c$, 334, 336
  $3 \times 3$ table, 336–337
*Tables of 105,000 Random Decimal Digits*, 232–233
Technical note, simulation, 252–254
Test concerning difference among means, statistic for, 360
Test concerning difference between means with $\sigma 1$ and $\sigma 2$ known, 310
Test concerning difference between two means, 308
Test concerning difference between two proportions, statistic for, 330
Test concerning standard deviation, statistic for, 318
Test concerning the equality of two standard deviations, statistic for, 321
Test of significance, *See* Significance tests
Tests concerning difference between means, $\sigma$'s unknown, 310
Tests concerning means, $\sigma$ unknown, statistic for, 304
  statistic for, 300
Tests concerning proportions, 327–328
  analysis of an $r \times c$ table, 333–342
  differences between proportions, 329–331
  large samples, 328–329
Tests of hypotheses, 262, 287–293
  based on count data, 326–350
    goodness of fit, 345–349
  means, 287–316
    differences between, 307–311
    operating characteristic curve, 292
    power function, 292
    Type I error, 288
    Type II error, 288
  significance tests, 293–297
  standard deviations, 317–325
  tests concerning means, 300–304
    $\sigma$ unknown, 304–306
Tests of randomness runs, 475–476
  above and below the median, 477–478

large samples, 476–477
*Theory of Games and Economic Behavior* (von Neumann/Morgenstern), 168
Theory of probability, *See* Probability theory
Theory of runs, 475
Third postulate of probability, 133, 139
$3 \times 3$ table, 336–337
$t$-interval, 269
TI-83 calculator, 25
Total sum of squares, 364, 433
Treatment effects, 359
Treatment mean square, 365
Treatment sum of squares, 364
Treatments, 364
Tree diagram, 101–102
$t$-statistic, 268
Two-factor experiments, 377
Two-sample $t$ tests, 311
Two-sample $z$ test, 310
Two-sided alternative, 295
Two-sided criterion, 297
Two-sided (tailed) test, 297
Two-variable data, summarizing, 37–41
Two-way analysis of variance, 377
  error sum of squares, 379
  with interaction, 382–384
  without interaction, 378–382, *See also* Analysis of variance (ANOVA)
Two-way experiments, 358
Type I error, 288
  specifying the probability of, 296
Type II error, 288, 452

**U**

$u$ test, 475
$U$ test, 465–469
  large samples, 469–471
Unbiased estimators, 77
Unions, 126
Upper boundary, 25
Upper class limits, 25
Upward slope, least-squares line, 403
U.S. Census, 2
U-shaped distributions, 91

**V**

Variance of the probability distribution, 201
Variance of the random variable, 201
Variance ratios, 321, 360
Variation, measures of, 43, 74–93
  defined, 74–75
  grouped data, description of, 86–87
  range, 75
  standard deviation, 75–79
    applications of, 79–82
Venn diagrams, 126–127, 135, 142
Vertical bar charts, 15

**W**

Weighted mean, 49–51
Wilcoxon rank-sum test, *See U* test
Wilcoxon signed-rank test, *See* Signed-rank test

**Z**

$z$ test, 300
$z$-interval, 267
$z$-scores, 210

Data Disk t/a Modern Elementary Statistics 12e
John E. Freund
Benjamin M. Perles
ISBN 0-13-187440-3
CD License Agreement
© 2007 Pearson Education, Inc.
Pearson Prentice Hall
Pearson Education, Inc.
Upper Saddle River, NJ 07458
All rights reserved.
Pearson Prentice Hall™ is a trademark of Pearson Education, Inc.

READ THIS LICENSE CAREFULLY BEFORE OPENING THIS PACKAGE. BY OPENING THIS PACKAGE, YOU ARE AGREEING TO THE TERMS AND CONDITIONS OF THIS LICENSE. IF YOU DO NOT AGREE, DO NOT OPEN THE PACKAGE. PROMPTLY RETURN THE UNOPENED PACKAGE AND ALL ACCOMPANYING ITEMS TO THE PLACE YOU OBTAINED THEM. THESE TERMS APPLY TO ALL LICENSED SOFTWARE ON THE DISK EXCEPT THAT THE TERMS FOR USE OF ANY SHAREWARE OR FREEWARE ON THE DISKETTES ARE AS SET FORTH IN THE ELECTRONIC LICENSE LOCATED ON THE DISK:

1. **GRANT OF LICENSE and OWNERSHIP:** The enclosed CD-ROM ("Software") is licensed, not sold, to you by Pearson Education, Inc. publishing as Pearson Prentice Hall ("We" or the "Company") in consideration of your adoption of the accompanying Company textbooks and/or other materials, and your agreement to these terms. You own only the disk(s) but we and/or our licensors own the Software itself. This license allows instructors and students enrolled in the course using the Company textbook that accompanies this Software (the "Course") to use and display the enclosed copy of the Software on up to one computer of an educational institution, for academic use only, so long as you comply with the terms of this Agreement. You may make one copy for back up only. We reserve any rights not granted to you.

2. **USE RESTRICTIONS:** You may not sell or license copies of the Software or the Documentation to others. You may not transfer, distribute or make available the Software or the Documentation, except to instructors and students in your school who are users of the adopted Company textbook that accompanies this Software in connection with the course for which the textbook was adopted. You may not reverse engineer, disassemble, decompile, modify, adapt, translate or create derivative works based on the Software or the Documentation. You may be held legally responsible for any copying or copyright infringement that is caused by your failure to abide by the terms of these restrictions.

3. **TERMINATION:** This license is effective until terminated. This license will terminate automatically without notice from the Company if you fail to comply with any provisions or limitations of this license. Upon termination, you shall destroy the Documentation and all copies of the Software. All provisions of this Agreement as to limitation and disclaimer of warranties, limitation of liability, remedies or damages, and our ownership rights shall survive termination.

4. **DISCLAIMER OF WARRANTY: THE COMPANY AND ITS LICENSORS MAKE NO WARRANTIES ABOUT THE SOFTWARE, WHICH IS PROVIDED "AS-IS." IF THE DISK IS DEFECTIVE IN MATERIALS OR WORKMANSHIP, YOUR ONLY REMEDY IS TO RETURN IT TO THE COMPANY WITHIN 30 DAYS FOR REPLACEMENT UNLESS THE COMPANY DETERMINES IN GOOD FAITH THAT THE DISK HAS BEEN MISUSED OR IMPROPERLY INSTALLED, REPAIRED, ALTERED OR DAMAGED. THE COMPANY DISCLAIMS ALL WARRANTIES, EXPRESS OR IMPLIED, INCLUDING WITHOUT LIMITATION, THE IMPLIED WARRANTIES OF MERCHANTABILITY AND FITNESS FOR A PARTICULAR PURPOSE. THE COMPANY DOES NOT WARRANT, GUARANTEE OR MAKE ANY REPRESENTATION REGARDING THE ACCURACY, RELIABILITY, CURRENTNESS, USE, OR RESULTS OF USE, OF THE SOFTWARE.**

5. **LIMITATION OF REMEDIES AND DAMAGES: IN NO EVENT, SHALL THE COMPANY OR ITS EMPLOYEES, AGENTS, LICENSORS OR CONTRACTORS BE LIABLE FOR ANY INCIDENTAL, INDIRECT, SPECIAL OR CONSEQUENTIAL DAMAGES ARISING OUT OF OR IN CONNECTION WITH THIS LICENSE OR THE SOFTWARE, INCLUDING, WITHOUT LIMITATION, LOSS OF USE, LOSS OF DATA, LOSS OF INCOME OR PROFIT, OR OTHER LOSSES SUSTAINED AS A RESULT OF INJURY TO ANY PERSON, OR LOSS OF OR DAMAGE TO PROPERTY, OR CLAIMS OF THIRD PARTIES, EVEN IF THE COMPANY OR AN AUTHORIZED**

REPRESENTATIVE OF THE COMPANY HAS BEEN ADVISED OF THE POSSIBILITY OF SUCH DAMAGES. SOME JURISDICTIONS DO NOT ALLOW THE LIMITATION OF DAMAGES IN CERTAIN CIRCUMSTANCES, SO THE ABOVE LIMITATIONS MAY NOT ALWAYS APPLY.

6. **GENERAL:** THIS AGREEMENT SHALL BE CONSTRUED IN ACCORDANCE WITH THE LAWS OF THE UNITED STATES OF AMERICA AND THE STATE OF NEW YORK, APPLICABLE TO CONTRACTS MADE IN NEW YORK, EXCLUDING THE STATE'S LAWS AND POLICIES ON CONFLICTS OF LAW, AND SHALL BENEFIT THE COMPANY, ITS AFFILIATES AND ASSIGNEES. This Agreement is the complete and exclusive statement of the agreement between you and the Company and supersedes all proposals, prior agreements, oral or written, and any other communications between you and the company or any of its representatives relating to the subject matter. If you are a U.S. Government user, this Software is licensed with "restricted rights" as set forth in subparagraphs (a)-(d) of the Commercial Computer-Restricted Rights clause at FAR 52.227-19 or in subparagraphs (c)(1)(ii) of the Rights in Technical Data and Computer Software clause at DFARS 252.227-7013, and similar clauses, as applicable. Should you have any questions concerning this agreement or if you wish to contact the Company for any reason, please contact in writing: Pearson Education, Inc., One Lake Street, Upper Saddle River, New Jersey 07458

************************************

## System Requirements
-Windows
Pentium II 300 MHz processor
Windows 98, NT, 2000, ME, or XP
64 MB RAM (128 MB RAM required for Windows XP)
4.3 MB available hard drive space (optional—for minimum QuickTime installation)
800 x 600 resolution
8x or faster CD-ROM drive
QuickTime 6.x
Sound card
-Macintosh
PowerPC G3 233 MHz or better
Mac OS 9.x or 10.x
64 MB RAM
10 MB available hard drive space for Mac OS 9, 19 MB on OS X (optional—if QuickTime installation is needed)
800 x 600 resolution
8x or faster CD-ROM drive
QuickTime 6.x

## Support Information
If you are having problems with this software, call (800) 677-6337 between 8:00 a.m. and 8:00 p.m. EST, Monday through Friday, and 5:00 p.m. through 12 a.m. EST on Sundays. You can also get support by filling out the web form located at: http://247.prenhall.com/mediaform
Our technical staff will need to know certain things about your system in order to help us solve your problems more quickly and efficiently. If possible, please be at your computer when you call for support. You should have the following information ready:

- Textbook ISBN
- CD-ROM ISBN
- corresponding product and title
- computer make and model
- Operating System (Windows or Macintosh) and Version
- RAM available
- hard disk space available
- Sound card? Yes or No
- printer make and model
- network connection
- detailed description of the problem, including the exact wording of any error messages.

NOTE: Pearson does not support and/or assist with the following:
- third-party software (i.e. Microsoft including Microsoft Office suite, Apple, Borland, etc.)
- homework assistance
- Textbooks and CD-ROMs purchased used are not supported and are non-replaceable. To purchase a new CD-ROM, contact Pearson Individual Order Copies at 1-800-282-0693.